Problems in Real Anal

Teodora-Liliana T. Rădulescu
Vicenţiu D. Rădulescu
Titu Andreescu

Problems in Real Analysis

Advanced Calculus on the Real Axis

Teodora-Liliana T. Rădulescu
Department of Mathematics
Fratii Buzesti National College
Craiova 200352
Romania
teodoraradulescu@yahoo.com

Vicenţiu D. Rădulescu
Simion Stoilow Mathematics Institute
Romanian Academy
Bucharest 014700
Romania
vicentiu.radulescu@math.cnrs.fr

Titu Andreescu
School of Natural Sciences and Mathematics
University of Texas at Dallas
Richardson, TX 75080
USA
titu.andreescu@utdallas.edu

ISBN: 978-0-387-77378-0   e-ISBN: 978-0-387-77379-7
DOI: 10.1007/978-0-387-77379-7
Springer Dordrecht Heidelberg London New York

Library of Congress Control Number: 2009926486

Mathematics Subject Classification (2000): 00A07, 26-01, 28-01, 40-01

© Springer Science+Business Media, LLC 2009
All rights reserved. This work may not be translated or copied in whole or in part without the written permission of the publisher (Springer Science+Business Media, LLC, 233 Spring Street, New York, NY 10013, USA), except for brief excerpts in connection with reviews or scholarly analysis. Use in connection with any form of information storage and retrieval, electronic adaptation, computer software, or by similar or dissimilar methodology now known or hereafter developed is forbidden.
The use in this publication of trade names, trademarks, service marks, and similar terms, even if they are not identified as such, is not to be taken as an expression of opinion as to whether or not they are subject to proprietary rights.

Printed on acid-free paper

Springer is part of Springer Science+Business Media (www.springer.com)

To understand mathematics means to be able to do mathematics. And what does it mean doing mathematics? In the first place it means to be able to solve mathematical problems.
—George Pólya (1887–1985)

We come nearest to the great when we are great in humility.
—Rabindranath Tagore (1861–1941)

Foreword

This carefully written book presents an extremely motivating and original approach, by means of problem-solving, to calculus on the real line, and as such, serves as a perfect introduction to real analysis. To achieve their goal, the authors have carefully selected problems that cover an impressive range of topics, all at the core of the subject. Some problems are genuinely difficult, but solving them will be highly rewarding, since each problem opens a new vista in the understanding of mathematics. This book is also perfect for self-study, since solutions are provided.

I like the care with which the authors intersperse their text with careful reviews of the background material needed in each chapter, thought-provoking quotations, and highly interesting and well-documented historical notes. In short, this book also makes very pleasant reading, and I am confident that each of its readers will enjoy reading it as much as I did. The charm and never-ending beauty of mathematics pervade all its pages.

In addition, this little gem illustrates the idea that one cannot learn mathematics without solving difficult problems. It is a world apart from the "computer addiction" that we are unfortunately witnessing among the younger generations of would-be mathematicians, who use too much ready-made software instead or their brains, or who stand in awe in front of computer-generated images, as if they had become the essence of mathematics. As such, it carries a very useful message.

One cannot help comparing this book to a "great ancestor," the famed *Problems and Theorems in Analysis*, by Pólya and Szegő, a text that has strongly influenced generations of analysts. I am confident that this book will have a similar impact.

Hong Kong, July 2008 *Philippe G. Ciarlet*

Preface

> *If I have seen further it is by standing on the shoulders of giants.*
> —Sir Isaac Newton (1642–1727), Letter to Robert Hooke, 1675

Mathematical analysis is central to mathematics, whether pure or applied. This discipline arises in various mathematical models whose dependent variables vary continuously and are functions of one or several variables. Real analysis dates to the mid-nineteenth century, and its roots go back to the pioneering papers by Cauchy, Riemann, and Weierstrass.

In 1821, Cauchy established new requirements of rigor in his celebrated *Cours d'Analyse*. The questions he raised are the following:

– What is a derivative really? Answer: a limit.
– What is an integral really? Answer: a limit.
– What is an infinite series really? Answer: a limit.

This leads to

– What is a limit? Answer: a number.

And, finally, the last question:

– What is a number?

Weierstrass and his collaborators (Heine, Cantor) answered this question around 1870–1872.

Our treatment in this volume is strongly related to the pioneering contributions in differential calculus by Newton, Leibniz, Descartes, and Euler in the seventeenth and eighteenth centuries, with mathematical rigor in the nineteenth century promoted by Cauchy, Weierstrass, and Peano. This presentation furthers modern directions in the integral calculus developed by Riemann and Darboux.

Due to the huge impact of mathematical analysis, we have intended in this book to build a bridge between ordinary high-school or undergraduate exercises and more difficult and abstract concepts or problems related to this field. We present in this volume an unusual collection of creative problems in elementary mathematical analysis. We intend to develop some basic principles and solution techniques and to offer a systematic illustration of how to organize the natural transition from problem-solving activity toward exploring, investigating, and discovering new results and properties.

The aim of this volume in elementary mathematical analysis is to introduce, through problems-solving, fundamental ideas and methods without losing sight of the context in which they first developed and the role they play in science and particularly in physics and other applied sciences. This volume aims at rapidly developing differential and integral calculus for real-valued functions of one real variable, giving relevance to the discussion of some differential equations and maximum principles.

The book is mainly geared toward students studying the basic principles of mathematical analysis. However, given its selection of problems, organization, and level, it would be an ideal choice for tutorial or problem-solving seminars, particularly those geared toward the Putnam exam and other high-level mathematical contests. We also address this work to motivated high-school and undergraduate students. This volume is meant primarily for students in mathematics, physics, engineering, and computer science, but, not without authorial ambition, we believe it can be used by anyone who wants to learn elementary mathematical analysis by solving problems. The book is also a must-have for instructors wishing to enrich their teaching with some carefully chosen problems and for individuals who are interested in solving difficult problems in mathematical analysis on the real axis. The volume is intended as a challenge to involve students as active participants in the course. To make our work self-contained, all chapters include basic definitions and properties. The problems are clustered by topic into eight chapters, each of them containing both sections of proposed problems with complete solutions and separate sections including auxiliary problems, their solutions being left to our readers. Throughout the book, students are encouraged to express their own ideas, solutions, generalizations, conjectures, and conclusions.

The volume contains a comprehensive collection of challenging problems, our goal being twofold: first, to encourage the readers to move away from routine exercises and memorized algorithms toward creative solutions and nonstandard problem-solving techniques; and second, to help our readers to develop a host of new mathematical tools and strategies that will be useful beyond the classroom and in a number of applied disciplines. We include representative problems proposed at various national or international competitions, problems selected from prestigious mathematical journals, but also some original problems published in leading publications. That is why most of the problems contained in this book are neither standard nor easy. The readers will find both classical topics of mathematical analysis on the real axis and modern ones. Additionally, historical comments and developments are presented throughout the book in order to stimulate further inquiry.

Traditionally, a rigorous first course or problem book in elementary mathematical analysis progresses in the following order:

> Sequences
> Functions $\Longrightarrow$ Continuity $\Longrightarrow$ Differentiability $\Longrightarrow$ Integration
> Limits

However, the historical development of these subjects occurred in reverse order:

$$\text{Cauchy (1821)} \Longleftarrow \text{Weierstrass (1872)} \Longleftarrow \begin{array}{c}\text{Newton (1665)}\\ \text{Leibniz (1675)}\end{array} \Longleftarrow \begin{array}{c}\text{Archimedes}\\ \text{Kepler (1615)}\\ \text{Fermat (1638)}\end{array}$$

This book brings to life the connections among different areas of mathematical analysis and explains how various subject areas flow from one another. The volume illustrates the richness of elementary mathematical analysis as one of the most classical fields in mathematics. The topic is revisited from the higher viewpoint of university mathematics, presenting a deeper understanding of familiar subjects and an introduction to new and exciting research fields, such as Ginzburg–Landau equations, the maximum principle, singular differential and integral inequalities, and nonlinear differential equations.

The volume is divided into four parts, ten chapters, and two appendices, as follows:

Part I. Sequences, Series, and Limits
    Chapter 1. Sequences
    Chapter 2. Series
    Chapter 3. Limits of Functions
Part II. Qualitative Properties of Continuous and Differentiable Functions
    Chapter 4. Continuity
    Chapter 5. Differentiability
Part III. Applications to Convex Functions and Optimization
    Chapter 6. Convex Functions
    Chapter 7. Inequalities and Extremum Problems
Part IV. Antiderivatives, Riemann Integrability, and Applications
    Chapter 8. Antiderivatives
    Chapter 9. Riemann Integrability
    Chapter 10. Applications of the Integral Calculus
Appendix A. Basic Elements of Set Theory
Appendix B. Topology of the Real Line

Each chapter is divided into sections. Exercises, formulas, and figures are numbered consecutively in each section, and we also indicate both the chapter and the section numbers. We have included at the beginning of chapters and sections quotations from the literature. They are intended to give the flavor of mathematics as a science with a long history. This book also contains a rich glossary and index, as well as a list of abbreviations and notation.

Key features of this volume:

- contains a collection of challenging problems in elementary mathematical analysis;
- includes incisive explanations of every important idea and develops illuminating applications of many theorems, along with detailed solutions, suitable cross-references, specific how-to hints, and suggestions;
- is self-contained and assumes only a basic knowledge but opens the path to competitive research in the field;
- uses competition-like problems as a platform for training typical inventive skills;
- develops basic valuable techniques for solving problems in mathematical analysis on the real axis;
- 38 carefully drawn figures support the understanding of analytic concepts;
- includes interesting and valuable historical account of ideas and methods in analysis;
- contains excellent bibliography, glossary, and index.

The book has elementary prerequisites, and it is designed to be used for lecture courses on methodology of mathematical research or discovery in mathematics. This work is a first step toward developing connections between analysis and other mathematical disciplines, as well as physics and engineering.

The background the student needs to read this book is quite modest. Anyone with elementary knowledge in calculus is well-prepared for almost everything to be found here. Taking into account the rich introductory blurbs provided with each chapter, no particular prerequisites are necessary, even if a dose of mathematical sophistication is needed. The book develops many results that are rarely seen, and even experienced readers are likely to find material that is challenging and informative.

Our vision throughout this volume is closely inspired by the following words of George Pólya [90] (1945) on the role of problems and discovery in mathematics: *Infallible rules of discovery leading to the solution of all possible mathematical problems would be more desirable than the philosopher's stone, vainly sought by all alchemists. The first rule of discovery is to have brains and good luck. The second rule of discovery is to sit tight and wait till you get a bright idea. Those of us who have little luck and less brain sometimes sit for decades. The fact seems to be, as Poincaré observed, it is the man, not the method, that solves the problem.*

Despite our best intentions, errors are sure to have slipped by us. Please let us know of any you find.

August 2008

Teodora-Liliana Rădulescu
Vicenţiu Rădulescu
Titu Andreescu

Acknowledgments

We acknowledge, with unreserved gratitude, the crucial role of Professors Catherine Bandle, Wladimir-Georges Boskoff, Louis Funar, Patrizia Pucci, Richard Stong, and Michel Willem, who encouraged us to write a problem book on this subject. Our colleague and friend Professor Dorin Andrica has been very interested in this project and suggested some appropriate problems for this volume. We warmly thank Professors Ioan Şerdean and Marian Tetiva for their kind support and useful discussions.

This volume was completed while Vicenţiu Rădulescu was visiting the University of Ljubljana during July and September 2008 with a research position funded by the Slovenian Research Agency. He would like to thank Professor Dušan Repovš for the invitation and many constructive discussions.

We thank Dr. Nicolae Constantinescu and Dr. Mirel Coşulschi for the professional drawing of figures contained in this book.

We are greatly indebted to the anonymous referees for their careful reading of the manuscript and for numerous comments and suggestions. These precious constructive remarks were very useful to us in the elaboration of the final version of this volume.

We are grateful to Ann Kostant, Springer editorial director for mathematics, for her efficient and enthusiastic help, as well as for numerous suggestions related to previous versions of this book. Our special thanks go also to Laura Held and to the other members of the editorial technical staff of Springer New York for the excellent quality of their work.

We are particularly grateful to copyeditor David Kramer for his guidance, thoroughness and attention to detail.

V. Rădulescu acknowledges the support received from the Romanian Research Council CNCSIS under Grant 55/2008 "Sisteme diferenţiale în analiza neliniară şi aplicaţii."

Contents

Foreword vii

Preface ix

Acknowledgments xiii

Abbreviations and Notation xix

Part I Sequences, Series, and Limits

1 **Sequences** 3
   1.1 Main Definitions and Basic Results 3
   1.2 Introductory Problems 7
   1.3 Recurrent Sequences 18
   1.4 Qualitative Results 30
   1.5 Hardy's and Carleman's Inequalities 45
   1.6 Independent Study Problems 51

2 **Series** 59
   2.1 Main Definitions and Basic Results 59
   2.2 Elementary Problems 66
   2.3 Convergent and Divergent Series 73
   2.4 Infinite Products 86
   2.5 Qualitative Results 89
   2.6 Independent Study Problems 110

3 **Limits of Functions** 115
   3.1 Main Definitions and Basic Results 115
   3.2 Computing Limits 118
   3.3 Qualitative Results 124
   3.4 Independent Study Problems 133

Part II Qualitative Properties of Continuous and Differentiable Functions

4 Continuity 139
    4.1   The Concept of Continuity and Basic Properties 139
    4.2   Elementary Problems 144
    4.3   The Intermediate Value Property 147
    4.4   Types of Discontinuities 151
    4.5   Fixed Points 154
    4.6   Functional Equations and Inequalities 163
    4.7   Qualitative Properties of Continuous Functions 169
    4.8   Independent Study Problems 177

5 Differentiability 183
    5.1   The Concept of Derivative and Basic Properties 183
    5.2   Introductory Problems 198
    5.3   The Main Theorems 218
    5.4   The Maximum Principle 235
    5.5   Differential Equations and Inequalities 238
    5.6   Independent Study Problems 252

Part III Applications to Convex Functions and Optimization

6 Convex Functions 263
    6.1   Main Definitions and Basic Results 263
    6.2   Basic Properties of Convex Functions and Applications 265
    6.3   Convexity versus Continuity and Differentiability 273
    6.4   Qualitative Results 278
    6.5   Independent Study Problems 285

7 Inequalities and Extremum Problems 289
    7.1   Basic Tools 289
    7.2   Elementary Examples 290
    7.3   Jensen, Young, Hölder, Minkowski, and Beyond 294
    7.4   Optimization Problems 300
    7.5   Qualitative Results 305
    7.6   Independent Study Problems 308

Part IV Antiderivatives, Riemann Integrability, and Applications

8 Antiderivatives 313
    8.1   Main Definitions and Properties 313
    8.2   Elementary Examples 315
    8.3   Existence or Nonexistence of Antiderivatives 317
    8.4   Qualitative Results 319
    8.5   Independent Study Problems 324

9 Riemann Integrability ... 325
- 9.1 Main Definitions and Properties ... 325
- 9.2 Elementary Examples ... 329
- 9.3 Classes of Riemann Integrable Functions ... 337
- 9.4 Basic Rules for Computing Integrals ... 339
- 9.5 Riemann Iintegrals and Limits ... 341
- 9.6 Qualitative Results ... 351
- 9.7 Independent Study Problems ... 367

10 Applications of the Integral Calculus ... 373
- 10.1 Overview ... 373
- 10.2 Integral Inequalities ... 374
- 10.3 Improper Integrals ... 390
- 10.4 Integrals and Series ... 402
- 10.5 Applications to Geometry ... 406
- 10.6 Independent Study Problems ... 409

Part V Appendix

A Basic Elements of Set Theory ... 417
- A.1 Direct and Inverse Image of a Set ... 417
- A.2 Finite, Countable, and Uncountable Sets ... 418

B Topology of the Real Line ... 419
- B.1 Open and Closed Sets ... 419
- B.2 Some Distinguished Points ... 420

Glossary ... 421

References ... 437

Index ... 443

Abbreviations and Notation

Abbreviations

We have tried to avoid using nonstandard abbreviations as much as possible. Other abbreviations include:

AMM	American Mathematical Monthly
GMA	Mathematics Gazette, Series A
MM	Mathematics Magazine
IMO	International Mathematical Olympiad
IMCUS	International Mathematics Competition for University Students
MSC	Miklós Schweitzer Competitions
Putnam	The William Lowell Putnam Mathematical Competition
SEEMOUS	South Eastern European Mathematical Olympiad for University Students

Notation

We assume familiarity with standard elementary notation of set theory, logic, algebra, analysis, number theory, and combinatorics. The following is notation that deserves additional clarification.

$\mathbb{N}$	the set of nonnegative integers ($\mathbb{N} = \{0, 1, 2, 3, \ldots\}$)
$\mathbb{N}^*$	the set of positive integers ($\mathbb{N}^* = \{1, 2, 3, \ldots\}$)
$\mathbb{Z}$	the set of integer real numbers ($\mathbb{Z} = \{\ldots, -3, -2, -1, 0, 1, 2, 3, \ldots\}$)
$\mathbb{Z}^*$	the set of nonzero integer real numbers ($\mathbb{Z}^* = \mathbb{Z} \setminus \{0\}$)
$\mathbb{Q}$	the set of rational real numbers $\left(\mathbb{Q} = \left\{\frac{m}{n}; m \in \mathbb{Z}, n \in \mathbb{N}^*, m \text{ and } n \text{ are relatively prime}\right\}\right)$
$\mathbb{R}$	the set of real numbers
$\mathbb{R}^*$	the set of nonzero real numbers ($\mathbb{R}^* = \mathbb{R} \setminus \{0\}$)
$\mathbb{R}_+$	the set of nonnegative real numbers ($\mathbb{R}_+ = [0, +\infty)$)
$\overline{\mathbb{R}}$	the completed real line $\left(\overline{\mathbb{R}} = \mathbb{R} \cup \{-\infty, +\infty\}\right)$

$\mathbb{C}$	the set of complex numbers
e	$\lim_{n\to\infty}\left(1+\frac{1}{n}\right)^n = 2.71828\ldots$
$\sup A$	the least upper bound of the set $A \subset \mathbb{R}$
$\inf A$	the greatest lower bound of the set $A \subset \mathbb{R}$
x_+	the positive part of the real number x ($x_+ = \max\{x,0\}$)
x_-	the negative part of the real number x ($x_- = \max\{-x,0\}$)
$\lvert x \rvert$	the modulus (absolute value) of the real number x ($\lvert x \rvert = x_+ + x_-$)
$\{x\}$	the fractional part of the real number x ($x = [x] + \{x\}$)
$\mathrm{Card}(A)$	cardinality of the finite set A
$\mathrm{dist}(x,A)$	the distance from $x \in \mathbb{R}$ to the set $A \subset \mathbb{R}$ ($\mathrm{dist}(x,A) = \inf\{\lvert x-a \rvert;\ a \in A\}$)
$\mathrm{Int}\,A$	the set of interior points of $A \subset \mathbb{R}$
$f(A)$	the image of the set A under a mapping f
$f^{-1}(B)$	the inverse image of the set B under a mapping f
$f \circ g$	the composition of functions f and g: $(f \circ g)(x) = f(g(x))$
$n!$	n factorial, equal to $n(n-1)\cdots 1$ \quad ($n \in \mathbb{N}^*$)
$(2n)!!$	$2n(2n-2)(2n-4)\cdots 4 \cdot 2$ \quad ($n \in \mathbb{N}^*$)
$(2n+1)!!$	$(2n+1)(2n-1)(2n-3)\cdots 3 \cdot 1$ \quad ($n \in \mathbb{N}^*$)
$\ln x$	$\log_e x$ ($x > 0$)
$x \nearrow x_0$	$x \to x_0 \in \mathbb{R}$ and $x < x_0$
$x \searrow x_0$	$x \to x_0 \in \mathbb{R}$ and $x > x_0$
$\limsup_{n\to\infty} x_n$	$\lim_{n\to\infty}\left(\sup_{k \geq n} x_k\right)$
$\liminf_{n\to\infty} x_n$	$\lim_{n\to\infty}\left(\inf_{k \geq n} x_k\right)$
$f^{(n)}(x)$	nth derivative of the function f at x
$C^n(a,b)$	the set of n-times differentiable functions $f:(a,b)\to\mathbb{R}$ such that $f^{(n)}$ is continuous on (a,b)
$C^\infty(a,b)$	the set of infinitely differentiable functions $f:(a,b)\to\mathbb{R}$ ($C^\infty(a,b) = \bigcap_{n=0}^{\infty} C^n(a,b)$)
Δf	the Laplace operator applied to the function $f: D \subset \mathbb{R}^N \to \mathbb{R}$ $\left(\Delta f = \dfrac{\partial^2 f}{\partial x_1^2} + \dfrac{\partial^2 f}{\partial x_2^2} + \cdots + \dfrac{\partial^2 f}{\partial x_N^2}\right)$
Landau's notation	$f(x) = o(g(x))$ as $x \to x_0$ if $f(x)/g(x) \to 0$ as $x \to x_0$
	$f(x) = O(g(x))$ as $x \to x_0$ if $f(x)/g(x)$ is bounded in a neighborhood of x_0
	$f \sim g$ as $x \to x_0$ if $f(x)/g(x) \to 1$ as $x \to x_0$
Hardy's notation	$f \prec\prec g$ as $x \to x_0$ if $f(x)/g(x) \to 0$ as $x \to x_0$
	$f \preceq g$ as $x \to x_0$ if $f(x)/g(x)$ is bounded in a neighborhood of x_0

Part I
Sequences, Series, and Limits

In the first three chapters of this volume we are mainly concerned with the theory of sequences and series of real numbers. Sequences describe wide classes of discrete processes arising in various applications. The theory of sequences is also viewed as a preliminary step in the attempt to model continuous phenomena in nature. In many ways, series of real numbers are much more interesting and important to analysis than finite sums. At the same time, the theory of series may be viewed as a sequel to the theory of sequences. We also develop the notion of limit of a function, which is one of the deepest concepts of mathematical analysis. This concept describes the behavior of a certain system, described by a function, as the variable approaches a certain value.

Chapter 1
Sequences

> As far as the laws of mathematics refer to reality, they are not certain; and as far as they are certain, they do not refer to reality.
> —Albert Einstein (1879–1955)

Abstract. In this chapter we study *real sequences*, a special class of functions whose domain is the set $\mathbb{N}$ of natural numbers and range a set of real numbers.

1.1 Main Definitions and Basic Results

> Hypotheses non fingo. ["I frame no hypotheses."]
> Sir Isaac Newton (1642–1727)

Sequences describe wide classes of discrete processes arising in various applications. The theory of sequences is also viewed as a preliminary step in the attempt to model continuous phenomena in nature. Since ancient times, mathematicians have realized that it is difficult to reconcile the discrete with the continuous. We understand counting $1, 2, 3, \ldots$ up to arbitrarily large numbers, but do we also understand moving from 0 to 1 through the continuum of points between them? Around 450 BC, Zeno thought not, because continuous motion involves infinity in an essential way. As he put it in his paradox of dichotomy:
There is no motion because that which is moved must arrive at the middle (of its course) before it arrives at the end.

Aristotle, *Physics*, Book VI, Ch. 9

A sequence of real numbers is a function $f : \mathbb{N} \to \mathbb{R}$ (or $f : \mathbb{N}^* \to \mathbb{R}$). We usually write a_n (or b_n, x_n, etc.) instead of $f(n)$. If $(a_n)_{n\geq 1}$ is a sequence of real numbers and if $n_1 < n_2 < \cdots < n_k < \cdots$ is an increasing sequence of positive integers, then the sequence $(a_{n_k})_{k\geq 1}$ is called a *subsequence* of $(a_n)_{n\geq 1}$.

A sequence of real numbers $(a_n)_{n\geq 1}$ is said to be *nondecreasing* (resp., *increasing*) if $a_n \leq a_{n+1}$ (resp., $a_n < a_{n+1}$), for all $n \geq 1$. The sequence $(a_n)_{n\geq 1}$ is called *nonincreasing* (resp., *decreasing*) if the above inequalities hold with "$\geq$" (resp., "$>$") instead of "$\leq$" (resp., "$<$").

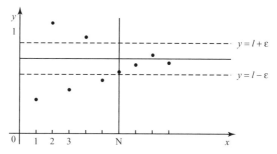

Fig. 1.1 Adapted from G.H. Hardy, *Pure Mathematics*, Cambridge University Press, 1952.

We recall in what follows some basic definitions and properties related to sequences. One of the main notions we need in the sequel is that of *convergence*. Let $(a_n)_{n\geq 1}$ be a sequence of real numbers. We say that $(a_n)_{n\geq 1}$ is a *convergent sequence* if there exists $\ell \in \mathbb{R}$ (which is called the *limit* of $(a_n)_{n\geq 1}$) such that for each neighborhood $\mathcal{N}$ of ℓ, we have $a_n \in \mathcal{N}$ for all $n \geq N$, where N is a positive integer depending on $\mathcal{N}$. In other words, $(a_n)_{n\geq 1}$ converges to ℓ if and only if for each $\varepsilon > 0$ there exists a natural number $N = N(\varepsilon)$ such that $|a_n - \ell| < \varepsilon$, for all $n \geq N$. In this case we write $\lim_{n\to\infty} a_n = \ell$ or $a_n \to \ell$ as $n \to \infty$. The concentration of a_n's in the strip $(\ell - \varepsilon, \ell + \varepsilon)$ for $n \geq N$ is depicted in Figure 1.1.

Example. Define $a_n = (1 + 1/n)^n$, $b_n = (1 + 1/n)^{n+1}$, $c_n = \sum_{k=0}^{n} 1/k!$. Then $(a_n)_{n\geq 1}$ and $(c_n)_{n\geq 1}$ are increasing sequences, while $(b_n)_{n\geq 1}$ is a decreasing sequence. However, all these three sequences have the same limit, which is denoted by e (e = 2.71828...). We will prove in the next chapter that e is an irrational number. Another important example is given by the following formula due to Stirling, which asserts that, asymptotically, $n!$ behaves like $n^n e^{-n} \sqrt{2\pi n}$. More precisely,

$$\lim_{n\to\infty} \frac{n!}{n^n e^{-n} \sqrt{2\pi n}} = 1.$$

Stirling's formula is important in calculating limits, because without this asymptotic property it is difficult to estimate the size of $n!$ for large n. In this capacity, it plays an important role in probability theory, when it is used in computing the probable outcome of an event after a very large number of trials. We refer to Chapter 9 for a complete proof of Stirling's formula.

We say that the sequence of real numbers $(a_n)_{n\geq 1}$ has limit $+\infty$ (resp., $-\infty$) if to each $\alpha \in \mathbb{R}$ there corresponds some positive integer N_0 such that $n \geq N_0$ implies $a_n > \alpha$ (resp., $a_n < \alpha$). We *do not* say that $(a_n)_{n\geq 1}$ converges in these cases.

An element $\alpha \in \overline{\mathbb{R}}$ is called an *accumulation point* of the sequence $(a_n)_{n\geq 1} \subset \mathbb{R}$ if there exists a subsequence $(a_{n_k})_{k\geq 1}$ such that $a_{n_k} \to \alpha$ as $n_k \to \infty$.

Any convergent sequence is bounded. The converse is not true (find a counterexample!), but we still have a partial positive answer, which is contained in the following result, which is due to Bernard Bolzano (1781–1848) and

1.1 Main Definitions and Basic Results

Karl Weierstrass (1815–1897). It seems that this theorem was revealed for the first time in Weierstrass's lecture of 1874.

Bolzano–Weierstrass Theorem. *A bounded sequence of real numbers has a convergent subsequence.*

However, some additional assumptions can guarantee that a bounded sequence converges. An important criterion for convergence is stated in what follows.

Monotone Convergence Theorem. *Let $(a_n)_{n\geq 1}$ be a bounded sequence that is monotone. Then $(a_n)_{n\geq 1}$ is a convergent sequence. If increasing, then $\lim_{n\to\infty} a_n = \sup_n a_n$, and if decreasing, then $\lim_{n\to\infty} a_n = \inf_n a_n$.*

In many arguments a central role is played by the following principle due to Cantor.

Nested Intervals Theorem. *Suppose that $I_n = [a_n, b_n]$ are closed intervals such that $I_{n+1} \subset I_n$, for all $n \geq 1$. If $\lim_{n\to\infty}(b_n - a_n) = 0$, then there is a unique real number x_0 that belongs to every I_n.*

A sequence $(a_n)_{n\geq 1}$ of real numbers is called a *Cauchy sequence* [Augustin Louis Cauchy (1789–1857)] if for every $\varepsilon > 0$ there is a natural number N_ε such that $|a_m - a_n| < \varepsilon$, for all $m, n \geq N_\varepsilon$. A useful result is that any Cauchy sequence is a bounded sequence. The following strong convergence criterion reduces the study of convergent sequences to that of Cauchy sequences.

Cauchy's Criterion. *A sequence of real numbers is convergent if and only if it is a Cauchy sequence.*

Let $(a_n)_{n\geq 1}$ be an arbitrary sequence of real numbers. The *limit inferior* of $(a_n)_{n\geq 1}$ (denoted by $\liminf_{n\to\infty} a_n$) is the supremum of the set X of $x \in \mathbb{R}$ such that there are at most a finite numbers of $n \in \mathbb{N}^*$ for which $a_n < x$. The *limit superior* of $(a_n)_{n\geq 1}$ (denoted by $\limsup_{n\to\infty} a_n$) is the infimum of the set Y of $y \in \mathbb{R}$ such that there are finitely many positive integers n for which $a_n > y$. Equivalent characterizations of these notions are the following:

(i) $\ell = \limsup_{n\to\infty} a_n$ ($\ell \in \overline{\mathbb{R}}$) if and only if whenever $\alpha < \ell$ the set $\{n \geq \mathbb{N}^*;\ a_n > \alpha\}$ is infinite, and whenever $\ell < \beta$ the set $\{n \geq \mathbb{N}^*;\ a_n > \beta\}$ is finite;

(ii) $\ell = \limsup_{n\to\infty} a_n$ if and only if $\ell = \inf_{m\geq 1} \sup_{n\geq m} a_n$.

Exercise. Formulate the above characterizations for $\liminf_{n\to\infty} a_n$.

The limit inferior and the limit superior of a sequence **always** exist, possibly in $\overline{\mathbb{R}}$. Moreover, the following relations hold:

$$\liminf_{n\to\infty} a_n \leq \limsup_{n\to\infty} a_n,$$

$$\liminf_{n\to\infty}(a_n + b_n) \geq \liminf_{n\to\infty} a_n + \liminf_{n\to\infty} b_n,$$

$$\limsup_{n\to\infty}(a_n + b_n) \leq \limsup_{n\to\infty} a_n + \limsup_{n\to\infty} b_n.$$

The existence of the limit of a sequence is closely related to "liminf" and "limsup." More precisely, the sequence $(a_n)_{n\geq 1}$ has a limit if and only if $\liminf_{n\to\infty} a_n = \limsup_{n\to\infty} a_n$.

A very useful result in applications concerns the following link between the ratio and the nth root of the terms of a sequence of positive numbers.

Theorem. *Let $(a_n)_{n\geq 1}$ be a sequence of positive numbers. Then*

$$\liminf_{n\to\infty} \frac{a_{n+1}}{a_n} \leq \liminf_{n\to\infty} \sqrt[n]{a_n} \leq \limsup_{n\to\infty} \sqrt[n]{a_n} \leq \limsup_{n\to\infty} \frac{a_{n+1}}{a_n}.$$

The above result implies, in particular, that if $\lim_{n\to\infty}(a_{n+1}/a_n) = \ell$, with $0 \leq \ell \leq \infty$, then $\lim_{n\to\infty} \sqrt[n]{a_n} = \ell$.

Example. The above theorem implies that $\lim_{n\to\infty} a_n = 1$, where $a_n = \sqrt[n]{n}$. We can give an elementary proof to the fact that $a_n > a_{n+1}$, for all $n \geq 3$. Indeed, define $b_n = (a_n/a_{n+1})^{n(n+1)}$, for all $n \geq 2$. Then $b_n > b_{n-1}$ for $n \geq 2$ is equivalent to $n^2 = a_n^{2n} > a_{n+1}^{n+1} a_{n-1}^{n-1} = n^2 - 1$. It remains to show that $b_n > 1$ for all $n \geq 3$. This follows from the fact that $(b_n)_{n\geq 2}$ is increasing, combined with $b_3 = 3^4/4^3 = 81/64 > 1$.

A sequence that is not convergent is called a *divergent* sequence. An important example of a divergent sequence is given by $a_n = 1 + 1/2 + 1/3 + \cdots + 1/n$, $n \geq 1$. Indeed, since $(a_n)_{n\geq 1}$ is increasing, it has a limit $\ell \in \mathbb{R} \cup \{+\infty\}$. Assuming that ℓ is finite, it follows that

$$\frac{1}{2} + \frac{1}{4} + \frac{1}{6} + \cdots = \frac{1}{2}\left(1 + \frac{1}{2} + \frac{1}{3} + \cdots\right) \to \frac{\ell}{2} \quad \text{as } n\to\infty.$$

This means that

$$\frac{1}{1} + \frac{1}{3} + \frac{1}{5} + \cdots \to \frac{\ell}{2} \quad \text{as } n\to\infty,$$

which is impossible because $1/1 > 1/2$, $1/3 > 1/4$, $1/5 > 1/6$, and so on. Consequently, $a_n \to +\infty$ as $n\to\infty$. The associated series $1 + 1/2 + 1/3 + 1/4 + \cdots$ is usually called the *harmonic series*. Furthermore, if $p < 1$ we have $1/n^p > 1/n$, for all $n \geq 2$. It follows that the sequence $(a_n)_{n\geq 1}$ defined by $a_n = \sum_{k=1}^n 1/k^p$ diverges, too. The situation is different if $p > 1$. In this case, the sequence $(a_n)_{n\geq 1}$ defined as above is convergent. Indeed,

$$\begin{aligned}
a_{2n+1} &= 1 + \left[\frac{1}{2^p} + \frac{1}{4^p} + \cdots + \frac{1}{(2n)^p}\right] + \left[\frac{1}{3^p} + \frac{1}{5^p} + \cdots + \frac{1}{(2n+1)^p}\right] \\
&< 1 + \left[\frac{1}{2^p} + \frac{1}{4^p} + \cdots + \frac{1}{(2n)^p}\right] + \left[\frac{1}{2^p} + \frac{1}{4^p} + \cdots + \frac{1}{(2n)^p}\right] \\
&= 1 + \frac{a_n}{2^p} + \frac{a_n}{2^p} < 1 + 2^{1-p} a_{2n+1},
\end{aligned}$$

because $a_n < a_{2n+1}$. Thus $(1 - 2^{1-p})a_{2n+1} < 1$. Since $p > 1$, we have $1 - 2^{1-p} > 0$. Hence $a_{2n+1} < (1 - 2^{1-p})^{-1}$ for all $n \geq 1$. So, the sequence $(a_n)_{n\geq 1}$ (for $p > 1$) is increasing and bounded above by $(1 - 2^{1-p})^{-1}$, which means that it converges. The corresponding limit $\lim_{n\to\infty} \sum_{k=1}^n 1/k^p = \sum_{n=1}^\infty 1/n^p$ is denoted by $\zeta(p)$ ($p > 1$) and is called the *Riemann zeta function* [Georg Friedrich Bernhard Riemann (1826–1866)]. This function is related to a celebrated mathematical conjecture (Hilbert's eighth problem). The Riemann hypothesis, formulated in 1859, is closely related to the frequency of prime numbers and is one of the seven grand challenges of mathematics ("Millennium Problems," as designated by the Clay Mathematics Institute; see http://www.claymath.org/millennium/).

The following result (whose "continuous" variant is l'Hôpital's rule for differentiable functions) provides us a method to compute limits of the indeterminate form $0/0$ or ∞/∞. This property is due to Otto Stolz (1859–1906) and Ernesto Cesàro (1859–1906).

Stolz–Cesàro Lemma. *Let $(a_n)_{n\geq 1}$ and $(b_n)_{n\geq 1}$ be two sequences of real numbers.*

(i) *Assume that $a_n \to 0$ and $b_n \to 0$ as $n \to \infty$. Suppose, moreover, that $(b_n)_{n\geq 1}$ is decreasing for all sufficiently large n and there exists*

$$\lim_{n\to\infty} \frac{a_{n+1}-a_n}{b_{n+1}-b_n} =: \ell \in \overline{\mathbb{R}}.$$

Then there exists $\lim_{n\to\infty} a_n/b_n$ and, moreover, $\lim_{n\to\infty} a_n/b_n = \ell$.

(ii) *Assume that $b_n \to +\infty$ as $n \to \infty$ and that $(b_n)_{n\geq 1}$ is increasing for all sufficiently large n. Suppose that there exists*

$$\lim_{n\to\infty} \frac{a_{n+1}-a_n}{b_{n+1}-b_n} =: \ell \in \overline{\mathbb{R}}.$$

Then there exists $\lim_{n\to\infty} a_n/b_n$ and, moreover, $\lim_{n\to\infty} a_n/b_n = \ell$.

1.2 Introductory Problems

> Nature not only suggests to us problems, she suggests their solution.
>
> Henri Poincaré (1854–1912)

We first prove with elementary arguments the following basic result.

Arithmetic–Geometric Means (AM–GM) Inequality. *For any positive numbers $a_1, a_2, \ldots, a_n$ we have*

$$\frac{a_1+a_2+\cdots+a_n}{n} \geq \sqrt[n]{a_1 a_2 \cdots a_n}.$$

Replacing a_k with $a_k \sqrt[n]{a_1 \cdots a_n}$ (for $1 \leq k \leq n$), we have $a_1 a_2 \cdots a_n = 1$, so it is enough to prove that $a_1 + a_2 + \cdots + a_n \geq n$. We argue by induction. For $n = 1$ the property is obvious. Passing from n to $n+1$, we can assume that $a_1 \leq 1 \leq a_2$. Hence $(1-a_1)(a_2-1) \geq 1$, that is, $a_1 + a_2 \geq 1 + a_1 a_2$. Next, we set $b_1 = a_1 a_2$ and, for any $k = 2, \ldots, n$, we set $b_k = a_{k+1}$. Hence $b_1 b_2 \cdots b_n = 1$. So, by the induction hypothesis,

$$a_1 a_2 + a_3 + \cdots + a_n \geq n.$$

It follows that
$$a_1+a_2+\cdots+a_{n+1} \geq a_1a_2+1+a_3+\cdots+a_{n+1} \geq n+1.$$

The AM–GM inequality also implies a relationship between the harmonic mean and the geometric mean: for any positive numbers $a_1, a_2, \ldots, a_n$ we have
$$\frac{n}{\frac{1}{a_1}+\frac{1}{a_2}+\cdots+\frac{1}{a_n}} \leq \sqrt[n]{a_1 a_2 \cdots a_n}.$$

The following easy exercise shows the importance of elementary monotony arguments for deducing the value of the limit of a convergent sequence.

1.2.1. Let $A_1, A_2, \ldots, A_k$ be nonnegative numbers. Compute
$$\lim_{n\to\infty} (A_1^n + A_2^n + \cdots + A_k^n)^{1/n}.$$

Solution. Without loss of generality, we may assume that $A_1 = \max\{A_1, \ldots, A_k\}$. Therefore $A_1^n \leq A_1^n + \cdots + A_k^n \leq kA_1^n$. It follows that
$$A_1 = \lim_{n\to\infty} (A_1^n)^{1/n} \leq \lim_{n\to\infty} (A_1^n + \cdots + A_k^n)^{1/n} \leq \lim_{n\to\infty} (kA_1^n)^{1/n} = A_1,$$
which shows that the limit sought is $A_1 := \min\{A_j;\ 1 \leq j \leq k\}$. □

Comments. We easily observe that the above result does not remain true if we do not assume that the numbers $A_1, \ldots, A_k$ are nonnegative. Moreover, in such a case it is possible that the sequence defined by $B_n := (A_1^n + \cdots + A_k^n)^{1/n}$ is not even convergent (give an example!).

We have already provided some basic examples of convergent and divergent sequences. The next exercise shows how, by means of monotony principles, we can construct further examples of such sequences.

1.2.2. Prove that $\lim_{n\to\infty} \sqrt[n]{n!} = +\infty$ and $\lim_{n\to\infty} \sqrt[n]{n!}/n = e^{-1}$.

Solution. We first observe that $(2n)! \geq \prod_{k=n}^{2n} k \geq n^{n+1}$. Hence
$$\sqrt[2n]{(2n)!} \geq \sqrt[2n]{n^{n+1}} \geq \sqrt{n} \quad \text{and} \quad \sqrt[2n+1]{(2n+1)!} \geq \sqrt[2n+1]{n^{n+1}} \geq \sqrt{n}.$$

For the last part we take into account that for any sequence $(a_n)_{n\geq 1}$ of positive numbers we have
$$\liminf_{n\to\infty} \frac{a_{n+1}}{a_n} \leq \liminf_{n\to\infty} \sqrt[n]{a_n} \leq \limsup_{n\to\infty} \sqrt[n]{a_n} \leq \limsup_{n\to\infty} \frac{a_{n+1}}{a_n}.$$

Taking $a_n = n!/n^n$ and using $\lim_{n\to\infty} (1+n^{-1})^{1/n} = e$, we conclude that $\lim_{n\to\infty} \sqrt[n]{n!}/n = e^{-1}$. An alternative argument is based on the Stolz–Cesàro lemma applied to $a_n = \log(n!/n^n)$ and $b_n = n$ and using again $\lim_{n\to\infty} (1+n^{-1})^{1/n} = e$. □

1.2 Introductory Problems

A nonobvious generalization of the above property is stated below. Our proof applies subtle properties of real-valued functions (see Chapter 5) but we strongly suggest that the reader refine the monotony arguments developed above.

Independent Study. *Prove that for real number $p \geq 0$ we have*

$$\lim_{n\to\infty} \frac{\left(1^{1^p} \cdot 2^{2^p} \cdots n^{n^p}\right)^{1/n^{p+1}}}{n^{1/(p+1)}} = e^{-1/(p+1)^2}.$$

Particular case:

$$\lim_{n\to\infty} \frac{\left(1^1 \cdot 2^2 \cdots n^n\right)^{1/n^2}}{n^{1/2}} = e^{-1/4}.$$

Hint. Use the mean value theorem. For instance, in the particular case $p = 2$, apply the Lagrange mean value theorem to the function $f(x) = (x^3 \ln x)/3 - x^3/9$ on the interval $[k, k+1]$, $1 \leq k \leq n$.

We have seen above that $\sqrt[n]{n!} \to \infty$ as $n \to \infty$. A natural question is to study the asymptotic behavior of the difference of two consecutive terms of this sequence. The next problem was published in 1901 and is due to the Romanian mathematician Traian Lalescu (1882–1929), who wrote one of the first treatises on integral equations.

1.2.3. *Find the limit of the sequence $(a_n)_{n\geq 2}$ defined by $a_n = \sqrt[n+1]{(n+1)!} - \sqrt[n]{n!}$.*

Solution. We can write $a_n = \sqrt[n]{n!}\,(b_n - 1)$, where $b_n = \sqrt[n+1]{(n+1)!}/\sqrt[n]{n!}$. Hence

$$a_n = \frac{\sqrt[n]{n!}}{n} \cdot \frac{b_n - 1}{\ln b_n} \cdot \ln b_n^n. \tag{1.1}$$

But $\lim_{n\to\infty} \sqrt[n]{n!}/n = e^{-1}$, so $b_n \to 1$ as $n \to \infty$. On the other hand,

$$\lim_{n\to\infty} b_n^n = \lim_{n\to\infty} \frac{(n+1)!}{n!} \cdot \frac{1}{\sqrt[n+1]{(n+1)!}} = e. \tag{1.2}$$

So, by (1.1) and (1.2), we obtain that $a_n \to e^{-1}$ as $n \to \infty$. $\square$

Remark. In the above solution we have used the property that if $b_n \to 1$ as $n \to \infty$ then $(b_n - 1)/\ln b_n \to 1$ as $n \to \infty$. This follows directly either by applying the Stolz–Cesàro lemma or after observing that

$$\lim_{n\to\infty} \frac{b_n - 1}{\ln b_n} = \lim_{n\to\infty} \frac{1}{\ln\left[1 + (b_n - 1)\right]^{1/(b_n - 1)}} = \frac{1}{\ln e} = 1.$$

Independent Study. *Let p be a nonnegative real number. Study the convergence of the sequence $(x_n)_{n\geq 1}$ defined by*

$$x_n = \left[1^{1^p} \cdot 2^{2^p} \cdots (n+1)^{(n+1)^p}\right]^{1/(n+1)^{p+1}} - \left(1^{1^p} \cdot 2^{2^p} \cdots n^{n^p}\right)^{1/n^{p+1}}.$$

We already know that the sequence $(s_n)_{n\geq 1}$ defined by $s_n = 1 + 1/2 + \cdots + 1/n$ diverges to $+\infty$. In what follows we establish the asymptotic behavior for s_n/n as $n\to\infty$. As a consequence, the result below implies that $\lim_{n\to\infty} n(1 - \sqrt[n]{n}) = \infty$.

1.2.4. Consider the sequence $(s_n)_{n\geq 1}$ defined by $s_n = 1 + 1/2 + \cdots + 1/n$. Prove that

(a) $n(n+1)^{1/n} < n + s_n$, for all integers $n > 1$;
(b) $(n-1)n^{-1/(n-1)} < n - s_n$, for all integers $n > 2$.

Solution. (a) Using the AM–GM inequality we obtain

$$\frac{n+s_n}{n} = \frac{(1+1) + \left(1+\frac{1}{2}\right) + \cdots + \left(1+\frac{1}{n}\right)}{n}$$

$$> \left[(1+1)\cdot\left(1+\frac{1}{2}\right)\cdots\left(1+\frac{1}{n}\right)\right]^{1/n} = (n+1)^{1/n}.$$

(b) By the definition of $(s_n)_{n\geq 1}$ we have

$$\frac{n-s_n}{n-1} = \frac{\left(1-\frac{1}{2}\right) + \cdots + \left(1-\frac{1}{n}\right)}{n-1} > \left[\left(1-\frac{1}{2}\right)\cdots\left(1-\frac{1}{n}\right)\right]^{1/(n-1)} = n^{-1/(n-1)}.$$

$\square$

The following exercise involves a second-order linear recurrence.

1.2.5. Let $\alpha \in (0,2)$. Consider the sequence defined by

$$x_{n+1} = \alpha x_n + (1-\alpha)x_{n-1}, \quad \text{for all } n \geq 1.$$

Find the limit of the sequence in terms of α, x_0, and x_1.

Solution. We have $x_n - x_{n-1} = (\alpha - 1)(x_n - x_{n-1})$. It follows that $x_n - x_{n-1} = (\alpha-1)^{n-1}(x_1 - x_0)$. Therefore

$$x_n - x_0 = \sum_{k=1}^{n}(x_k - x_{k-1}) = (x_1 - x_0)\sum_{k=1}^{n}(\alpha-1)^{k-1}.$$

Now, using the assumption $\alpha \in (0,2)$, we deduce that

$$\lim_{n\to\infty} x_n = \frac{(1-\alpha)x_0 + x_1}{2-\alpha}. \quad \square$$

Next, we discuss a first-order quadratic recurrence in order to establish a necessary and sufficient condition for convergence in terms of the involved real parameter.

1.2.6. Let a be a positive number. Define the sequence $(x_n)_{n\geq 0}$ by

$$x_{n+1} = a + x_n^2, \quad \text{for all } n \geq 0, \ x_0 = 0.$$

1.2 Introductory Problems

Find a necessary and sufficient condition such that the sequence is convergent.

Solution. If $\lim_{n \to \infty} x_n = \ell$ then $\ell = a + \ell^2$, that is,

$$\ell = \frac{1 \pm \sqrt{1-4a}}{2}.$$

So, necessarily, $a \leq 1/4$.

Conversely, assume that $0 < a \leq 1/4$. From $x_{n+1} - x_n = x_n^2 - x_{n-1}^2$ it follows that the sequence (x_n) is increasing. Moreover,

$$x_{n+1} = a + x_n^2 < \frac{1}{4} + \frac{1}{4} = \frac{1}{2},$$

provided that $x_n < 1/2$. This shows that the sequence is bounded, so it converges. $\square$

The above result is extended below to larger classes of quadratic nonlinearities. The close relationship between boundedness, monotony, and convergence is pointed out and a complete discussion is developed in the following exercise.

1.2.7. Let $f(x) = 1/4 + x - x^2$. For any $x \in \mathbb{R}$, define the sequence $(x_n)_{n \geq 0}$ by $x_0 = x$ and $x_{n+1} = f(x_n)$. If this sequence is convergent, let x_∞ be its limit.

(a) Show that if $x = 0$, then the sequence is bounded and increasing, and compute its limit $x_\infty = \ell$.
(b) Find all possible values of ℓ and the corresponding real numbers x such that $x_\infty = \ell$.

Solution. (a) We have

$$f(x) = \frac{1}{2} - \left(x - \frac{1}{2}\right)^2,$$

so $x_n \leq 1/2$, for all $n \geq 1$. This inequality also shows that $(x_n)_{n \geq 0}$ is increasing. Passing to the limit, we obtain

$$\ell = \frac{1}{2} - \left(\ell - \frac{1}{2}\right)^2.$$

Since all the terms of the sequence are positive, we deduce that $\ell = 1/2$.

(b) By the definition of f it follows that

$$f(x) \leq x, \quad \text{for all } x \leq -\frac{1}{2},$$

and

$$f(x) \leq -\frac{1}{2}, \quad \text{for all } x \geq \frac{3}{2}.$$

So, the sequence diverges both if $x \leq -1/2$ and for $x \geq 3/2$.

We now prove that if $x \in (-1/2, 3/2)$, then the sequence converges and its limit equals $1/2$. Indeed, in this case we have

$$\left| f(x) - \frac{1}{2} \right| < \left| x - \frac{1}{2} \right|.$$

It follows that

$$\left| x_{n+1} - \frac{1}{2} \right| < \left| x - \frac{1}{2} \right|^n \to 0 \quad \text{as } n \to \infty. \quad \square$$

The next exercise gives an example of a convergent sequence defined by means of an integer-valued function. We invite the reader to establish more properties of the function $\langle \cdot \rangle : [0, \infty) \to \mathbb{N}$.

1.2.8. For any integer $n \geq 1$, let $\langle n \rangle$ be the closest integer to $\sqrt{n}$. Compute

$$\lim_{n \to \infty} \sum_{j=1}^{n} \frac{2^{\langle j \rangle} + 2^{-\langle j \rangle}}{2^j}.$$

Solution. Since $(k - 1/2)^2 = k^2 - k + 1/4$ and $(k + 1/2)^2 = k^2 + k + 1/4$, it follows that $\langle n \rangle = k$ if and only if $k^2 - k + 1 \leq n \leq k^2 + k$. Hence

$$\lim_{n \to \infty} \sum_{j=1}^{n} \frac{2^{\langle j \rangle} + 2^{-\langle j \rangle}}{2^j} = \sum_{k=1}^{\infty} \sum_{\langle j \rangle = k} \frac{2^{\langle j \rangle} + 2^{-\langle j \rangle}}{2^j} = \sum_{k=1}^{\infty} \sum_{n=k^2-k+1}^{k^2+k} \frac{2^k + 2^{-k}}{2^n}$$

$$= \sum_{k=1}^{\infty} (2^k + 2^{-k})(2^{-k^2+k} - 2^{-k^2-k}) = \sum_{k=1}^{\infty} \left(2^{-k(k-2)} - 2^{-k(k+2)} \right)$$

$$= \sum_{k=1}^{\infty} 2^{-k(k-2)} - \sum_{k=3}^{\infty} 2^{-k(k-2)} = 3. \quad \square$$

We give below a characterization of the sequences having a certain growth property. As in many cases, monotony arguments play a central role.

1.2.9. *(i) Let $(a_n)_{n \geq 1}$ be a sequence of real numbers such that $a_1 = 1$ and $a_{n+1} > 3a_n/2$ for all $n \geq 1$. Prove that the sequence $(b_n)_{n \geq 1}$ defined by*

$$b_n = \frac{a_n}{(3/2)^{n-1}}$$

either has a finite limit or tends to infinity.

(ii) Prove that for all $\alpha > 1$ there exists a sequence $(a_n)_{n \geq 1}$ with the same properties such that

$$\lim_{n \to \infty} \frac{a_n}{(3/2)^{n-1}} = \alpha.$$

International Mathematical Competition for University Students, 2003

Solution. (i) Our hypothesis $a_{n+1} > 3a_n/2$ is equivalent to $b_{n+1} > b_n$, and the conclusion follows immediately.

(ii) For any $\alpha > 1$ there exists a sequence $1 = b_1 < b_2 < \cdots$ that converges to α. Choosing $a_n = (3/2)^{n-1} b_n$, we obtain the required sequence $(a_n)_{n \geq 1}$. □

Qualitative properties of a sequence of positive integers are established in the next example.

1.2.10. Consider the sequence $(a_n)_{n \geq 1}$ defined by $a_n = n^2 + 2$.
(i) Find a subsequence $(a_{n_k})_{k \geq 1}$ such that if $i < j$ then a_{n_i} is a divisor of a_{n_j}.
(ii) Find a subsequence such that any two terms are relatively prime.

Solution. (i) For any integer $k \geq 1$, take $n_k = 2^{(3^k+1)/2}$. Then $a_{n_k} = 2(2^{3^n} + 1)$ and the conclusion follows.

(ii) Consider the subsequence $(b_n)_{n \geq 1}$ defined by $b_1 = 3$, $b_2 = b_1^2 + 2$, and, for any integer $n \geq 3$, $b_n = (b_1 \cdots b_{n-1})^2 + 2$. □

The Stolz–Cesàro lemma is a powerful instrument for computing limits of sequences (always keep in mind that it gives only a **sufficient** condition for the existence of the limit!). A simple illustration is given in what follows.

1.2.11. *The sequence of real numbers $(x_n)_{n \geq 1}$ satisfies $\lim_{n \to \infty}(x_{2n} + x_{2n+1}) = 315$ and $\lim_{n \to \infty}(x_{2n} + x_{2n-1}) = 2003$. Evaluate $\lim_{n \to \infty}(x_{2n}/x_{2n+1})$.*

Harvard–MIT Mathematics Tournament, 2003

Solution. Set $a_n = x_{2n}$ and $b_n = x_{2n+1}$ and observe that

$$\frac{a_{n+1} - a_n}{b_{n+1} - b_n} = \frac{(x_{2n+2} + x_{2n+1}) - (x_{2n+1} + x_{2n})}{(x_{2n+3} + x_{2n+2}) - (x_{2n+2} + x_{2n+1})} \longrightarrow \frac{2003 - 315}{315 - 2003} = -1,$$

as $n \to \infty$. Thus, by the Stolz–Cesàro lemma, the required limit equals -1. □

Remark. We observe that the value of $\lim_{n \to \infty}(x_{2n}/x_{2n+1})$ does **not** depend on the values of $\lim_{n \to \infty}(x_{2n} + x_{2n+1})$ and $\lim_{n \to \infty}(x_{2n} + x_{2n-1})$ but only on the convergence of these two sequences.

We refine below the asymptotic behavior of a sequence converging to zero. The proof relies again on the Stolz–Cesàro lemma, and the method can be extended to large classes of recurrent sequences.

1.2.12. *Let $(a_n)_{n \geq 1}$ be a sequence of real numbers such that $\lim_{n \to \infty} a_n \sum_{k=1}^{n} a_k^2 = 1$. Prove that $\lim_{n \to \infty}(3n)^{1/3} a_n = 1$.*

I. J. Schoenberg, Amer. Math. Monthly, Problem 6376

Solution. Set $s_n = \sum_{k=1}^{n} a_k^2$. Then the condition $a_n s_n \to 1$ implies that $s_n \to \infty$ and $a_n \to 0$ as $n \to \infty$. Hence we also have that $a_n s_{n-1} \to 1$ as $n \to \infty$. Therefore

$$s_n^3 - s_{n-1}^3 = a_n^2 (s_n^2 + s_n s_{n-1} + s_{n-1}^2) \to 3 \quad \text{as } n \to \infty. \tag{1.3}$$

Thus, by the Stolz–Cesàro lemma, $s_n^3/n \to 3$ as $n \to \infty$, or equivalently, $\lim_{n \to \infty} n^{-2/3} s_n^2 = 3^{2/3}$. So, by (1.3), we deduce that $\lim_{n \to \infty}(3n)^{1/3} a_n = 1$. □

We study in what follows a sequence whose terms are related to the coefficients of certain polynomials.

1.2.13. Fix a real number $x \neq -1 \pm \sqrt{2}$. Consider the sequence $(s_n)_{n \geq 1}$ defined by $s_n = \sum_{k=0}^{n} a_k x^k$ such that $\lim_{n \to \infty} s_n = 1/(1 - 2x - x^2)$. Prove that for any integer $n \geq 0$, there exists an integer m such that $a_n^2 + a_{n+1}^2 = a_m$.

Solution. We have

$$\frac{1}{1 - 2x - x^2} = \frac{1}{2\sqrt{2}} \left[\frac{\sqrt{2}+1}{1 - (1+\sqrt{2})x} + \frac{\sqrt{2}-1}{1 - (1-\sqrt{2})x} \right]$$

and

$$\frac{1}{1 + (1 \pm \sqrt{2})x} = \lim_{n \to \infty} \sum_{k=0}^{n} (1 \pm \sqrt{2})^k x^k.$$

Therefore

$$a_n = \frac{1}{2\sqrt{2}} \left[(\sqrt{2}+1)^{n+1} - (1 - \sqrt{2})^{n+1} \right].$$

A straightforward computation shows that $a_n^2 + a_{n+1}^2 = a_{2n+2}$. □

The following (not easy!) problem circulated in the folklore of contestants in Romanian mathematical competitions in the 1980s. It gives us an interesting property related to bounded sequences of **real** numbers. Does the property given below remain true if b_n are not real numbers?

1.2.14. Suppose that $(a_n)_{n \geq 1}$ is a sequence of real numbers such that $\lim_{n \to \infty} a_n = 1$ and $(b_n)_{n \geq 1}$ is a bounded sequence of real numbers. If k is a positive integer such that $\lim_{n \to \infty}(b_n - a_n b_{n+k}) = \ell$, prove that $\ell = 0$.

Solution. (Călin Popescu). Let $b = \liminf_{n \to \infty} b_n$ and $B = \limsup_{n \to \infty} b_n$. Since $(b_n)_{n \geq 1}$ is bounded, both b and B are finite. Now there are two subsequences (b_{p_r}) and (b_{q_r}) of $(b_n)_{n \geq 1}$ such that $b_{p_r} \to b$ and $b_{q_r} \to B$ as $r \to \infty$. Since $a_n \to 1$ and $b_n - a_n b_{n+k} \to \ell$ as $n \to \infty$, it follows that the subsequences (b_{p_r+k}) and (b_{q_r+k}) of $(b_n)_{n \geq 1}$ tend to $b - \ell$ and $B - \ell$, respectively, as $r \to \infty$. Consequently, $b - \ell \geq b$ and $B - \ell \leq B$, hence $\ell = 0$. □

Elementary trigonometry formulas enable us to show in what follows that a very simple sequence diverges. A deeper property of this sequence will be proved in Problem 1.4.26.

1.2.15. Set $a_n = \sin n$, for any $n \geq 1$. Prove that the sequence $(a_n)_{n \geq 1}$ is divergent.

Solution. Arguing by contradiction, we assume that the sequence $(a_n)_{n \geq 1}$ is convergent. Let $a = \lim_{n \to \infty} \sin n$. Using the identity

$$\sin(n+1) = \sin n \cos 1 + \cos n \sin 1$$

we deduce that the sequence $(\cos n)$ converges and, moreover,

$$a = a \cos 1 + b \sin 1,$$

where $b = \lim_{n\to\infty} \cos n$. Using now

$$\cos(n+1) = \cos n \cos 1 - \sin n \sin 1$$

we deduce that

$$b = b\cos 1 - a\sin 1.$$

These two relations imply $a = b = 0$, a contradiction, since $a^2 + b^2 = 1$. □

Under what assumption on the function f one can deduce that a sequence $(a_n)_{n\geq 1}$ converges, provided the sequence $(f(a_n))_{n\geq 1}$ is convergent? The next exercise offers a sufficient condition such that this happens.

1.2.16. Let $(a_n)_{n\geq 1}$ be a sequence of real numbers such that $a_n \geq 1$ for all n and the sequence $(a_n + a_n^{-1})_{n\geq 1}$ converges. Prove that the sequence $(a_n)_{n\geq 1}$ is convergent.

Solution. Let $a = \liminf_{n\to\infty} a_n$ and $A = \limsup_{n\to\infty} a_n$. Then both a and A exist and are finite. Indeed, if $A = +\infty$ then we obtain a contradiction from $a_n + a_n^{-1} > a_n$ and our hypothesis that the sequence $(a_n + a_n^{-1})_{n\geq 1}$ is bounded (since it is convergent).

Arguing by contradiction, assume that $A > a$. Choose subsequences $(a_{n_k})_{k\geq 1}$ and $(a_{m_k})_{k\geq 1}$ such that $a_{n_k} \to A$ as $k \to \infty$ and $a_{m_k} \to a$ as $k \to \infty$. Therefore $a_{n_k} + 1/a_{n_k} \to A + 1/A$ as $k \to \infty$ and $a_{m_k} + 1/a_{m_k} \to a + 1/a$ as $k \to \infty$. But the sequence $(a_n + a_n^{-1})_{n\geq 1}$ is convergent to some limit ℓ. It follows that

$$\ell = A + \frac{1}{A} = a + \frac{1}{a}.$$

Thus, $(A - a)((Aa - 1) = 0$. Hence either $A = a$ or $Aa = 1$ and both are impossible since $A > a \geq 1$. □

1.2.17. Given a sequence $(a_n)_{n\geq 1}$ such that $a_n - a_{n-2} \to 0$ as $n \to \infty$, show that

$$\lim_{n\to\infty} \frac{a_n - a_{n-1}}{n} = 0.$$

Solution. For $\varepsilon > 0$, let n_0 be sufficiently large that $|a_n - a_{n-2}| < \varepsilon$ for all $n \geq n_0$. We have

$$a_n - a_{n-1} = (a_n - a_{n-2}) - (a_{n-1} - a_{n-3}) + (a_{n-2} - a_{n-4})$$
$$- \cdots + \{(a_{n_0+2} - a_{n_0}) - (a_{n_0+1} - a_{n_0-1})\}.$$

Thus

$$|a_n - a_{n-1}| \leq (n - n_0)\varepsilon + |a_{n_0+1} - a_{n_0-1}|$$

and so $(a_n - a_{n-1})/n$ tends to zero as $n \to \infty$. □

1.2.18. Let

$$S_n = \sum_{k=1}^{n} \left(\sqrt{1 + \frac{k}{n^2}} - 1\right).$$

Show that $\lim_{n\to\infty} S_n = 1/4$.

Solution. We first observe that for all $x > -1$,

$$\frac{x}{2+x} < \sqrt{1+x} - 1 < \frac{x}{2}.$$

Hence, setting $x = k/n^2$, we obtain

$$\frac{k}{2n^2+k} < \sqrt{1+\frac{k}{n^2}} - 1 < \frac{k}{2n^2}.$$

Therefore

$$\sum_{k=1}^{n} \frac{k}{2n^2+k} < S_n < \frac{1}{2n^2}\sum_{k=1}^{n} k.$$

We have

$$\frac{1}{2n^2}\sum_{k=1}^{n} k = \frac{n(n+1)}{4n^2} \to \frac{1}{4} \quad \text{as } n \to \infty.$$

On the other hand,

$$\lim_{n\to\infty}\left\{\frac{1}{2n^2}\sum_{k=1}^{n} k - \sum_{k=1}^{n}\frac{k}{2n^2+k}\right\} = \lim_{n\to\infty}\sum_{k=1}^{n}\frac{k^2}{2n^2(2n^2+k)}.$$

But

$$\sum_{k=1}^{n}\frac{k^2}{2n^2(2n^2+k)} < \sum_{k=1}^{n}\frac{k^2}{4n^4} = \frac{n(n+1)(2n+1)}{24n^4}.$$

We deduce that

$$\lim_{n\to\infty}\left\{\frac{1}{2n^2}\sum_{k=1}^{n} k - \sum_{k=1}^{n}\frac{k}{2n^2+k}\right\} = 0$$

and

$$\lim_{n\to\infty}\sum_{k=1}^{n}\frac{k}{2n^2+k} = \frac{1}{4},$$

hence the desired conclusion. □

1.2.19. Let $(x_n)_{n\geq 1}$ be a sequence and set $y_n = x_{n-1} + 2x_n$ for all $n \geq 2$. Suppose that the sequence $(y_n)_{n\geq 2}$ converges. Show that the sequence $(x_n)_{n\geq 1}$ converges.

Solution. Let $\tilde{y} = \lim_{n\to\infty} y_n$ and set $\tilde{x} = \tilde{y}/3$. We show that $\tilde{x} = \lim_{n\to\infty} x_n$. For $\varepsilon > 0$ there is a positive integer n_0 such that for all $n \geq n_0$, $|y_n - \tilde{y}| < \varepsilon/2$. Hence

$$\varepsilon/2 > |y_n - \tilde{y}| = |x_{n-1} + 2x_n - 3\tilde{x}| = |2(x_n - \tilde{x}) + (x_{n-1} - \tilde{x})|$$
$$\geq 2|x_n - \tilde{x}| - |x_{n-1} - \tilde{x}|.$$

Thus, $|x_n - \tilde{x}| < \varepsilon/2 + (1/2)|x_{n-1} - \tilde{x}|$, which can be iterated to give

$$|x_{n+m} - \tilde{x}| < \frac{\varepsilon}{4}\left(\sum_{i=1}^{m} 2^{-i}\right) + 2^{-(m+1)}|x_{n-1} - \tilde{x}| < \frac{\varepsilon}{2} + 2^{-(m+1)}|x_{n-1} - \tilde{x}|.$$

1.2 Introductory Problems

By taking m large enough, $2^{-(m+1)}|x_{n-1} - \tilde{x}| < \varepsilon/2$. Thus for all sufficiently large k, $|x_k - \tilde{x}| < \varepsilon$. □

1.2.20. (i) Let $(a_n)_{n \geq 1}$ be a bounded sequence of real numbers such that $a_n \neq 0$ for all $n \geq 1$. Show that there is a subsequence $(b_n)_{n \geq 1}$ of $(a_n)_{n \geq 1}$ such that the sequence $(b_{n+1}/b_n)_{n \geq 1}$ converges.

(ii) Let $(a_n)_{n \geq 1}$ be a sequence of real numbers such that for every subsequence $(b_n)_{n \geq 1}$ of $(a_n)_{n \geq 1}$, $\lim_{n \to \infty} |b_{n+1}/b_n| \leq 1$. Prove that $(a_n)_{n \geq 1}$ has at most two limit points. Moreover, if these limit points are not equal and if one of them is t, then the other is $-t$.

Solution. (i) We distinguish two cases.

CASE 1: there exists $\varepsilon > 0$ such that for infinitely many k, $|a_k| \geq \varepsilon$. Let $(b_n)_{n \geq 1}$ be the subsequence of $(a_n)_{n \geq 1}$ consisting of those a_k with $|a_k| \geq \varepsilon$. Then $|b_{n+1}/b_n| \leq (\sup_k |a_k|)/\varepsilon$ for all n.

CASE 2: for every $\varepsilon > 0$ there is an integer N_ε such that for all $k \geq N_\varepsilon$, $|a_k| < \varepsilon$. Then there is a subsequence $(b_n)_{n \geq 1}$ of $(a_n)_{n \geq 1}$ such that $|b_{n+1}| < |b_n|$.

In both cases, the sequence of ratios $(b_{n+1}/b_n)_{n \geq 1}$ is bounded, and thus by the Bolzano–Weierstrass theorem there exists a convergent subsequence.

(ii) By hypothesis it follows that the sequence $(a_n)_{n \geq 1}$ is bounded. Now suppose that $(a_n)_{n \geq 1}$ has at least two distinct limit points s and t, where $s \neq -t$. Then there are subsequences $(b_p)_{p \geq 1}$ and $(c_q)_{q \geq 1}$ of $(a_n)_{n \geq 1}$ converging to s and t respectively. Let $(d_j)_{j \geq 1}$ be the following subsequence of $(a_n)_{n \geq 1}$:

$$d_j = \begin{cases} b_p & \text{if } j \text{ is odd (that is, } d_1 = b_1, d_3 = b_2, \ldots), \\ c_q & \text{if } j \text{ is even (that is, } d_2 = c_1, d_4 = c_2, \ldots). \end{cases}$$

Then $\lim_{j \to \infty} |d_{j+1}/d_j|$ does not exist. This contradiction concludes the proof. □

1.2.21. Let $(a_n)_{n \geq 1}$ be a sequence of real numbers such that $\lim_{n \to \infty}(2a_{n+1} - a_n) = \ell$. Prove that $\lim_{n \to \infty} a_n = \ell$.

Solution. We first show that the sequence $(a_n)_{n \geq 1}$ is bounded. Since the sequence $(2a_{n+1} - a_n)_{n \geq 1}$ is bounded (as a convergent sequence), there exists $M > 0$ such that $|a_1| \leq M$ and for all $n \geq 1$, $|2a_{n+1} - a_n| \leq M$. We prove by induction that $|a_n| \leq M$ for all n. Indeed, suppose that $|a_n| \leq M$. Then

$$|a_{n+1}| = \left| \frac{a_n + (2a_{n+1} - a_n)}{2} \right| \leq \frac{1}{2}(|a_n| + |2a_{n+1} - a_n|) \leq M.$$

This concludes the induction and shows that $(a_n)_{n \geq 1}$ is bounded.

Taking lim sup in

$$a_{n+1} = \frac{a_n + (2a_{n+1} - a_n)}{2},$$

we obtain

$$\limsup_{n \to \infty} a_n \leq \frac{\limsup_{n \to \infty} a_n + \ell}{2}.$$

This yields $\limsup_{n\to\infty} a_n \leq \ell$. Similarly, we deduce that $\liminf_{n\to\infty} a_n \geq \ell$. We conclude that $\lim_{n\to\infty} a_n = \ell$. □

1.2.22. Prove that a countably infinite set of positive real numbers with a finite nonzero limit point can be arranged in a sequence $(a_n)_{n\geq 1}$ such that $(a_n^{1/n})_{n\geq 1}$ is convergent.

P. Orno, Math. Magazine, Problem 1021

Solution. Let $(x_n)_{n\geq 1}$ denote the real numbers of concern and let a be a nonzero finite limit point. Choose A such that $1/A < a < A$ and let a_1 denote the x_n of smallest subscript that lies in the interval $(1/A, A)$. Assuming $a_1, a_2, \ldots, a_{k-1}$ to have been chosen, let a_k be the x_n of smallest subscript different from the $n-1$ already chosen lying in the interval $(A^{-\sqrt{n}}, A^{\sqrt{n}})$. Since a is a limit point, such an x_n can be found, and since $A^{\sqrt{n}} \to \infty$ and $A^{-\sqrt{n}} \to 0$, we deduce that every x_n eventually is included in an interval of the form $(A^{-\sqrt{n}}, A^{\sqrt{n}})$ and will therefore eventually become part of the sequence $(a_n)_{n\geq 1}$. For each n we have $1/A^{\sqrt{n}} < a_n < A^{\sqrt{n}}$ and therefore $(1/A)^{1/\sqrt{n}} < a_n^{1/n} < A^{1/\sqrt{n}}$. But $\lim_{n\to\infty} A^{1/\sqrt{n}} = \lim_{n\to\infty} (1/A)^{1/\sqrt{n}} = 1$ and therefore $\lim_{n\to\infty} a_n^{1/n} = 1$. □

1.3 Recurrent Sequences

Mathematics is trivial, but I can't do
my work without it.
<hr>Richard Feynman (1918–1988)

Recurrent sequences are widely encountered in nature, and they should be seen as a major step toward the *discretization* of various *continuous* models. One of the most famous recurrent sequences goes back to Leonardo Fibonacci (1170–1250). About 1202, Fibonacci formulated his famous rabbit problem, which led to the Fibonacci sequence $1, 1, 2, 3, 5, 8, 13, \ldots$. The terms of this sequence have beautiful properties, mainly related to the *golden ratio*. At the same time, it appears in several applications in biology, including leaves and petal arrangements, branching plants, rabbit colonies, and bees' ancestors (see Figure 1.2). It seems that the Fibonacci recurrence $F_n = F_{n-1} + F_{n-2}$ was first written down by Albert Girard around 1634 and solved by de Moivre in 1730. Bombelli studied the equation $y_n = 2 + 1/y_{n-1}$ in 1572, which is similar to the equation $z_n = 1 + 1/z_{n-1}$ satisfied by ratios of Fibonacci numbers, in order to approximate $\sqrt{2}$. Fibonacci also gave a rough definition for the concept of continued fractions that is intimately associated with *difference equations*, which are now intensively applied in the modeling of continuous phenomena. A more precise definition was formulated by Cataldi around 1613. The method of recursion was significantly advanced with the invention of mathematical induction by Maurolico in the sixteenth century and with its development by Fermat and Pascal in

1.3 Recurrent Sequences

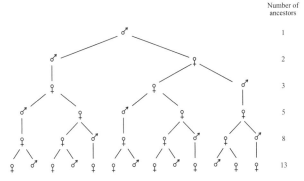

Fig. 1.2 Bees have Fibonacci-number ancestors.

the seventeenth century. Sir Thomas Harriet (1560–1621) invented the calculus of finite differences, and Henry Briggs (1556–1630) applied it to the calculation of logarithms. It was rediscovered by Leibniz around 1672. Sir Isaac Newton, Leonhard Euler (1707–1783), Joseph-Louis Lagrange (1736–1813), Carl Friedrich Gauss (1777–1855), and many others used this calculus to study interpolation theory. The theory of finite differences was developed largely by James Stirling (1692–1770) in the early eighteenth century. Recurrence relations were extended to the study of several sequences. A celebrated example is related to the cooperative recurrences $x_n = (x_{n-1} + y_{n-1})/2$ and $y_n = \sqrt{x_{n-1}y_{n-1}}$, which were associated by Lagrange with the evaluation of elliptic integrals.

In this section we are interested in the study of recurrent sequences that are not necessarily linear.

The first example gives an interesting connection with Euler's indicator function.

1.3.1. Consider the sequence $(a_n)_{n \geq 1}$ defined by $a_1 = 1$, $a_2 = 2$, $a_3 = 24$, and

$$a_n = \frac{6a_{n-1}^2 a_{n-3} - 8a_{n-1}a_{n-2}^2}{a_{n-2}a_{n-3}} \quad \text{for all } n \geq 4.$$

Show that for all n, a_n is an integer that is multiple of n.

Solution. We have
$$\frac{a_n}{a_{n-1}} = 6\frac{a_{n-1}}{a_{n-2}} - 8\frac{a_{n-2}}{a_{n-3}}.$$

Set $b_n = a_n/a_{n-1}$. Thus $b_2 = 2$ and $b_3 = 12$. It follows that $b_n = 2^{n-1}(2^{n-1}-1)$ and $a_n = 2^{n(n-1)/2} \prod_{i=1}^{n-1}(2^i - 1)$. To prove that a_n is a multiple of n, let $n = 2^k m$, where m is odd. Then $k \leq n \leq n(n-1)/2$ and there exists $i \leq m-1$ such that m is a divisor of $2^i - 1$ [for $i = \varphi(m)$, where φ denotes Euler's function]. Consequently, a_n is a multiple of n. □

In certain cases a sequence defined by a nonlinear recurrence relation may be described in a linear way. We give below an example.

1.3.2. Define the sequence $(a_n)_{n\geq 1}$ by $a_1 = a_2 = 1$ and $a_n = a_{n-1}^2 + 2/a_{n-2}$, for any $n \geq 3$. Prove that for all n, a_n is an integer.

Solution. The recurrence relation is of second order, so we try to find a_n of the form $a_n = \alpha q_1^n + \beta q_2^n$. From $a_1 = a_2 = 1$ we obtain $\alpha = q_2 - 1/q_1(q_2 - q_1)$, $\beta = 1 - q_1/q_2(q_2 - q_1)$. Substituting in the recurrence relation, we obtain

$$(\alpha q_1^n + \beta q_2^n)(\alpha q_1^{n-2} + \beta q_2^{n-2}) = \alpha^2 q_1^{2n-2} + \beta^2 q_2^{2n-2} + 2\alpha\beta q_1^{n-1} q_2^{n-1} + 2.$$

Hence
$$\alpha\beta q_1^{n-2} q_2^{n-2} (q_1 - q_2)^2 = 2, \quad \text{for all } n.$$

Therefore
$$(q_2 - 1)(1 - q_1) q_1^{n-3} q_2^{n-3} = 2, \quad \text{for all } n,$$

so $q_1 q_2 = 1$ and $q_1 + q_2 = 4$. It follows that q_1 and q_2 are the roots of the second-order equation $q^2 - 4q + 1 = 0$. This is the characteristic (secular) equation of the sequence defined by

$$a_n = 4a_{n-1} - a_{n-2}, \quad a_1 = a_2 = 1.$$

Thus we obtain that if $a_{n-1} \in \mathbb{N}$ and $a_{n-2} \in \mathbb{N}$ then $a_n \in \mathbb{N}$. □

Recurrent sequences may be also applied to finding coefficients in various expansions, as illustrated in the following exercise.

1.3.3. Consider the expression

$$\left(\cdots\left(\left((x-2)^2 - 2\right)^2 - 2\right)^2 - \cdots - 2\right)^2 \quad (n \text{ squares}).$$

Find the coefficient of x^2.

Solution. Let a_n be the coefficient of x^2, and b_n the coefficient of x. We observe that for any n, the term not containing x in the above development is 4. It follows that

$$a_n = 4a_{n-1} + b_{n-1}^2 \quad \text{and} \quad b_n = 4b_{n-1},$$

where $a_1 = 1$ and $b_1 = -4$. We first obtain that $b_n = -4^n$. Substituting in the recurrence relation corresponding to a_n, we obtain $a_n = 4a_{n-1} + 4^{2n-2}$, which implies $a_n = 4^{n-1}(4^n - 1)/3$. □

An interesting additive decomposition of positive integers in terms of Fibonacci numbers is presented in what follows.

1.3.4. Prove that any positive integer N may be written as the sum of distinct and nonconsecutive terms of the Fibonacci sequence.

Solution. Let $(F_n)_{n\geq 1}$ be the Fibonacci sequence. Then $F_1 = F_2 = 1$ and $F_{n+2} = F_{n+1} + F_n$, for all $n \geq 1$. Let us assume that $F_n \leq N < F_{n+1}$. So, $0 \leq N - F_n < F_{n-1}$. It follows that there exists $s < n - 1$ such that $F_s \leq N - F_n < F_{s+1}$. Hence $0 \leq N - F_n - F_s < F_{s-1}$ and $s - 1 < n - 2$. We thus obtain that N may be written

1.3 Recurrent Sequences

as $N = F_n + F_s + F_p + \cdots + F_r$, where the consecutive subscripts $n, s, p, \ldots, r$ are nonconsecutive numbers. □

We need in what follows elementary differentiability properties of polynomials.

1.3.5. *For any real number a and for any positive integer n define the sequence $(x_k)_{k \geq 0}$ by $x_0 = 0$, $x_1 = 1$, and*

$$x_{k+2} = \frac{cx_{k+1} - (n-k)x_k}{k+1}, \quad \text{for all } k \geq 0.$$

Fix n and let c be the largest real number such that $x_{n+1} = 0$. Find x_k in terms of n and k, $1 \leq k \leq n$.

Solution. We first observe that x_{n+1} is a polynomial of degree n in c. Thus, it is enough to find n values of c such that $x_{n+1} = 0$. We will prove that these values are $c = n - 1 - 2r$, for $r = 0, 1, \ldots, n - 1$. In this case, x_k is the coefficient of t^{k-1} in the polynomial $f(t) = (1-t)^r(1+t)^{n-1-r}$. This property follows after observing that f satisfies the identity

$$\frac{f'(t)}{f(t)} = \frac{n-1-r}{1+t} - \frac{r}{1-t},$$

that is,

$$(1-t^2)f'(t) = f(t)[(n-1-r)(1-t) - r(1+t)] = f(t)[(n-1-2r) - (n-1)t].$$

Identifying the coefficients of t^k in both sides, we obtain

$$(k+1)x_{k+2} - (k-1)x_k = (n-1-2r)x_{k+1} - (n-1)x_k.$$

In particular, the largest c is $n - 1$, and $x_k = C_{n-1}^{k-1}$, for $k = 1, 2, \ldots, n$. □

The next problem is devoted to the study of the normalized *logistic equation*, which is a successful model of many phenomena arising in genetics and mathematical biology. In its simplest form, the logistic equation is a formula for approximating the evolution of an animal population over time. The unknown a_n in the following recurrent sequence represents the number of animals after the nth year. It is easy to observe that the sequence $(a_n)_{n \geq 1}$ converges to zero. The interesting part of the problem is to deduce the first- and second-order decay terms of this sequence.

1.3.6. *Consider the sequence $(a_n)_{n \geq 1}$ satisfying $a_1 \in (0, 1)$ and $a_{n+1} = a_n(1 - a_n)$, for all $n \geq 1$. Show that $\lim_{n \to \infty} na_n = 1$ and $\lim_{n \to \infty} n(1 - na_n)/\ln n = 1$.*

Solution. We first prove that $a_n < 1/(n+1)$, for all $n \geq 2$. For this purpose we use the fact that the mapping $f(x) = x(1-x)$ is increasing on $(0, 1/2]$ and decreasing on $[1/2, 1)$. Thus, $a_2 = a_1(1 - a_1) \leq 1/4 < 1/3$. Let us now assume that $a_n < 1/(n+1)$. Hence

$$a_{n+1} = a_n(1 - a_n) \leq \frac{1}{n+1}\left(1 - \frac{1}{n+1}\right) = \frac{n}{(n+1)^2} < \frac{1}{n+2}.$$

We observe that the sequence $(a_n)_{n\geq 1}$ is decreasing, because $a_{n+1}-a_n=-a_n^2<0$. Moreover, $(a_n)_{n\geq 1}$ is bounded and $\lim_{n\to\infty}a_n=0$. Let $c_n=1/a_n$. We compute $\lim_{n\to\infty}c_n/n=\lim_{n\to\infty}1/(na_n)$. We have

$$\lim_{n\to\infty}\frac{c_{n+1}-c_n}{(n+1)-n}=\lim_{n\to\infty}\frac{a_n-a_{n+1}}{a_na_{n+1}}=\lim_{n\to\infty}\frac{a_n^2}{a_na_n(1-a_n)}=\lim_{n\to\infty}\frac{1}{1-a_n}=1.$$

So, by the Stolz–Cesàro lemma, $\lim_{n\to\infty}c_n/n=1$. Therefore $\lim_{n\to\infty}na_n=1$.

We provide the following alternative proof to this result. Let us first observe that

$$(n+1)a_{n+1}=na_n+a_n-(n+1)a_n^2=na_n+a_n(1-(n+1)a_n). \tag{1.4}$$

To see that $(na_n)_{n\geq 1}$ is increasing, we need to show that $1-(n+1)a_n\geq 0$. From the graph of $y=x(1-x)$ we note that $a_2\leq 1/4$ and $a_n\leq a\leq 1/2$ imply $a_{n+1}\leq a(1-a)$. So, by induction,

$$(n+1)a_n\leq (n+1)\frac{1}{n}\left(1-\frac{1}{n}\right)=1-\frac{1}{n^2}\leq 1.$$

Furthermore, $na_n<(n+1)a_n\leq 1$, and so $(na_n)_{n\geq 1}$ is bounded above by 1. Thus na_n converges to a limit ℓ with $0<na_n<\ell\leq 1$. Now summing (1.4) from 2 to n, we obtain

$$1\geq(n+1)a_{n+1}=2a_2+a_2(1-3a_2)+a_3(1-4a_3)+\cdots+a_n(1-(n+1)a_n). \tag{1.5}$$

If $\ell\neq 1$ then $(1-(n+1)a_n)\geq (1-\ell)/2$ for all large n, and thus relation (1.5) shows that the series $\sum_{n=1}^\infty a_n$ is convergent. However, $na_n\geq a_1$ and so $\sum_{n=1}^\infty a_n\geq a_1\sum_{n=1}^\infty 1/n$. But $\sum_{n=1}^\infty 1/n$ is divergent. This contradiction shows that $\ell=1$.

For the second part, set $b_n=1/a_n$. Then

$$b_{n+1}=\frac{b_n^2}{b_n-1}=b_n+1+\frac{1}{b_n-1}. \tag{1.6}$$

Using this recurrence relation, a standard induction argument implies that $b_n\geq n$, for all $n\geq 0$. Thus, by (1.6),

$$b_n\leq b_{n-1}+1+\frac{1}{n-1}\leq b_{n-2}+2+\frac{1}{n-1}+\frac{1}{n-2}\leq\cdots$$
$$\leq b_1+n+\sum_{k=1}^n\frac{1}{k}\leq n+\ln n+C, \tag{1.7}$$

where C is a real constant [we have used here the fact that $1+1/2+\cdots+1/n-\ln n$ tends to a **finite** limit]. This in turns proves the existence of a constant C_0 such that for all integer $n\geq 1$,

$$b_n\geq b_{n-1}+1+\frac{1}{n-1+\ln n+C}\geq\cdots$$
$$\geq b_1+n+\sum_{k=1}^{n-1}\frac{1}{k+\ln(k+1)+C}\geq n+\ln n+C_0. \tag{1.8}$$

1.3 Recurrent Sequences

It follows from (1.7) and (1.8) that

$$\frac{n(1-na_n)}{\ln n} = \frac{n(b_n-n)}{b_n \ln n} \leq \frac{n(\ln n + C)}{(n+\ln n + C_0)\ln n} = 1 + o(1) \quad \text{as } n \to \infty$$

and

$$\frac{n(1-na_n)}{\ln n} = \frac{n(b_n-n)}{b_n \ln n} \geq \frac{n(\ln n + C_0)}{(n+\ln n + C)\ln n} = 1 + o(1) \quad \text{as } n \to \infty.$$

This concludes the proof. □

Independent Study. *Find a sequence* $(x_n)_{n\geq 1}$ *such that*

$$\lim_{n\to\infty} x_n \left(1 - \frac{n(1-na_n)}{\ln n}\right) = 1.$$

The following problem is difficult, and it offers a curious comparison property with respect to the Fibonacci sequence.

1.3.7. *Fix $x_1 \in [0,1)$ and define the sequence $(x_n)_{n\geq 1}$ by $x_{n+1} = 0$ if $x_n = 0$ and $x_{n+1} = \frac{1}{x_n} - [1/x_n]$. Prove that for all integers $n \geq 1$,*

$$\sum_{k=1}^{n} x_k < \sum_{k=1}^{n} \frac{F_k}{F_{k+1}},$$

where $(F_n)_{n\geq 1}$ is the Fibonacci sequence.

USA, Proposed to the 33rd International Mathematical Olympiad, 1992

Solution. Consider the function $f(x) = (x+1)^{-1}$, and for any positive integer n, define the mapping $g_n(x) = x + f(x) + f^2(x) + \cdots + f^n(x)$, where f^n is the nth iterate of f, that is, $f^n = f \circ \cdots \circ f$ (n times). We easily check the following properties:

(i) for all $0 \leq x < y \leq 1$ we have $0 < f(x) - f(y) < y - x$;
(ii) g_n is increasing in $[0,1]$;
(iii) $F_1/F_2 = 1$, $f(F_n/F_{n+1}) = F_{n+1}/F_{n+2}$ and $g_{n-1}(1) = \sum_{k=1}^{n} F_k/F_{k+1}$, for all positive integers n.

We just remark that for proving (i) we use

$$g_n(x) - g_n(y) = (x-y) + [f(x) - f(y)] + \cdots + [f^n(x) - f^n(y)]$$

combined with the fact that any difference is less in modulus than and of opposite sign with respect to the previous one [see (i)].

If for some $2 \leq k < n$ we have $x_k = 0$, then by definition, $x_n = 0$, and we conclude the proof by induction with respect to the first $n-1$ terms of the sequence. If this does not occur, then for all $2 \leq k \leq n$, we may write $x_{k-1} = (a_k + x_k)^{-1}$, where $a_k > 0$ is the integer part of x_{k-1}^{-1}. Hence

$$\sum_{k=1}^{n} x_k = x_n + \cfrac{1}{a_n + x_n} + \cfrac{1}{a_{n-1} + \cfrac{1}{a_n + x_n}} + \cdots + \cfrac{1}{a_2 + \cfrac{1}{a_3 + \cfrac{1}{\cdots a_{n-1} + \cfrac{1}{a_n + x_n}}}}.$$

We prove by induction with respect to k that for any fixed $x_n \in [0, 1)$, the right-hand side of the above equality is maximum if and only if $a_k = 1$, for all k. We first observe that a_2 must be 1, since a_2 appears only in the last expression. Next, let us assume that $k > 2$ and that for any values of $x_n, a_n, a_{n-1}, \ldots, a_{k+1}$, the above expression attains its maximum for $a_{k-1} = a_{k-2} = \cdots = a_2 = 1$. Observe that only the last $k-1$ terms on the right-hand side contain a_k, and moreover, their sum is

$$g_{k-2}\left(\cfrac{1}{a_k + \cfrac{1}{a_{k+1} + \cfrac{1}{\cdots a_{n-1} + \cfrac{1}{a_n + x_n}}}}\right).$$

By (ii), the function g_{k-2} is increasing, so its maximum is achieved for $a_k = 1$. Using now (ii) and (iii), we obtain

$$\sum_{k=1}^{n} x_k \leq x_n + \cfrac{1}{1 + x_n} + \cfrac{1}{1 + \cfrac{1}{1 + x_n}} + \cdots + \cfrac{1}{1 + \cfrac{1}{1 + \cfrac{1}{\cdots 1 + \cfrac{1}{1 + x_n}}}}$$

$$= g_{n-1}(x) < g_{n-1}(1) = \sum_{k=1}^{n} \frac{F_k}{F_{k+1}}.$$

This concludes our proof. □

A simple linear recurrence generates a sequence of perfect squares, as shown below.

1.3.8. *A sequence of integers $(a_n)_{n \geq 1}$ is given by the conditions $a_1 = 1$, $a_2 = 12$, $a_3 = 20$, and $a_{n+3} = 2a_{n+2} + 2a_{n+1} - a_n$ for every $n \geq 1$. Prove that for every positive integer n, the number $1 + 4a_n a_{n+1}$ is a perfect square.*

Problem M1174*, Kvant

Solution. Define the sequence $(b_n)_{n \geq 1}$ by $b_n = a_{n+2} - a_{n+1} - a_n$, for any $n \geq 1$. Observe that $(b_n)_{n \geq 1}$ fulfills the same recurrence relation as the sequence $(a_n)_{n \geq 1}$. The idea is to prove that for all n, we have $1 + 4a_n a_{n+1} = b_n^2$. We argue by induction and we first observe that this equality holds for $n = 1$. Next, assuming that $1 + 4a_{n-1}a_n = b_{n-1}^2$, we prove that $1 + 4a_n a_{n+1} = b_n^2$. Indeed, since

$$b_n = (2a_{n+1} + 2a_n - a_{n-1}) - a_{n+1} - a_n = a_{n+1} + a_n - a_{n-1}$$
$$= (a_{n+1} - a_n - a_{n-1}) + 2a_n = b_{n-1} + 2a_n,$$

1.3 Recurrent Sequences

we obtain

$$\begin{aligned}1+4a_{n-1}a_n &= b_{n-1}^2 = (b_n-2a_n)^2 = b_n^2 - 4a_nb_n + 4a_n^2 \\ &= b_n^2 - 4a_n(a_{n+1}+a_n-a_{n-1}) + 4a_n^2 \\ &= b_n^2 - 4a_na_{n+1} - 4a_n^2 + 4a_na_{n-1} + 4a_n^2 \\ &= b_n^2 - 4a_na_{n+1} + 4a_na_{n-1}.\end{aligned}$$

Subtracting $4a_na_{n-1}$, we obtain $1+4a_na_{n+1} = b_n^2$, which completes the proof. □

We always have to pay attention whether the models we study indeed occur in reality. We show in what follows that there is a unique sequence of positive integers with a prescribed property.

1.3.9. *Prove that there exists a unique sequence* $(x_n)_{n\geq 0}$ *of positive integers such that* $x_1=1$, $x_2=1$, *and* $x_{n+1}^3+1 = x_nx_{n+2}$ *for all* $n\geq 1$.

Solution. A direct computation shows that $x_3=2$ and $x_4=9$. Next, for any $n\geq 4$ we have

$$x_n^3+1 = \frac{(x_{n-1}^3+1)^3}{x_{n-2}^3}+1 = \frac{(x_{n-1}^3+1)^3}{x_{n-1}x_{n-3}-1}+1$$

$$= \frac{x_{n-1}(x_{n-1}^8+3x_{n-1}^5+3x_{n-1}^2+x_{n-3})}{x_{n-1}x_{n-3}-1}.$$

Hence

$$(x_n^3+1)(x_{n-1}x_{n-3}-1) = x_{n-1}(x_{n-1}^8+3x_{n-1}^5+3x_{n-1}^2+x_{n-3}).$$

We assume, by induction, that $x_1,\ldots,x_n$ are integers. Observing that $x_{n-1}x_{n-3}-1$ and x_{n-1} are relatively prime, it follows that x_{n-1} is a divisor of x_n^3+1. Therefore x_{n+1} is an integer. □

Stirling's formula is a powerful tool for proving asymptotic estimates. We give below an example that involves an elementary recurrence.

1.3.10. *The sequence* $(a_n)_{n\geq 1}$ *is defined by* $a_1=1$ *and* $a_{n+1}=n/a_n$, *for all integers* $n\geq 1$. *Evaluate*

$$\lim_{n\to\infty} n^{-1/2}\left(\frac{1}{a_1}+\frac{1}{a_2}+\cdots+\frac{1}{a_n}\right).$$

W. W. Chao, Amer. Math. Monthly, Problem E 3356

Solution. We show that the limit is $\sqrt{2/\pi}+\sqrt{\pi/2}$. We first observe that $a_na_{n+1}=n$ and $a_{n-1}a_n=n-1$, for all $n\geq 2$. Subtracting these yields $a_{n+1}-a_{n-1}=a_n^{-1}$. Hence $a_2^{-1}+\cdots+a_n^{-1} = a_n+a_{n+1}-a_1-a_2$. It follows that

$$\lim_{n\to\infty} \frac{a_1^{-1}+a_2^{-1}+\cdots+a_n^{-1}}{\sqrt{n}} = \lim_{n\to\infty}\left(\frac{a_n}{\sqrt{n}}+\frac{a_{n+1}}{\sqrt{n}}\right). \qquad (1.9)$$

Since $a_{n+1} = n/a_n = a_{n-1}n/(n-1)$, we can iterate to obtain

$$a_{2n+1} = \frac{2^{2n}(n!)^2}{(2n)!} \quad \text{and} \quad a_{2n+2} = \frac{(2n+1)!}{2^{2n}(n!)^2}.$$

By Stirling's formula, we have $a_{2n+1} \sim \sqrt{\pi/2} \cdot \sqrt{2n}$ and $a_{2n+2} \sim \sqrt{2/\pi} \cdot \sqrt{2n}$. Hence the two contributions of the right-hand side of (1.9) approach $\sqrt{2/\pi}$ and $\sqrt{\pi/2}$. □

Elementary inequalities imply that the following recurrent sequence diverges like $2\sqrt{n}$. As a consequence, a quite precise estimate for a high-order term is obtained.

1.3.11. Let $(a_n)_{n \geq 0}$ be the sequence defined by

$$a_n = a_{n-1} + \frac{1}{a_{n-1}}, \quad \text{for all } n \geq 1, \, a_0 = 5.$$

Prove that $45 < a_{1000} < 45.1$.

Solution. We first observe that for all $n \geq 1$,

$$a_n^2 = a_{n-1}^2 + 2 + \frac{1}{a_{n-1}^2} > a_{n-1}^2 + 2.$$

It follows that $a_n^2 > a_0^2 + 2n$ and, in particular, $a_{1000}^2 > 2025 = 45^2$.

For the reverse inequality, we write

$$a_n^2 < a_{n-1}^2 + 2 + \frac{1}{a_n a_{n-1}} = a_{n-1}^2 + 2 + \frac{1}{a_n - a_{n-1}}\left(\frac{1}{a_{n-1}} - \frac{1}{a_n}\right)$$

$$= a_{n-1}^2 + 2 + a_{n-1}\left(\frac{1}{a_{n-1}} - \frac{1}{a_n}\right) < a_{n-1}^2 + 2 + a_n\left(\frac{1}{a_{n-1}} - \frac{1}{a_n}\right).$$

By addition we obtain, for all $n \geq 1$,

$$a_n^2 < a_0^2 + 2n + a_n\left(\frac{1}{a_0} - \frac{1}{a_n}\right).$$

Therefore

$$a_{1000}^2 < 25 + 2000 + \frac{a_{1000}}{5} - 1,$$

and a straightforward computation shows that $a_{1000} < 45.1$. □

Bounded sequences are **not** necessarily convergent. We give below a sufficient condition that this happens. Monotonicity arguments again play a central role.

1.3.12. *Let $(a_n)_{n \geq 0}$ be a bounded sequence of real numbers satisfying*

$$a_{n+2} \leq \frac{1}{2}(a_n + a_{n+1}), \quad \text{for all } n \geq 0.$$

1.3 Recurrent Sequences

(a) *Prove that the sequence $(A_n)_{n\geq 0}$ defined by $A_n = \max\{a_n, a_{n+1}\}$ is convergent.*
(b) *Deduce that the sequence $(a_n)_{n\geq 0}$ is convergent.*

Solution. (a) We first observe that

$$a_{n+2} \leq \frac{2\max\{a_n, a_{n+1}\}}{2} = A_n.$$

Since $a_{n+1} \leq \max\{a_n, a_{n+1}\} = A_n$, it follows that $\max\{a_{n+1}, a_{n+2}\} = A_{n+1} \leq A_n$, which shows that $(A_n)_{n\geq 0}$ is nonincreasing. Since $(a_n)_{n\geq 0}$ is bounded, it follows that the sequence $(A_n)_{n\geq 0}$ is also bounded, so it converges.

(b) By the boundedness of $(a_n)_{n\geq 0}$, it follows that there exists a convergent subsequence (a_{n_p}). Moreover, since $(A_n)_{n\geq 0}$ is nonincreasing, we can assume that $a_{n_1} \geq a_{n_2} \geq \cdots \geq a_{n_p} \geq \cdots$. Set $\ell = \lim_{p\to\infty} a_{n_p}$. Obviously, $a_{n_p} \geq \ell$.

Set $\ell_1 = \lim_{n\to\infty} A_n$. We prove that $\ell \geq \ell_1$. Indeed, arguing by contradiction, it follows that for any $\varepsilon > 0$, there exists a positive integer $N'(\varepsilon)$ such that

$$a_{n_p-1} < \ell_1 + \varepsilon, \; a_{n_p} < \ell_1 - \varepsilon \quad \text{for all } n_p \geq N'(\varepsilon).$$

On the other hand, $a_{n_p+1} \leq 1/2(a_{n_p} + a_{n_p-1}) < \ell_1$ and $A_{n_p} = \max\{a_{n_p}, a_{n_p+1}\} < \ell_1$, which is impossible because the sequence $(A_n)_{n\geq 0}$ is nonincreasing and converges to ℓ_1. By $\ell = \lim_{p\to\infty} a_{n_p}$ we obtain that for all $\varepsilon > 0$, there exists a positive integer $N_1 = N_1(\varepsilon) \in \mathbb{N}$ such that $a_{n_p} - \ell < \varepsilon$, for all $n_p \geq N_1(\varepsilon)$. From $\ell_1 \leq \ell$ we obtain N_2 such that for all $m \geq N_2$ there exist n_p and n_{p+1} satisfying $a_{n_{p+1}} \leq a_m < a_{n_p}$. Set $N = \max\{N_1, N_2\}$. Hence

$$a_n - \ell \leq a_{n_p} - \ell < \varepsilon, \quad \text{for all } n \geq N,$$

that is, $\ell = \lim_{n\to\infty} a_n$.

An easier alternative argument is the following. Fix $\varepsilon > 0$. There is a positive integer N_1 such that for $n \geq N_1$ we have $\ell \leq A_n < \ell + \varepsilon$. Hence $a_n \leq A_n < \ell + \varepsilon$. Suppose $a_n < \ell - 3\varepsilon$. Then $a_{n+1} = A_n < \ell + \varepsilon$. Thus the recurrence gives

$$a_{n+2} \leq \frac{a_n + a_{n-1}}{2} < \ell - \varepsilon \quad \text{and} \quad a_{n+3} \leq \frac{a_{n+1} + a_{n+2}}{2} < \ell.$$

Hence $A_{n+2} < \ell$, a contradiction. Thus for $n \geq N_1$ we have $\ell - 3\varepsilon \leq a_n < \ell + \varepsilon$. Since $\varepsilon > 0$ was arbitrary, it follows that $\lim_{n\to\infty} a_n = \ell$. □

Recurrent sequences may be useful for solving functional equations, as illustrated in the following exercise.

1.3.13. *Prove that there exists a unique function $f : [0,\infty) \to [0,\infty)$ such that $(f \circ f)(x) = 6x - f(x)$ and $f(x) > 0$, for all $x > 0$.*

Solution. Fix $x \geq 0$ and define the sequence $(x_n)_{n\geq 0}$ by $x_0 = x$ and $x_n = f(x_{n-1})$, for all integers $n \geq 1$. By hypothesis, we have $x_n \geq 0$ and

$$x_{n+2} + x_{n+1} - 6x_n = (f \circ f)(x_n) + f(x_n) - 6x_n = 0, \quad \text{for all } n \geq 0.$$

The roots of the associated characteristic equation $r^2 + r - 6 = 0$ are $r_1 = 2$ and $r_2 = -3$. Thus, there exist real numbers a and b such that for all $n \geq 0$,

$$x_n = a \cdot 2^n + b \cdot (-3)^n. \tag{1.10}$$

We intend to prove that $b = 0$. We argue by contradiction and assume that $b \neq 0$. In this case, by (1.10), it follows that for any n sufficiently large (we choose n **even** if $b < 0$, resp. **odd** if $b > 0$) we have $x_n < 0$, contradiction. So, $b = 0$ and hence $x_1 = f(x) = 2x_0 = 2x$. Therefore the unique function satisfying our hypotheses is $f(x) = 2x$.

We recall that if $(a_n)_{n \geq 1}$ and $(b_n)_{n \geq 1}$ are sequences of positive real numbers, then we use the notation
$$a_n \sim b_n \quad \text{as } n \to \infty$$
if $\lim_{n \to \infty} a_n / b_n = 1$. □

1.3.14. *Let $(a_n)_{n \geq 1}$ be a sequence of real numbers such that $e^{a_n} + na_n = 2$ for all positive integers n.*

Compute $\lim_{n \to \infty} n(1 - na_n)$.

Teodora–Liliana Rădulescu, Mathematical Reflections, No. 2 (2006)

Solution. For any integer $n \geq 1$, define $f_n(x) = e^x + nx - 2$. Then f_n increases on $\mathbb{R}$, as a sum of increasing functions. So, by $f_n(0) = -1 < 0$ and $f_n(\ln 2) = n \ln 2 > 0$, we conclude that for all integers $n \geq 1$, there exists a unique $a_n \in (0, \ln 2)$ such that $f_n(a_n) = 0$. Next, we observe that

$$f_{n+1}(a_n) = e^{a_n} + na_n - 2 + a_n = f_n(a_n) + a_n = a_n > 0.$$

Since $f_{n+1}(a_{n+1}) = 0$ and f_n increases for any $n \geq 1$, we deduce that $a_n > a_{n+1}$, for all $n \geq 1$. So, there exists $\ell := \lim_{n \to \infty} a_n \in [0, \ln 2)$. If $\ell > 0$ then the recurrence relation $e^{a_n} + na_n = 2$ yields the contradiction $+\infty = 2$. Thus, $\ell = 0$ and $\lim_{n \to \infty} na_n = 1$, that is, $a_n \sim n^{-1}$ as $n \to \infty$. Using again the recurrence relation, we obtain

$$1 - na_n = e^{a_n} - 1 \sim e^{1/n} - 1 \quad \text{as } n \to \infty.$$

Since $e^x - 1 = x + o(x)$ as $x \to 0$ we deduce that

$$1 - na_n \sim \frac{1}{n} \quad \text{as } n \to \infty.$$

It follows that $\lim_{n \to \infty} n(1 - na_n) = 1$. □

With similar arguments as above (which are based on the asymptotic expansion of e^x around $x = 0$) we can prove that $\lim_{n \to \infty} n[n(1 - na_n) - 1] = 1/2$ (exercise!).

In some cases, connections with trigonometry may be useful for computing limits. How can one solve the following problem without using trigonometric formulas?

1.3.15. *Let $-1 < a_0 < 1$, and define recursively $a_n = [(1 + a_{n-1})/2]^{1/2}$, for all $n > 0$. Find the limits, as $n \to \infty$, of $b_n = 4^n(1 - a_n)$ and $c_n = a_1 a_2 \cdots a_n$.*

M. Golomb, Amer. Math. Monthly, Problem E 2835

1.3 Recurrent Sequences

Solution. Let $\varphi \in (0, \pi)$ be such that $\cos \varphi = a_0$. Using the formula $\cos^2 x = (1 + \cos 2x)/2$ we find that $a_n = \cos(\varphi/2^n)$. Thus, since $\sin^2 x = (1 - \cos 2x)/2$ and $\lim_{x \to 0} \sin x / x = 1$, we obtain

$$b_n = 4^n \left(1 - \cos \frac{\varphi}{2^n}\right) = 2^{2n+1} \sin^2 \frac{\varphi}{2^{n+1}} \longrightarrow \frac{\varphi^2}{2} \quad \text{as } n \to \infty.$$

Next, we observe that

$$\sin \frac{\varphi}{2^n} c_n = \cos \frac{\varphi}{2} \cos \frac{\varphi}{2^2} \cdots \cos \frac{\varphi}{2^n} \sin \frac{\varphi}{2^n}.$$

Hence

$$c_n = \frac{\sin \varphi}{\sin(\varphi/2^n)} 2^n \longrightarrow \frac{\sin \varphi}{\varphi} \quad \text{as } n \to \infty. \quad \square$$

Convergent linear combinations of the terms of a given sequence do not imply that the sequence is convergent, too! We give below a complete response to a problem of this type.

1.3.16. *Let $(a_n)_{n \geq 1}$ be a sequence of real numbers and $-1 < \alpha < 1$ such that the sequence $(a_{n+1} + \alpha a_n)_{n \geq 1}$ is convergent. Prove that the sequence $(a_n)_{n \geq 1}$ converges, too.*

Solution. Set $b_n = a_{n+1} + \alpha a_n$ and $\ell := \lim_{n \to \infty} b_n$. We will prove that $a_n \to \ell/(1+\alpha)$ as $n \to \infty$. We first observe that if we set $a'_n = a_n - \ell/(1+\alpha)$ and $b'_n = b_n - \ell$, then the same recurrence relation is fulfilled (that is, $b'_n = a'_{n+1} + \alpha a'_n$), but $b'_n \to 0$ as $n \to \infty$. This shows that we can assume, without loss of generality, that $\ell = 0$. It remains to prove that $a_n \to 0$ as $n \to \infty$. Since the assertion is obvious for $\alpha = 0$, we assume in what follows that $\alpha \neq 0$. Set $a_n = (-\alpha)^n x_n$. Then $b_n = (-\alpha)^{n+1} x_{n+1} + \alpha(-\alpha)^n x_n$ and hence, for all $n \geq 1$,

$$x_{n+1} - x_n = -\left(-\frac{1}{\alpha}\right)^n b_n.$$

It follows that for all $n \geq 1$,

$$x_n = x_1 - \sum_{k=1}^{n-1} \left(-\frac{1}{\alpha}\right)^k b_k.$$

Since $x_n = (-1/\alpha)^n a_n$, we obtain, for all $n \geq 1$,

$$a_n = (-\alpha)^n x_1 - (-\alpha)^n \sum_{k=1}^{n-1} \left(-\frac{1}{\alpha}\right)^k b_k.$$

But $(-\alpha)^n x_1 \to 0$ as $n \to \infty$. So, it remains to prove that $(-\alpha)^n \sum_{k=1}^{n-1} (-1/\alpha)^k b_k \to 0$ as $n \to \infty$. We first observe that by the Stolz–Cesàro lemma,

$$\lim_{n \to \infty} \frac{\sum_{k=1}^{n-1} (-1/\alpha)^k b_k}{\sum_{k=1}^{n-1} (-1/\alpha)^k} = \lim_{n \to \infty} \frac{(-1/\alpha)^n b_n}{(-1/\alpha)^n} = \lim_{n \to \infty} b_n = 0.$$

Hence

$$\sum_{k=1}^{n-1}\left(-\frac{1}{\alpha}\right)^k b_k = o\left(\sum_{k=1}^{n-1}\left(-\frac{1}{\alpha}\right)^k\right) = o\left(\left(-\frac{1}{\alpha}\right)^n\right) \quad \text{as } n\to\infty.$$

Therefore

$$(-\alpha)^n \sum_{k=1}^{n-1}\left(-\frac{1}{\alpha}\right)^k b_k = (-\alpha)^n \cdot o\left(\left(-\frac{1}{\alpha}\right)^n\right) = o(1) \quad \text{as } n\to\infty.$$

This concludes the proof. □

Remarks. 1. The above statement is sharp, in the sense that the property does **not** remain true if $\alpha \notin (-1,1)$. Give counterexamples!

2. A "continuous" variant of the above "discrete" property is the following.

Assume that f is a real-valued continuously differentiable function on some interval $[a,\infty)$ such that $f'(x) + \alpha f(x)$ tends to zero as x tends to infinity, for some $\alpha > 0$. Prove that $f(x)$ tends to zero as x tends to infinity.

A complete proof of this result will be given in Chapter 10.

1.4 Qualitative Results

> A mathematician who is not also a poet will never be a complete mathematician.
>
> Karl Weierstrass (1815–1897)

The following inequalities due to Niels Henrik Abel (1802–1829) are useful in many applications (see, e.g., Carleman's inequality in the next section). We illustrate here this result to deduce an interesting inequality involving positive numbers. We point out that a very important international prize was created by the Norwegian Academy of Science and Letters in order to celebrate the Abel centenary in 2002. This is a kind of Nobel Prize awarded to mathematicians, and the first laureates of the Abel Prize were Jean-Pierre Serre (2003), Sir Michael Francis Atiyah and Isadore M. Singer (2004), Peter D. Lax (2005), Lennart Carleson (2006), Srinivasa S.R. Varadhan (2007), and John Griggs Thompson and Jacques Tits (2008).

1.4.1. (**Abel's inequalities**). (a) Let $a_1,\ldots,a_n$ be real numbers. Set $m = \min(a_1, a_1+a_2,\ldots,a_1+a_2+\cdots+a_n)$, $M = \max(a_1, a_1+a_2,\ldots,a_1+a_2+\cdots+a_n)$.
Prove that if $b_1 \geq b_2 \geq \cdots \geq b_n \geq 0$, then

$$Mb_1 \geq a_1 b_1 + a_2 b_2 + \cdots + a_n b_n \geq m b_1.$$

(b) Assume that $x_1 \geq x_2 \geq \cdots \geq x_n \geq 0$, $y_1 \geq y_2 \geq \cdots \geq y_n \geq 0$, and $x_1 \geq y_1$, $x_1 + x_2 \geq y_1 + y_2, \ldots, x_1 + x_2 + \cdots + x_n \geq y_1 + y_2 + \cdots + y_n$.

Prove that

$$x_1^k + x_2^k + \cdots + x_n^k \geq y_1^k + y_2^k + \cdots + y_n^k \quad \text{for all } n, k \geq 1.$$

Solution. (a) Define $s_1 = a_1$, $s_2 = a_1 + a_2, \ldots, s_n = a_1 + a_2 + \cdots + a_n$. Then

$$E := a_1 b_1 + a_2 b_2 + \cdots + a_n b_n$$
$$= s_1 b_1 + (s_2 - s_1) b_2 + \cdots + (s_{n-1} - s_{n-2}) b_{n-1} + (s_n - s_{n-1}) b_n$$
$$= s_1(b_1 - b_2) + s_2(b_2 - b_3) + \cdots + s_{n-1}(b_{n-1} - b_n) + b_n s_n.$$

But $M \geq s_i \geq m$ and $b_1 - b_2 \geq 0$, $b_2 - b_3 \geq 0, \ldots, b_{n-1} - b_n \geq 0$. Therefore

$$Mb_1 = M(b_1 - b_2 + b_2 - b_3 + \cdots + b_{n-1} - b_n) + Mb_n \geq E$$
$$\geq m(b_1 - b_2 + b_2 - b_3 + \cdots + b_{n-1} - b_n) + mb_n = mb_1.$$

Consequently, if $a_1 \geq 0$, $a_1 + a_2 \geq 0, \ldots, a_1 + a_2 + \cdots + a_n \geq 0$ and $b_1 \geq b_2 \geq \cdots \geq b_n \geq 0$, then $a_1 b_1 + a_2 b_2 + \cdots + a_n b_n \geq 0$.

(b) Since $x_1 \geq x_2 \geq 0$ and $y_1 \geq y_2 \geq 0$, we have

$$b_1 := x_1^{k-1} + x_1^{k-2} y_1 + \cdots + x_1 y_1^{k-2} + y_1^{k-1}$$
$$\geq x_2^{k-1} + x_2^{k-2} y_2 + \cdots + x_2 y_2^{k-2} + y_2^{k-1} =: b_2.$$

Analogously we obtain $b_1 \geq b_2 \geq \cdots \geq b_n \geq 0$.

Define $a_1 = x_1 - y_1$, $a_2 = x_2 - y_2, \ldots, a_n = x_n - y_n$. We have $a_1 \geq 0$, $a_1 + a_2 \geq 0, \ldots, a_1 + a_2 + \cdots + a_n \geq 0$. So, by (a), it follows that $a_1 b_1 + a_2 b_2 + \cdots + a_n b_n \geq 0$, that is,

$$x_1^k - y_1^k + x_2^k - y_2^k + \cdots + x_n^k - y_n^k \geq 0.$$

Hence

$$x_1^k + x_2^k + \cdots + x_n^k \geq y_1^k + y_2^k + \cdots + y_n^k. \quad \square$$

Alternative proof of (b). Our reasoning relies on the following auxiliary result.
Lemma. *If $y_1 \geq y_2$ and $x_1 - y_1 \geq x_2 - y_2$, then $x_1^k - y_1^k \geq x_2^k - y_2^k$.*
Indeed,

$$x_1^k - y_1^k = (y_1 + (x_1 - y_1))^k - y_1^k = \sum_{i=1}^{k} C_k^i y_1^{k-i} (x_1 - y_1)^i$$
$$\geq \sum_{i=1}^{k} C_k^i y_2^{k-i} (x_2 - y_2)^i = x_2^k - y_2^k,$$

which concludes the proof of the lemma.

Using now the lemma and our hypotheses, we obtain

$$x_1^k - y_1^k \geq y_n^k - (y_n - x_1 + y_1)^k,$$
$$x_2^k - y_2^k \geq (y_n - x_1 + y_1)^k - (y_n - x_1 - x_2 + y_1 + y_2)^k,$$

$$x_{n-1}^k - y_{n-1}^k \ldots \geq \left(y_n - \sum_{i=1}^{n-2} x_i + \sum_{i=1}^{n-2} y_i\right)^k - \left(y_n - \sum_{i=1}^{n-1} x_i + \sum_{i=1}^{n-1} y_i\right)^k.$$

After summing these relations, we obtain

$$\sum_{i=1}^{n-1}(x_i^k - y_i^k) \geq y_n^k - \left(y_n - \sum_{i=1}^{n-1} x_i + \sum_{i=1}^{n-1} y_i\right)^k \geq y_n^k - x_n^k.$$

It follows that $\sum_{i=1}^n x_i^k \geq \sum_{i=1}^n y_i^k$.

The following problem shows that a subadditivity condition guarantees the existence of the limit for a certain sequence. This property does not remain true if the subadditivity assumption is partially fulfilled.

1.4.2. *(i) Assume that the sequence $(x_n)_{n\geq 1}$ satisfies*

$$x_{m+n} \leq x_m + x_n \quad \text{for all } m, n \geq 1.$$

Show that the sequence $(x_n/n)_{n\geq 1}$ has a limit and, moreover,

$$\lim_{n\to\infty} \frac{x_n}{n} = \inf_{n\geq 1} \frac{x_n}{n}.$$

(ii) Let k be an arbitrary positive integer. Assume that $(x_n)_{n\geq 1}$ is a sequence of nonnegative real numbers satisfying

$$x_{m+n} \leq x_m + x_n \quad \text{for all } m, n \geq 1 \text{ with } |m - n| \leq k.$$

Prove that $\lim_{n\to\infty} x_n/n$ does not necessarily exist.

Solution. (i) Set $\alpha = \inf_{n\geq 1} x_n/n$. It is enough to show that $\limsup_{n\to\infty} x_n/n \leq \alpha$. Arguing by contradiction, there exists $\delta \in \mathbb{R}$ such that

$$\alpha < \delta < \limsup_{n\to\infty} \frac{x_n}{n}.$$

Let $k \in \mathbb{N}$ be such that $x_k/k < \delta$. For any $n > k$ we may write $n = n(k) \cdot k + r(k)$, where $n(k) \in \mathbb{N}$ and $0 \leq r(k) < k$. It follows that

$$\frac{x_n}{n} \leq \frac{n(k)x_k + x_{r(k)}}{n(k)k + r(k)} = \frac{x_k}{k + \frac{r(k)}{n(k)}} + \frac{x_{r(k)}}{n(k)k + r(k)} \leq \frac{x_k}{k + \frac{r(k)}{n(k)}} + \frac{\beta}{n},$$

where $\beta = \sup_{0 \leq m \leq k} x_m$. Since $\lim_{n\to\infty} n(k) = \infty$, it follows that

1.4 Qualitative Results

$$\limsup_{n\to\infty} \frac{x_n}{n} \leq \frac{x_k}{k} < \delta,$$

contradiction. Hence $\limsup_{n\to\infty} x_n/n \leq \alpha$.

(ii) We construct the following example of a sequence $(x_n)_{n\geq 1}$ such that $\lim_{n\to\infty} x_n/n$ does not exist. Set $T := \{2^m;\ m \in \mathbb{N}\}$. For each fixed integer $k \geq 0$ define the sequence $(x_n)_{n\geq 1}$ by

$$x_n = \begin{cases} n & \text{if dist}(n, T) \leq k, \\ 0 & \text{otherwise.} \end{cases}$$

It is clear that $\liminf_{n\to\infty} x_n/n = 0$ and $\limsup_{n\to\infty} x_n/n = 1$, so $\lim_{n\to\infty} x_n/n$ does not exist. It suffices to prove that this sequence has the required properties. Indeed, if not, there exist integers $m, n \geq 1$ such that

$$|m - n| \leq k \quad \text{and} \quad x_{m+n} > x_m + x_n. \tag{1.11}$$

This means that either $x_n = 0$ or $x_m = 0$, say $x_n = 0$. Thus, for some positive integer q,

$$2^{q-1} + k < n < 2^q - k,$$

which implies

$$2^q + k < n + (n-k) \leq n + m \leq n + (n+k) < 2^{q+1} - k.$$

This forces $x_{n+m} = 0$. Thus, we contradict our assumption (1.11). □

The next three exercises give applications of sequences in number theory.

1.4.3. *The finite sequence $a_0, a_1, \ldots, a_{n-1}$ has the following property: for any $i \in \{0, 1, \ldots, n-1\}$, a_i denotes the number of occurrences of i in the terms of the sequence. [For instance, if $n = 4$, then an example is $a_0 = 1$, $a_1 = 2$, $a_2 = 1$, $a_3 = 0$.] Prove that if $n \geq 7$, then the sequence is uniquely defined by $n - 4, 2, 1, 0, \ldots, 0, 1, 0, 0, 0$.*

Solution. We first prove that $a_0 + a_1 + \cdots + a_{n-1} = n$. Indeed, by the definition of a_i ($0 \leq i \leq n-1$), it follows that $a_0 + \cdots + a_{n-1}$ equals the number of terms of the sequence, that is, n.

Since a_0 denotes the number of terms equal to 0, we deduce that $n - a_0$ signifies the number of terms different from 0, while $n - a_0 - 1$ represents the number of nonzero terms other than a_0 (it is obvious that $a_0 \neq 0$, since if not, then $a_0 \geq 1$, contradiction). By $a_0 + a_1 + \cdots + a_{n-1} = n$, it follows that the sum of all $n - a_0 - 1$ nonzero terms other than a_0 is equal to $n - a_0$. Thus, $n - a_0 - 2$ of them equal 1 and exactly one is equal to 2.

If $a_0 = 1$, then $a_1 = n - a_0 - 2 + 1 = n - 2$, $a_2 = 1$, $a_{n-2} = 1$, so $a_0 + a_1 + a_2 + a_{n-2} = n + 1 > n$, contradiction. Hence $a_0 \neq 1$. Analogously, $a_0 \neq 2$, so $a_0 \geq 3$.

Set $a_0 = k$. It follows that $a_1 = n - k - 2$, $a_2 = 1$, $a_{n-k-2} = 1$. Therefore $n - k - 2 = a_1 = a_2 + a_{n-k-2} = 1 + 1$, so $k = n - 4 = a_0$. Thus, $a_1 = 2$, $a_2 = 1$,

$a_{n-4} = 1$, $a_0 \geq 3$. We deduce that $n \geq 7$. It follows that the unique sequence satisfying our hypotheses is $n-4$, 2, 1, 0,..., 0, 1, 0, 0, 0. □

1.4.4. Consider an infinite arithmetic progression of natural numbers.

(a) Prove that if this progression contains cubes of integers then it contains infinitely many such terms.
(b) Give an example of an infinite arithmetic progression of positive integers such that no term is the cube of an integer.

Solution. (a) Let $(a_n)_{n\geq 1}$ be an infinite arithmetic progression of natural numbers whose common difference is d. The idea is to show that if $a_k = q_1^3$ is a cube, then $(q_1 + md)^3$, with $m \geq 1$, are all cubes in the arithmetic progression. Indeed, assume that $a_k = q_1^3$, with $q_1 \in \mathbb{N}$. We find $q_2 \in \mathbb{N}$ such that there exists $m_1 \in \mathbb{N}$ satisfying

$$a_k + m_1 d = q_2^3 \iff q_1^3 + m_1 d = q_2^3 \iff m_1 d = (q_2 - q_1)(q_1^2 + q_1 q_2 + q_2^2).$$

Set $q_2 = q_1 + d$. We observe that in this case $m_1 \in \mathbb{N}$, that is, $a_{k+m_1} = (q_1 + d)^3$. With the same argument for a_{k+m_1} we find a term a_{k+m_2}, with $m_2 > m_1$, that is also a cube, and so on. This shows that in fact, there are infinitely many terms with this property.

(b) Let a be an odd positive integer. We show that the arithmetic progression defined by $a_1 = 2a$ and $d = 4a^2$ does not contain cubes of integers. Indeed, if $2a + 4a^2 k = q^3$, then $2a + 4a^2 k = 8q_1^3$. Hence $a + 2a^2 k = 4q_1^3$, impossible because the left-hand side is odd. □

An easier example consists in choosing an arithmetic progression with all members congruent to 2 (mod 7).

1.4.5. Let p_k be the kth prime number. Define the sequence $(u_n)_{n\geq 1}$ by $u_n = p_1 + \cdots + p_n$. Prove that between u_n and u_{n+1} there is at least one perfect square.

Solution. We prove that if $m^2 \leq u_n$ then $(m+1)^2 - m^2 < u_{n+1} - u_n = p_{n+1}$. Indeed, observing that

$$2\sqrt{u_n} + 1 \geq 2m + 1 = (m+1)^2 - m^2,$$

we have only to show that $p_{n+1} > 2\sqrt{u_n} + 1$. This inequality may be rewritten as

$$\left(\frac{p_{n+1} - 1}{2}\right)^2 > p_1 + \cdots + p_n.$$

The above inequality is true for $n = 4$ and then it is justified by induction, using $p_{n+2} \geq p_{n+1} + 1$. Let us now assume that there exists $n \in \mathbb{N}$ such that between u_n and u_{n+1} there is no perfect square. This means that there exists $m \in \mathbb{N}$ such that $m^2 \leq u_n$ and $(m+1)^2 \geq u_{n+1}$. It follows that $(m+1)^2 - m^2 \geq u_{n+1} - u_n$, which contradicts the above relation. □

An interesting representation formula is proved in the next problem: any real number is a linear combination with integer coefficients of the terms of a given sequence converging to zero! It is also argued that this last assumption is sharp.

1.4 Qualitative Results   35

1.4.6. *Let $(a_n)_{n\geq 1}$ be a sequence of real numbers converging to 0 and containing infinitely many nonzero terms. Prove that for any real number x there exists a sequence of integers $(\lambda_n(x))_{n\geq 1}$ such that $x = \lim_{n\to\infty} \sum_{k=1}^{n} \lambda_k(x) a_k$.*

Is the assumption $\lim_{n\to\infty} a_n = 0$ necessary?

Solution. We can assume that $a_n \neq 0$, for all n. Fix arbitrarily a positive number x. We choose $\lambda_n(x)$ such that $\lambda_n(x) a_n \geq 0$, for all $n \in \mathbb{N}$.

We first observe that there exist $n_1 \in \mathbb{N}$ and $\lambda_{n_1}(x) \in \mathbb{Z}$ such that $\lambda_{n_1}(x) a_{n_1} \leq x$ and $|x - \lambda_{n_1}(x) a_{n_1}| < |a_{n_1}|$. We point out that the existence of n_1 and $\lambda_{n_1}(x)$ follows from $\lim_{n\to\infty} a_n = 0$. For the same reasons, there exist $n_2 \in \mathbb{N}$, $n_2 > n_1$, and $\lambda_{n_2}(x) \in \mathbb{Z}$ such that $\lambda_{n_2}(x) a_{n_2} \leq x - \lambda_{n_1}(x) a_{n_1}$ and $|x - \lambda_{n_1}(x) a_{n_1} - \lambda_{n_2}(x) a_{n_2}| < |a_{n_2}|$. Thus, we obtain an increasing sequence of positive integers (n_k) and a sequence of integers $(\lambda_{n_k}(x))_{n_k}$ such that

$$\lambda_{n_{k+1}}(x) a_{n_{k+1}} \leq x - \lambda_{n_1}(x) a_{n_1} - \cdots - \lambda_{n_k}(x) a_{n_k}$$

and

$$|x - \lambda_{n_1}(x) a_{n_1} - \cdots - \lambda_{n_{k+1}}(x) a_{n_{k+1}}| < |a_{n_{k+1}}|.$$

If $n \notin \{n_1, n_2, \ldots, n_k, \ldots\}$ then we take $\lambda_n(x) = 0$. Since $\lim_{k\to\infty} a_{n_k} = 0$, we obtain $x = \lim_{n\to\infty} \sum_{k=1}^{n} \lambda_k(x) a_k$.

If $x = 0$, then we take $\lambda_n(0) = 0$, and if $x < 0$, then we choose $\lambda_n(x) = -\lambda_n(-x)$, for all n.

If the sequence (a_n) does not converge to 0, then the result is not true. Indeed, it suffices to take $a_n = 1$, for all $n \geq 1$. If $\lambda_n \in \mathbb{Z}$ and if the sequence defined by $\sum_{k=1}^{n} \lambda_k(x) a_k$ is convergent, then its limit cannot belong ro $\mathbb{R} \setminus \mathbb{Z}$. □

The following is an interesting application of sequences in a problem of interest for many people. The statement is in connection with (easy!) IMO-type problems.

1.4.7. *A chess player plays at least one game every day and at most 12 games every week. Prove that there exists a sequence of consecutive days in which he plays exactly 20 games.*

Solution. Let a_n be the number of games played in the first n days. By hypothesis, $21 \leq a_{21} \leq 36$. Among the numbers $a_1, a_2, \ldots, a_{21}$ there exist a_i and a_j ($1 \leq i < j \leq 21$) giving the same remainder on division by 20. Hence

$$1 \leq a_j - a_i \leq 35,$$

and 20 divides $a_j - a_i$. It follows that $a_j - a_i = 20$. Consequently, in the days $i+1, i+2, \ldots, j$, he played exactly 20 games. □

The method used for proving the next problem can be easily extended to deduce similar properties of positive integers. Try to formulate some related properties!

1.4.8. *Prove that among any 39 consecutive positive integers there exists at least one having the sum of its digits a multiple of 11.*

Solution. Let s_n denote the sum of digits of the positive integer n. If the last digit of n is not 9, then $s_{n+1} = s_n + 1$. If the last k digits of n are 9, then

$s_{n+1} = s_n - 9k + 1$. Next, we take into account that in a sequence of fewer than 100 consecutive positive integers there exists at most one number that ends with at least two digits 9. The most unfavorable case for our problem is the following: $s_n, s_n+1, \ldots, s_n+8, s_n, s_n+1, \ldots, s_n+9, s_n-9k+10, s_n-9k+11, \ldots, s_n-9k+19, s_n-9k+11, s_n-9k+12, \ldots, s_n-9k+20$, where $n-1$ ends with exactly one 0, $n+18$ has the last k digits equal to 9, and

$$-9k \equiv 1 \pmod{11} \quad \text{and} \quad s_{n-1} \equiv 0 \pmod{11}.$$

In the above scheme, which contains the sum of digits of 39 consecutive positive integers, we observe that only the number $s_{n+38} = s_n - 9k + 20$ is a multiple of 11. □

The existence of the maximum in the next example follows by standard compactness arguments. However, it is much more difficult to find the explicit value of the maximum.

1.4.9. *Let $n \geq 2$ be an integer, and $a_1, \ldots, a_n$ real numbers belonging to $[-1, 1]$. Establish for what values of $a_1, \ldots, a_n$ the expression*

$$E = \left| a_1 - \frac{a_2 + \cdots + a_n}{n} \right| + \cdots + \left| a_n - \frac{a_1 + \cdots + a_{n-1}}{n} \right|$$

achieves its maximum.

Solution. By virtue of the symmetry of E, we can assume that $a_1 \geq a_2 \geq \cdots \geq a_n$. Define $s_k = \sum_{i=1}^k a_i$ ($k = 1, \ldots, n$). It follows that

$$E = \sum_{i=1}^n \frac{1}{n} |(n+1)a_i - s_n|.$$

Let α be the number of terms a_i such that $(n+1)a_i \geq s_n$. Hence

$$E = \frac{1}{n}[(n+1)s_\alpha - \alpha s_n - (n+1)(s_n - s_\alpha) + (n-\alpha)s_n]$$
$$= \frac{1}{n}[2(n+1)s_\alpha - s_n(2\alpha + 1)].$$

Assume α is fixed. Since $-1 \leq a_i \leq 1$, it follows that $s_n \geq s_\alpha + \alpha - n$, $s_\alpha \leq \alpha$. Therefore

$$E \leq \frac{1}{n}[(2n+2)s_\alpha - (s_\alpha + \alpha - n)(2\alpha + 1)],$$

and thus

$$E \leq \frac{1}{n}[(2n - 2\alpha + 1)\alpha - (\alpha - n)(2\alpha + 1)] = \frac{1}{n}(-4\alpha^2 + 4n\alpha + n).$$

This means that for any fixed α, we have $\max E = (-4\alpha^2 + 4n\alpha + n)/n$, which is achieved for $a_i = 1$ ($1 \leq i \leq \alpha$) and $a_j = -1$ ($j = \alpha+1, \ldots, n$). If α is varying, then by monotony arguments, we obtain that the maximum of E is attained for

1.4 Qualitative Results

$\alpha = n/2$ (if n is even) and $\alpha = (n+1)/2$ or $\alpha = (n-1)/2$ (if n is odd), that is, for $\alpha = [n+1/2]$. Therefore

$$\max E = -\frac{4}{n}\left[\frac{n+1}{2}\right]^2 + 4\left[\frac{n+1}{2}\right] + 1. \quad \square$$

The solution of the next problem relies deeply on the order structure of real numbers combined with a certain basic property of the set of positive integers. What is this property and why does the result not remain true if the three sequences are in $\mathbb{R}$?

1.4.10. *Prove that for any sequences of positive integers $(a_n)_{n\geq 0}$, $(b_n)_{n\geq 0}$, $(c_n)_{n\geq 0}$ there exist different positive integers p and q such that $a_p \geq a_q$, $b_p \geq b_q$, and $c_p \geq c_q$.*

Solution. Let $q \in \mathbb{N}$ be such that $c_q = \min_n c_n$. If there exists $p \in \mathbb{N}$, $p \neq q$, such that $a_p \geq a_q$ and $b_p \geq b_q$, then we conclude the proof. Let us now assume that for all $p \in \mathbb{N}$, $p \neq q$, we have either $a_p < a_q$ or $b_p < b_q$. This means that there are infinitely many indices $p \neq q$ such that $a_p < a_q$ or $b_p < b_q$. We can assume, without loss of generality, that $a_p < a_q$, for infinitely many $p \in \mathbb{N}$. Since between 1 and a_q there are finitely many integers, it follows (by Dirichlet's principle) that between 1 and a_q there exists a number that appears infinitely many times in the sequence $(a_n)_{n\geq 0}$. In other words, the sequence $(a_n)_{n\geq 0}$ contains a constant subsequence (a_{n_k}): $a_{n_1} = a_{n_2} = \cdots = a_{n_k} = \cdots$.

Next, we consider the corresponding subsequences (b_{n_k}) and (c_{n_k}). Let $j \in \mathbb{N}$ be such that $c_{n_j} = \min_k c_{n_k}$. If there is some $i \in \mathbb{N}$ such that $b_{n_i} \geq b_{n_j}$, then the proof is concluded because $a_{n_i} = a_{n_j}$. So, we assume that $b_{n_i} < b_{n_j}$, for all $i \in \mathbb{N}$. In particular, the subsequence (b_{n_k}) is bounded, that is, it contains a constant subsequence (b_{m_k}).

Consider now the subsequence (c_{m_k}). The proof is concluded after observing that choosing $r, s \in \mathbb{N}$ such that $c_{m_r} \geq c_{m_s}$, we have $b_{m_r} = b_{m_s}$ and $a_{m_r} = a_{m_s}$. $\quad \square$

Kepler's equation is an important tool for understanding major phenomena arising in celestial mechanics. This contribution is due to the German mathematician Johannes Kepler (1571–1630), who formulated the three laws of planetary motion, a major step in the formulation of the laws of motion and universal gravitation by Sir Isaac Newton. The unknown of the following equation denotes the *eccentric anomaly*, which appears in astrodynamics and is related to Kepler's second law: *A line joining a planet and the sun sweeps out equal areas during equal intervals of time*.

1.4.11. *For any $0 < \varepsilon < 1$ and $a \in \mathbb{R}$, consider the Kepler equation $x - \varepsilon \sin x = a$. Define $x_0 = a$, $x_1 = a + \varepsilon \sin x_0, \ldots, x_n = a + \varepsilon \sin x_{n-1}$. Prove that there exists $\xi = \lim_{n \to \infty} x_n$ and that ξ is the unique solution of Kepler's equation.*

Solution. We first observe that the sequence $(x_n)_{n\geq 0}$ is bounded. This follows from

$$|x_n| = |a + \varepsilon \sin x_{n-1}| \leq |a| + \varepsilon < |a| + 1.$$

We prove that the sequence $(x_n)_{n\geq 0}$ converges using Cauchy's criterion. More precisely, we show that for any $\delta > 0$, there exists $N(\delta) \in \mathbb{N}$ such that for all $n \geq N(\delta)$ and for any $p \in \mathbb{N}$ we obtain $|x_n - x_{n+p}| < \delta$.

We have

$$|x_n - x_{n+p}| = \varepsilon |\sin x_{n-1} - \sin x_{n+p-1}|$$
$$= 2\varepsilon \left|\sin \frac{x_{n-1} - x_{n+p-1}}{2}\right| \cdot \left|\cos \frac{x_{n-1} + x_{n+p-1}}{2}\right|$$
$$\leq 2\varepsilon \left|\sin \frac{x_{n-1} - x_{n+p-1}}{2}\right| \leq 2\varepsilon^2 \cdot \frac{|x_{n-1} - x_{n+p-1}|}{2}$$
$$= \varepsilon |x_{n-1} - x_{n+p-1}|.$$

But $|x_0 - x_p| = \varepsilon |\sin x_{p-1}| \leq \varepsilon$, $|x_1 - x_{p+1}| \leq \varepsilon \cdot \varepsilon = \varepsilon^2$. This implies that

$$|x_{n-1} - x_{n+p-1}| \leq \varepsilon^n < \delta,$$

provided that $n > \log \delta / \log \varepsilon$ [we have used here our assumption $\varepsilon \in (0,1)$]. Set $\xi = \lim_{n \to \infty} x_n$. It follows that $\xi - \varepsilon \sin \xi = a$, which shows that ξ is a solution of Kepler's equation.

In order to prove the uniqueness, let ξ_1, ξ_2 be distinct solutions of Kepler's equation. We have

$$|\xi_1 - \xi_2| = \varepsilon |\sin \xi_1 - \sin \xi_2| = 2\varepsilon \left|\sin \frac{\xi_1 - \xi_2}{2}\right| \cdot \left|\cos \frac{\xi_1 + \xi_2}{2}\right|$$
$$\leq 2\varepsilon \left|\sin \frac{\xi_1 - \xi_2}{2}\right| < 2 \left|\frac{\xi_1 - \xi_2}{2}\right| = |\xi_1 - \xi_2|. \quad \square$$

This contradiction shows that the Kepler equation has a unique solution.

The following quadratic *Diophantine equation* (that is, the unknowns are positive integers) is a special case of the Pell equation $x^2 - Dy^2 = m$ (where D is a nonsquare positive integer) and m is an integer. It seems that the English mathematician John Pell (1610–1685) has nothing to do with this equation. According to Lenstra [69], Euler mistakenly attributed to Pell a solution method that had been found by another English mathematician, William Brouncker (1620–1684). An exposition for solving the Pell equation may be found even in Euler's *Algebra* [25, Abschn. 2, Cap. 7].

Usually, Pell-type equations of this form are solved by finding the continued fraction of $\sqrt{D}$. This method is also due to Euler (see [84, Chap. 7]). Our proof is based on an iterative process. The idea of computing by recursion is as old as counting itself. It occurred in primitive form in the efforts of the Babylonians as early as 2000 B.C. to extract roots and in more explicit form around 450 B.C. in the Pythagoreans' study of figurative numbers, since in modern notation the triangular numbers satisfy the difference equation $t_n = t_{n-1} + n$, the square numbers the equation $s_n = s_{n-1} + 2n - 1$, and so forth. The Pythagoreans also used a system of difference equations $x_n = x_{n-1} + 2y_{n-1}$, $y_n = x_{n-1} + y_{n-1}$ to generate large solutions of Pell's equation $x^2 - 2y^2 = 1$, and thereby approximations of $\sqrt{2}$. In his attempts to compute the circumference of a circle, Archimedes (about 250 B.C.) employed equations of the form $P_{2n} = 2p_n P_n/(p_n + P_n)$, $p_{2n} = \sqrt{p_n P_{2n}}$ to compute the perimeters P_n and p_n of the circumscribed polygon of n sides and the inscribed polygon of n sides, respectively. Other familiar ancient discoveries about recurrence include the Euclidean algorithm and Zeno's paradox. Euclid also studied geometric series, although the general form of the sum was not obtained until around 1593 by Vieta.

1.4.12. *Prove that the equation $x^2 - 2y^2 = 7$ has infinitely many integer solutions.*

Solution. We first observe that $x_0 = 3$, $y_0 = 1$ is a solution of the above equation. We prove that if (x_n, y_n) is a solution then (x_{n+1}, y_{n+1}) is a solution, too, where $x_{n+1} = 3x_n + 4y_n$ and $y_{n+1} = 2x_n + 3y_n$. Indeed,

$$x_{n+1}^2 - 2y_{n+1}^2 = 9x_n^2 + 16y_n^2 + 24x_ny_n - 8x_n^2 - 18y_n^2 - 24x_ny_n = x_n^2 - 2y_n^2 = 7.$$

This shows that the equation has infinitely many distinct positive solutions in $\mathbb{Z} \times \mathbb{Z}$. □

A representation property in terms of a given sequence of positive integers is stated below.

1.4.13. *Let a be an arbitrary positive integer. Show that there exists at least one positive integer that can be expressed in two distinct ways as sums of distinct terms of the form $1^a, 2^a, \ldots, n^a, \ldots$.*

Solution. We first prove that $1^a + 2^a + \cdots + n^a < 2^n - 1$, for any n sufficiently large. Indeed, we easily observe with an induction argument that $1^a + 2^a + \cdots + n^a$ is a polynomial of degree $a + 1$ in n. The inequality $1^a + 2^a + \cdots + n^a < 2^n - 1$, for n large enough, follows now easily after taking into account that the exponential function grows faster at infinity than any polynomial function.

We observe that the number of sums of distinct terms in the finite sequence 1^a, $2^a, \ldots, n^a$ equals $2^n - 1$, which coincides with the number of nonempty subsets of a set that contains n elements. The largest of these sums, that is, $1^a + 2^a + \cdots + n^a$, is less than $2^n - 1$. It follows that all the $2^n - 1$ sums are less than $2^n - 1$, too. Since all these sums are positive integers, it follows that at least two of them must coincide. □

An elementary property of positive integers is used to find the sum of digits of a huge number!

1.4.14. *Let $S(x)$ denote the sum of digits of the positive integer x. Let n be a multiple of 9 having fewer than 10^{10} digits. Prove that $S(S(S(n))) = 9$.*

Solution. We use the property that the difference between any positive integer and the sum of its digits is a multiple of 9. Since $n \equiv 0 \pmod 9$, we have $S(n) - n \equiv 0 \pmod 9$, $S(S(n)) - S(n) \equiv 0 \pmod 9$, $S(S(S(n))) - S(S(n)) \equiv 0 \pmod 9$. Therefore $S(S(S(n))) \equiv 0 \pmod 9$. Since n has fewer than 10^{10} digits, it follows that $S(n) \leq 9 \cdot 10^{10}$, so $S(S(n)) \leq 9 \cdot 10 + 8 = 98$. Hence $S(S(S(n))) \leq 8 + 9 = 17$. Now, using $S(S(S(n))) \equiv 0 \pmod 9$, we deduce that $S(S(S(n))) = 9$. □

Independent Study. The above property is related to the following problem proposed to the International Mathematical Olympiad in 1975. Find its solution!

Let A be the sum of the decimal digits of 4444^{4444}, and let B be the sum of the decimal digits of A. Find the sum of the decimal digits of B.

We find below all possible values of a sum.

1.4.15. *Let A a positive real number. Find all the possible values of $\sum_{j=0}^{\infty} x_j^2 := \lim_{n\to\infty} \sum_{j=0}^{n} x_j^2$, where $x_0, x_1, \ldots, x_n, \ldots$ are positive numbers satisfying $\sum_{j=0}^{\infty} x_j = A$.*

Solution. We prove that the given expression may take any value in $(0, A^2)$. We first observe that

$$0 < \left(\sum_{j=0}^{n} x_j\right)^2 = \sum_{j=0}^{n} x_j^2 + 2 \sum_{0 \leq j < k \leq n} x_j x_k.$$

This implies that $\sum_{j=0}^{n} x_j^2 \leq A^2 - 2x_0 x_1$. Passing at the limit as $n \to \infty$ we obtain $\sum_{j=0}^{\infty} x_j^2 \leq A^2 - 2x_0 x_1 < A^2$.

Next, we prove that the expression can achieve any value in $(0, A^2)$. Indeed, let $(x_n)_{n \geq 0}$ be a geometric progression of ratio d. In this case, $\sum_{j=0}^{\infty} x_j = x_0/(1-d)$ and

$$\sum_{j=0}^{\infty} x_j^2 = \frac{x_0^2}{1 - d^2} = \frac{1-d}{1+d} \left(\sum_{j=0}^{\infty} x_j\right)^2.$$

If d grows from 0 to 1, then $(1-d)/(1+d)$ decreases from 1 to 0. Consequently, if $(x_n)_{n \geq 0}$ is a geometric progression of positive numbers satisfying $\sum_{j=0}^{\infty} x_j = A$, then $\sum_{j=0}^{\infty} x_j^2$ takes values between 0 and A^2. □

The terms of a sequence defined by a polynomial recurrence satisfy an interesting property. Our solution relies on a fundamental property of polynomials with integer coefficients.

1.4.16. *Let $f(x)$ be a polynomial with integer coefficients. Define the sequence $(a_n)_{n \geq 0}$ by $a_0 = 0$ and $a_{n+1} = f(a_n)$, for all $n \geq 0$. Prove that if there exists $m \geq 1$ such that $a_m = 0$, then either $a_1 = 0$ or $a_2 = 0$.*

Solution. We first recall that if $f(x)$ is a polynomial with integer coefficients then $m - n$ is a divisor of $f(m) - f(n)$, for any different integers m and n. In particular, if $b_n = a_{n+1} - a_n$, then b_n is a divisor of b_{n+1}, for any n. On the other hand, $a_0 = a_m = 0$, so $a_1 = a_{m+1}$, which implies $b_0 = b_m$. If $b_0 = 0$, then $a_0 = a_1 = \cdots = a_m$, and the proof is concluded. If not, then $|b_0| = |b_1| = |b_2| = \cdots$, so $b_n = \pm b_0$, for all n.

Using $b_0 + \ldots + b_{m-1} = a_m - a_0 = 0$ we deduce that half of the integers $b_0, \ldots, b_{m-1}$ are positive. In particular, there exists an integer $0 < k < m$ such that $b_{k-1} = -b_k$, that is, $a_{k-1} = a_{k+1}$. It follows that $a_n = a_{n+2}$, for all $n \geq k - 1$. Taking $m = n$, we deduce that $a_0 = a_m = a_{m+2} = f(f(a_0)) = a_2$. □

For the next exercise the reader must be familiar with the properties of the scalar product of vectors.

1.4.17. *Find all vectors $\mathbf{x}, \mathbf{y} \neq 0$ such that the sequence $a_n = |\mathbf{x} - n\mathbf{y}|$ $(n \geq 1)$ is (a) increasing; (b) decreasing.*

Solution. We have

$$a_n^2 = |\mathbf{x}|^2 - 2n(\mathbf{x}\cdot\mathbf{y}) + n^2|\mathbf{y}|^2 = |\mathbf{y}|^2\left[\left(n - \frac{\mathbf{x}\cdot\mathbf{y}}{|\mathbf{y}|^2}\right)^2\right] + \left[|\mathbf{x}|^2 - \frac{(\mathbf{x}\cdot\mathbf{y})^2}{|\mathbf{y}|^4}\right].$$

The nonconstant term of these quantities contains $(n-c)^2$. This shows that the sequence is never decreasing and the sequence is increasing if and only if $c \leq 3/2$. Consequently, the sequence is increasing if and only if

$$\frac{\mathbf{x}\cdot\mathbf{y}}{|\mathbf{y}|^2} \leq \frac{3}{2}. \quad \square$$

An interesting convergent rearrangement is possible, provided a sequence of positive real numbers has a finite nonzero accumulation point.

1.4.18. *Let $(a_n)_{n\geq 1}$ be a sequence of positive real numbers that has a finite nonzero accumulation point. Prove that $(a_n)_{n\geq 1}$ can be arranged in a sequence $(x_n)_{n\geq 1}$ so that $(x_n^{1/n})_{n\geq 1}$ is convergent.*

Solution. Let $a \in (0,\infty)$ be an accumulation point of $(a_n)_{n\geq 1}$ and choose $\alpha > 0$ such that $\alpha^{-1} < a < \alpha$. Let x_1 denote the a_n of smallest subscript that lies in the interval (α^{-1}, α). Assume that $x_1, x_2, \ldots, x_{k-1}$ have been chosen. Then we define $x_k = a_n$, where n is the smallest subscript different from the $k-1$ already chosen lying in the interval $(\alpha^{-\sqrt{n}}, \alpha^{\sqrt{n}})$. Such a subscript does exist, since a is an accumulation point. Moreover, by $\alpha^{\sqrt{n}} \to \infty$ and $\alpha^{-\sqrt{n}} \to 0$ as $n\to\infty$, it follows that every a_n belongs to a certain interval $(\alpha^{-\sqrt{n}}, \alpha^{\sqrt{n}})$ and will therefore become part of the sequence $(x_n)_{n\geq 1}$. For each n we have $\alpha^{-\sqrt{n}} < x_n < \alpha^{\sqrt{n}}$ and therefore $(1/\alpha)^{1/\sqrt{n}} < x_n < \alpha^{1/\sqrt{n}}$. Since $\lim_{n\to\infty}\alpha^{1/\sqrt{n}} = \lim_{n\to\infty}(1/\alpha)^{1/\sqrt{n}} = 1$, we conclude that $\lim_{n\to\infty} x_n^{1/n} = 1$. $\quad \square$

We are now concerned with the maximum possible values of the terms of a sequence satisfying a certain assumption.

1.4.19. *Let $(a_n)_{n\geq 0}$ be a sequence of nonnegative numbers with $a_0 = 1$ and satisfying $\sum_{k=n}^{\infty} a_k \leq ca_n$, for all $n \geq 0$, where $c > 1$ is a prescribed constant. Find the maximum of a_n in terms of c.*

Solution. For any integer $n \geq 1$, the maximum of a_n is $(c-1)^n/c^{n-1}$. Indeed, since $a_n \geq 0$, we have

$$ca_{n-1} \geq \sum_{k=n-1}^{\infty} a_k \geq a_{n-1} + a_n,$$

so $a_{n-1} \geq a_n/(c-1)$. Similarly, we obtain

$$ca_{n-2} \geq a_{n-2} + a_{n-1} + a_n \geq a_{n-2} + \left(\frac{1}{c-1} + 1\right)a_n,$$

so $a_{n-2} \geq ca_n/(c-1)^2$. With a similar argument we obtain, by induction,

$$a_{n-k} \geq c^{k-1} a_n/(c-1)^k, \quad \text{for all } 1 \leq k \leq n.$$

For $k = n$ we have $1 = a_0 \geq c^{n-1} a_n/(c-1)^n$, so $a_n \leq (c-1)^n/c^{n-1}$.

For $n \geq 1$ fixed, this upper bound is achieved by the sequence $(a_k)_{k \geq 0}$ defined by $a_k = 0$ for $k > n$, $a_n = (c-1)^n/c^{n-1}$, and $a_k = (c-1)^k/c^k$, provided that $1 \leq k < n$. It follows that $\sum_{k=m}^{\infty} a_k \leq ca_m$ for $m \geq n$, while for $m < n$ we have

$$\sum_{k=m}^{\infty} a_k = a_n + \sum_{k=m}^{n-1} \frac{(c-1)^k}{c^k} = \frac{(c-1)^n}{c^{n-1}} - c\left[\frac{(c-1)^n}{c^n} - \frac{(c-1)^m}{c^m}\right] = ca_m. \quad \square$$

We know what monotone sequences are. What about monotone sequences starting with a certain rank? More precisely, we are concerned with sequences of real numbers $(x_n)_{n \geq 0}$ such that $(x_n)_{n \geq N}$ is monotone for some positive integer N.

1.4.20. Characterize the sequences of real numbers $(a_n)_{n \geq 0}$ for which there exists a permutation σ of $\mathbb{N}$ such that the sequence $(a_{\sigma(n)})_{n \geq 0}$ is monotone starting with a certain rank.

Solution. We shall restrict to the case in which the sequence $(a_{\sigma(n)})_{n \geq 0}$ is increasing starting with a certain rank. We first observe that such a sequence has a limit, not necessarily finite. Indeed, let $\ell \in \mathbb{R} \cup \{+\infty\}$ be the limit of $(a_{\sigma(n)})_{n \geq 0}$ and fix a neighborhood V of ℓ. So, there exists a certain rank N such that for all $n \geq N$, $a_{\sigma(n)} \in V$. Set $M = \max_{0 \leq k \leq n-1} \sigma(k)$. Then, for all integer $m \geq M$ we have $\sigma^{-1}(m) \geq N$, that is, $a_m \in V$.

We first show that if $\ell = +\infty$ then there exists a permutation with the required property. Indeed, in this case, there exists $a_{n_0} = \min_{n \geq 0} a_n$. Set $\sigma(0) = n_0$. Define recurrently the permutation σ by

$$a_{\sigma(k)} = \min\{a_m; m \neq \sigma(j), \text{ for all } j < k\}.$$

Then σ is one-to-one and the limit of $(a_{\sigma(n)})_{n \geq 0}$ is $+\infty$. It remains to show that σ is onto. Indeed, if not, there is some integer $p \in \mathbb{N}$ such that $p \notin \sigma(\mathbb{N})$. In this case, we consider the smallest integer i with $a_p < a_{\sigma(i)}$, which contradicts the definition of $\sigma(i)$.

Next, we assume that $\ell \in \mathbb{R}$. In this case we show that there exists a permutation with the required property if and only if one of the following conditions holds: either

(i) $a_n = \ell$ for any n large enough

or

(ii) $a_n < \ell$ for all n.

Indeed, these conditions are sufficient. This is obvious if (i) holds. In case (ii) we argue to an argument similar to used if $\ell = +\infty$.

Conversely, let us now assume that $(a_{\sigma(n)})_{n \geq 0}$ is increasing starting with a certain rank N. If the above condition (i) does not hold, then $(a_{\sigma(n)})_{n \geq 0}$ is not stationary. Thus the set $\{n; a_{\sigma(n)} \geq \ell\}$ is finite, that is, condition (ii) holds. $\square$

Sometimes functional equations may be solved by using recurrent sequences. The example below can be easily extended to larger classes of related problems.

1.4.21. Determine all bijections f from $[0,1]$ to $[0,1]$ satisfying $f(2x - f(x)) = x$ for all x in $[0,1]$.

K.-W. Lih, Amer. Math. Monthly, Problem E 2893

Solution. By hypothesis, $f^{-1}(x) = 2x - f(x)$, so that $f(x) - x = x - f^{-1}(x)$, for all $x \in [0,1]$. Fix $x_0 \in [0,1]$ and define inductively the sequence $(x_n)_{n \geq 0}$ by $x_n = f(x_{n-1})$, for all $n \geq 1$. Setting $x = x_{n-1}$ in the above equation, we obtain $x_n - x_{n-1} = x_{n-1} - x_{n-2}$. It follows that $x_n - x_{n-1} = x_1 - x_0$ for all $n \geq 1$. Hence $x_n - x_0 = n(x_1 - x_0)$. Since $|x_n - x_0| \leq 1$, we conclude that $|x_1 - x_0| < 1/n$ for all $n \geq 1$. Therefore $|x_1 - x_0| = 0$ and, consequently, $f(x_0) = x_0$. This argument shows that the only solution is given by $f(x) = x$, for all $x \in [0,1]$.

We now discuss a sufficient condition for a nondecreasing sequence $(a_n)_{n \geq 1}$ such that $(a_n/n)_{n \geq 1}$ converges. It is easy to observe that $(a_n)_{n \geq 1}$ should have at most linear growth. Is this enough? □

1.4.22. Let $(a_n)_{n \geq 1}$ be a nondecreasing sequence of nonnegative numbers. Assume that $a_{mn} \geq m a_n$ for all m, n, and also $\sup(a_n/n) = \ell < \infty$. Prove that the sequence $(a_n/n)_{n \geq 1}$ converges to ℓ.

J. M. Borden, Amer. Math. Monthly, Problem E 2860

Solution. If $\ell = 0$ the proof is immediate. If $\ell > 0$, given $\varepsilon > 0$, there exists a positive integer N such that $\ell - \varepsilon/2 < a_N/N \leq \ell$. For every m, it now follows that $\ell - \varepsilon/2 < a_{mN}/(mN) \leq \ell$. Fix $\varepsilon > 0$ and suppose $n \geq 2\ell N/\varepsilon$. Then we have $mN \leq n < (m+1)N$ for some $m \geq 2\ell/\varepsilon - 1$ (hence $m/m+1 \geq 1 - \varepsilon/2\ell$). Then we have

$$\frac{a_n}{n} \geq \frac{a_{mN}}{(m+1)N} = \frac{a_{mN}}{mN} \cdot \frac{m}{m+1}$$
$$> \left(\ell - \frac{\varepsilon}{2}\right) \cdot \frac{m}{m+1} \geq \left(\ell - \frac{\varepsilon}{2}\right)\left(1 - \frac{\varepsilon}{2\ell}\right) > \ell - \varepsilon.$$

Hence $\lim_{n \to \infty} a_n/n = \ell$. □

A linear combination of two consecutive terms of any sequence converges to zero. What about the coefficients of such a combination?

1.4.23. Find all positive real numbers a and b such that the limit of every sequence $(x_n)_{n \geq 1}$ satisfying $\lim_{n \to \infty}(ax_{n+1} - bx_n) = 0$ equals zero.

Solution. We prove that this property holds if and only if $a > b$. Indeed, if $a < b$ then the sequence $x_n = (b/a)^n$ satisfies the hypothesis but does not converge to 0. If $a = b$, a counterexample is offered by the sequence $x_n = 1 + 1/2 + \cdots + 1/n$.

Next, we assume that $a > b$. Let ℓ_- (resp., ℓ_+) be the limit inferior (resp., limit superior) of the given sequence. Now we pass to the "lim sup" in our hypothesis, using the relation $\limsup_{n \to \infty}(a_n + b_n) \geq \limsup_{n \to \infty} a_n + \limsup_{n \to \infty} b_n$. It follows that

$$0 = \lim_{n\to\infty}(ax_{n+1} - bx_n) \geq \limsup_{n\to\infty}(ax_{n+1}) + \limsup_{n\to\infty}(-bx_n)$$
$$= a\ell_+ - \liminf_{n\to\infty}(bx_n) = a\ell_+ - b\ell_-.$$

Thus, $\ell_+ \leq (b/a)\ell_-$. Since $\ell_- \leq \ell_+$, it follows that $\ell_+ \leq (b/a)\ell_+$, and so $\ell_- \leq \ell_+ \leq 0$. Similarly, passing at the "lim inf," we obtain $\ell_- \geq (b/a)\ell_+$. Since $\ell_+ \geq \ell_-$, we deduce that $\ell_- \geq (b/a)\ell_-$. Thus, $\ell_+ \geq \ell_- \geq 0$. Therefore $\ell_+ = \ell_- = 0$, and the sequence converges to 0. □

1.4.24. Let $(a_n)_{n\geq 0}$ be a sequence of nonnegative numbers such that $a_{2k} - a_{2k+1} \leq a_k^2$, $a_{2k+1} - a_{2k+2} \leq a_k a_{k+1}$ for any $k \geq 0$ and $\limsup_{n\to\infty} na_n < 1/4$. Prove that $\limsup_{n\to\infty} \sqrt[n]{a_n} < 1$.

Solution. Let $c_l = \sup_{n\geq 2^l}(n+1)a_n$ for $l \geq 0$. We show that $c_{l+1} \leq 4c_l^2$. Indeed, for any integer $n \geq 2^{l+1}$ there exists an integer $k \geq 2^l$ such that $n = 2k$ or $n = 2k+1$. In the first case we have

$$a_{2k} - a_{2k+1} \leq a_k^2 \leq \frac{c_l^2}{(k+1)^2} \leq \frac{4c_l^2}{2k+1} - \frac{4c_l^2}{2k+2},$$

while in the second case we obtain

$$a_{2k+1} - a_{2k+2} \leq a_k a_{k+1} \leq \frac{c_l^2}{(k+1)(k+2)} \leq \frac{4c_l^2}{2k+2} - \frac{4c_l^2}{2k+3}.$$

Hence a sequence $(a_n - 4c_l^2(n+1)^{-1})_{n\geq 2^{l+1}}$ is nondecreasing and its terms are nonpositive, since it converges to 0. Therefore $a_n \leq 4c_l^2/n+1$ for $n \geq 2^{l+1}$, meaning that $c_{l+1}^2 \leq 4c_l^2$. This implies that a sequence $((4c_l)^{2^{-l}})_{l\geq 0}$ is nonincreasing and therefore bounded from above by some number $q \in (0,1)$, since all its terms except finitely many are less than 1. Hence $c_l \leq q^{2^l}$ for l large enough. For any n between 2^l and 2^{l+1} there is $a_n \leq c_l/n+1 \leq q^{2^l} \leq (\sqrt{q})^n$, yielding $\limsup_{n\to\infty} \sqrt[n]{a_n} \leq \sqrt{q} < 1$, which ends the proof. □

Leopold Kronecker (1823–1891) was a German mathematician with deep contributions in number theory, algebra, and analysis. One of his famous quotations is the following: "God made the integers; all else is the work of man." The next density property has a crucial importance for the understanding of the set of real numbers.

1.4.25. (**Kronecker's Theorem**). *Let α be an irrational real number. Prove that the set $A = \{m + n\alpha;\ m, n \in \mathbb{Z}\}$ is dense in $\mathbb{R}$.*

Solution. We recall that a set $A \subset \mathbb{R}$ is dense in $\mathbb{R}$ if for all $\varepsilon > 0$ and for any $x \in \mathbb{R}$, there exists $a \in A$ such that $|x - a| < \varepsilon$. In other words, the set $A \subset \mathbb{R}$ is dense in $\mathbb{R}$ if for all $x \in \mathbb{R}$ there exists a sequence $(a_n)_{n\geq 1}$ in A such that $a_n \to x$ as $n \to \infty$.

We first prove that there exists a sequence $(a_n)_{n\geq 1}$ in A such that $a_n \to 0$ as $n \to \infty$. For this purpose, consider the numbers $\{\alpha\}, \{2\alpha\}, \ldots, \{(n+1)\alpha\}$, where $n \geq 1$ is an arbitrary integer. Since $\alpha \in \mathbb{R} \setminus \mathbb{Q}$, it follows that $\{j\alpha\} \neq \{k\alpha\}$, for any $1 \leq j < k \leq n+1$. Next, we observe that

$$\{\{\alpha\}, \{2\alpha\}, \ldots, \{(n+1)\alpha\}\} \subset [0,1] = \left[0, \frac{1}{n}\right) \cup \left[\frac{1}{n}, \frac{2}{n}\right) \cup \cdots \cup \left[\frac{n-1}{n}, 1\right).$$

So, by the pigeonhole principle[1] (Dirichlet's principle), there exist two elements, say $\{p\alpha\}$ and $\{q\alpha\}$ ($1 \leq p < q \leq n+1$), that belong to the same interval $[k/n, (k+1)/n)$. This means that $|\{p\alpha\} - \{q\alpha\}| < 1/n$. Now setting

$$a_n = \{p\alpha\} - \{q\alpha\} = (p-q)\alpha + [q\alpha] - [p\alpha] \in A,$$

we deduce that $a_n \to 0$ as $n \to \infty$. So, since $a_n \neq 0$, we obtain that the set $B := \{ka_n;\ k \in \mathbb{Z},\ n \in \mathbb{N}^*\}$ is dense in $\mathbb{R}$. Since $B \subset A$, this implies that A is dense in $\mathbb{R}$, too. $\square$

We have proved in Exercise 1.1.15. that the sequence $(\sin n)_{n\geq 1}$ is divergent. The next result yields much more information about the behavior of this sequence.

1.4.26. *Prove that the set of accumulation points of the sequence $(\sin n)_{n\geq 1}$ is the interval $[-1,1]$.*

Solution. Fix arbitrarily $y \in [-1,1]$ and take $x \in [-\pi/2, \pi/2]$ such that $\sin x = y$. By Kronecker's density theorem, there exists a sequence $(a_n)_{n\geq 1}$, with $a_n = p_n + 2\pi q_n$ ($p_n, q_n \in \mathbb{Z}$), such that $a_n \to x$ as $n \to \infty$. In particular, this implies that $p_n \neq 0$ for infinitely many n, and hence we assume that $p_n \neq 0$ for all $n \geq 1$. By $a_n \to x$ as $n \to \infty$ we obtain $\sin a_n = \sin p_n \to \sin x = y$ as $n \to \infty$. We distinguish two cases:

(i) $p_n > 0$ for infinitely many $n \geq 1$. In this situation the proof is concluded.
(ii) $p_n > 0$ for finitely many $n \geq 1$. We assume, without loss of generality, that $p_n < 0$ for all $n \geq 1$. Set $b_n = \pi - a_n = -p_n - (2q_n - 1)\pi \to \pi - x$ as $n \to \infty$. So, $-p_n > 0$ and $\sin b_n \to \sin(\pi - x) = \sin x = y$ as $n \to \infty$. $\square$

1.5 Hardy's and Carleman's Inequalities

> In great mathematics there is a very high degree of unexpectedness, combined with inevitability and economy.
>
> Godfrey H. Hardy (1877–1947),
> *A Mathematician's Apology*

We are concerned in this section with two important classical results that have many applications in various domains, such as the treatment of various classes of nonlinear singular problems arising in mathematical physics.

---

[1] The pigeonhole principle asserts that if $n+1$ pigeons are placed in n pigeonholes, then some pigeonhole contains at least two of the pigeons.

To illustrate Hardy's and Carleman's inequalities, let us assume that $(a_n)_{n\geq 1}$ is a sequence of nonnegative real numbers and let $p > 1$ be an arbitrary real number. Set $s_n := \sum_{k=1}^n a_k$, $S_n := \sum_{k=1}^n a_k^p$, $m_n := \sum_{k=1}^n (a_1 a_2 \ldots a_k)^{1/k}$, and $M_n := \sum_{k=1}^n \left(1/k \sum_{j=1}^k a_j\right)^p$. Then the following properties hold:

(i) if the sequence $(s_n)_{n\geq 1}$ is convergent, then the sequence $(m_n)_{n\geq 1}$ converges, too;
(ii) if the sequence $(m_n)_{n\geq 1}$ is divergent, then the sequence $(s_n)_{n\geq 1}$ diverges, too;
(iii) if the sequence $(S_n)_{n\geq 1}$ is convergent, then the sequence $(M_n)_{n\geq 1}$ converges, too;
(iv) if the sequence $(M_n)_{n\geq 1}$ is divergent, then the sequence $(S_n)_{n\geq 1}$ diverges, too.

More precise statements of these results will be given below.

In [40], the British mathematician Godfrey H. Hardy (1877–1947) proved an important inequality, related to mixed means of nonnegative real numbers. Our first purpose in this section is to give a proof of this classical result.

1.5.1. (**Hardy's Inequality, 1920**). *Assume that $p > 1$ is a real number and $(a_n)_{n\geq 1}$ is a sequence of nonnegative real numbers. Then*

$$\sum_{n=1}^\infty \left(\frac{1}{n}\sum_{k=1}^n a_k\right)^p \leq \left(\frac{p}{p-1}\right)^p \sum_{n=1}^\infty a_n^p,$$

with equality if and only if $a_n = 0$ for every $n \geq 1$. Moreover, the constant $p^p(p-1)^{-p}$ is the best possible.

Solution. Our arguments rely upon the following mixed means inequality, which is due to Mond and Pečarić [78]:

$$\left[\frac{1}{n}\sum_{k=1}^n \left(\frac{a_1+\cdots+a_k}{k}\right)^p\right]^{1/p} \leq \frac{1}{n}\sum_{k=1}^n \left(\frac{1}{k}\sum_{i=1}^k a_i^p\right)^{1/p},$$

with equality if and only if $a_1 = \cdots = a_n$. The above inequality can be written, equivalently,

$$\sum_{k=1}^n \left(\frac{a_1+\cdots+a_k}{k}\right)^p \leq n^{1-p}\left[\sum_{k=1}^n \left(\frac{1}{k}\sum_{i=1}^k a_i^p\right)^{1/p}\right]^p. \quad (1.12)$$

We also assume the following inequality:

$$\sum_{k=1}^n \left(\frac{1}{k}\right)^{1/p} < \frac{p}{p-1} n^{1-1/p}, \quad (1.13)$$

for all integers $n \geq 1$ and any real number $p > 1$. We will give in Chapter 6 proofs of inequalities (1.12), and (1.13).

1.5 Hardy's and Carleman's Inequalities

Set $S_n := \sum_{j=1}^{n} a_j^p$. Thus, by (1.12), (1.13), and the observation that $S_n \geq \sum_{j=1}^{k} a_j^p$ for all $1 \leq k \leq n$, we obtain

$$\sum_{k=1}^{n} \left(\frac{a_1 + \cdots + a_k}{k}\right)^p \leq n^{1-p} S_n \left[\sum_{k=1}^{n} \left(\frac{1}{k}\right)^{1/p}\right]^p$$

$$\leq n^{1-p} S_n \frac{p^p}{(p-1)^p} n^{p-1} = \frac{p^p}{(p-1)^p} \sum_{k=1}^{n} a_k^p. \tag{1.14}$$

We also deduce that equality holds in the above inequality if and only if both $a_1 = \cdots = a_n$ and $S_n = \sum_{j=1}^{k} a_j^p$ for all $1 \leq k \leq n$, that is, if and only if $a_1 = \cdots = a_n = 0$.
Taking $n \to \infty$ in (1.14), we obtain Hardy's inequality.

In order to show that $p^p(p-1)^{-p}$ is the best constant in Hardy's inequality, we take the sequence $(a_n)_{n \geq 1}$ defined by $a_n = n^{-1/p}$ if $n \leq N$ and $a_n = 0$ elsewhere, where N is a fixed positive integer. A straightforward computation shows that for every $\varepsilon \in (0,1)$ there exists a positive integer $N(\varepsilon)$ such that

$$\sum_{n=1}^{\infty} \left(\frac{1}{n} \sum_{k=1}^{n} a_k\right)^p > (1-\varepsilon) \left(\frac{p}{p-1}\right)^p \sum_{n=1}^{\infty} a_n^p,$$

for the above choice of $(a_n)_{n \geq 1}$ and for all $N \geq N(\varepsilon)$. This justifies that the constant $p^p(p-1)^{-p}$ in Hardy's inequality cannot be replaced with a smaller one.

We point out that the integral version of the above discrete Hardy's inequality is

$$\int_0^{\infty} \left[\frac{1}{x} \int_0^x f(t) \, dt\right]^p dx \leq \left(\frac{p}{p-1}\right)^p \int_0^{\infty} f^p(x) \, dx, \tag{1.15}$$

for all real numbers $p > 1$ and any continuous function $f : [0, \infty) \to [0, \infty)$ such that $\int_0^{\infty} f^p(x) \, dx := \lim_{x \to \infty} \int_0^x f^p(t) \, dt$ exists and is finite. Equality holds in (1.15) if and only if $f \equiv 0$, and the constant $p^p(p-1)^{-p}$ is the best possible. We give a proof of Hardy's integral inequality in Chapter 8 of this work.

The purpose of the next exercise is to give a proof of Carleman's inequality, which has important applications in the theory of quasianalytic functions. This inequality was presented in 1922 by the Swedish mathematician Torsten Carleman (1892–1949) on the occasion of the Scandinavian Congress of Mathematics in Helsinki. We point out that the continuous version of Carleman's inequality is

$$\int_0^{\infty} \exp\left(\frac{1}{x} \int_0^x \ln f(t) \, dt\right) dx < e \int_0^{\infty} f(x) \, dx,$$

where $f(t)$ is a positive function. This is sometimes called Knopp's inequality, even though it seems that George Pólya (1887–1985) was the first to discover this inequality. □

1.5.2. Let E denote the set of all real sequences $A = (A_n)_{n \geq 0}$ satisfying $A_0 = 1$, $A_n \geq 1$ and $(A_n)^2 \leq A_{n+1} A_{n-1}$, for all $n \geq 1$.

(a) Check that the sequence A defined by $A_n = n!$ is an element of E.

Fix $A \in E$. Define the sequence $(\lambda_n)_{n \geq 0}$ and $(\mu_n)_{n \geq 0}$ by $\lambda_0 = \mu_0 = 1$, $\lambda_n = A_{n-1}/A_n$ and $\mu_n = (A_n)^{-1/n}$ for all integers $n \geq 1$.

(b) Prove that the sequence (λ_n) is decreasing and $(\lambda_n)^n \leq \lambda_1 \lambda_2 \cdots \lambda_n$ for any integer $n \geq 0$.
(c) Show that the sequence (μ_n) is decreasing.
(d) Prove that for all $n \in \mathbb{N}$, and for any $j \in [0,n] \cap \mathbb{N}$, $A_{n+1}/A_{n+1-j} \geq A_n/A_{n-j}$. Deduce that for all $n \in \mathbb{N}$ and for any $j \in [0,n] \cap \mathbb{N}$, $A_j A_{n-j} \leq A_n$.
(e) Establish the inequality $\lambda_n \leq \mu_n$.

Let (a_n) and (c_n) be sequences of positive numbers such that the sequence defined by $s_n = \sum_{k=1}^n a_k$ is convergent. Let $u_n = (a_1 a_2 \cdots a_n)^{1/n}$ and $b_n = (c_1 c_2 \cdots c_n)^{-1/n}$ for all integers $n \geq 1$.

(f) Show that $u_n \leq \frac{b_n}{n} \sum_{k=1}^n a_k c_k$.
(g) We assume, additionally, that $\sum_{k=1}^n \frac{b_k}{k}$ is convergent and define, for any $k \geq 1$, $B_k = \lim_{n \to \infty} \sum_{p=k}^n \frac{b_p}{p}$. Prove that $\sum_{p=1}^n u_p \leq \sum_{k=1}^n B_k c_k a_k$.
(h) (**Carleman's inequality**). Let $c_n = (n+1)^n/n^{n-1}$. Deduce that the sequence $(\sum_{k=1}^n u_k)_n$ is convergent and $\sum_{n=1}^\infty (a_1 a_2 \ldots a_n)^{1/n} < e \sum_{n=1}^\infty a_n$.

Solution. (a) Set $A = (n!)$. Then $A_n^2/A_{n+1}A_{n-1} = n/n+1 \leq 1$, so $A \in E$.

(b) From $A_n^2 \geq A_{n+1} A_{n-1}$ we deduce that $\lambda_{n+1} \leq \lambda_n$. Fix $n \geq 1$. We have $0 < \lambda_n \leq \lambda_k$ for all $k = 1, \ldots, n$. Therefore
$$0 < \lambda_n^n \leq \lambda_1 \lambda_2 \ldots \lambda_n.$$

(c) We may write
$$\left(\frac{\mu_{n+1}}{\mu_n}\right)^{n(n+1)} = \frac{A_n^{n+1}}{A_{n+1}^n} = A_{n+1} \lambda_{n+1}^{n+1}.$$
The inequality proved in (b) can be rewritten as $A_n \lambda_n^n \leq 1$, since $\lambda_1 \lambda_2 \cdots \lambda_n = 1/A_n$. Hence
$$\left(\frac{\mu_{n+1}}{\mu_n}\right)^{n/(n+1)} \leq 1 \iff \frac{\mu_{n+1}}{\mu_n} \leq 1,$$
which means that the sequence (μ_n) is decreasing.

(d) Let j and n be integers such that $0 \leq j \leq n$. The sequence (λ_m) is decreasing, so $\lambda_{n+1} \leq \lambda_{n-j+1}$, that is, $A_n/A_{n+1} \leq A_{n-j}/A_{n-j+1}$. It follows that $A_{n+1}/A_{n-j+1} \geq A_n/A_{n-j}$.

The sequence $(A_n/A_{n-j})_{n \geq j}$ is increasing, so $A_n/A_{n-j} \geq A_j/A_0$. Since $A_0 = 1$, it follows that $A_n \geq A_{n-j} A_j$.

(e) Using (b), we have $\lambda_n^n \leq \lambda_1 \lambda_2 \ldots \lambda_n = 1/A_n = \mu_n^n$, that is, $\lambda_n \leq \mu_n$.

1.5 Hardy's and Carleman's Inequalities

(f) Let $n \geq 1$. We have $u_n/b_n = (\alpha_1 \alpha_2 \ldots \alpha_n)^{1/n}$, where $\alpha_i = a_i c_i$ for all $1 \leq i \leq n$. By the AM–GM inequality we deduce that

$$\frac{u_n}{b_n} \leq \frac{1}{n}(a_1 c_1 + a_2 c_2 + \cdots + a_n c_n).$$

(g) Observe that $b_n/n = B_n - B_{n+1}$ and set $S_n = a_1 c_1 + \cdots + a_n c_n$. We have

$$\sum_{k=1}^{n} u_k \leq \sum_{k=1}^{n} (B_k - B_{k+1}) S_k.$$

Applying Abel's inequality, we obtain

$$\sum_{k=1}^{n}(B_k - B_{k+1})S_k = \sum_{k=1}^{n} B_k S_k - \sum_{k=2}^{n+1} B_k S_{k-1} = \sum_{k=1}^{n} B_k a_k c_k - B_{n+1} S_n.$$

Since $B_{n+1} S_n \geq 0$, we have

$$\sum_{k=1}^{n} u_k \leq \sum_{k=1}^{n} a_k B_k c_k.$$

(h) Set $c_n = (n+1)^n / n^{n-1}$. Thus $c_n = \gamma_{n+1}/\gamma_n$, where $\gamma_n = n^{n-1}$. It follows that $c_1 c_2 \cdots c_n = (n+1)^n$.

Thus we see that $b_n = 1/(n+1)$ satisfies

$$\frac{b_n}{n} = \frac{1}{n(n+1)} = \frac{1}{n} - \frac{1}{n+1}.$$

Therefore

$$\lim_{n \to \infty} \sum_{k=1}^{n} \frac{b_k}{k} < +\infty$$

and

$$\frac{b_n}{n} + \frac{b_{n+1}}{n+1} + \frac{b_{n+2}}{n+2} + \cdots + \frac{b_{n+k}}{n+k} \to \frac{1}{n},$$

as $k \to \infty$, for all $n \geq 1$. Thus we see that $B_n = 1/n$ satisfies $B_n c_n = (1 + 1/n)^n < e$. Using (g), it follows that $\sum_{k=1}^{n} u_k \leq e \sum_{k=1}^{n} a_k$. We deduce that the sequence defined by $\sum_{k=1}^{n} u_k$ is convergent and, moreover,

$$\lim_{n \to \infty} \sum_{k=1}^{n} (a_1 a_2 \cdots a_k)^{1/k} < e \lim_{n \to \infty} \sum_{k=1}^{n} a_k. \tag{1.16}$$

Carleman proved that e is the best constant in (1.16), that is, Carleman's inequality is no longer valid if we replace e with a smaller constant. We give in what follows the following alternative proof to Carleman's inequality, based on the arithmetic–

geometric means inequality. We first observe that for all positive integers k,

$$\frac{(k+1)^k}{k!} = \left(1+\frac{1}{1}\right)\left(1+\frac{1}{2}\right)^2 \cdots \left(1+\frac{1}{k}\right)^k < e^k.$$

Hence

$$\sum_{n=1}^{\infty} a_n = \sum_{n=1}^{\infty} na_n \sum_{k=n}^{\infty} \frac{1}{k(k+1)} = \sum_{k=1}^{\infty} \frac{1}{k(k+1)} \sum_{n=1}^{k} na_n$$

$$= \sum_{k=1}^{\infty} \frac{a_1 + 2a_2 + \cdots + ka_k}{k(k+1)} > \sum_{k=1}^{\infty} \frac{1}{k+1} \left(k! \prod_{n=1}^{k} a_n\right)^{1/k}$$

$$= \sum_{k=1}^{\infty} \left[\frac{k!}{(k+1)^k}\right]^{1/k} \left(\prod_{n=1}^{k} a_n\right)^{1/k} \geq \frac{1}{e} \sum_{k=1}^{\infty} \left(\prod_{n=1}^{k} a_n\right)^{1/k}.$$

Strict inequality holds, since we cannot have equality at the same time in all terms of the inequality. This can occur only if $a_k = c/k$ for some $c > 0$ and all $k \geq 1$. Such a choice is not possible, since $\sum_{k=1}^{\infty} a_k$ is convergent.

Using Hardy's inequality we can give the following proof of Carleman's inequality. We first apply Hardy's inequality for a_k replaced by $a_k^{1/p}$. Hence

$$\sum_{n=1}^{\infty} \left(\frac{a_1^{1/p} + \cdots + a_n^{1/p}}{n}\right)^p < \left(\frac{p}{p-1}\right)^p \sum_{n=1}^{\infty} a_n. \qquad (1.17)$$

Since $a^b = e^{b \ln a}$ (for $a, b > 0$), it follows that

$$\left(\frac{a_1^{1/p} + \cdots + a_n^{1/p}}{n}\right)^p = \exp\left(\frac{\ln(a_1^{1/p} + \cdots + a_n^{1/p}) - \ln n}{1/p}\right) \to \sqrt[n]{a_1 \cdots a_n}, \quad (1.18)$$

as $p \to \infty$. We point out that we have used above a basic result that asserts that

$$\frac{\ln(a_1^x + \cdots + a_n^x) - \ln n}{x} \quad \text{tends to } \ln(a_1 \cdots a_n)^{1/n} \text{ as } x \to 0,$$

where $n \geq 1$ is a fixed integer (see Chapter 3 for a proof of this result).

Since $(p/p-1)^p$ increases to e as $p \to \infty$, we deduce from (1.17) and (1.18) that

$$a_1 + \sqrt{a_1 a_2} + \sqrt[3]{a_1 a_2 a_3} + \cdots + \sqrt[n]{a_1 a_2 \cdots a_n} + \cdots < e(a_1 + a_2 + \cdots + a_n + \cdots),$$

which is just Carleman's inequality.

The following alternative proof of Carleman's inequality is due to George Pólya (1887–1985) (see [109]). We first observe that the arithmetic–geometric means inequality implies

$$\sqrt[n]{a_1 a_2 \cdots a_n} \leq \frac{\alpha_1 a_1 + \alpha_2 a_2 + \cdots + \alpha_n a_n}{n \sqrt[n]{\alpha_1 \alpha_2 \cdots \alpha_n}},$$

for all positive numbers $\alpha_1, \ldots, \alpha_n$. Choosing $\alpha_n = (n+1)^n / n^{n-1}$ and observing that $\sqrt[n]{\alpha_1 \alpha_2 \cdots \alpha_n} = n+1$, we obtain

$$\sqrt[n]{a_1 a_2 \cdots a_n} \leq \frac{1}{n(n+1)} \sum_{k=1}^{n} \frac{(k+1)^k}{k^{k-1}} a_k.$$

Set

$$x_{m,n} = \begin{cases} \frac{\alpha_m a_m}{n(n+1)} & \text{if } 1 \leq m \leq n, \\ 0 & \text{if } m > n. \end{cases}$$

It follows that

$$\sum_{n=1}^{\infty} \sqrt[n]{a_1 a_2 \cdots a_n} \leq \sum_{n=1}^{\infty} \left[\frac{1}{n(n+1)} \sum_{m=1}^{n} \alpha_m a_m \right]$$

$$= \sum_{n=1}^{\infty} \left(\sum_{m=1}^{\infty} x_{m,n} \right) = \sum_{m=1}^{\infty} \left(\sum_{n=1}^{\infty} x_{m,n} \right)$$

$$= \sum_{m=1}^{\infty} \left(\sum_{n=m}^{\infty} \frac{\alpha_m a_m}{n(n+1)} \right) = \sum_{m=1}^{\infty} \left[\alpha_m a_m \sum_{n=m}^{\infty} \left(\frac{1}{n} - \frac{1}{n+1} \right) \right]$$

$$= \sum_{m=1}^{\infty} \frac{\alpha_m a_m}{m} < e \sum_{m=1}^{\infty} a_m,$$

since $\alpha_m < em$, for all $m \geq 1$. This concludes the proof of Carleman's inequality. $\square$

1.6 Independent Study Problems

> Mathematics is the only good metaphysics.
>
> Lord Kelvin (1824–1907)

1.6.1. Assume that $a_1 \geq a_2 \geq \cdots \geq a_n \geq \cdots \geq 0$ and the sequence $s_n = \sum_{k=1}^{n} \varepsilon_k a_k$ is convergent, where $\varepsilon_k = \pm 1$. Prove that $\lim_{n \to \infty} (\varepsilon_1 + \varepsilon_2 + \cdots + \varepsilon_n) a_n = 0$.

1.6.2. Suppose that the sequence of real numbers $(a_n)_{n \geq 1}$ is defined such that

$$a_m + a_n - 1 < a_{m+n} < a_m + a_n + 1, \quad \text{for all } m, n \geq 1.$$

Prove that $\lim_{n \to \infty} a_n / n = \omega$ exists, is finite, and

$$\omega n - 1 < a_n < \omega n + 1, \quad n \geq 1.$$

1.6.3. Consider n arithmetic progressions of positive integers having distinct differences bigger than 1. Prove that these progressions cannot cover the set of positive integers.

1.6.4. Let $(a_n)_{n\geq 1}$ be a sequence of positive numbers such that a_1, a_2, a_3 are in arithmetic progression, a_2, a_3, a_4 are in geometric progression, a_3, a_4, a_5 are in arithmetic progression, a_4, a_5, a_6 are in geometric progression, and so on.

Find an expression for a_n in terms for a_1 and a_2.

1.6.5. Let d be a real number. For any integer $m \geq 0$, define the sequence $\{a_m(j)\}_{j\geq 0}$ by

$$a_m(0) = \frac{d}{2^m}, \quad a_m(j+1) = (a_m(j))^2 + 2a_m(j), \quad \text{for all } j \geq 0.$$

Show that $\lim_{n\to\infty} a_n(n)$ exists.

1.6.6. Compute

$$\lim_{n\to\infty} \sum_{k=0}^{n} \operatorname{arccot}(k^2 + k + 1).$$

1.6.7. For any integer $n \geq 1$, let a_n denote the number of zeros in the base-3 representation of n. Find all positive numbers x such that the sequence $(s_n)_{n\geq 1}$ defined by

$$s_n = \sum_{k=1}^{n} \frac{x^{a_n}}{n^3}$$

is convergent.

1.6.8. Let $x_1 = 0.8$ and $y_1 = 0.6$. Define

$$x_{n+1} = x_n \cos y_n - y_n \sin y_n, \quad y_{n+1} = x_n \sin y_n + y_n \cos y_n, \quad \text{for all } n \geq 1.$$

Study the convergence of these sequences and, in case of positive answers, find the corresponding limits.

1.6.9. For all integers $n \geq 1$, define

$$x_n = \min\{|a - b\sqrt{3}|; \, a, b \in \mathbb{N}, \, a + b = n\}.$$

Find the smallest positive number p such that $x_n \leq p$, for all $n \geq 1$.

1.6.10. Let $(a_n)_{n\geq 1}$ be a sequence of positive numbers such that the sequence $(b_n)_{n\geq 1}$ defined by $b_n = \sum_{k=1}^{n} a_k$ is convergent. Set

$$x_n = \sum_{k=1}^{n} a_k^{k/(k+1)}, \quad n \geq 1.$$

Prove that the sequence $(x_n)_{n\geq 1}$ is convergent.

1.6 Independent Study Problems

1.6.11. Is the number $\sqrt{2}$ the limit of a sequence of the form $n^{1/3} - m^{1/3}$ ($n,m = 0,1,\ldots$)?

1.6.12. For any integer $k \geq 0$, let $S(k) = k - m^2$, where m is the largest integer such that $m^2 \leq k$. Define the sequence $(a_n)_{n\geq 0}$ by

$$a_0 = A, \quad a_{n+1} = a_n + S(a_n), \quad \text{for all } n \geq 0.$$

Find all positive integers A such that the above sequence is eventually constant.

1.6.13. For any positive integer n, set $a_n = 0$ (resp., $a_n = 1$), provided that the number of digits 1 in the binary representation of n is even (resp., odd). Prove that there do not exist positive integers k and m such that

$$a_{k+j} = a_{k+m+j} = a_{k+2m+j}, \quad \text{for all } 0 \leq j \leq m-1.$$

1.6.14. Let $(a_n)_{n\geq 1}$ be a sequence of positive numbers such that $a_n \leq a_{2n} + a_{2n+1}$, for all n. Set $s_n = \sum_{k=1}^n a_k$. Prove that the sequence $(s_n)_{n\geq 1}$ is divergent.

1.6.15. Prove that if $a_n \neq 0$, for all $n = 1, 2, \ldots$, and $\lim_{n\to\infty} a_n = 0$, then for any real number x there exists a sequence of integers $(m_n)_{n\geq 1}$ such that $x = \lim_{n\to\infty} \prod_{j=1}^n m_j a_j$.
Hint. Construct by induction the sequence of integers (m_n) such that

$$m_n a_n \leq \frac{x}{\prod_{j=1}^{n-1} m_j a_j} < (m_n + 1) a_n.$$

1.6.16. Prove that if $(\sqrt{5} - 1)/2 = \lim_{p\to\infty} \sum_{k=1}^p 2^{-n_k}$, where n_k are positive integers, then $n_k \leq 5 \cdot 2^{k-2} - 1$.
Hint. Show that $n_{k+1} \leq 2n_k + 2$.

1.6.17. (**Erdős–Szekeres Theorem**). Prove that every finite sequence of real numbers consisting of $n^2 + 1$ terms contains a monotone subsequence of $n + 1$ terms.
Hint. Associate to any term of the sequence an ordered pair consisting of the length of maximal decreasing and increasing subsequences starting with this term.

1.6.18. Prove that there is an enumeration $(x_n)_{n\geq 1}$ of the rational numbers in $(0,1)$ such that the sequence $\left(\sum_{k=1}^n \prod_{i=1}^k x_i\right)_{n\geq 1}$ is divergent.
Hint. Let (b_n) be the sequence $(n/(n+1))_{n\geq 1}$ and let $(a_n)_{n\geq 1}$ denote the sequence of rational numbers in $(0,1)$ excepting those contained by (b_n). Let $(c_n)_{n\geq 1}$ be an arbitrary sequence such that $\sum_{k=1}^n c_k \to \infty$ (we can choose, e.g., $c_n = 1/n$). Take k_1, $k_2, \ldots$ such that

$$a_1 + a_1 b_1 + a_1 b_1 b_2 + \cdots + a_1 b_1 b_2 \cdots b_{k_1} \geq c_1,$$
$$a_1 b_1 b_2 \cdots b_{k_1}(a_2 + a_2 b_{k_1+1} + \cdots + a_2 b_{k_1+1} \cdots b_{k_2}) \geq c_2,$$

and so on.

1.6.19. Prove that any sequence of real numbers $(a_n)_{n\geq 1}$ satisfying $a_{n+1} = |a_n| - a_{n-1}$ is periodic.
Answer. We have $a_{n+9} = a_n$, for all n.

1.6.20. Let $(K_n)_{n\geq 1}$ be a sequence of squares of areas a_n. Prove that the plane can be covered by the elements of $(K_n)_{n\geq 1}$, provided that $\lim_{n\to\infty} \sum_{k=1}^n a_k = \infty$.
Hint. It is enough to show that the unit square can be covered with these squares.

1.6.21. Let $p_j \in [0,1]$, $j = 1,\ldots,n$. Prove that

$$\inf_{0\leq x\leq 1} \sum_{j=1}^n \frac{1}{|x-p_j|} \leq 8n\left(1 + \frac{1}{3} + \cdots + \frac{1}{2n-1}\right).$$

Hint. We split the interval $[0,1]$ into $2n$ intervals of the same length. Next, we choose x in an interval that does not contain any of the numbers p_j.

1.6.22. Let u_n be the unique positive root of the polynomial $x^n + x^{n-1} + \cdots + x - 1$. Prove that $\lim_{n\to\infty} u_n = 1/2$.

1.6.23. Let a_n be the sequence defined by $a_0 = 0$, $a_1 = 1$, $a_2 = 0$, $a_3 = 1$, and for all $n \geq 1$,

$$a_{n+3} = \frac{(n^2+n+1)(n+1)}{n} a_{n+2} + (n^2+n+1)a_{n+1} - \frac{n+1}{n} a_n.$$

Prove that a_n is the square of an integer, for any $n \geq 0$.

1.6.24. The sequence of positive numbers (a_n) satisfies

$$\cos a_{n+m} = \frac{\cos a_n + \cos a_m}{1 + \cos a_n \cos a_m} \quad \text{for all } n, m \geq 0.$$

Such a sequence is said to be *minimal* provided that $0 \leq a_n \leq \pi$, for all n. Find all minimal sequences with the above property.

1.6.25. Compute $\lim_{n\to\infty} (2k^{1/n} - 1)^n$ and $\lim_{n\to\infty} (2n^{1/n} - 1)/n^2$.
Answer. The limits equal k^2, resp. 0.

1.6.26. Let (a_n) and (b_n) be sequences of positive numbers such that $\lim_{n\to\infty} a_n^n = a$, $\lim_{n\to\infty} b_n^n = b$, $a, b \in (0, \infty)$. Let p and q be positive numbers such that $p + q = 1$. Compute $\lim_{n\to\infty} (pa_n + qb_n)^n$.
Hint. We have $x_n^n \to x$ if and only if $n(x_n - 1) \to \ln x$.

1.6.27. Compute $\lim_{n\to\infty} \sum_{k=1}^{n^2} n/(n^2 + k^2)$.

1.6.28. Study the convergence of the sequences

$$a_n = \sqrt{1!\sqrt{2!\cdots\sqrt{n!}}}, \quad b_n = \sqrt{2\sqrt{3\cdots\sqrt{n}}}.$$

1.6 Independent Study Problems

1.6.29. Prove that

$$\lim_{n\to\infty} \left(1+\frac{1}{n}\right)\left(1+\frac{2}{n}\right)^{1/2}\cdots\left(1+\frac{n}{n}\right)^{1/n} = e^{\pi^2/12}.$$

1.6.30. Let $(a_n)_{n\geq 1}$ be an arithmetic progression of positive numbers. Compute

$$\lim_{n\to\infty} \frac{n(a_1\cdots a_n)^{1/n}}{a_1+\cdots+a_n}.$$

Answer. The limit equals 1 if $(a_n)_{n\geq 1}$ is a constant sequence and $2/e$, otherwise.

1.6.31. Let $(a_n)_{n\geq 1}$ be a sequence of positive numbers such that the series $\sum_{n\geq 1} a_n$ converges. Let a be the sum of this series. Prove that there exists an increasing sequence of integers $(n_k)_{k\geq 1}$ with $n_1 = 1$ and satisfying, for all positive integer k, $\sum_{n\geq n_k} a_n \leq 4^{1-k}a$.

1.6.32. Let $(a_n)_{n\geq 1}$ be a sequence of real numbers. Prove that if the sequence $(2a_{n+1}+\sin a_n)_{n\geq 1}$ is convergent then $(a_n)_{n\geq 1}$ converges, too.

V. Vâjâitu

1.6.33. Prove that for any $x \in [0,\infty)$ there exist sequences of nonnegative integers $(a_n)_{n\geq 0}$ and $(b_n)_{n\geq 0}$ such that

$$\lim_{n\to\infty}\left(\frac{1}{a_n+1}+\frac{1}{a_n+2}+\cdots+\frac{1}{a_n+b_n}\right) = x.$$

1.6.34. Prove that the sequence $(a_n)_{n\geq 2}$ defined by $a_n = \log n!$ does not contain infinite arithmetic progressions.

A. Iuga

1.6.35. The sequence $(a_n)_{n\geq 1}$ satisfies the relations $a_1 = a_2 = 1$ and $a_{n+2} = a_n + a_{n+1}^{-1}$, for all positive integers n. Find the general term a_n.
Answer. $a_n = (n-1)!!/(n-2)!!$.

1.6.36. Let α be an irrational number, let $(k_n)_{n\geq 1}$ be an increasing sequence of positive integers, and let $(x_n)_{n\geq 1}$ be a sequence of real numbers defined by $x_n = \cos 2\pi k_n \alpha$, for all $n \geq 1$. For any natural number N we denote by $s(N)$ the number of terms of the sequence $(k_n)_{n\geq 1}$ that do not exceed N.

(a) Prove that if $(x_n)_{n\geq 1}$ is convergent, then $\lim_{N\to\infty} s(N)/N = 0$.
(b) Let $a \geq 2$ be a positive integer and let $k_n = a^n$, for all $n \geq 1$. Prove that $\lim_{N\to\infty} s(N)/N = 0$ but the sequence $(x_n)_{n\geq 1}$ does not converge.

Marian Tetiva

1.6.37. Let $(a_n)_{n\geq 1}$ be a sequence of real numbers in $(0,1)$ such that for all $n \geq 1$, $(1-a_n)a_{n+1} > 1/4$. Prove that $\lim_{n\to\infty} a_n = 1/2$.

H.S. Wall, Amer. Math. Monthly, Problem E598

1.6.38. Let $a_n = \sum_{1 \leq k \leq n}(k/n)^n$.

(i) Find the limit $\lim_{n\to\infty} a_n$.
(ii) Determine the character of monotonicity of the sequence $(a_n)_{n \geq 1}$.
(iii) Is it true that $a_n \leq 2$ for any $n \geq 1$?

Hint. (i) The main contribution is given by the last terms:

$$\sum_{n-\sqrt[3]{n} \leq k \leq n} \left(\frac{k}{n}\right)^k = \sum_{0 \leq j \leq \sqrt[3]{n}} \left(1 - \frac{j}{n}\right)^{n-j}$$

$$= \sum_{0 \leq j \leq \sqrt[3]{n}} e^{-j}\left(1 + O\left(\frac{j^2}{n}\right)\right) = \sum_{j \geq 0} e^{-j} + o(1).$$

The sum of the remaining terms is infinitesimally small, since $\max_{1 < k < n-\sqrt[3]{n}}(k/n)^k = 4/n^2$.

1.6.39. Suppose that the nondecreasing sequence of positive numbers $(a_n)_{n \geq 1}$ has finite limit ℓ. Prove that the sequence $(b_n)_{n \geq 1}$ defined by $b_n = \sum_{k=1}^{n}(\ell - a_k)$ converges if and only if $a_1 a_2 \cdots a_n / \ell^n \to C$ for some $C > 0$.

1.6.40. Define the sequence $(a_n)_{n \geq 2}$ by $a_n = \sqrt{1 + \sqrt{2 + \cdots + \sqrt{n}}}$. Prove that $a_n \to \ell \in \mathbb{R}$ and $\sqrt{n} \cdot \sqrt[n]{\ell - a_n} \to \sqrt{e}/2$.

1.6.41. Suppose that the sequence $(a_n)_{n \geq 1}$ of real numbers is such that $a_{n+1} - a_n \to 0$. Prove that the set of limits of its convergent subsequences is the interval with endpoints $\liminf_{n\to\infty} a_n$ and $\limsup_{n\to\infty} a_n$.

1.6.42. Suppose that the sequence $(a_n)_{n \geq 1}$ of real numbers is such that $a_{n+1} + a_n \to 0$. Prove that the set of limits of convergent subsequences of this sequence is either infinite or contains at most two points.

1.6.43. Suppose that the sequence $(a_n)_{n \geq 1}$ of nonnegative numbers is such that

$$a_n \leq \frac{a_{n-1} + a_{n-2}}{n^2} \quad \text{for all } n \geq 3.$$

Prove that $a_n = O(1/n!)$.

Hint. Show that the inequalities $a_{n-1} \leq c/(n-1)!$ and $a_{n-2} \leq c/(n-2)!$ imply the inequality $a_n \leq c/n!$. The constant c is chosen so that the last inequality holds for $n = 1, 2$.

1.6.44. Consider the sequence $(a_n)_{n \geq 1}$ of real numbers such that $a_{m+n} \leq a_m + a_n$ for all $m, n \geq 1$.

Prove that for any n we have

$$\sum_{k=1}^{n} \frac{a_k}{k} \geq a_n.$$

1.6.45. *A sequence of positive integers* $(a_n)_{n\geq 1}$ *contains each positive integer exactly once. Assume that if* $m \neq n$ *then*

$$\frac{1}{1998} < \frac{|a_m - a_n|}{|m - n|} < 1998.$$

Prove that $|a_n - n| < 2000000$ *for all* n.

Russian Mathematical Olympiad, 1998

1.6.46. *Let* $(a_n)_{n\geq 1}$ *be a sequence of real numbers such that* $a_1 = 1$ *and* $a_{n+1} = a_n + \sqrt[3]{a_n}$ *for* $n \geq 1$. *Prove that there exist real numbers* p *and* q *such that* $\lim_{n \to \infty} a_n/(pn^q) = 1$.

1.6.47. *Find all functions* $f; [0, \infty) \to [0, \infty)$ *satisfying, for all* $x \geq 0$, $(f \circ f)(x) = 6x - f(x)$.

1.6.48. *Let* $(a_n)_{n\geq 1}$ *be a sequence of positive real numbers such that for all* $n \geq 1$,

$$a_{n+1} \leq a_n + \frac{1}{(n+1)^2}.$$

Prove that the sequence $(a_n)_{n\geq 1}$ *is convergent.*

Chapter 2
Series

Knowledge must precede application.
—Max Planck (1858–1947), *The Nature of Light*

Abstract. In this chapter we investigate whether we can "add" infinitely many real numbers. In other words, if $(a_n)_{n\geq 1}$ is a sequence of real numbers, then we ask whether we can give a meaning to a symbol such as "$a_1 + a_2 + \cdots + a_n + \cdots$" or $\sum_{n=1}^{\infty} a_n$. We are mainly concerned with series of positive numbers, alternating series, but also with arbitrary series of real numbers, viewed as a natural generalization of the concept of finite sums of real numbers. This leads us to discuss deeper techniques of testing the behavior of infinite series as regards convergence. The theory of series may be viewed as a sequel to the theory of sequences developed in Chapter 1.

2.1 Main Definitions and Basic Results

> ...and I already see a way for finding the sum of this row $\frac{1}{1} + \frac{1}{4} + \frac{1}{9} + \frac{1}{16}$ etc.
>
> Johann Bernoulli (1667–1759), Letter to his brother, 1691

The concept of *series* had a continuous development in the last few centuries, due to its major applications in various fields of modern analysis. Roughly speaking, a series is represented by the sum of a sequence of terms. We are concerned only with series of real or complex numbers. In such a way, a series appears as a sum of the type
$$a_1 + a_2 + \cdots + a_n + \cdots,$$
where a_n are either real or complex numbers. We give the main definitions and properties in the case of series of *real* numbers. As soon as it is possible (paying attention, e.g., to order properties, which are no longer valid in $\mathbb{C}$!) the same definitions and properties extend automatically to the series of complex numbers. Important advances to a rigorous theory of series are due to Gauss (1812) and Bolzano (1817).

The first book containing a systematic treatment of infinite series is Cauchy's famous *Analyse Algébrique* [16], 1821.

Assuming that $(a_n)_{n\geq 1}$ is a sequence of real numbers, we define the sequence $(S_n)_{n\geq 1}$ by

$$S_n = a_1 + a_2 + \cdots + a_n, \quad \text{for any } n \geq 1.$$

Then S_n are called *partial sums* of the series $a_1 + a_2 + \cdots + a_n + \cdots$. We say that a series $\sum_{n=1}^{\infty} a_n$ *converges* if the sequence $(S_n)_{n\geq 1}$ is convergent. In such a case we have

$$\sum_{n=1}^{\infty} a_n := \lim_{n \to \infty} S_n.$$

In particular, this shows that if a series $\sum_{n=1}^{\infty} a_n$ converges then the sequence $(a_n)_{n\geq 1}$ tends to zero as n goes to infinity. If $(S_n)_{n\geq 1}$ diverges, then we say that the series $\sum_{n=1}^{\infty} a_n$ is a *divergent* series.

Leonhard Euler found several examples of convergent series, in relationship to the numbers $\sqrt{2}$, π, and e. For instance, he proved that

$$\sqrt{2} = \frac{7}{5}\left(1 + \frac{1}{100} + \frac{1\cdot 3}{100 \cdot 200} + \frac{1\cdot 3\cdot 5}{100 \cdot 200 \cdot 300} + \cdots\right),$$

$$e = 1 + \frac{1}{1!} + \frac{1}{2!} + \cdots + \frac{1}{n!} + \cdots,$$

$$\frac{\pi^2}{6} = 1 + \frac{1}{2^2} + \frac{1}{3^2} + \cdots + \frac{1}{n^2} + \cdots,$$

$$\frac{\pi^4}{90} = 1 + \frac{1}{2^4} + \frac{1}{3^4} + \cdots + \frac{1}{n^4} + \cdots,$$

$$\frac{\pi^6}{945} = 1 + \frac{1}{2^6} + \frac{1}{3^6} + \cdots + \frac{1}{n^6} + \cdots,$$

$$\frac{\pi^8}{9450} = 1 + \frac{1}{2^8} + \frac{1}{3^8} + \cdots + \frac{1}{n^8} + \cdots.$$

We note that the exact value of the sum of the series $\sum_{n=1}^{\infty} 1/n^3$ is not known. Furthermore, using the formula

$$\frac{\pi}{6} = \frac{1}{\sqrt{3}} - \frac{1}{3(\sqrt{3})^3} + \cdots + (-1)^n \frac{1}{(2n+1)(\sqrt{3})^{2n+1}} + \cdots,$$

Euler was able to compute the first 127 decimals of π, his only mistake being at the 113rd decimal! The only holiday celebrating a number is dedicated to π, and this happens in one of the most famous universities in the world! Indeed, Harvard University celebrates the *Pi Day* on 14 March (03/14) at precisely 3:14 PM A related convergent series associated with π was discovered by Leibniz, who showed that

$$\frac{\pi}{4} = 1 - \frac{1}{3} + \frac{1}{5} - \frac{1}{7} + \cdots.$$

2.1 Main Definitions and Basic Results

A useful convergence criterion is due to Cauchy, which asserts that a series $\sum_{n=1}^{\infty} a_n$ converges if and only if for any $\varepsilon > 0$, there exists a positive integer $N = N(\varepsilon)$ such that for all integers $n > m > N$,

$$|a_{m+1} + \cdots + a_n| < \varepsilon.$$

Using Cauchy's criterion one deduces that the harmonic series $\sum_{n=1}^{\infty} 1/n$ diverges. Indeed, we observe that the partial sums satisfy, for all $n \geq 1$,

$$|S_{2n} - S_n| > \frac{1}{2},$$

which shows that Cauchy's criterion is not fulfilled. The divergence of the harmonic series was first established by the Italian mathematician Mengoli in 1650.

A direct consequence, based on the remark that $n^{1/p} \leq n$ provided $p < 1$, implies that the series $\sum_{n=1}^{\infty} 1/n^p$ diverges, for any $p \leq 1$. In this argument we have applied the **first comparison test**, which asserts that if $\sum_{n=1}^{\infty} a_n$ and $\sum_{n=1}^{\infty} b_n$ are series of *positive* numbers and $a_n \leq b_n$ for all $n \geq 1$, then the following assertions are true:

(i) if $\sum_{n=1}^{\infty} b_n$ converges, then $\sum_{n=1}^{\infty} a_n$ converges, too;
(ii) if $\sum_{n=1}^{\infty} a_n$ diverges, then $\sum_{n=1}^{\infty} b_n$ diverges, too.

A **second comparison test** is the following. Assume that $\sum_{n=1}^{\infty} a_n$ and $\sum_{n=1}^{\infty} b_n$ are two series of positive numbers such that $\sum_{n=1}^{\infty} a_n$ is convergent and $\sum_{n=1}^{\infty} b_n$ is divergent. Given a series $\sum_{n=1}^{\infty} x_n$ of positive numbers, we have:

(i) If the inequality $x_{n+1}/x_n \leq a_{n+1}/a_n$ is true for all $n \geq 1$, then $\sum_{n=1}^{\infty} x_n$ is convergent.
(ii) If the inequality $x_{n+1}/x_n \geq b_{n+1}/b_n$ is true for all $n \geq 1$, then $\sum_{n=1}^{\infty} x_n$ is divergent.

Many series of nonnegative real numbers have terms that decrease monotonically. For such series the following test due to Cauchy is usually helpful.

Cauchy's Condensation Criterion. Suppose that $a_1 \geq a_2 \geq \cdots \geq 0$. Then the series $\sum_{n=1}^{\infty} a_n$ is convergent if and only if the series

$$\sum_{n=0}^{\infty} 2^n a_{2^n} = a_1 + 2a_2 + 4a_4 + 8a_8 + \cdots$$

is convergent.

A consequence of the above condensation theorem is that both series $\sum_{n=1}^{\infty} 1/n^p$ and $\sum_{n=1}^{\infty} 1/n(\log n)^p$ converge if and only if $p > 1$. With similar arguments we can show that the series

$$\sum_{n=3}^{\infty} \frac{1}{n \log n \log \log n}$$

diverges, whereas

$$\sum_{n=3}^{\infty} \frac{1}{n \log n (\log \log n)^2}$$

converges.

Two tests that are closely related are stated in what follows.

Root Test. Given a series $\sum_{n=1}^{\infty} a_n$, define $\ell = \limsup_{n\to\infty} \sqrt[n]{|a_n|} \in [0,+\infty]$. Then the following properties are true:

(i) If $\ell < 1$ then the series $\sum_{n=1}^{\infty} a_n$ is convergent.
(ii) If $\ell > 1$ then the series $\sum_{n=1}^{\infty} a_n$ is divergent.
(iii) If $\ell = 1$ then the test is inconclusive. However, if $\sqrt[n]{|a_n|} \geq 1$ for infinitely many distinct values of n, then the series $\sum_{n=1}^{\infty} a_n$ diverges.

The following ratio test of Jean le Rond d'Alembert (1717–1783) for convergence of series depends on the limit of the ratio a_{n+1}/a_n. If the limit is 1, the test fails.

Ratio Test. Let $\sum_{n=1}^{\infty} a_n$ be a series such that $a_n \neq 0$ for all n. Then the following properties are true:

(i) The series $\sum_{n=1}^{\infty} a_n$ converges if $\limsup_{n\to\infty} |a_{n+1}/a_n| < 1$.
(ii) The series $\sum_{n=1}^{\infty} a_n$ diverges if there exists $m \in \mathbb{N}$ such that $|a_{n+1}/a_n| \geq 1$ for all $n \geq m$.
(iii) If $\liminf_{n\to\infty} |a_{n+1}/a_n| \leq 1 \leq \limsup_{n\to\infty} |a_{n+1}/a_n|$ then the test is inconclusive.

In most cases it is easier to apply the ratio test than the root test. However, given a sequence $(a_n)_{n\geq 1}$ of positive numbers, then

$$\liminf_{n\to\infty} \frac{a_{n+1}}{a_n} \leq \liminf_{n\to\infty} \sqrt[n]{|a_n|} \leq \limsup_{n\to\infty} \sqrt[n]{|a_n|} \leq \liminf_{n\to\infty} \frac{a_{n+1}}{a_n}.$$

These inequalities show the following:

(i) if the ratio test implies convergence, so does the root test;
(ii) if the root test is inconclusive, so is the ratio test.

Consider the sequence $(a_n)_{n\geq 1}$ defined by $a_n = (1/2)^{n/2}$ if n is even and $a_n = 2(1/2)^{(n-1)/2}$ if n is odd. Then $\liminf_{n\to\infty} a_{n+1}/a_n = 1/4$ and $\limsup_{n\to\infty} a_{n+1}/a_n = 2$, so the ratio test is inconclusive for the series $\sum_{n=1}^{\infty} a_n$. At the same time, since $\lim_{n\to\infty} \sqrt[n]{|a_n|} = 1/\sqrt{2} < 1$, the root test implies that the series $\sum_{n=1}^{\infty} a_n$ is convergent.

The following test was given by the German mathematician Ernst Kummer [62] in 1835 (and published in the celebrated Crelle's Journal) when he was 23, though with a restrictive condition that was first observed by Ulysse Dini [23] in 1867. Later it was rediscovered several times and gave rise to a violent contention on questions of priority. Kummer's test gives very powerful sufficient conditions for convergence or divergence of a positive series.

Kummer's Test. Let $(a_n)_{n\geq 1}$ and $(b_n)_{n\geq 1}$ be two sequences of positive numbers. Suppose that the series $\sum_{n=1}^{\infty} 1/b_n$ diverges and let $x_n = b_n - (a_{n+1}/a_n)b_{n+1}$. Then the series $\sum_{n=1}^{\infty} a_n$ converges if there is some $h > 0$ such that $x_n \geq h$ for all n (equivalently, if $\liminf_{n\to\infty} x_n > 0$) and diverges if $x_n \leq 0$ for all n (which is the case if, e.g., $\limsup_{n\to\infty} x_n > 0$).

2.1 Main Definitions and Basic Results

In many cases the following corollaries of Kummer's test may be useful.

Raabe's Test. Let $(a_n)_{n\geq 1}$ be a sequence of positive numbers. Then the series $\sum_{n=1}^{\infty} a_n$ converges if $a_{n+1}/a_n \leq 1 - r/n$ for all n, where $r > 1$ (equivalently, if $\liminf_{n\to\infty} n(1 - a_{n+1}/a_n) > 1$) and diverges if $a_{n+1}/a_n \geq 1 - 1/n$ for all n (which is the case if, e.g., $\limsup_{n\to\infty} n(1 - a_{n+1}/a_n) < 1$).

Gauss's Test. Let $(a_n)_{n\geq 1}$ be a sequence of positive numbers such that for some constants $r \in \mathbb{R}$ and $p > 1$, we have

$$\frac{a_{n+1}}{a_n} = 1 - \frac{r}{n} + O\left(\frac{1}{n^p}\right) \quad \text{as } n\to\infty.$$

Then the series $\sum_{n=1}^{\infty} a_n$ converges if $r > 1$ and diverges if $r \leq 1$.

Abel's summation formula establishes that given a pair of real sequences $(a_n)_{n\geq 1}$ and $(b_n)_{n\geq 1}$, then

$$\sum_{j=m}^{n} a_j b_j = \sum_{j=m}^{n-1} S_j(b_j - b_{j+1}) + S_n b_n - S_{m-1} b_n,$$

provided $1 \leq m < n$, where $S_n = \sum_{j=1}^{n} a_j$ and $S_0 = 0$. Direct consequences of this elementary summation formula are formulated in the following results.

Dirichlet's Test. Let $\sum_{n=1}^{\infty} a_n$ be a series of real numbers whose partial sums $s_n = \sum_{k=1}^{n} a_k$ form a bounded sequence. If $(b_n)_{n\geq 1}$ is a decreasing sequence of nonnegative numbers converging to 0, then the series $\sum_{n=1}^{\infty} a_n b_n$ converges.

Abel's Test. Let $\sum_{n=1}^{\infty} a_n$ be a convergent series of real numbers. Then for any bounded monotone sequence $(b_n)_{n\geq 1}$, the series $\sum_{n=1}^{\infty} a_n b_n$ is also convergent.

If $(a_n)_{n\geq 1}$ is a sequence of *positive* numbers, then the series $(-1)^n a_n$ (or $(-1)^{n+1} a_n$) is called an *alternating series*. For such series one of the most useful convergence tests is the following result.

Leibniz's Test. Let $(a_n)_{n\geq 1}$ be a decreasing sequence of positive numbers. Then the alternating series $\sum_{n=1}^{\infty} (-1)^n a_n$ is convergent.

A simple application of Leibniz's test shows that the series $\sum_{n=1}^{\infty} (-1)^{n+1} 1/n$ converges (we will see later that its sum equals $\ln 2$).

Let $(x_n)_{n\geq 1}$ be a sequence of real numbers. An infinite product $\prod_{n=1}^{\infty}(1 + x_n)$ is said to be *convergent* if there is a number $n_0 \in \mathbb{N}$ such that $\lim_{n\to\infty} \prod_{k=n_0}^{N}(1 + x_k)$ exists and is different from zero; otherwise, the product is called *divergent*.

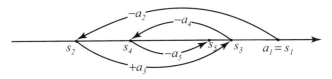

Fig. 2.1 Illustration of Leibniz's test for alternating series.

We observe from this definition that if an infinite product $\prod_{n=1}^{\infty}(1+x_n)$ is convergent then $x_n \to 0$ as $n \to \infty$. An infinite product $\prod(1+x_n)$ is convergent if and only if $\sum_{n=n_0}^{\infty} \log(1+x_n)$ is convergent for some $n_0 \in \mathbb{N}$. As a result, there is the following simple correspondence between the unconditional convergence of series and that of products: a product $\prod_{k=1}^{\infty}(1+x_k)$ converges for arbitrary arrangements of its factors if and only if the series $\sum_{n=1}^{\infty}|x_n|$ is convergent; in this case the product is called *absolutely (or unconditionally) convergent*.

There is no equally simple correspondence between the *conditional* convergence of the product $\prod_{n=1}^{\infty}(1+x_n)$ and that of the series $\sum_{n=1}^{\infty} x_n$. For instance, the series

$$\left(1-\frac{1}{\sqrt{2}}\right)\left(1+\frac{1}{\sqrt{2}}\right)\left(1-\frac{1}{\sqrt{3}}\right)\left(1+\frac{1}{\sqrt{3}}\right)\cdots$$

diverges to zero $\left(\text{since } \prod_{n=2}^{N}(1-1/n) = 1/N\right)$, while

$$-\frac{1}{\sqrt{2}} + \frac{1}{\sqrt{2}} - \frac{1}{\sqrt{3}} + \frac{1}{\sqrt{3}} - + \cdots$$

is convergent. We also observe that the product

$$\prod_{n=2}^{\infty}\left(1-\frac{1}{\sqrt[4]{n}}\right)\left(1+\frac{1}{\sqrt[4]{n}}\right)\left(1+\frac{1}{\sqrt{n}}\right)\left(1+\frac{1}{n}\right)$$

is convergent $\left(\text{since } \prod_{n=2}^{N}(1-1/n^2) = (N+1)/2N\right)$, whereas

$$\sum x_n = -\frac{1}{\sqrt[4]{2}} + \frac{1}{\sqrt[4]{2}} + \frac{1}{\sqrt{2}} + \frac{1}{2} - \frac{1}{\sqrt[4]{3}} + \cdots$$

and even $\sum(x_n - (1/2)x_n^2)$ (observe that $\log(1+x_n) = x_n - (1/2)x_n^2 + O(x_n^3)$) diverges.

The following two infinite products for π, due respectively to John Wallis (1616–1703) and François Viète (1540–1603), are well known:

$$\frac{\pi}{2} = \frac{2 \cdot 2}{1 \cdot 3} \frac{4 \cdot 4}{3 \cdot 5} \frac{6 \cdot 6}{5 \cdot 7} \cdots \frac{2n \cdot 2n}{(2n-1) \cdot (2n+1)} \cdots,$$

$$\frac{\pi}{2} = \frac{2}{\sqrt{2}} \frac{2}{\sqrt{2+\sqrt{2}}} \frac{2}{\sqrt{2+\sqrt{2+\sqrt{2}}}} \frac{2}{\sqrt{2+\sqrt{2+\sqrt{2+\sqrt{2}}}}} \cdots.$$

While both formulas are beautiful, the convergence of Wallis's formula is painfully slow. Viète's representation formula is much better, although Viète himself was able to approximate π to only 9 digits past the decimal point in 1593.

Other important infinite products related to trigonometric functions were found by Euler (1748):

2.1 Main Definitions and Basic Results

$$\sin x = x\left(1 - \frac{x^2}{\pi^2}\right)\left(1 - \frac{x^2}{4\pi^2}\right)\left(1 - \frac{x^2}{9\pi^2}\right)\cdots,$$

$$\cos x = \left(1 - \frac{4x^2}{\pi^2}\right)\left(1 - \frac{4x^2}{9\pi^2}\right)\left(1 - \frac{4x^2}{25\pi^2}\right)\cdots.$$

Cauchy's Criterion for Infinite Products. Let $x_n > -1$ for all n. If $\lim_{n\to\infty} \sum_{k=1}^{n} x_k$ exists then so does $\lim_{n\to\infty} \prod_{k=1}^{n}(1+x_k)$. Moreover, the limit is zero if and only if $\sum_{n=1}^{\infty} x_n^2 = \infty$.

Cauchy explicitly attributes this theorem to Gaspard-Gustave Coriolis, who is well known for his work in mechanics. The following test is useful in many situations.

Coriolis Test. If $(x_n)_{n\geq 1}$ is a sequence of real numbers such that $\sum_{n=1}^{\infty} x_n$ and $\sum_{n=1}^{\infty} x_n^2$ are convergent, then $\prod_{n=1}^{\infty}(1+x_n)$ converges.

As can be seen from the second of the examples mentioned above, the Coriolis conditions are not necessary for convergence. In fact, it seems impossible to find a simple necessary and sufficient condition for the convergence of the product $\prod_{n=1}^{\infty}(1+x_n)$ in terms of that of series like $\sum_{n=1}^{\infty} x_n$. Indeed, if any such convergent product is given, we can choose an arbitrary sequence (y_n) converging to zero and conclude that

$$\prod_{n=1}^{\infty}(1+x_n)(1+y_n)\left(1 - \frac{y_n}{1+y_n}\right)$$

is convergent, too, while

$$\sum_{n=1}^{\infty}\left(x_n + y_n - \frac{y_n}{1+y_n}\right) = \sum\left(x_n + \frac{y_n^2}{1+y_n}\right)$$

may be strongly divergent.

One of the best-known number series is the celebrated harmonic series $\sum_{n=1}^{\infty} 1/n$. We have already observed that this series diverges. We are now interested to estimate the divergence rate of the harmonic series. For any positive integer n, consider its partial sum $H_n = \sum_{k=1}^{n} 1/k$. We will see in Chapter 9 that the sequence $(a_n)_{n\geq 1}$ defined by $a_n = H_n - \ln n$ converges to a positive number γ, called Euler's constant. Moreover, the following asymptotic estimate holds:

$$\sum_{k=1}^{n} \frac{1}{k} = \gamma + \ln n + \frac{1}{2n} - \frac{1}{12n^2} + \frac{1}{120n^4} - \frac{1}{252n^6} + \frac{1}{240n^8} + O(n^{-9}) \quad \text{as } n\to\infty.$$

The symbol γ was first used by Mascheroni, and the approximate value of this constant is

$$\gamma = 0.5772156649015328606065120900824024 31042\ldots.$$

The following mnemonic for remembering the value of γ is due to Ward [115]:

These numbers proceed to a limit Euler's subtle mind discerned.

An interesting open problem is that it is not known whether γ is a rational, irrational, or transcendental number. There is a huge literature on Euler's constant, especially due to its impact on other fields of mathematics. Some interesting properties of this number are listed below:

$$\lim_{n\to\infty} \left(\frac{(H_n - \ln n)^\gamma}{\gamma^{H_n - \ln n}} \right)^{2n} = \frac{e}{\gamma}.$$

We also observe that $\gamma = \lim_{n\to\infty} [H_n - \ln(n+\alpha)]$, for any $\alpha \in \mathbb{R}$. An interesting problem is to establish what choice of α makes this limit converge as fast as possible. Partial answers were given in [44] as follows: (i) the standard case $\alpha = 0$ gives an error of about $O(n^{-1})$; (ii) $\alpha = 2^{-1}$ gives the better error $O(n^{-2})$, as $n\to\infty$.

It is striking to observe the exceedingly slow growth of the partial sums H_n. Indeed, in [44] it is showed that the least n for which $H_n \geq 100$ is a 44-digit number (taking nearly a whole line of text). Another interesting tidbit (which follows with a quite elementary argument related to the prime decomposition of integers) is that H_n is never an integer, except for $n = 1, 2, 6$.

We also point out that Euler's constant γ and the sequence $(H_n)_{n \geq 1}$ of partial sums of the harmonic series are closely related to the celebrated Euler's gamma function, which is defined by

$$\Gamma(t) = \int_0^\infty x^{t-1} e^{-x} dx, \quad \text{for all } t > 0.$$

More precisely (see [44]), for any integer $n \geq 2$,

$$\frac{\Gamma'(n)}{\Gamma(n)} = H_{n-1} - \gamma.$$

The gamma function is an extension of the factorial to positive **real** numbers, in the sense that for any positive integer n, $\Gamma(n) = (n-1)!$. The above formula implies that the value of the derivative of Γ on positive integers is given by

$$\Gamma'(n) = (n-1)! \left(1 + \frac{1}{2} + \cdots + \frac{1}{n-1} - \gamma \right) \quad \text{for all } n \geq 2.$$

2.2 Elementary Problems

> In mathematics the art of proposing a question must be held of higher value than solving it.
>
> Georg Cantor (1845–1918)

By means of simple inequalities we deduce below lower and upper estimates for the sum of series $\sum_{n=1}^\infty 1/n^2$. We recall that the exact value of this sum equals $\pi^2/6 = 1.644934\ldots$. This result was obtained by Euler in 1735 and is the starting point of

2.2 Elementary Problems

a celebrated paper written by Bernhard Riemann (1826–1866) in 1859, in which he defines the *zeta function* ζ by

$$\zeta(x) = \sum_{n=1}^{\infty} 1/n^x, \quad \text{for any } x > 1.$$

We have already observed that this series diverges if $x \leq 1$. The series $\sum_{n=1}^{\infty} 1/n^x$ converges for any $x > 1$, as a consequence of Gauss's test. The Riemann zeta function can be also expressed in terms of an infinite product by means of Euler's formula

$$\zeta(x) = \prod_{n=1}^{\infty} \frac{1}{1 - 1/p_n^x} \quad \text{for all } x > 1,$$

where $(p_n)_{n \geq 1}$ is the sequence of prime numbers ($p_1 = 2, p_2 = 3, p_3 = 5, \ldots$).

The Riemann zeta function satisfies

$$\zeta(x) \to 1 \quad \text{as } x \to \infty$$

and

$$\frac{1}{x-1} \leq \zeta(x) \leq \frac{x}{x-1}, \quad \text{for any } x > 1.$$

Thus, the following additional asymptotic estimates hold:

$$\zeta(x) \to \infty \quad \text{as } x \searrow 1,$$
$$(x-1)\zeta(x) \to 1 \quad \text{as } x \searrow 1.$$

The problem of finding the exact value of $\zeta(x)$, for any real number $x > 1$, is also known as the *Basel problem*. The problem of evaluating $\zeta(2) = \sum_{n=1}^{\infty} 1/n^2$ in closed form has an interesting history, which apparently began in 1644 when Pietro Mengoli (1625–1686) asked for the sum of the reciprocals of the squares. The city of Basel was the hometown of the famous brothers Jakob and Johann Bernoulli, who made serious but unsuccessful attempts to find $\zeta(2)$. In 1689, Jakob Bernoulli wrote, "If somebody should succeed in finding what till now withstood our efforts and communicate it to us we shall be much obliged to him." The prodigious mathematician Leonhard Euler found for the first time the value of $\zeta(2)$ in 1735; see [26]. Since then, many different ways of evaluating this sum have been discovered. Fourteen proofs are collected in [17], and two of them have been included in [1]. Euler also obtained the more general formula

$$\zeta(2n) = (2\pi)^{2n} \frac{|B_{2n}|}{2(2n)!},$$

where B_n are the Bernoulli numbers, defined by $B_0 = 1$ and

$$\sum_{k=0}^{n} \binom{n}{k} B_k = 0 \quad \text{for any } n \geq 2.$$

Sadly, Jakob Bernoulli did not live to see young Euler's triumphant discovery and its surprising connection with Bernoulli numbers.

2.2.1. *Prove that* $29/18 < \sum_{n=1}^{\infty} 1/n^2 < 31/18$.

Solution. We have

$$\frac{29}{18} = 1 + \frac{1}{4} + \frac{1}{9} + \sum_{n=4}^{\infty} \frac{1}{n(n+1)} < \sum_{n=1}^{\infty} \frac{1}{n^2}$$
$$< 1 + \frac{1}{4} + \frac{1}{9} + \sum_{n=4}^{\infty} \frac{1}{n(n-1)} = \frac{61}{36} < \frac{31}{18}. \quad \square$$

We now consider a series whose terms are defined by a recurrence relation. In this case we are able not only to justify that the series converges, but also to find the value of its sum.

2.2.2. *Let $(a_n)_{n \geq 1}$ be the sequence of real numbers defined by $a_n = 1/2$ and, for any $n \geq 1$,*

$$a_{n+1} = \frac{a_n^2}{a_n^2 - a_n + 1}.$$

Prove that the series $\sum_{n=1}^{\infty} a_n$ converges and find its sum.

Solution. Writing $b_n = 1/a_n$, we obtain $b_1 = 2$ and $b_{n+1} = b_n^2 - b_n + 1 = b_n(b_n - 1) + 1$, for any positive integer n. We deduce that $(b_n)_{n \geq 1}$ is an increasing sequence that diverges to $+\infty$ and satisfies

$$\frac{1}{b_n} = \frac{1}{b_n - 1} - \frac{1}{b_{n+1} - 1}.$$

By addition we deduce that

$$a_1 + \cdots + a_n = \frac{1}{b_1 - 1} - \frac{1}{b_{n+1} - 1} = 1 - \frac{1}{b_{n+1} - 1}.$$

This relation shows that the series $\sum_{n=1}^{\infty} a_n$ converges and its sum equals 1. $\quad \square$

2.2.3. *Prove that*

$$\frac{1}{3} + \frac{1}{3 \cdot 7} + \frac{1}{3 \cdot 7 \cdot 47} + \frac{1}{3 \cdot 7 \cdot 47 \cdot 2207} + \cdots = \frac{3 - \sqrt{5}}{2},$$

where each factor in the denominator is equal to the square of the preceding factor diminished by 2.

Solution. We determine λ such that

$$\frac{y_0 - \lambda}{2} = \frac{1}{y_0} + \frac{1}{y_0 y_1} + \frac{1}{y_0 y_1 y_2} + \cdots,$$

where $y_0 = 3$, $y_1 = 7$, $y_2 = 47$, and in general, $y_n = y_{n-1}^2 - 2$.

2.2 Elementary Problems

We notice that

$$\frac{y_1 - \lambda y_0}{2} = \frac{1}{y_1} + \frac{1}{y_1 y_2} + \cdots;$$

$$\frac{y_2 - \lambda y_0 y_1}{2} = \frac{1}{y_2} + \frac{1}{y_2 y_3} + \cdots;$$

$$\frac{y_n - \lambda y_0 y_1 \cdots y_{n-1}}{2} = \frac{1}{y_n} + \frac{1}{y_n y_{n+1}} + \cdots.$$

Letting $n \to \infty$, we observe that

$$\lambda = \lim_{n \to \infty} \frac{y_n}{y_0 y_1 \cdots y_{n-1}}.$$

Since $y_n = y_{n-1}^2 - 2$, we deduce $y_n^2 - 4 = y_{n-1}^2(y_{n-2}^2 - 4)$. Hence

$$y_n^2 - 4 = y_0^2 y_1^2 y_2^2 \cdots y_{n-1}^2 (y_0^2 - 4),$$

and so

$$\lim_{n \to \infty} \frac{y_n}{y_0 y_1 \cdots y_{n-1}} = \sqrt{y_0^2 - 4} = \sqrt{3^2 - 4} = \sqrt{5} = \lambda. \quad \square$$

2.2.4. Show that

$$\sum_{n=1}^{\infty} \left(n \log \frac{2n+1}{2n-1} - 1 \right) = \frac{1}{2}(1 - \log 2).$$

Solution. Set

$$s_n = \sum_{k=1}^{n} \left(k \log \frac{2k+1}{2k-1} - 1 \right) = \log \frac{1}{1} \cdot \frac{1}{3} \cdot \frac{1}{5} \cdot \frac{1}{7} \cdots \frac{1}{2n-1} (2n+1)^n - n.$$

Therefore

$$s_n = \log \frac{2^{n-1}(n-1)!(2n+1)^n}{(2n-1)!} - n.$$

Using Stirling's formula, we deduce that the sequence $(s_n)_{n \geq 1}$ tends to the same limit as the sequence $(t_n)_{n \geq 1}$, where

$$t_n = \log \frac{2^{n-1} \sqrt{2\pi(n-1)}(n-1)^{n-1} e^{1-n}(2n+1)^n}{\sqrt{2\pi(2n-1)}(2n-1)^{2n-1} e^{1-2n}},$$

as $n \to \infty$. But

$$t_n = \frac{1}{2} \log \frac{n-1}{2n-1} + \log \left(\frac{2n-2}{2n-1} \right)^{n-1} \left(\frac{2n+1}{2n-1} \right)^n \to -\frac{1}{2} \log 2 + \frac{1}{2} \quad \text{as } n \to \infty.$$

Indeed,
$$\left(\frac{2n-2}{2n-1}\right)^{n-1} = \left(1 - \frac{1}{2n-1}\right)^{n-1} \to \frac{1}{\sqrt{e}}$$
and
$$\left(\frac{2n+1}{2n-1}\right)^{n} = \left(1 + \frac{2}{2n-1}\right)^{n} \to e,$$
as $n \to \infty$. This completes the proof. $\square$

2.2.5. Evaluate
$$\sum_{n=1}^{\infty} \frac{n}{n^4 + n^2 + 1}.$$

Solution. The series $\sum_{n=1}^{\infty} n/(n^4 + n^2 + 1)$ converges because for all $n \geq 1$,
$$0 < \frac{n}{n^4 + n^2 + 1} < \frac{1}{n^3} \quad \text{and} \quad \sum_{n=1}^{\infty} \frac{1}{n^3} \quad \text{converges.}$$

On the other hand, we have
$$\frac{n}{n^4 + n^2 + 1} = \frac{n}{(n^2+1)^2 - n^2} = \frac{n}{(n^2-n+1)(n^2+n+1)}$$
$$= \frac{1/2}{(n^2-n+1)} - \frac{1/2}{(n^2+n+1)}.$$

Set
$$a_n = \frac{1/2}{(n^2-n+1)} \Rightarrow a_{n+1} = \frac{1/2}{(n^2+n+1)}.$$

Therefore,
$$S_N = \sum_{n=1}^{N} \frac{n}{n^4+n^2+1} = \sum_{n=1}^{N} (a_n - a_{n+1})$$
$$= (a_1 - a_2) + \cdots + (a_N - a_{N+1}) = a_1 - a_{N+1}$$
$$= \frac{1}{2} - \frac{1/2}{N^2+N+1}.$$

We conclude that
$$\sum_{n=1}^{\infty} \frac{n}{n^4+n^2+1} = \lim_{N \to \infty} S_N = \frac{1}{2}. \quad \square$$

2.2.6. Suppose that $(a_n)_{n \geq 1}$ and $(b_n)_{n \geq 1}$ are decreasing sequences of positive numbers that converge to 0 and both series $\sum_{n=1}^{\infty} a_n$ and $\sum_{n=1}^{\infty} b_n$ diverge. Does the series $\sum_{n=1}^{\infty} \min(a_n, b_n)$ always diverge?

Solution. We first construct positive but nonmonotone sequences $(\alpha_n)_{n \geq 1}$ and $(\beta_n)_{n \geq 1}$ such that $\sum_{n=1}^{\infty} \alpha_n = \sum_{n=1}^{\infty} \beta_n = +\infty$ (exercise!). The required monotone

2.2 Elementary Problems

series are obtained by considering the series

$$\alpha_1 + \cdots + \frac{\alpha_k}{n_k} + \cdots + \frac{\alpha_k}{n_k} + \cdots$$

and

$$\beta_1 + \cdots + \frac{\beta_k}{n_k} + \cdots + \frac{\beta_k}{n_k} + \cdots$$

(the terms of the form α_k/n_k and β_k/n_k are repeated n_k times) for a suitable choice of the sequence $(n_k)_{k \geq 1}$.

A direct application of the ratio test implies that the series

$$1 + \frac{1}{1!} + \frac{1}{2!} + \cdots + \frac{1}{n!} + \cdots$$

is convergent. Denote by e its sum, that is, $e = \sum_{n=0}^{\infty} 1/n!$. Since $1/n! \leq 1/2(3^{n-2})$ for all $n \geq 3$ and $\sum_{n=3}^{\infty} 1/3^{n-2} = 1/2$, we deduce that for any $n \geq 3$,

$$a_n := \sum_{k=0}^{n} \frac{1}{k!} < 1 + 1 + \frac{1}{2} + \sum_{k=3}^{\infty} \frac{1}{2 \cdot 3^{k-2}} = \frac{11}{4}.$$

This shows that $e \leq 11/4 < 3$; hence $2 < e < 3$. □

The next exercise establishes further qualitative properties of the number e, which is also called the *base of the natural logarithm*.

2.2.7. (i) *Prove that* $e = \lim_{n \to \infty}(1 + 1/n)^n$.

(ii) *Show that the number e is irrational.*

Solution. (i) Set $b_n = (1 + 1/n)^n$. Thus, by Newton's binomial,

$$b_n = 1 + 1 + \frac{1}{2!}\left(1 - \frac{1}{n}\right) + \cdots + \frac{1}{n!}\left(1 - \frac{1}{n}\right)\left(1 - \frac{2}{n}\right) \cdots \left(1 - \frac{n-1}{n}\right).$$

It follows that $b_n \leq a_n$, so that

$$\limsup_{n \to \infty} b_n \leq e.$$

Next, for any fixed positive integers m and $n \geq m$, we have

$$b_n \geq 1 + 1 + \frac{1}{2!}\left(1 - \frac{1}{n}\right) + \cdots + \frac{1}{m!}\left(1 - \frac{1}{n}\right) \cdots \left(1 - \frac{m-1}{n}\right).$$

Letting $n \to \infty$ we obtain

$$a_m := \sum_{k=0}^{m} \frac{1}{k!} \leq \liminf_{n \to \infty} b_n.$$

Since m is arbitrary, it follows that

$$e \leq \liminf_{n \to \infty} b_n.$$

Thus, $\limsup_{n\to\infty} b_n \leq e \leq \liminf_{n\to\infty} b_n$, and this concludes the proof of (i).

(ii) If $e = p/q$ with p, q positive integers, then

$$q!e - \left(\sum_{k=0}^{q} \frac{1}{k!}\right) q! = \frac{1}{q+1} + \frac{1}{(q+1)(q+2)} + \cdots.$$

The left-hand side of this relation is an integer, while the right-hand side satisfies

$$0 < \frac{1}{q+1} + \frac{1}{(q+1)(q+2)} + \cdots < \frac{1}{q+1} + \frac{1}{(q+1)^2} + \cdots = \frac{1}{q} < 1.$$

This contradiction proves that e is indeed irrational. □

2.2.8. *Show that the integer nearest to $n!/e$ ($n \geq 2$) is divisible by $n-1$ but not by n.*

Solution. We have

$$e^{-1} = \sum_{k=0}^{\infty} \frac{(-1)^k}{k!}.$$

Therefore

$$\frac{n!}{e} = n!e^{-1} = n! \sum_{k=0}^{n} \frac{(-1)^k}{k!} + n! \sum_{k=n+1}^{\infty} \frac{(-1)^k}{k!}.$$

The first term is obviously an integer, and the second term is bounded by

$$\left| n! \sum_{k=n+1}^{\infty} \frac{(-1)^k}{k!} \right| \leq n! \cdot \frac{1}{(n+1)!} = \frac{1}{n+1} \leq \frac{1}{3}, \quad \text{since } n \geq 2.$$

Therefore $n! \sum_{k=0}^{n} (-1)^k/k!$ is the nearest integer to $n!/e$. This integer is not divisible by n, since

$$n! \sum_{k=0}^{n} \frac{(-1)^k}{k!} = n \cdot \left[(n-1)! \sum_{k=0}^{n-1} \frac{(-1)^k}{k!} \right] + (-1)^n.$$

This number is divisible by $(n-1)$ because

$$n! \sum_{k=0}^{n} \frac{(-1)^k}{k!} = n(n-1) \left[(n-2)! \sum_{k=0}^{n-2} \frac{(-1)^k}{k!} \right] + (-1)^{n-1} \cdot n + (-1)^n$$

$$= (n-1) \left\{ n \left[(n-2)! \sum_{k=0}^{n-2} \frac{(-1)^k}{k!} \right] + (-1)^{n-1} \right\}$$

and since the terms inside square brackets are obviously integers. □

2.2.9. *Find the sum of the series*

$$\sum_{n=0}^{\infty} \arctan \frac{1}{n^2 + n + 1}.$$

2.3 Convergent and Divergent Series

Solution. For any integer $n \geq 0$ we have

$$\arctan \frac{1}{n^2+n+1} = \arctan \frac{(n+1)-n}{n^2+n+1} = \arctan(n+1) - \arctan n.$$

Therefore

$$\sum_{n=0}^{N} \arctan \frac{1}{n^2+n+1} = \arctan(N+1),$$

which shows that the given series converges to $\pi/2$. □

2.2.10. *Fix $a > 0$ and denote by $\mathscr{A}$ the class of all sequences $(a_n)_{n \geq 1}$ of positive real numbers such that $\sum_{n=1}^{\infty} a_n = a$. Find $\{\sum_{n=1}^{\infty} a_n^2; (a_n)_{n \geq 1} \in \mathscr{A}\}$.*

Solution. Let $(a_n)_{n \geq 1} \in \mathscr{A}$. Then

$$a^2 = \sum_{n=1}^{\infty} a_n^2 + 2 \sum_{1 \leq m < n} a_m a_n,$$

which implies $0 < \sum_{n=1}^{\infty} a_n^2 < a^2$. Conversely, we prove that for any $b \in (0, a^2)$, there is a sequence $(a_n)_{n \geq 1} \in \mathscr{A}$ such that $b = \sum_{n=1}^{\infty} a_n^2$. For this purpose we set

$$r = \frac{a^2+b}{a^2-b} \quad \text{and} \quad \lambda = \frac{(r-1)a}{r}.$$

For any $n \geq 1$, let $a_n = \lambda/r^{n-1}$. Then $(a_n)_{n \geq 1} \in \mathscr{A}$ and $\sum_{n=1}^{\infty} a_n^2 = b$. Consequently, $\{\sum_{n=1}^{\infty} a_n^2; (a_n)_{n \geq 1} \in \mathscr{A}\} = (0, a^2)$. □

2.3 Convergent and Divergent Series

> It is not enough that we do our best;
> sometimes we have to do what is
> required.
>
> Sir Winston Churchill (1874–1965)

2.3.1. *Prove that the series*

$$\sum_{n=3}^{\infty} \frac{1}{(\ln \ln n)^{\ln n}}$$

is convergent.

Solution. We have

$$(\ln \ln n)^{\ln n} = e^{\ln n \cdot \ln \ln \ln n} = (e^{\ln n})^{\ln \ln \ln n} = n^{\ln \ln \ln n}.$$

But
$$\ln\ln\ln e^{e^{e^{e^2}}} = \ln\ln e^{e^2} = \ln e^2 = 2.$$
Thus, for any $n > e^{e^{e^2}}$, we have $n^{\ln\ln\ln n} > n^2$, hence
$$\frac{1}{(\ln\ln n)^{\ln n}} = \frac{1}{n^{\ln\ln\ln n}} < \frac{1}{n^2}. \quad \square$$

Applying the first comparison test, we deduce that the given series converges.

2.3.2. Study the convergence of the series
$$\sum_{n=1}^{\infty} \frac{a^n}{\sqrt[n]{n!}},$$
where a is a given positive number.

Solution. Since $\sqrt[n]{n!} \geq 1$ for any positive integer n, we deduce that for all $n \geq 1$,
$$\frac{a^n}{\sqrt[n]{n!}} \leq a^n.$$

Thus, by the first comparison test, the given series converges provided $a < 1$.

Next, using the inequality
$$\sqrt[n]{n!} \leq \sqrt[n]{n^n} = n,$$
we obtain, for all $n \geq 1$,
$$\frac{a^n}{n} \leq \frac{a^n}{\sqrt[n]{n!}}.$$

But the series $\sum_{n=1}^{\infty} a^n/n$ diverges for any $a \geq 1$. Indeed, for $a = 1$ it coincides with the harmonic series, while the sequence of its terms diverges to $+\infty$ provided $a > 1$. So, by the second comparison test, the given series diverges for all $a \geq 1$. $\quad \square$

2.3.3. Study the convergence of the series
$$\sum_{n=1}^{\infty} a^n \left(1 + \frac{1}{n}\right)^n,$$
where a is a given positive number.

Solution. Set $x_n = a^n(1 + 1/n)^n$. Thus, $\lim_{n\to\infty} \sqrt[n]{x_n} = a$. Hence, by the root test, the given series converges if $a < 1$ and diverges for any $a > 1$. If $a = 1$, then the series diverges because $x_n \to e \neq 0$ as $n \to \infty$. $\quad \square$

2.3.4. Let a be a positive number. Prove that the series
$$\sum_{n=1}^{\infty} \frac{a^n}{\sqrt{n!}}$$
is convergent.

2.3 Convergent and Divergent Series

Solution. Set $x_n = a^n/\sqrt{n!}$. We have

$$\lim_{n\to\infty} \frac{x_{n+1}}{x_n} = \lim_{n\to\infty} \frac{a^{n+1}\sqrt{n!}}{a^n\sqrt{(n+1)!}} = 0 < 1. \quad \square$$

Applying the ratio test, we deduce that the given series is convergent.

2.3.5. Let $(a_n)_{n\geq 1}$ be a sequence of positive numbers converging to a and assume that b is a positive number. Study the convergence of the series

$$\sum_{n=1}^{\infty} \frac{n!b^n}{(b+a_1)(b+a_2)\cdots(b+a_n)}.$$

Solution. Applying the Raabe test, we compute

$$\lim_{n\to\infty} n\left(1 - \frac{x_{n+1}}{x_n}\right) = \lim_{n\to\infty} \frac{na_{n+1}}{(n+1)b} = \frac{a}{b}.$$

We deduce that the given series converges if $b < a$ and that it diverges when $b > a$. If $a = b$, the behavior of the series strongly depends on the convergence rate of the sequence $(a_n)_{n\geq 1}$. $\quad \square$

2.3.6. Let a, b, c be real numbers such that $-c$ is not a natural number. Study the convergence of the hypergeometric series

$$\sum_{n=1}^{\infty} \frac{a(a+1)\cdots(a+n-1)b(b+1)\cdots(b+n-1)}{n!c(c+1)\cdots(c+n-1)}.$$

Solution. Denote by x_n the general term of the series. A straightforward computation (we leave the details to the reader) yields

$$\frac{x_{n+1}}{x_n} = 1 - \frac{c+1-a-b}{n} + O\left(\frac{1}{n^2}\right) \quad \text{as } n\to\infty.$$

The Gauss test enables us to conclude that the hypergeometric series converges if and only if $c > a + b$. $\quad \square$

2.3.7. For all $n \geq 1$, set

$$s_n = \sum_{k=1}^{n} a_k \quad \text{and} \quad \sigma_n = \sum_{k=1}^{n}\left(1 - \frac{k}{n+1}\right)a_k.$$

Assume that the series $\sum_{n=1}^{\infty} |s_n - \sigma_n|^\alpha$ converges, for some $\alpha > 0$. Prove that the series $\sum_{n=1}^{\infty} a_n$ is convergent.

Solution. Since $s_n - \sigma_n \to 0$ as $n\to\infty$, it is sufficient to prove that the sequence $(\sigma_n)_{n\geq 1}$ converges. We have

$$\sigma_n - \sigma_{n-1} = \frac{1}{n(n+1)}\sum_{k=1}^{n} ka_k = \frac{s_n - \sigma_n}{n}.$$

Summing for $n = 1, 2, \ldots, N$, we obtain

$$\sigma_N = \sum_{n=1}^{N} \frac{s_n - \sigma_n}{n}.$$

If $\alpha \leq 1$, the convergence is immediate. If $\alpha > 1$, Hölder's inequality yields

$$\sum_{n=p}^{q} \frac{|s_n - \sigma_n|}{n} \leq \left\{ \sum_{n=p}^{q} |s_n - \sigma_n|^\alpha \right\}^{1/\alpha} \left\{ \sum_{n=p}^{q} n^{-\beta} \right\}^{1/\beta},$$

where $1/\alpha + 1/\beta = 1$. Since both factors on the right-hand side converge to 0, so does the left-hand side. Thus, the sequence $(\sigma_n)_{n\geq 1}$ is Cauchy, hence convergent. □

2.3.8. Let $(a_n)_{n\geq 1}$ be a sequence of real numbers such that the series $\sum_{n=1}^{\infty} a_n^4$ is convergent. Prove that the series $\sum_{n=1}^{\infty} a_n^5$ is convergent.

Solution. The hypothesis implies that $a_n^4 \to 0$ as $n \to \infty$. Thus, there exists a positive integer N such that $a_n^4 < 1$ for all $n \geq N$; hence $|a_n^5| < a_n^4$, provided $n \geq N$. So, by the first comparison test, the series $\sum_{n=1}^{\infty} a_n^5$ is convergent. □

2.3.9. Let $(a_n)_{n\geq 1}$ be a sequence of positive numbers such that $a_n \notin \{0, 1\}$ and the series $\sum_{n=1}^{\infty} a_n$ is convergent. Prove that the series $\sum_{n=1}^{\infty} a_n/(1-a_n)$ is convergent.

Solution. An argument similar to that developed in the previous exercise implies that the series $\sum_{n=1}^{\infty} a_n^2$ is convergent. Since the series $\sum_{n=1}^{\infty} a_n$ converges, it follows that $a_n \to 0$ as $n \to \infty$ and there exists an integer N such that $1 - a_n > 0$ and $a_n/(1-a_n) < 2a_n + 2a_n^2$ for all $n \geq N$. Thus, by the first comparison test, the series $\sum_{n=1}^{\infty} a_n/(1-a_n)$ is convergent. □

2.3.10. Let $(a_n)_{n\geq 1}$ be a decreasing sequence of positive numbers. Prove that if there is a positive number k such that $a_n \geq k/n$ for infinitely many n, then the series $\sum_{n=1}^{\infty} a_n$ is divergent.

Solution. Fix $\varepsilon > 0$ such that $\varepsilon \leq k/3$. For N arbitrarily large, select $n > 3N$ such that $a_n \geq k/n$. Let m be a positive integer such that $N < m < 2N$. Thus, $m/n < 2/3$. We have

$$|a_{m+1} + \cdots + a_n| > (n-m)a_n \geq \frac{(n-m)k}{n} = k\left(1 - \frac{m}{n}\right) > \frac{k}{3} \geq \varepsilon.$$

So, in view of Cauchy's criterion, the series $\sum_{n=1}^{\infty} a_n$ is divergent. □

2.3.11. Let $(a_n)_{n\geq 1}$ be a sequence of nonnegative numbers such that $a_{2n} - a_{2n+1} \leq a_n^2$, $a_{2n+1} - a_{2n+2} \leq a_n a_{n+1}$ for any $n \geq 1$ and $\limsup_{n\to\infty} na_n < 1/4$. Prove that the series $\sum_{n=1}^{\infty} a_n$ is convergent.

Solution. We prove that $\limsup_{n\to\infty} \sqrt[n]{a_n} < 1$. Then, by the root test, we obtain the conclusion.

Let $c_j = \sup_{n \geq 2^j}(n+1)a_n$ for $j \geq 1$. We show that $c_{j+1} \leq 4c_j^2$. Indeed, for any integer $n \geq 2^{j+1}$ there exists an integer $k \geq 2^j$ such that either $n = 2k$ or $n = 2k+1$.

2.3 Convergent and Divergent Series

In the first case we have

$$a_{2k} - a_{2k+1} \leq a_k^2 \leq \frac{c_j^2}{(k+1)^2} \leq \frac{4c_j^2}{2k+1} - \frac{4c_j^2}{2k+2},$$

whereas in the second case we obtain

$$a_{2k+1} - a_{2k+2} \leq a_k a_{k+1} \leq \frac{c_j^2}{(k+1)(k+2)} \leq \frac{4c_j^2}{2k+2} - \frac{4c_j^2}{2k+3}.$$

It follows that the sequence $(a_n - 4c_j^2/n + 1)_{n \geq 2^{j+1}}$ is nondecreasing and its terms are nonpositive, since it converges to 0. Therefore $a_n \leq 4c_j^2/(n+1)$ for $n \geq 2^{j+1}$, that is, $c_{j+1}^2 \leq 4c_j^2$. This implies that the sequence $((4c_j)^{2^{-j}})_{j \geq 0}$ is nonincreasing and therefore bounded from above by some number $q \in (0,1)$, since all its terms except finitely many are less than 1. Hence $c_j \leq q^{2^j}$ for l large enough. For any n between 2^j and 2^{j+1} we have

$$a_n \leq \frac{c_j}{n+1} \leq q^{2^j} \leq (\sqrt{q})^n.$$

Hence $\limsup \sqrt[n]{a_n} \leq \sqrt{q} < 1$, which concludes the proof. □

2.3.12. If $a, b > 1$ prove that

$$\sum_{n=1}^{\infty} \frac{1}{\sqrt{n^a n^b (n^a + n^b)}} \leq \frac{1}{4} (\zeta(a) + \zeta(b) + \zeta(a+b)),$$

where ζ denotes the Riemann zeta function.

Solution. It is sufficient to prove that

$$\frac{1}{\sqrt{n^a n^b (n^a + n^b)}} \leq \frac{1}{4} \left(\frac{1}{n^a} + \frac{1}{n^b} + \frac{1}{n^{a+b}} \right) = \frac{n^a + n^b + 1}{4n^{a+b}}.$$

But by the AM–GM inequality, $n^a + n^b + 1 \geq 2\sqrt{n^a + n^b}$ and $n^a + n^b \geq 2\sqrt{n^a n^b}$. Hence

$$4\sqrt{n^a n^b} \leq \sqrt{n^a + n^b} (n^a + n^b + 1),$$

which concludes the proof. □

2.3.13. Let $(p_n)_{n \geq 1}$ be a sequence of positive real numbers. Show that if the series $\sum_{n=1}^{\infty} 1/p_n$ converges, then so does the series

$$\sum_{n=1}^{\infty} \frac{n^2}{(p_1 + p_2 + \cdots + p_n)^2} p_n.$$

Solution. Put $q_n = p_1 + p_2 + \cdots + p_n$ and $q_0 = 0$. The main idea is to find an estimate of

$$S_m = \sum_{n=1}^{m} \left(\frac{n}{q_n} \right)^2 (q_n - q_{n-1})$$

in terms of $T = \sum_{n=1}^{\infty} 1/p_n$. For this purpose we observe that

$$S_m \leq \frac{1}{p_1} + \sum_{n=2}^{m} \frac{n^2}{q_n q_{n-1}}(q_n - q_{n-1}) = \frac{1}{p_1} + \sum_{n=2}^{m} \frac{n^2}{q_{n-1}} - \sum_{n=2}^{m} \frac{n^2}{q_n}$$

$$= \frac{1}{p_1} + \sum_{n=1}^{m-1} \frac{(n+1)^2}{q_n} - \sum_{n=2}^{m} \frac{n^2}{q_n} \leq \frac{5}{p_1} + 2\sum_{n=2}^{m} \frac{n}{q_n} + \sum_{n=2}^{m} \frac{1}{q_n}.$$

Next, by the Cauchy–Schwarz inequality,

$$\left(\sum_{n=1}^{m} \frac{n}{q_n}\right)^2 \leq \left(\sum_{n=2}^{m} \frac{n^2}{q_n^2} p_n\right)\left(\sum_{n=2}^{m} \frac{1}{p_n}\right).$$

We deduce that $S_m \leq 5/p_1 + 2\sqrt{S_m T} + T$. But this inequality implies

$$\sqrt{S_m} \leq T + \sqrt{2T + 5/p_1}. \quad \square$$

2.3.14. *Suppose that $(a_n)_{n \geq 1}$ is a sequence of real numbers such that*

$$a_n = \sum_{k=1}^{\infty} a_{n+k}^2 \quad \text{for } n = 1, 2, \ldots.$$

Show that if the series $\sum_{n=1}^{\infty} a_n$ converges, then $a_n = 0$ for all n.

Solution. We first observe that $0 \leq a_{n+1} \leq a_n$ for all $n \geq 1$. If $\sum_{j=1}^{\infty} a_j$ converges, we take $k \geq 1$ such that $\sum_{j=k+1}^{\infty} a_j < 1$. Then

$$a_{k+1} \leq a_k = \sum_{j=k+1}^{\infty} a_j^2 \leq a_{k+1}\left(\sum_{j=k+1}^{\infty} a_j\right) \leq a_{k+1}.$$

Hence $a_k = a_{k+1}$ and so $a_{k+1} = 0$, implying $a_j = 0$ for $j > k$ since $a_{k+1} = \sum_{j=k+1}^{\infty} a_j^2$. By induction we deduce that $a_j = 0$ for all $j < k+1$. $\quad \square$

2.3.15. *Let $(a_n)_{n \geq 1}$ be an increasing sequence of positive numbers such that $a_n \to \infty$ as $n \to \infty$. Then $\sum_{n=1}^{\infty} (a_{n+1} - a_n)/a_{n+1}$ diverges.*

Solution. We first observe that

$$\sum_{k=1}^{n-1} \frac{a_{k+1} - a_k}{a_{k+1}} \geq \frac{a_n - a_1}{a_n} \to 1 \quad \text{as } n \to \infty.$$

Pick n_1 so large that $(a_{n_1} - a_1)/a_{n_1} > 1/2$. In general, given n_r, pick n_{r+1} so large that $(a_{n_{r+1}} - a_{n_r})/a_{n_{r+1}} > 1/2$, which is possible since $(a_n - a_{n_r})/a_n \to 1$ as $n \to \infty$.

2.3 Convergent and Divergent Series

Now

$$\sum_{k=1}^{n_r-1} (a_{k+1} - a_k)/a_{k+1} = \sum_{k=1}^{n_1-1} (a_{k+1} - a_k)/a_{k+1} + \sum_{k=n_1}^{n_2-1} (a_{k+1} - a_k)/a_{k+1} + \cdots$$
$$+ \sum_{k=n_{r-1}}^{n_r-1} (a_{k+1} - a_k)/a_{k+1}$$
$$\geq (a_{n_1} - a_1)/a_{n_1} + (a_{n_2} - a_{n_1})/a_{n_2} + \cdots + (a_{n_r} - a_{n_{r-1}})/a_{n_r}$$
$$> r/2 \to \infty \text{ as } r \to \infty.$$

Thus, since each term of $\sum_{k=1}^{\infty} (a_{k+1} - a_k)/a_{k+1}$ is positive, the series diverges.

The well known Raabe's test for convergence or divergence of a positive series $\sum_{n=1}^{\infty} a_n$ rests on knowing the behavior of an associated sequence

$$\mathscr{R}_n := n\left(1 - \frac{a_{n+1}}{a_n}\right).$$

As established in Knopp [59], the series $\sum_{n=1}^{\infty} a_n$ converges, provided $\liminf_{n \to \infty} \mathscr{R}_n > 1$. If $\mathscr{R}_n \leq 1$ for all sufficiently large n, then the series diverges.

A distinct associated sequence is

$$\hat{\mathscr{R}}_n := n\left(\frac{a_n}{a_{n+1}} - 1\right).$$

If $\liminf_{n \to \infty} \hat{\mathscr{R}}_n > 1$, then the series $\sum_{n=1}^{\infty} a_n$ converges. If $\hat{\mathscr{R}}_n \leq 1$ for all sufficiently large n, then the series diverges.

These two versions of Raabe's test are not equivalent. Indeed, consider the series

$$\sum_{n=1}^{\infty} \frac{(n-1)! \, n! \, 4^n}{(2n)! \sqrt{n}}.$$

Since

$$\frac{a_n}{a_{n+1}} = \left(1 + \frac{1}{2n}\right)\left(1 + \frac{1}{n}\right)^{1/2} = \left(1 + \frac{1}{2n}\right)\left(1 + \frac{\frac{1}{2}}{n} - \frac{\frac{1}{8}}{n^2} + o\left(\frac{1}{n^2}\right)\right),$$

we have

$$\hat{\mathscr{R}}_n = 1 + \frac{\frac{1}{8}}{n} + o\left(\frac{1}{n^2}\right) \to 1^+ \quad \text{as } n \to \infty,$$

which means that the test is inconclusive. On the other hand,

$$\frac{a_{n+1}}{a_n} = \left(1 - \frac{1}{2n+1}\right)\left(1 - \frac{1}{n+1}\right)^{1/2} = 1 - \frac{1}{n} + \frac{\frac{3}{8}}{n^2} + o\left(\frac{1}{n^2}\right).$$

Therefore

$$\mathscr{R}_n = 1 - \frac{\frac{3}{8}}{n} + o\left(\frac{1}{n^2}\right) \to 1^- \quad \text{as } n \to \infty,$$

which means that the series diverges. However, removing the first term of the above investigated series, we obtain

$$\sum_{n=2}^{\infty} \frac{(n-1)!n!4^n}{(2n)!\sqrt{n}} = \sum_{n=1}^{\infty} \frac{n!(n+1)!4^{n+1}}{(2n+2)!\sqrt{n+1}}.$$

For this last series we find that

$$\frac{a_n}{a_{n+1}} = \left(1 + \frac{\frac{1}{2}}{n+1}\right)\left(1 + \frac{1}{n+1}\right)^{1/2} = 1 + \frac{1}{n+1} + \frac{\frac{1}{8}}{(n+1)^2} + o\left(\frac{1}{n^2}\right),$$

and hence

$$\hat{\mathscr{R}}_n = 1 - \frac{\frac{7}{8}}{n} + o\left(\frac{1}{n^2}\right) \to 1^- \quad \text{as } n \to \infty.$$

We conclude the divergence of the series by the second form of Raabe's test.

As in [95], we introduce the following "parametrized" associated sequence

$$\hat{\mathscr{R}}_n^{(k)} := (n-k)\left(\frac{a_n}{a_{n+1}} - 1\right),$$

where k runs over all nonnegative integers. □

2.3.16. Let $\sum_{n=1}^{\infty} a_n$ be a series of positive terms.

(i) If $\hat{\mathscr{R}}_n^{(k)} \leq 1$ for a nonnegative integer k and for all sufficiently large indices n, then the series diverges.

(ii) If

$$n\left(\frac{a_n}{a_{n+1}} - 1\right) = 1 + O\left(\frac{1}{n}\right),$$

then the series diverges.

Solution. (i) Indeed, the inequality $\hat{\mathscr{R}}_n^{(k)} \leq 1$ implies

$$\frac{a_{n+1}}{a_n} \geq \frac{\frac{1}{n+1-k}}{\frac{1}{n-k}}$$

for all sufficiently large n. Thus since the series $\sum_{n>k} 1/n - k$ diverges, by the ratio comparison test the series $\sum a_n$ diverges as well.

(ii) The equality implies that

$$\frac{a_n}{a_{n+1}} - 1 \leq \frac{1}{n} + \frac{M}{n^2}$$

for some positive integer M and for all sufficiently large n.

2.3 Convergent and Divergent Series

Thus
$$\hat{\mathscr{R}}_n^{(M+1)} = (n-M-1)\left(\frac{1}{n}+\frac{M}{n^2}\right) = 1 - \frac{1}{n} + O\left(\frac{1}{n^2}\right) \to 1^- \quad \text{as } n\to\infty,$$

and hence, by (i), the series $\sum_{n=1}^{\infty} a_n$ diverges. □

2.3.17. Assume that $a_n > 0$ for each n, and that $\sum_{n=1}^{\infty} a_n$ converges. Prove that the series $\sum_{n=1}^{\infty} a_n^{(n-1)/n}$ converges as well.

Solution. Applying the AM–GM inequality, we have, for any $n \geq 2$,
$$a_n^{(n-1)/n} = (a_n^{1/2} a_n^{1/2} \cdot a_n^{n-2})^{\frac{1}{n}} \leq \frac{2\sqrt{a_n} + (n-2)a_n}{n}.$$

But $2\sqrt{a_n}/n \leq 1/n^2 + a_n$ (because $2xy \leq x^2 + y^2$), and $(n-2)a_n/n \leq a_n$ (because $(n-2)/n \leq 1$).

Therefore, $0 < a_n^{(n-1)/n} \leq 1/n^2 + 2a_n$, for each $n \geq 1$. Finally, the comparison test shows that $\sum_{n=1}^{\infty} a_n^{(n-1)/n}$ converges, since $\sum_{n=1}^{\infty} 1/n^2 + 2\sum_{n=1}^{\infty} a_n$ clearly converges.

An alternative solution is based on the observation that each term a_n satisfies either the inequality $0 < a_n \leq 1/2^n$ or $1/2^n < a_n$. In the first case, $a_n^{(n-1)/n} \leq 1/2^{n-1}$. In the second one,
$$a_n^{(n-1)/n} = \frac{a_n}{a_n^{1/n}} \leq 2a_n.$$

Therefore, in both cases,
$$0 < a_n^{(n-1)/n} \leq \frac{1}{2^n} + 2a_n.$$

The conclusion is now immediate since $\sum_{n=1}^{\infty} 1/2^n$ converges, and so does $\sum_{n=1}^{\infty} 2a_n$.

In the next exercise we give a full description of Bertrand's series. □

2.3.18. Let α and β be real numbers and consider the Bertrand series
$$\sum_{n=2}^{\infty} \frac{1}{n^\alpha (\ln n)^\beta}.$$

Prove the following properties:
(i) *the Bertrand series converges if and only if either $\alpha > 1$ or $\alpha = 1$ and $\beta > 1$;*
(ii) *the Bertrand series diverges if and only if either $\alpha < 1$ or $\alpha = 1$ and $\beta \leq 1$.*

Solution. The easiest cases correspond to $\alpha > 1$ and $\alpha < 1$. We postpone to Chapter 10 the proof in the other cases.

We first assume that $\alpha > 1$. Set $\gamma = (1+\alpha)/2$. For every $n \geq 2$ we have
$$n^\gamma u_n = n^{(1-\alpha)/2} (\ln n)^{-\beta}.$$

But $\lim_{n\to\infty} n^{\frac{1-\alpha}{2}} (\ln n)^{-\beta} = 0$, since $(1-\alpha)/2 < 0$. Hence for any $\varepsilon > 0$, there exists a positive integer N such that $n^\gamma u_n \leq \varepsilon$, for any $n \geq N$. In particular, for $\varepsilon = 1$, we deduce that $u_n \leq n^\gamma$, provided n is large enough. Since $\gamma > 1$, the series $\sum_{n=1}^\infty 1/n^\gamma$ converges, and using the comparison test we deduce that the series $\sum_{n=1}^\infty u_n$ converges as well.

In the case $\beta > 0$, using the fact that the mapping $t \longmapsto 1/t^\beta$ is decreasing, we can conclude the convergence of the above series with the following elementary arguments. For every $n \geq 2$ we have

$$n^\alpha u_n = \frac{1}{(\ln n)^\beta} \leq \frac{1}{(\ln 2)^\beta}.$$

We have

$$u_n \leq \frac{M}{n^\alpha},$$

where $M = (\ln 2)^{-\beta}$. But the series $\sum_{n=1}^\infty 1/n^\alpha$ converges because $\alpha > 1$. Using again the comparison test we conclude the convergence of the series having the general term u_n.

Next, we assume that $\alpha < 1$. For any $n \geq 2$ we have

$$nu_n = \frac{n^{1-\alpha}}{(\ln n)^\beta}.$$

But $1 - \alpha > 0$; hence

$$\lim_{n \to \infty} \frac{n^{1-\alpha}}{(\ln n)^\beta} = \infty.$$

Thus, there exist $M > 0$ and $N \in \mathbb{N}$ such that for all $n \in \mathbb{N}$,

$$n \geq N \Rightarrow nu_n \geq M,$$

that is, $u_n \geq M/n$ for any $n \geq N$. Since the harmonic series diverges, the comparison test implies that the series $\sum_{n=1}^\infty u_n$ also diverges.

We can illuminate the relationship between the root and ratio tests with the following simple calculation:

$$\lim_{n \to \infty} |a_n|^{1/n} = \lim_{n \to \infty} \left| \frac{a_n}{a_{n-1}} \cdot \frac{a_{n-1}}{a_{n-2}} \cdots \frac{a_2}{a_1} \cdot a_1 \right|^{1/n} = \lim_{n \to \infty} \left| \frac{a_n}{a_{n-1}} \cdots \frac{a_2}{a_1} \right|^{1/n}.$$

The right-hand side is the limit of the geometric means of the first n consecutive ratios of the series. In other words, while the ratio test depends on the behavior (in the limit) of each consecutive ratio, the root test considers only the average behavior of these ratios. Clearly, if all the consecutive ratios get small, then their average value will get small as well. The converse is false, which is why the root test is stronger. Thus, for example, the ratio test fails on the rearranged geometric series

$$\frac{1}{2} + 1 + \frac{1}{8} + \frac{1}{4} + \frac{1}{32} + \frac{1}{16} + \cdots, \tag{2.1}$$

2.3 Convergent and Divergent Series

since the consecutive ratios alternate in value between 2 and $1/8$. However, the geometric mean of the first $2n$ consecutive ratios is

$$\left| 2^n \cdot \frac{1}{8^n} \right|^{1/2n} = \frac{1}{2},$$

so the root test shows that the series converges. □

The following convergence test may be viewed as an *arithmetic mean test*.

2.3.19. Assume that $(a_n)_{n \geq 1}$ is a sequence of positive numbers such that

$$\lim_{n \to \infty} \frac{1}{n} \left(\frac{a_n}{a_{n-1}} + \frac{a_{n-1}}{a_{n-2}} + \cdots + \frac{a_2}{a_1} \right) < 1.$$

Prove that the series $\sum_{n=1}^{\infty} a_n$ converges.

Solution. By the arithmetic–geometric means inequality, the arithmetic mean of n consecutive ratios of a series is always larger than their geometric mean. So, by the root test, the series $\sum_{n=1}^{\infty} a_n$ converges.

The arithmetic mean test is stronger than the ratio test but weaker than the root test. However, in some cases it may be easier to compute the arithmetic mean of the consecutive ratios than it is to compute their geometric mean. For example, consider the series $\sum_{n=1}^{\infty} a_n$, where the a_n's are defined inductively by

$$a_0 = 1, \quad a_n = a_{n-1} \cdot \frac{\log(1 + 1/(n+1))}{\log(n+1)\log(n+2)}, \quad \forall n \geq 1.$$

Then

$$\lim_{n \to \infty} \left| \frac{a_{n+1}}{a_n} \right| = 1,$$

so the ratio test fails. To apply the root test, we need to evaluate

$$\lim_{n \to \infty} \left| \prod_{k=1}^{n} \frac{\log(1 + 1/(k+1))}{\log(k+1)\log(k+2)} \right|^{1/n}.$$

The arithmetic mean test, however, requires only the following calculation:

$$\lim_{n \to \infty} \frac{1}{n} \sum_{k=1}^{n} \left| \frac{\log(1 + 1/(k+1))}{\log(k+1)\log(k+2)} \right| = \lim_{n \to \infty} \frac{1}{n} \sum_{k=1}^{n} \left(\frac{1}{\log(k+1)} - \frac{1}{\log(k+2)} \right)$$

$$= \lim_{n \to \infty} \frac{1}{n} \left(\frac{1}{\log 2} - \frac{1}{\log(n+2)} \right)$$

$$= 0.$$

Hence the series converges. □

An example of a series for which the arithmetic mean test fails is series (2.1) above. In this case the arithmetic mean of the the consecutive ratios converges to $17/16$. □

2.3.20. Suppose that the series $\sum_{n=1}^{\infty} a_n$ diverges and $\lambda_n \to +\infty$. Does the series $\sum_{n=1}^{\infty} \lambda_n a_n$ always diverge? What happens if $(\lambda_n)_{n \geq 1}$ is an unbounded increasing sequence?

Solution. Consider the counterexample $\lambda_n = (1 + (-1)^n (\ln n)/\sqrt{n})^{-1} \ln n$, $a_n \lambda_n = (-1)^n/\sqrt{n}$.

Assuming now that $(\lambda_n)_{n \geq 1}$ is an increasing sequence of positive numbers such that $\lambda_n \to +\infty$ as $n \to \infty$, then the series $\sum_{n=1}^{\infty} \lambda_n a_n$ diverges, since otherwise the series $\sum_{n=1}^{\infty} a_n$ would converge by the Dirichlet test. □

2.3.21. Suppose that the series $\sum_{n=1}^{\infty} a_n$ converges. Can the series $\sum_{n=1}^{\infty} a_n^3$ diverge? Can the series $\sum_{n=1}^{\infty} a_n^p$ diverge for any $p = 2, 3, \ldots$?

Solution. In the first case, take $a_n = \cos(2n\pi/3)/\ln(1+n)$. In the more general framework, consider the sequence $(a_n)_{n \geq 5}$ defined by $a_{3n} = 2/\ln n$, $a_{3n-1} = a_{3n+1} = -1/\ln n$, for all $n \geq 2$.

We know that the series $\sum_{n=1}^{\infty} 1/n^p$ converges if and only if $p > 1$. Suppose that we sum the series for all n that can be written without using the numeral 9. Suppose, for example, that the key for 9 on the typewriter is broken and we type all the natural numbers we can. Call the summation $\sum'$. □

2.3.22. Prove that the series $\sum' 1/n^p$ converges if and only if $p > \log_{10} 9$.

Solution. There are

$$(9-1) \text{ terms with } \frac{1}{10^p} < \frac{1}{n^p} \leq \frac{1}{1^p},$$

$$(9^2 - 9) \text{ terms with } \frac{1}{100^p} < \frac{1}{n^p} \leq \frac{1}{10^p},$$

$$(9^k - 9^{k-1}) \text{ terms with } \frac{1}{10^{kp}} < \frac{1}{n^p} \leq \frac{1}{10^{(k-1)p}}.$$

Then

$$\frac{8}{10^p} + \frac{8 \cdot 9}{10^{2p}} + \frac{8 \cdot 9^2}{10^{3p}} + \cdots < \sum' \frac{1}{n^p} < \frac{8}{1^p} + \frac{8 \cdot 9}{10^p} + \frac{8 \cdot 9^2}{10^{2p}} + \cdots.$$

The ratio of both geometric series is $9/10^p$. Then the series converges if and only if $9/10^p < 1$, that is, if $p > \log_{10} 9$. □

2.3.23. Suppose that the series of positive real numbers $\sum_{n=1}^{\infty} a_n$ converges. Prove that the series $\sum_{n=1}^{\infty} a_n^{\log n/(1+\log n)}$ also converges.

Solution. Define $I = \left\{ n;\ a_n^{\log n/(1+\log n)} \leq e^2 a_n \right\}$ and $J = \mathbb{N} \setminus I$. If $n \in J$, then $a_n^{\log n} > (en)^2 a_n^{1+\log n}$, so $a_n < (en)^{-2}$. Therefore

$$\sum_{n=1}^{\infty} a_n^{\log n/(1+\log n)} \leq \sum_{n \in I} e^2 a_n + \sum_{n \in J} (en)^{-2} \leq e^2 \sum_{n=1}^{\infty} a_n + e^{-2} \sum_{n=1}^{\infty} n^{-2} < +\infty.$$

2.3 Convergent and Divergent Series

We give in what follows an elementary proof (see [49]) of the identity

$$\frac{\pi^2}{6} = 1 + \frac{1}{2^2} + \frac{1}{3^2} + \cdots + \frac{1}{n^2} + \cdots. \quad \square$$

2.3.24. Prove that

$$\sum_{n=1}^{\infty} \frac{1}{n^2} = \frac{\pi^2}{6}.$$

Solution. The main idea is to apply repeatedly the following identity:

$$\frac{1}{\sin^2 x} = \frac{1}{4\sin^2 \frac{x}{2} \cos^2 \frac{x}{2}} = \frac{1}{4}\left[\frac{1}{\sin^2 \frac{x}{2}} + \frac{1}{\cos^2 \frac{x}{2}}\right] = \frac{1}{4}\left[\frac{1}{\sin^2 \frac{x}{2}} + \frac{1}{\sin^2 \frac{\pi+x}{2}}\right].$$

Hence

$$1 = \frac{1}{\sin^2 \frac{\pi}{2}} = \frac{1}{4}\left[\frac{1}{\sin^2 \frac{\pi}{4}} + \frac{1}{\sin^2 \frac{3\pi}{4}}\right]$$

$$= \frac{1}{16}\left[\frac{1}{\sin^2 \frac{\pi}{8}} + \frac{1}{\sin^2 \frac{3\pi}{8}} + \frac{1}{\sin^2 \frac{5\pi}{8}} + \frac{1}{\sin^2 \frac{7\pi}{8}}\right] = \cdots \quad (2.2)$$

$$= \frac{1}{4^n}\sum_{k=0}^{2^n-1} \frac{1}{\sin^2 \frac{(2k+1)\pi}{2^{n+1}}} = \frac{2}{4^n}\sum_{k=0}^{2^{n-1}-1} \frac{1}{\sin^2 \frac{(2k+1)\pi}{2^{n+1}}}.$$

Taking the termwise limit as $n \to \infty$ and using $\lim_{N\to\infty} N\sin(x/N) = x$ for $N = 2^n$ and $x = (2k+1)\pi/2$ yields

$$1 = \frac{8}{\pi^2}\sum_{k=0}^{\infty} \frac{1}{(2k+1)^2}. \quad (2.3)$$

However, taking the limit termwise requires some care, as the example

$$1 = \frac{1}{2} + \frac{1}{2} = \frac{1}{4} + \frac{1}{4} + \frac{1}{4} + \frac{1}{4} = \cdots \to 0 + 0 + 0 + \cdots$$

shows. In our case, the implication (2.2) $\Rightarrow$ (2.3) is justified because the kth term in the sum (2.2) is bounded by $2/(2k+1)^2$ (independently of n), since $\sin x > 2x/\pi$ holds for $0 < x < \pi/2$.

To conclude the proof we observe that

$$S := \sum_{n=1}^{\infty} \frac{1}{n^2} = \sum_{n=\text{odd}} \frac{1}{n^2} + \sum_{n=\text{even}} \frac{1}{n^2} = \frac{\pi^2}{8} + \frac{1}{4}\sum_{k=1}^{\infty} \frac{1}{k^2} = \frac{\pi^2}{8} + \frac{S}{4},$$

which yields $S = \pi^2/6$. $\square$

2.4 Infinite Products

> I recognize the lion by his paw. [After reading an anonymous solution to a problem that he realized was Newton's solution.]
>
> Jakob Bernoulli (1654–1705)

If a real sequence $(x_n)_{n\in\mathbb{N}}$ satisfies the Coriolis conditions, then necessarily not only $\prod_{n=1}^{\infty}(1+x_n)$, but also $\prod_{n=1}^{\infty}(1+cx_n)$ for every $c \in \mathbb{R}$ is convergent. It turns out that this, or even the convergence of $\prod_{n=1}^{\infty}(1+c_1 x_n)$, $\prod_{n=1}^{\infty}(1+c_2 x_n)$ for any two different nonzero real numbers c_1, c_2, is equivalent to the Coriolis conditions.

There remains a "pathological" special case of convergence of a real product that is characterized by the following properties: the product $\prod_{n=1}^{\infty}(1+x_n)$ converges, while the series $\sum_{n=1}^{\infty} x_n = \sum x_n^2$ diverges. In this case the balance of factors is destroyed by any scaling of the deviations from unity: $\prod_{n=1}^{\infty}(1+cx_n)$ diverges for $c \in \mathbb{R} \setminus \{0, 1\}$.

The following exercise summarizes properties of this type.

2.4.1. Let $(x_n)_{n \geq 1}$ be a sequence of real numbers.

(i) If any two of the four expressions

$$\prod_{n=1}^{\infty}(1+x_n), \quad \prod_{n=1}^{\infty}(1-x_n), \quad \sum_{n=1}^{\infty} x_n, \quad \sum_{n=1}^{\infty} x_n^2$$

are convergent, then this holds also for the remaining two.

(ii) If $\sum_{n=1}^{\infty} x_n$ is convergent and $\sum_{n=1}^{\infty} x_n^2$ is divergent, then $\prod_{n=1}^{\infty}(1+x_n)$ diverges to zero.

(iii) If $\sum_{n=1}^{\infty} x_n^2$ is convergent and $\sum_{n=1}^{\infty} x_n$ is divergent, then

$$\prod_{k=1}^{n}(1+x_k)/\exp\left(\sum_{k=1}^{n} x_k\right)$$

tends to a finite limit for $n \to \infty$.

(iv) If $\prod_{n=1}^{\infty}(1+x_n)$ is convergent and $\sum_{n=1}^{\infty} x_n^2$ is divergent, then $\sum_{n=1}^{\infty} x_n = \infty$.

(v) If $\prod_{n=1}^{\infty}(1+cx_n)$ is convergent for two different values $c \in \mathbb{R} \setminus \{0\}$, then the product is convergent for every $c \in \mathbb{R}$.

Solution. (i) If any two of the four expressions are convergent, then there is a positive integer n_0 such that $|x_n| \leq 1/2$ for $n \geq n_0$. Thus, for $n_0 \leq n_1 \leq n_2$,

$$\sum_{n=n_1}^{n_2} \log(1+x_n) = \sum_{n=n_1}^{n_2}\left(x_n - \frac{v_n}{2}x_n^2\right) = \sum_{n=n_1}^{n_2} x_n - \frac{\tilde{v}}{2}\sum_{n=n_1}^{n_2} x_n^2, \quad (2.4)$$

where $v_n, \tilde{v} \in \left(\frac{4}{9}, 4\right)$ by Taylor's formula.

2.4 Infinite Products

Thus, by Cauchy's criterion for series, if any two of the three expressions $\prod_{n=1}^{\infty}(1+x_n)$, $\sum_{n=1}^{\infty} x_n$, $\sum_{n=1}^{\infty} x_n^2$ are convergent, then so is the third one.

Since the convergence of $\sum_{n=1}^{\infty} x_n$ and $\sum_{n=1}^{\infty} x_n^2$ is not affected by changing the sign of each x_n, there remains only to be shown that the convergence of $\prod_{n=1}^{\infty}(1+x_n)$ and $\prod_{n=1}^{\infty}(1-x_n)$ implies that of $\sum_{n=1}^{\infty} x_n^2$.

But if $\prod_{n=1}^{\infty}(1+x_n)$ and $\prod_{n=1}^{\infty}(1-x_n)$ are convergent, then so is the product

$$\prod_{n=1}^{\infty}(1+x_n)(1-x_n) = \prod_{n=1}^{\infty}(1-x_n^2),$$

and hence $\sum_{n=1}^{\infty} x_n^2$ converges.

The assertions (ii), (iii), and (iv) follow directly from relation (2.4).

(v) Without loss of generality we assume that $\prod_{n=1}^{\infty}(1+x_n)$ and $\prod_{n=1}^{\infty}(1+c_0 x_n)$ are convergent, where $c_0 \in \mathbb{R} \setminus \{0,1\}$. Then with $|x_n|, |c_0 x_n| < 1$ for $n \geq n_0$,

$$\left(\prod_{n=n_0}^{\infty}(1+x_n)\right)^{c_0} = \prod_{n=n_0}^{\infty}(1+x_n)^{c_0}.$$

Thus, the infinite product

$$\prod_{n=n_0}^{\infty} \frac{(1+x_n)^{c_0}}{1+c_0 x_n}$$

converges. But

$$\frac{(1+x_n)^{c_0}}{1+c_0 x_n} = 1 + x_n^2 \cdot \frac{c_0(c_0-1)}{2}(1+\varepsilon_n) \quad (n \geq n_0)$$

as $\varepsilon_n \to 0$ for $n \to \infty$. This implies the convergence of $\sum_{n=1}^{\infty} x_n^2$, and hence by (i), the conclusion follows.

Instructive examples are furnished by the products

$$\prod_{n=1}^{\infty}\left(1 + c \cdot \binom{\alpha}{n}\right) \quad \text{for } \alpha, c \in \mathbb{R}, \tag{2.5}$$

and

$$\prod_{n=1}^{\infty}\left(1 + c \cdot \frac{\alpha^{2n}}{\sqrt{n}-1/2}\right)\left(1 - c \cdot \frac{\alpha^{2n+1}}{\sqrt{n}+1/2}\right) \quad \text{for } \alpha, c \in \mathbb{R}. \tag{2.6}$$

In order to discuss (2.5), the convergence properties of the series

$$\sum_{n=1}^{\infty}\binom{\alpha}{n} \quad \text{and} \quad \sum_{n=1}^{\infty}\binom{\alpha}{n}^2$$

need to be investigated. To do this it is most convenient to use the Gauss test.

Using this criterion and the fact that $|\binom{\alpha}{n}| \searrow 0$ as $n \to \infty$ for $-1 < \alpha < 0$, whereas $|\binom{\alpha}{n}| \to \infty$ as $n \to \infty$ for $\alpha < -1$, we deduce the following properties:

- if $\alpha \geq 0$, then the product (2.5) is absolutely convergent for every $c \in \mathbb{R}$;
- if $-1/2 < \alpha < 0$, then the product (2.5) is conditionally convergent for $c \in \mathbb{R} \setminus \{0\}$;
- if $-1 < \alpha \leq -1/2$, then the product (2.5) diverges to zero for $c \in \mathbb{R} \setminus \{0\}$;
- if $\alpha = -1$, then the product (2.5) diverges to zero for $0 < |c| < \sqrt{2}$, and is indefinitely divergent for $|c| \geq \sqrt{2}$;
- if $\alpha < -1$, then the product (2.5) is indefinitely divergent for $c \in \mathbb{R} \setminus \{0\}$. We now discuss the infinite product (2.6).
- If $|\alpha| < 1$, then the product (2.6) obviously is absolutely convergent;
- if $|\alpha| > 1$, then the product (2.6) is divergent for $c \in \mathbb{R} \setminus \{0\}$;
- if $\alpha = 1$, then the product (2.6) is convergent for $c = 1$, since

$$\left(1 + \frac{1}{\sqrt{n} - 1/2}\right)\left(1 - \frac{1}{\sqrt{n} + 1/2}\right) = 1$$

for all $n \in \mathbb{N}$. Thus, from $\sum x_n^2 = \infty$ and the above exercise, we conclude that for every $c \in \mathbb{R} \setminus \{0, 1\}$, the infinite product (2.6) is divergent. □

2.4.2. Establish the following formula of Viète:

$$\frac{2}{\pi} = \sqrt{\frac{1}{2}} \cdot \sqrt{\frac{1}{2} + \frac{1}{2}\sqrt{\frac{1}{2}}} \cdot \sqrt{\frac{1}{2} + \frac{1}{2}\sqrt{\frac{1}{2} + \frac{1}{2}\sqrt{\frac{1}{2}}}} \cdots.$$

Solution. Let $t \neq 0$. We commence by showing that

$$\lim_{n \to \infty} \cos \frac{t}{2} \cos \frac{t}{2^2} \cdots \cos \frac{t}{2^n} = \frac{\sin t}{t}. \tag{2.7}$$

Indeed,

$$\sin t = 2^n \cos \frac{t}{2} \cos \frac{t}{2^2} \cdots \cos \frac{t}{2^n} \sin \frac{t}{2^n}.$$

Hence

$$\cos \frac{t}{2} \cos \frac{t}{2^2} \cdots \cos \frac{t}{2^n} = \frac{\sin t}{2^n \sin t/2^n} = \frac{\sin t}{t} \cdot \frac{t/2^n}{\sin t/2^n}.$$

But

$$\frac{\sin t/2^n}{t/2^n} \to 1 \quad \text{as } n \to \infty.$$

From (2.7) we obtain, for $t = \pi/2$,

$$\frac{2}{\pi} = \lim_{n \to \infty} \cos \frac{\pi}{4} \cos \frac{\pi}{8} \cdots \cos \frac{\pi}{2^{n+1}}.$$

Since

$$\cos \frac{\pi}{4} = \frac{1}{\sqrt{2}} \quad \text{and} \quad \cos \frac{\theta}{2} = \sqrt{\frac{1}{2} + \frac{1}{2}\cos \theta},$$

our conclusion follows. □

2.5 Qualitative Results

> We are servants rather than masters in mathematics.
>
> ―――――――
> Charles Hermite (1822–1901)

A direct application of the definition implies that for any *convergent* series $\sum_{n=1}^{\infty} a_n$ of real numbers,

$$\liminf_{n \to \infty} na_n \leq 0 \leq \limsup_{n \to \infty} na_n.$$

This means that to prove $\lim_{n \to \infty} na_n = 0$, it is sufficient to show that the sequence $(na_n)_{n \geq 1}$ is convergent. We apply this result to obtain in what follows a property that is an extension of a very old result of Olivier ([85], 1827).

2.5.1. *Let $\sum_{n=1}^{\infty} a_n$ be a convergent series of real numbers, and suppose that there exists a real constant α such that for every nonnegative integer n (we define $a_0 = 0$),*
$$na_n \geq (n + \alpha) a_{n-1}.$$
Prove that the sequence $(na_n)_{n \geq 1}$ converges to 0.

Solution. Define, for any positive integer n, $r_n = na_n$ and $s_n = -(1+\alpha)\sum_{k=1}^{n-1} a_k$. Set $b_n = r_n + s_n$. We can rewrite the hypothesis as

$$na_n \geq (n-1)a_{n-1} + (1+\alpha)a_{n-1}, \quad \text{for all } n \geq 2.$$

Thus, for all integer $n \geq 2$, we have $b_n \geq b_{n-1}$, and so the sequence $(b_n)_{n \geq 1}$ is nondecreasing. Using the convergence of the series $\sum_{n=1}^{\infty} a_n$ combined with the fact that $r_n = b_n - s_n$, we deduce that the sequence $(r_n)_{n \geq 1}$ is convergent. □

2.5.2. *(Cesàro's lemma) Consider a sequence of positive numbers $(a_n)_{n \geq 1}$. Then the following assertions are equivalent.*

(i) *The series $\sum_{n \geq 1} a_n$ diverges.*
(ii) *For any sequence of real numbers $(b_n)_{n \geq 1}$ that admits a limit ℓ in $\overline{\mathbb{R}}$, the sequence $(a_1 b_1 + \cdots + a_n b_n / a_1 + \cdots + a_n)_{n \geq 1}$ tends to ℓ, too.*

Solution. (i) $\Longrightarrow$ (ii) Let us first assume that $\ell \in \mathbb{R}$. Fix $\varepsilon > 0$. Then there exists a positive integer $N = N_\varepsilon$ such that for any $n > N$, $|b_n - \ell| \leq \varepsilon$. For any $n > N$ we set $c_n := \left| \frac{a_1 b_1 + \cdots + a_n b_n}{a_1 + \cdots + a_n} - \ell \right|$. Thus, using the convergence of $(b_n)_{n \geq 1}$ combined with our hypothesis (i),

$$c_n \leq \frac{1}{a_1 + \cdots + a_n} \sum_{k=1}^{N} |a_k(b_k - \ell)| + \frac{1}{a_1 + \cdots + a_n} \sum_{k=N+1}^{n} a_k |b_k - \ell|$$

$$\leq \frac{C}{a_1 + \cdots + a_n} + \frac{a_{N+1} + \cdots + a_n}{a_1 + \cdots + a_n} \varepsilon \leq 2\varepsilon,$$

provided that $n > N_0 \geq N$ and C does not depend on n.

If $\ell = +\infty$ we use a similar argument, based on the appropriate definition.

(ii) $\implies$ (i) Let $L := \sum_{n=1}^{\infty} a_n \in (0, +\infty]$. We argue by contradiction and assume that $L \in \mathbb{R}$. Consider the sequence $(b_n)_{n \geq 1}$ defined by $b_1 = 1$ and $b_n = 0$, for all $n \geq 2$. Then, by (ii), we obtain $0 = 1/L$, a contradiction. This implies that $L = +\infty$, so the series $\sum_{n \geq 1} a_n$ diverges. $\square$

The following elementary result has a crucial role in the study of the Ishikawa iteration process with errors for nonlinear strongly accretive operators in Banach spaces.

2.5.3. Let $(a_n)_{n \geq 0}$, $(b_n)_{n \geq 0}$, and $(c_n)_{n \geq 0}$ be sequences of nonnegative numbers satisfying, for all integers $n \geq 0$,

$$a_{n+1} \leq (1 - \lambda_n) a_n + b_n + c_n$$

with $\lambda_n \in [0, 1]$, $\sum_{n=0}^{\infty} \lambda_n = \infty$, $b_n = o(\lambda_n)$ as $n \to \infty$, and $\sum_{n=0}^{\infty} c_n < \infty$. Prove that the sequence $(a_n)_{n \geq 0}$ converges to zero.

Solution. Since $b_n = o(\lambda_n)$ as $n \to \infty$, there exists a sequence $(d_n)_{n \geq 0}$ such that $b_n = \lambda_n d_n$. Our inequality in the hypothesis enables us to prove by induction that for all integers $0 \leq k < n$,

$$0 \leq a_{n+1} \leq a_k \prod_{j=k}^{n}(1-\lambda_j) + \sum_{j=k}^{n}\left[\lambda_j \prod_{i=j+1}^{n}(1-\lambda_i)\right] d_j + \sum_{j=k}^{n} c_j \prod_{i=j+1}^{n}(1-\lambda_i). \quad (2.8)$$

Using the inequality $\ln(1+x) \leq x$ for all $x > -1$, we obtain

$$\prod_{j=k}^{n}(1-\lambda_j) \leq e^{-\sum_{j=k}^{n}\lambda_j} \longrightarrow 0 \quad \text{as } n \to \infty.$$

Next, for any integer $0 \leq k < n$,

$$\sum_{j=k}^{n} \lambda_j \prod_{i=j+1}^{n}(1-\lambda_i) = 1 - (1-\lambda_n)(1-\lambda_{n-1})\cdots(1-\lambda_k) \leq 1.$$

Fix $\varepsilon > 0$. Since $\lim_{n \to \infty} d_n = 0$ and $\sum_{n=0}^{\infty} c_n < \infty$, there exists a natural number $N_0 = N_0(\varepsilon)$ such that $d_j < \varepsilon$ for all $j \geq N_0$ and $\sum_{j=N_0}^{\infty} c_j < 2\varepsilon$. Thus, by (2.8),

$$0 \leq \liminf_{n \to \infty} a_n \leq \limsup_{n \to \infty} a_n \leq 2\varepsilon.$$

Letting $\varepsilon \to 0$, we obtain $\lim_{n \to \infty} a_n = 0$. $\square$

2.5.4. Suppose $\sum_{n=1}^{\infty} a_n$ converges. Do the following series have to converge as well?

(i) $a_1 + a_2 + a_4 + a_3 + a_8 + a_7 + a_6 + a_5 + a_{16} + a_{15} + \cdots + a_9 + a_{32} + \cdots$;
(ii) $a_1 + a_2 + a_3 + a_4 + a_5 + a_7 + a_6 + a_8 + a_9 + a_{11} + a_{13} + a_{15} + a_{10} + a_{12} + a_{14} + a_{16} + a_{17} + a_{19} + \cdots$.

International Mathematics Competition for University Students, 1997

2.5 Qualitative Results

Solution. (i) Yes. Let $S = \sum_{n=1}^{\infty} a_n$, $S_n = \sum_{k=1}^{n} a_k$. Fix $\varepsilon > 0$ and a positive integer $N = N(\varepsilon)$ such that $|S_n - S| < \varepsilon$ for all $n > N$. The partial sums of the permuted series have the form

$$T_{2^{n-1}+k} = S_{2^{n-1}} + S_{2^n} - S_{2^n-k}, \quad \text{for all } 0 \le k < 2^{n-1}.$$

It follows that if $2^{n-1} > N$ then $|T_{2^{n-1}+k} - S| < 3\varepsilon$, which shows that the permuted series converges.

(ii) No. Indeed, consider $a_n = (-1)^{n-1}/\sqrt{n}$. Then

$$T_{3 \cdot 2^{n-2}} = S_{2^{n-1}} + \sum_{k=2^{n-2}}^{2^{n-1}-1} \frac{1}{\sqrt{2k+1}}$$

and

$$T_{3 \cdot 2^{n-2}} - S_{2^{n-1}} \ge 2^{n-2} \frac{1}{\sqrt{2^n}} \longrightarrow \infty \text{ as } n \to \infty.$$

So, by hypothesis, $T_{3 \cdot 2^{n-2}} \to \infty$.

A nice elementary result, whose proof is based on the *uniform boundedness principle* in functional analysis (see [37]), asserts that if for a given series $\sum_{n=1}^{\infty} b_n^2$ of positive numbers, the series $\sum_{n=1}^{\infty} a_n b_n$ converges for each sequence of real numbers $(a_n)_{n \ge 1}$ such that $\sum_{n=1}^{\infty} a_n^2$ converges, then the series $\sum_{n=1}^{\infty} b_n^2$ converges, too. This property can be also viewed as a kind of reciprocal of the Cauchy–Schwarz inequality. We give a constructive proof of this result in what follows. □

2.5.5. Suppose that the series $\sum_{n=1}^{\infty} b_n^2$ of positive numbers diverges. Prove that there exists a sequence $(a_n)_{n \ge 1}$ of real numbers such that

$$\sum_{n=1}^{\infty} a_n^2 < \infty \quad \text{and} \quad \sum_{n=1}^{\infty} |a_n b_n| = \infty.$$

G. Helmberg

Solution. Suppose first that there exists a subsequence $(n_k)_{k \ge 1}$ of $\mathbb{N}$ for which $|b_{n_k}| \ge 1$ for all $k \ge 1$. Set

$$a_n = \begin{cases} 1/k & \text{if } n = n_k, \\ 0 & \text{otherwise}. \end{cases}$$

Thus, we obtain $\sum_{n=1}^{\infty} a_n^2 = \sum_{k=1}^{\infty} 1/k^p < \infty$ and $\sum_{n=1}^{\infty} |a_n b_n| = \sum_{k=1}^{\infty} |b_{n_k}|/k = \infty$.

Without loss of generality we may therefore suppose that $0 < |b_k| < 1$ for all $k \ge 1$. For any $n \ge 1$ define the sequence $(\beta_{n,k})_{k \ge 1}$ as follows:

$$\beta_{n,k} = \begin{cases} b_k & \text{if } 1 \le k \le n, \\ 0 & \text{otherwise}. \end{cases}$$

Next, fix $\varepsilon \in (0,1)$ and set

$$\alpha_k = \frac{|b_k|}{\left(\sum_{j=1}^k \beta_{k,j}^2\right)^{1-\varepsilon/2}}.$$

Then

$$\sum_{k=1}^\infty |a_k b_k| = \sum_{k=1}^\infty \frac{b_k^2}{\left(\sum_{j=1}^k \beta_{k,j}^2\right)^{1-\varepsilon/2}} \geq \sum_{k=1}^n \frac{b_k^2}{\left(\sum_{j=1}^k \beta_{n,j}^2\right)^{1-\varepsilon/2}} = \left(\sum_{k=1}^n b_k^2\right)^{\varepsilon/2} \longrightarrow \infty$$

as $n \to \infty$.

It remains to show that $\sum_{k=1}^\infty a_k^2 < \infty$. Since $\lim_{n\to\infty} \sum_{k=1}^n b_k^2 = \infty$ and $0 < |b_k| < 1$, it is possible to define an increasing subsequence $(n_j)_{j\geq 1}$ of $\mathbb{N}^*$ by requiring that

$$\sum_{k=1}^{n_j-1} b_k^2 < \ell \leq \sum_{k=1}^{n_j} b_k^2.$$

As a consequence we obtain the following inequalities:

$$\sum_{k=1}^{n_j} b_k^2 < \ell + 1;$$

$$\sum_{k=n_j+1}^{n_{j+1}} b_k^2 = \sum_{k=1}^{n_{j+1}} b_k^2 - \sum_{k=1}^{n_j} b_k^2 < \ell + 2 - \ell = 2;$$

$$\sum_{k=n_j+1}^{n_{j+1}} a_k^2 = \sum_{k=n_j+1}^{n_{j+1}} \frac{b_k^2}{\left(\sum_{i=1}^k \beta_{k,i}^2\right)^{2-\varepsilon}} \leq \sum_{k=n_j+1}^{n_{j+1}} \frac{b_k^2}{\left(\sum_{i=1}^{n_j} \beta_{n_j,i}^2\right)^{2-\varepsilon}} \leq \frac{2}{\left(\sum_{i=1}^{n_j} \beta_{n_j,i}^2\right)^{2-\varepsilon}}.$$

Accordingly,

$$\sum_{k=1}^\infty a_k^2 \leq \sum_{k=1}^{n_1} |a_k| + 2\sum_{i=1}^\infty \frac{1}{\left(\sum_{i=1}^{n_j} \beta_{n_j,i}^2\right)^{2-\varepsilon}} < \frac{2}{b_1^4} + 2\sum_{i=1}^\infty \frac{1}{\ell^{2-\varepsilon}} < \infty. \quad \square$$

The next result asserts, in particular, that if $(a_n)_{n\geq 1}$ is a sequence of positive numbers, then:

◇ if the series $\sum_{n=1}^\infty a_n^{-1}$ converges, then the series $\sum_{n=1}^\infty n(a_1 + a_2 + \cdots + a_n)^{-1}$ converges, too;

◇ if the series $\sum_{n=1}^\infty n(a_1 + a_2 + \cdots + a_n)^{-1}$ diverges, then the series $\sum_{n=1}^\infty a_n^{-1}$ diverges, too.

2.5.6. Prove that for any sequence of positive numbers $(a_n)_{n\geq 1}$ we have

2.5 Qualitative Results

$$\sum_{n=1}^{\infty}\frac{n}{a_1+a_2+\cdots+a_n} \leq 2\sum_{n=1}^{\infty}\frac{1}{a_n}.$$

Putnam Competition, 1964

Solution. We first observe that the Cauchy–Schwarz inequality implies

$$(1+2+\cdots+n)^2 \leq (a_1+a_2+\cdots+a_n)\left(\frac{1^2}{a_1}+\frac{2^2}{a_2}+\cdots+\frac{n^2}{a_n}\right).$$

Hence

$$\frac{2n+1}{a_1+\cdots+a_n} \leq 4\frac{2n+1}{n^2(n+1)^2}\sum_{k=1}^{n}\frac{k^2}{a_k},$$

$$\sum_{n=1}^{\infty}\frac{2n+1}{a_1+\cdots+a_n} \leq 4\sum_{n=1}^{\infty}\frac{2n+1}{n^2(n+1)^2}\sum_{k=1}^{n}\frac{k^2}{a_k} = 4\sum_{k=1}^{n}\frac{k^2}{a_k}\sum_{n=k}^{\infty}\frac{2n+1}{n^2(n+1)^2}.$$

But

$$\sum_{n=k}^{\infty}\frac{2n+1}{n^2(n+1)^2} = \sum_{n=k}^{\infty}\left(\frac{1}{n^2}-\frac{1}{(n+1)^2}\right) = \frac{1}{k^2}.$$

It follows that

$$\frac{2n+1}{a_1+\cdots+a_n} \leq 4\sum_{k=1}^{n}\frac{1}{a_k},$$

and the conclusion follows. □

2.5.7. Let $\sum_{n=1}^{\infty}a_n$ be a convergent series of real numbers. Does the series $\sum_{n=1}^{\infty}a_n|a_n|$ converge, too?

Solution. For any integer $n \geq 0$, define $a_{3n+1} = a_{3n+2} = 1/\sqrt{n+1}$ and $a_{3n+3} = -2/\sqrt{n+1}$. Then the series $\sum_{n=1}^{\infty}a_n$ converges, but $\sum_{n=1}^{\infty}a_n|a_n|$ diverges. □

2.5.8. Let $(a_n)_{n\geq 1}$ be a sequence of real numbers such that $0 < a_n \leq a_{2n}+a_{2n+1}$, for all integers $n \geq 1$. Prove that the series $\sum_{n=1}^{\infty}a_n$ diverges.

Putnam Competition, 1994

Solution. We argue by contradiction and assume that the series converges. Then

$$\sum_{n=1}^{\infty}a_n \leq \sum_{n=1}^{\infty}(a_{2n}+a_{2n+1}) = \sum_{n=2}^{\infty}a_n < \sum_{n=1}^{\infty}a_n,$$

contradiction. □

An immediate consequence of the following problem is that both the harmonic series $\sum_{n=1}^{\infty}(n)^{-1}$ and the Riemann series $\sum_{n=2}^{\infty}(\ln n)^{-1}$ are divergent.

2.5.9. Let $(a_n)_{n\geq 1}$ be a sequence of positive numbers such that $a_n < a_{n+1}+a_{n^2}$, for all $n \geq 1$. Prove that the series $\sum_{n=1}^{\infty}a_n$ is divergent.

Solution. We have: $a_2 < a_3 + a_4 < a_4 + a_9 + a_5 + a_{16} < \cdots$. Let I_k denote the set of positive subscripts that appear at the kth step in the above iteration, that is, $I_1 = \{2\}, I_2 = \{3,4\}, I_3 = \{4,5,9,16\}$, etc. We observe that the length of I_k is less than or equal to 2^{k-1}. Moreover, if $|I_k| = 2^{k-1}$ for all $k \geq 1$, then $\sum_{p \in I_k} a_p > a_2$, which implies that all remainders of $\sum_{n=1}^{\infty} a_n$ are greater than a_2, and so the series $\sum_{n=1}^{\infty} a_n$ diverges.

In order to prove that $|I_k| = 2^{k-1}$ for all $k \geq 1$, we argue by contradiction and denote by m the least positive integer m such that $i + 1 = j^2$, for some $i, j \in I_m$. We observe that the numbers $j^2 - 1, j^2 - 2, \ldots, j^2 - 2j + 2$ are not squares. On the other hand, $I_k = (I_{k-1} + 1) \cup I_{k-1}^2$, where $I_{k-1} + 1 := \{q+1; q \in I_{k-1}\}$ and $I_{k-1}^2 = \{q^2; q \in I_{k-1}\}$. For all $k < m$, the definition of m implies that $(I_{k-1} + 1) \cap I_{k-1}^2 = \emptyset$. Therefore $j^2 - 2 \in I_{m-1}$, $j^2 - 3 \in I_{m-2}$, $\ldots$, $j^2 - 2j + 2 \in I_{m-2j+3}$. So, $m - 2j + 3 \geq 1$, that is, $m \geq 2j - 2$. Since, obviously, $j \geq m + 1$ for all $j \in I_m$, we obtain $m \geq 2j - 2 \geq 2(m+1) - 2 = 2m$, contradiction. $\square$

2.5.10. (a) *Find a sequence of distinct complex numbers $(z_n)_{n \geq 1}$ and a sequence of nonzero real numbers $(\alpha_n)_{n \geq 1}$ such that the series $\sum_{n=1}^{\infty} \alpha_n |z - z_n|^{-1}$ either converges to a positive number or diverges to $+\infty$ for almost all complex numbers z, but not all α_n are positive.*

(b) *Let $(z_n)_{n \geq 1}$ be a sequence of distinct complex numbers. Assume that $\sum_{n=1}^{\infty} \alpha_n$ is an absolutely convergent series of real numbers such that $\sum_{n=1}^{\infty} \alpha_n |z - z_n|^{-1}$ converges to a nonnegative number, for almost all $z \in \mathbb{C}$. Prove that α_n are nonnegative for all $n \geq 1$.*

T.-L. Rădulescu and V. Rădulescu, Amer. Math. Monthly, Problem 11304

Solution. (a) We prove that the series

$$-\frac{1}{|z|} + \frac{1}{|z + \frac{1}{2}|} + \frac{1}{|z - \frac{1}{2}|} + \cdots + \frac{1}{|z + \frac{1}{n}|} + \frac{1}{|z - \frac{1}{n}|} + \cdots \quad (2.9)$$

diverges to $+\infty$ for all $z \in \mathbb{C} \setminus \{0, \pm\frac{1}{2}, \cdots, \pm\frac{1}{n}, \ldots\}$.

Indeed, we first observe that for any fixed $z \in \mathbb{C} \setminus \{0, \pm\frac{1}{2}, \ldots, \pm\frac{1}{n}, \ldots\}$, the above series has the same nature as the series $-1 + 1 + 1 + 1 + \cdots$, which diverges. Next, we observe that

$$-\frac{1}{|z|} + \frac{1}{|z + \frac{1}{n}|} \geq 0 \quad \text{if and only if} \quad \operatorname{Re} z \leq -\frac{1}{2n}$$

and

$$-\frac{1}{|z|} + \frac{1}{|z - \frac{1}{n}|} \geq 0 \quad \text{if and only if} \quad \operatorname{Re} z \geq \frac{1}{2n}.$$

The above relations show that for any $z \in \mathbb{C} \setminus \{0, \pm\frac{1}{2}, \ldots, \pm\frac{1}{n}, \ldots\}$ with $\operatorname{Re} z \neq 0$ there exists $N \in \mathbb{N}$ such that $-\frac{1}{|z|} + \sum_{k=1}^{N} \left(\frac{1}{|z + \frac{1}{k}|} + \frac{1}{|z - \frac{1}{k}|} \right) > 0$. It remains to prove that this is also true if $z = iy$, $y \in \mathbb{R} \setminus \{0\}$. For this purpose

2.5 Qualitative Results

we observe that

$$\frac{1}{|z+\frac{1}{2}|}+\frac{1}{|z-\frac{1}{2}|}+\cdots+\frac{1}{|z+\frac{1}{n}|}+\frac{1}{|z-\frac{1}{n}|}$$
$$=\frac{2}{\sqrt{y^2+\frac{1}{4}}}+\cdots+\frac{2}{\sqrt{y^2+\frac{1}{n^2}}}\geq\frac{2n-2}{\sqrt{y^2+\frac{1}{4}}}\geq\frac{1}{|y|}=\frac{1}{|z|},$$

provided $2|y| \geq (4n^2 - 8n + 3)^{-1}$. In conclusion, the series (2.9) diverges to $+\infty$ for all $z \in \mathbb{C} \setminus \{0, \pm\frac{1}{2}, \ldots, \pm\frac{1}{n}, \ldots\}$.

Another example of series with the above properties is

$$-\frac{1}{|z|}+\sum_{n=2}^{\infty}\frac{1}{n}\left(\frac{1}{|z+\ln n|}+\frac{1}{|z-\ln n|}\right), \quad z \in \mathbb{C} \setminus \{0, \pm\ln 2; \pm\ln 3, \ldots\}.$$

(b) It is sufficient to focus on an arbitrary term of the sequence, say α_1, and to show that $\alpha_1 \geq 0$. We can assume, without loss of generality, that $z_1 = 0$. Fix arbitrarily $\varepsilon \in (0,1)$. Since $\sum_{n=1}^{\infty}|\alpha_n| < \infty$, there exists a positive integer N such that $\sum_{i=N+1}^{\infty}|\alpha_i| < \varepsilon$. Next, we choose $r > 0$ small enough that $|a_i| > r/\varepsilon$, for all $i \in \{2, \ldots, N\}$. Set

$$f(z) = \sum_{n=1}^{\infty}\frac{\alpha_n}{|z-z_n|}.$$

It follows that

$$0 \leq \int_{B_r(0)} f(z)dz = \alpha_1 \int_{B_r(0)} \frac{dz}{|z|} + \sum_{i=2}^{N}\alpha_i \int_{B_r(0)} \frac{dz}{|z-z_i|} + \sum_{i=N+1}^{\infty}\alpha_i \int_{B_r(0)} \frac{dz}{|z-z_i|}$$
$$\leq \alpha_1 \int_{B_r(0)} \frac{dz}{|z|} + \sum_{i=2}^{N}|\alpha_i| \int_{B_r(0)} \frac{dz}{|z-z_i|} + \sum_{i=N+1}^{\infty}|\alpha_i| \int_{B_r(0)} \frac{dz}{|z-z_i|}$$
$$= 2\pi r\alpha_1 + \sum_{i=2}^{N}|\alpha_i| \int_{B_r(0)} \frac{dz}{|z-z_i|} + \sum_{i=N+1}^{\infty}|\alpha_i| \int_{B_r(0)} \frac{dz}{|z-z_i|}$$
$$\leq 2\pi r\alpha_1 + \sum_{i=2}^{N}|\alpha_i| \int_{B_r(0)} \frac{dz}{|z-z_i|} + \varepsilon \sup_{i \geq N+1} \int_{B_r(0)} \frac{dz}{|z-z_i|}. \quad (2.10)$$

For every $i \in \{2, \ldots, N\}$ we have $|z - z_i| \geq |z_i| - |z| \geq \frac{r}{\varepsilon} - r = r(1-\varepsilon)/\varepsilon$, so

$$\int_{B_r(0)} \frac{dz}{|z-z_i|} \leq \frac{\varepsilon}{r(1-\varepsilon)} \int_{B_r(0)} dz = \frac{\varepsilon \pi r}{1-\varepsilon}. \quad (2.11)$$

If $i \geq N+1$ we distinguish two cases: either $|z_i| \geq 2r$ or $|z_i| < 2r$. In the first situation we deduce that $|z - z_i| \geq r$, for any $z \in B_r(0)$. Thus

$$\int_{B_r(0)} \frac{dz}{|z-z_i|} \leq \frac{1}{r}\int_{B_r(0)} dz = \pi r.$$

If $|z_i| < 2r$ then
$$\int_{B_r(0)} \frac{dz}{|z-z_i|} \leq \int_{B_{4r}(z_i)} \frac{dz}{|z-z_i|} = 8\pi r.$$

The above two relations show that
$$\sup_{i \geq N+1} \int_{B_r(0)} \frac{dz}{|z-z_i|} \leq 8\pi r. \tag{2.12}$$

Using (2.10), (2.11), and (2.12), we obtain
$$0 \leq 2\pi r \alpha_1 + \sum_{i=2}^{N} |\alpha_i| \cdot \frac{\varepsilon \pi r}{1-\varepsilon} + 8\varepsilon \pi r.$$

Dividing by r and letting $\varepsilon \to 0$ we deduce that $\alpha_1 \geq 0$. □

Open problem. For part (a) of this problem, we have **not** been able to find an example of a series $\sum_{n=1}^{\infty} \alpha_n |z-z_n|^{-1}$ that converges to a positive number for almost all complex numbers z, but with not all α_n positive. It might be possible that such a series does not exist and the unique situation that can occur is that under our assumptions described in (a), the series $\sum_{n=1}^{\infty} \alpha_n |z-z_n|^{-1}$ always diverges to $+\infty$.

2.5.11. Suppose that $a_1 > a_2 > \cdots$ and $\lim_{n \to \infty} a_n = 0$. Define
$$S_n = \sum_{j=n}^{\infty} (-1)^{j-n} a_j = a_n - a_{n+1} + a_{n+2} - \cdots.$$

Prove that the series $\sum_{n=1}^{\infty} S_n^2$, $\sum_{n=1}^{\infty} a_n S_n$, and $\sum_{n=1}^{\infty} a_n^2$ converge or diverge together.

W. Trench, Amer. Math. Monthly, Problem 10624

Solution. By the Leibniz alternating series test, S_n exists and satisfies $0 < S_n < a_n$. Thus
$$\sum_{n=1}^{\infty} S_n^2 < \sum_{n=1}^{\infty} a_n S_n < \sum_{n=1}^{\infty} a_n^2.$$

So it suffices to show that finiteness of $\sum_{n=1}^{\infty} S_n^2$ implies finiteness of $\sum_{n=1}^{\infty} a_n^2$. To prove it, we use $S_n = a_n - S_{n+1}$ and the inequality $(x+y)^2 \leq 2(x^2+y^2)$ to infer
$$\sum_{n=1}^{\infty} a_n^2 = \sum_{n=1}^{\infty} (S_n + S_{n+1})^2 \leq \sum_{n=1}^{\infty} 2(S_n^2 + S_{n+1}^2) = 2\sum_{n=1}^{\infty} S_n^2 + 2\sum_{n=1}^{\infty} S_{n+1}^2 < 4\sum_{n=1}^{\infty} S_n^2.$$

Using Stirling's formula
$$\lim_{n \to \infty} \frac{n!}{n^n e^{-n} \sqrt{2\pi n}} = 1$$

we deduce that if $a_n = (2n)!/4^n (n!)^2$, then $a_n = O(1/\sqrt{n})$ as $n \to \infty$. On the other hand, $a_{n+1}/a_n = (2n+1)/(2n+2) < 1$, so $(a_n)_{n \geq 1}$ is a decreasing sequence of

2.5 Qualitative Results

positive numbers converging to 0. Thus, by the Leibniz alternating series test, the series $\sum_{n=1}^{\infty}(-1)^n a_n$ is convergent. At an elementary level, however, the convergence of the series $\sum_{n=1}^{\infty}(-1)^n a_n$ may be a little more difficult to obtain, the hard part being to show that $a_n \to 0$ as $n \to \infty$. For this purpose one may use the following simple result. □ **2.5.12.** Suppose $(a_n)_{n \geq 1}$ is a decreasing sequence of positive numbers and for each natural number n, define $b_n = 1 - a_{n+1}/a_n$. Then the sequence $(a_n)_{n \geq 1}$ converges to zero if and only if the series $\sum_{n=1}^{\infty} b_n$ diverges.

Solution. We first observe that for $M > N$,

$$0 \leq \frac{a_N - a_{M+1}}{a_N} = \frac{1}{a_N} \sum_{n=N}^{M} (a_n - a_{n+1}) \leq \sum_{n=N}^{M} b_n \leq \frac{1}{a_M} \sum_{n=N}^{M} (a_n - a_{n+1}) = \frac{a_N - a_{M+1}}{a_M}.$$

Now, if $\sum_{n=1}^{\infty} b_n$ converges and $a_n \to 0$, then letting $M \to \infty$ gives $1 \leq \sum_{n=N}^{\infty} b_n$, which is impossible. Conversely, if $a_n \to \alpha > 0$ then let $N = 1$ and let $M \to \infty$. The inequality above yields $\sum_{n=1}^{\infty} b_n \leq (a_1 - \alpha)/\alpha$, and so the series converges.

Returning now to the above example, we see that $b_n = 1/(2n+2)$ for any $n \geq 1$, so the divergence of the series $\sum_{n=1}^{\infty} b_n$ implies that $a_n \to 0$ as $n \to \infty$. The same technique gives an easy proof of the convergence of such series as

$$\sum_{n=1}^{\infty} \frac{(-1)^n n^n}{e^n n!}$$

and the series of binomial coefficients

$$\sum_{n=[\alpha]+1}^{\infty} \binom{\alpha}{n} \quad \text{with } \alpha > -1. \quad \square$$

2.5.13. Let $\sum_{n=1}^{\infty} a_n$ be a convergent series of positive numbers and set $r_n = \sum_{k=n}^{\infty} a_k$. Prove that the series $\sum_{n=1}^{\infty} a_n/r_n$ diverges.

O. Kellogg

Solution. By definition, $r_n = a_n + r_{n+1}$. Hence

$$\frac{r_{n+1}}{r_n} = 1 - \frac{a_n}{r_n},$$

$$\frac{r_{n+2}}{r_n} = \left(1 - \frac{a_n}{r_n}\right)\left(1 - \frac{a_{n+1}}{r_{n+1}}\right) > 1 - \frac{a_n}{r_n} - \frac{a_{n+1}}{r_{n+1}}.$$

By induction we establish that for all positive numbers n and p we have

$$\frac{r_{n+p}}{r_n} => 1 - \frac{a_n}{r_n} - \frac{a_{n+1}}{r_{n+1}} - \cdots - \frac{a_{n+p}}{r_{n+p}}.$$

Since the series $\sum_{n=1}^{\infty} a_n$ is convergent, we deduce that $r_{n+p} \to 0$ as $p \to \infty$. Fix $\delta \in (0,1)$. Hence, to any positive integer n, there corresponds $p \in \mathbb{N}$ such that

$$\frac{a_n}{r_n} + \frac{a_{n+1}}{r_{n+1}} + \cdots + \frac{a_{n+p}}{r_{n+p}} > 1 - \frac{r_{n+p}}{r_n} > \delta.$$

It follows that the series $\sum_{n=1}^{\infty} a_n/r_n$ diverges.

Let $\sum_{n=1}^{\infty} a_n$ be an arbitrary series. Each increasing sequence $n_1, n_2, n_3, \ldots$ of positive integers determines a series

$$a_{n_1} + a_{n_2} + a_{n_3} + \cdots,$$

which is called a *subseries* of the given series. The following exercise establishes an interesting property of some subseries of a divergent series. □

2.5.14. Let $(a_n)_{n\geq 1}$ be a sequence of positive numbers such that $\lim_{n\to\infty} a_n = 0$ and the series $\sum_{n=1}^{\infty} a_n$ diverges. Let a be a positive number. Then there is a subseries of $\sum_{n=1}^{\infty} a_n$ that converges to a.

Solution. Let n_1 be the least integer such that $a_n < a/2$ for all $n \geq n_1$. Let m_1 be the greatest integer for which

$$s_1 = a_{n_1} + a_{n_1+1} + \cdots + a_{m_1} < a.$$

Then $a/2 < s_1 < a$. Let n_2 be the least integer such that $n_2 > m_1$ and $a_n < (a-s_1)/2$, provided $n \geq n_2$. Let m_2 denote the greatest integer for which

$$s_2 = s_1 + a_{n_2} + a_{n_2+1} + \cdots + a_{m_2} < a.$$

Then $a - a/2^2 < s_2 < a$. Let n_3 be the least integer such that $n_3 > m_2$ and $a_n < (a-s_2)/2$, provided $n \geq n_3$. Let m_3 denote the greatest integer for which

$$s_3 = s_2 + a_{n_3} + a_{n_3+1} + \cdots + a_{m_3} < a.$$

Then $a - a/2^3 < s_3 < a$. Continuation of the construction yields the sequences of integers $(n_k)_{k\geq 1}$ and $(m_k)_{k\geq 1}$ such that

$$n_1 < m_1 < n_2 < m_2 < n_3 < m_3 < \cdots,$$
$$s_q = s_{q-1} + a_{n_q} + a_{n_q+1} + \cdots + a_{m_q} \quad \text{for all } q \geq 2,$$

and $a - a/2^q < s_q < a$. It follows that the required subseries converging to a is

$$a_{n_1} + a_{n_1+1} + \cdots + a_{m_1} + a_{n_2} + a_{n_2+1} + \cdots + a_{m_2} + a_{n_3} + a_{n_3+1} + \cdots. \quad \square$$

2.5.15. Let $(\lambda_n)_{n\geq 1}$ be a sequence of positive numbers. Set $a_1 = 1$, $a_{n+1} = a_n + \lambda_n a_n^{-1}$ for $n \geq 1$. Prove that $\lim_{n\to\infty} a_n$ exists if and only if the series $\sum_{n=1}^{\infty} \lambda_n$ converges.

T. Davison

Solution. We first observe that $a_{n+1} > a_n \geq 1$ for all $n \geq 1$. Assume that $\lim_{n\to\infty} a_n = a$. Hence, for all n, we have $a_n \leq a$. Therefore

$$\sum_{n=1}^{N} (a_{n+1} - a_n) = a_{N+1} - a_1 = \sum_{n=1}^{N} \frac{\lambda_n}{a_n} \geq \frac{1}{a} \sum_{n=1}^{N} \lambda_n.$$

2.5 Qualitative Results

It follows that $\sum_{n=1}^{\infty} \lambda_n \leq a(a-1) < \infty$.

Conversely, assume that the series $\sum_{n=1}^{\infty} \lambda_n$ converges. Then

$$\sum_{n=1}^{N}(a_{n+1} - a_n) = a_{N+1} - a_1 = \sum_{n=1}^{N} \frac{\lambda_n}{a_n} \geq \sum_{n=1}^{N} \lambda_n.$$

Therefore $\lim_{n \to \infty} a_n = 1 + \sum_{n=1}^{\infty} \lambda_n$.

Since the harmonic series $\sum_{n=1}^{\infty} 1/n$ is divergent, the partial sums $S_n = \sum_{k=1}^{n} 1/k$ are larger than every real number, provided n is big enough. A remarkable property is that S_n is never an integer, for all $n \geq 2$. Another striking result related to the harmonic series is the following. □

2.5.16. *Prove that every positive rational number is the sum of a finite number of distinct terms of the harmonic series.*

Solution. Let a/b be any given positive rational number and assume without loss of generality that $a/b > 1$. Let n_0 be uniquely determined by

$$S_{n_0} = 1 + \frac{1}{2} + \cdots + \frac{1}{n_0} \leq \frac{a}{b} < 1 + \frac{1}{2} + \cdots + \frac{1}{n_0} + \frac{1}{n_0 + 1}.$$

Let $a_1/b_1 = a/b - S_{n_0}$ and assume that $a_1/b_1 > 0$. Consider the positive integer n_1 uniquely determined by $n_1 < b_1/a_1 \leq n_1 + 1$ and set $a_2/b_2 = a_1/b_1 - 1/(n_1 + 1)$. Let n_2 be determined by $n_2 < b_2/a_2 \leq n_2 + 1$. Define $a_3/b_3 = a_2/b_2 - 1/(n_2 + 1)$ and so on. Since $a_1 > a_2 > a_3 > \cdots$, we deduce that after finitely many steps we obtain $a_m = 1$, that is,

$$\frac{1}{b_m} = \frac{a_m}{b_m} = \frac{a_{m-1}}{b_{m-1}} - \frac{1}{n_{m-1} + 1}.$$

Therefore

$$\frac{a}{b} = S_{n_0} + \frac{1}{n_1 + 1} + \frac{1}{n_2 + 1} + \cdots + \frac{1}{n_{m-1} + 1} + \frac{1}{b_m}. \quad \square$$

2.5.17. *Let $(a_n)_{n \geq 1}$ be a sequence of real numbers such that $a_n \geq 0$ and the series $\sum_{n=1}^{\infty} a_n^2$ is convergent. Define $S_n = \sum_{k=1}^{n} a_k$. Prove that the series $\sum_{n=1}^{\infty} (S_n/n)^2$ is convergent.*

Solution. We first observe that

$$\left(\frac{S_n}{n}\right)^2 = \left(a_n + \frac{S_n}{n} - a_n\right)^2 \leq 2a_n^2 + 2\left(\frac{S_n}{n} - a_n\right)^2$$

$$= 4a_n^2 + 2\left(\frac{S_n}{n}\right)^2 - 4\frac{a_n S_n}{n}.$$

Hence

$$\sum_{n=1}^{N}\left(\frac{S_n}{n}\right)^2 \leq 4\sum_{n=1}^{N} a_n^2 + 2\sum_{n=1}^{N}\left(\frac{S_n}{n}\right)^2 - 4\sum_{n=1}^{N} \frac{a_n S_n}{n} \qquad (2.13)$$

for each n. Moreover,
$$-2a_n S_n = -(S_n^2 - S_{n-1}^2) - a_n^2 \leq -(S_n^2 - S_{n-1}^2).$$

Therefore
$$-2 \sum_{n=1}^{N} \frac{a_n S_n}{n} \leq -\sum_{n=1}^{N} \frac{S_n^2 - S_{n-1}^2}{n}$$
$$= -\frac{S_1^2}{1 \cdot 2} - \frac{S_2^2}{2 \cdot 3} - \cdots - \frac{S_{N-1}^2}{(N-1) \cdot N} - \frac{S_N^2}{N}$$
$$\leq -\sum_{n=1}^{N} \frac{1}{n(n+1)} S_n^2.$$

By substituting this estimate into (2.13) we obtain
$$\sum_{n=1}^{N} \left(\frac{S_n}{n}\right)^2 \leq 4 \sum_{n=1}^{N} a_n^2 + 2 \sum_{n=1}^{N} \left(\frac{S_n}{n}\right)^2 - 2 \sum_{n=1}^{N} \frac{1}{n(n+1)} S_n^2$$
$$= 4 \sum_{n=1}^{N} a_n^2 + 2 \sum_{n=1}^{N} \frac{1}{n^2(n+1)} S_n^2,$$

which yields
$$\sum_{n=1}^{N} \left(1 - \frac{2}{n+1}\right) \left(\frac{S_n}{n}\right)^2 \leq 4 \sum_{n=1}^{N} a_n^2.$$

This concludes the proof. $\square$

2.5.18. *Does there exist a bijective map $\pi : \mathbb{N} \to \mathbb{N}$ such that*
$$\sum_{n=1}^{\infty} \frac{\pi(n)}{n^2} < \infty?$$

International Mathematical Competition for University Students, 1999

Solution. We argue that the series $\sum_{n=1}^{\infty} \pi(n)/n^2$ is divergent for *any* bijective map $\pi : \mathbb{N} \to \mathbb{N}$. Indeed, let π be a permutation of $\mathbb{N}$. For any $n \in \mathbb{N}$, the numbers $\pi(1), \ldots, \pi(n)$ are distinct positive integers; hence
$$\pi(1) + \cdots + \pi(n) \geq 1 + \cdots + n = \frac{n(n+1)}{2}.$$

Therefore
$$\sum_{n=1}^{\infty} \frac{\pi(n)}{n^2} = \sum_{n=1}^{\infty} (\pi(1) + \cdots + \pi(n)) \left(\frac{1}{n^2} - \frac{1}{(n+1)^2}\right)$$
$$\geq \sum_{n=1}^{\infty} \frac{n(n+1)}{2} \cdot \frac{2n+1}{n^2(n+1)^2} = \sum_{n=1}^{\infty} \frac{2n+1}{2n(n+1)} \geq \sum_{n=1}^{\infty} \frac{1}{n+1} = \infty.$$

2.5 Qualitative Results

An alternative argument is based on the observation that for any positive integer N we have

$$\sum_{n=N+1}^{3N} \frac{\pi(n)}{n^2} > \frac{1}{9}.$$

Indeed, only N of the $2N$ integers $\pi(N+1), \ldots, \pi(3N)$ are at most N, so that at least N of them are strictly greater than N. Hence

$$\sum_{n=N+1}^{3N} \frac{\pi(n)}{n^2} \geq \frac{1}{(3N)^2} \sum_{n=N+1}^{3N} \pi(n) > \frac{1}{9N^2} \cdot N \cdot N = \frac{1}{9}. \quad \square$$

2.5.19. Let $a_0 = b_0 = 1$. For each $n \geq 1$ let

$$a_n = \sum_{k=0}^{\infty} \frac{k^n}{k!}, \quad b_n = \sum_{k=0}^{\infty} (-1)^k \frac{k^n}{k!}.$$

Show that $a_n \cdot b_n$ is an integer.

Solution. We prove by induction on n that a_n/e and $b_n e$ are integers, for any $n \geq 1$.

From the power series of e^x, we deduce that $a_1 = e^1 = e$ and $b_n = e^{-1} = 1/e$. Suppose that for some $n \geq 1$, $a_0, a_1, \ldots, a_n$ (resp., $b_0, b_1, \ldots, b_n$) are all integer multiples of e (resp., $1/e$). Then, by the Newton's binomial,

$$a_{n+1} = \sum_{k=0}^{\infty} \frac{(k+1)^{n+1}}{(k+1)!} = \sum_{k=0}^{\infty} \frac{(k+1)^n}{k!}$$

$$= \sum_{k=0}^{\infty} \sum_{m=0}^{n} \binom{n}{m} \frac{k^m}{k!} = \sum_{m=0}^{n} \binom{n}{m} \sum_{k=0}^{\infty} \frac{k^m}{k!}$$

$$= \sum_{m=0}^{n} \binom{n}{m} a_m$$

and

$$b_{n+1} = \sum_{k=0}^{\infty} (-1)^{k+1} \frac{(k+1)^{n+1}}{(k+1)!} = -\sum_{k=0}^{\infty} (-1)^k \frac{(k+1)^n}{k!}$$

$$= -\sum_{k=0}^{\infty} (-1)^k \sum_{m=0}^{n} \binom{n}{m} \frac{k^m}{k!} = -\sum_{m=0}^{n} \binom{n}{m} \sum_{k=0}^{\infty} (-1)^k \frac{k^m}{k!}$$

$$= -\sum_{m=0}^{n} \binom{n}{m} b_m.$$

The numbers a_{n+1} and b_{n+1} are expressed as linear combinations of the previous elements with integer coefficients, which concludes the proof. $\quad \square$

The following result is closely related to the divergence of the harmonic series and establishes an interesting property that holds for all sequences of positive real numbers.

2.5.20. Let $(a_n)_{n\geq 1}$ be a sequence of positive numbers. Show that

$$\limsup_{n\to\infty} \left(\frac{a_1+a_{n+1}}{a_n}\right)^n \geq e.$$

Solution. Arguing by contradiction, we deduce that there exists a positive integer N such that for all $n \geq N$,

$$\left(\frac{a_1+a_{n+1}}{a_n}\right)^n < \left(1+\frac{1}{n}\right)^n.$$

This inequality may be rewritten as

$$\frac{a_1}{n+1} < \frac{a_n}{n} - \frac{a_{n+1}}{n+1} \quad \text{for all } n \geq N.$$

Summing these inequalities, we obtain

$$a_1 \left(\frac{1}{N+1} + \cdots + \frac{1}{n}\right) < \frac{a_N}{N} - \frac{a_{n+1}}{n+1} < \frac{a_N}{N},$$

which is impossible, since the harmonic series is divergent. □

The following result is optimal, in the sense that (under the same hypotheses) there is no function f with $f(n)/n \to \infty$ and $f(n)a_n \to 0$ as $n \to \infty$.

2.5.21. Let $(a_n)_{n\geq 1}$ be a sequence of real numbers such that the series $\sum_{n=1}^{\infty} a_n$ converges and $a_n \geq a_{n+1} \geq 0$ for all $n \geq 1$. Prove that $na_n \to 0$ as $n \to \infty$.

Solution. Assume that m and n are positive integers with $2m < n$. Since the series $\sum_{n=1}^{\infty} a_n$ converges, we may assume that $r_m := \sum_{j=m+1}^{\infty} a_j < \varepsilon$, where $\varepsilon > 0$ is fixed arbitrarily. Then

$$(n-m)a_n \leq a_{m+1} + \cdots + a_n \leq r_m := \sum_{j=m+1}^{\infty} a_j.$$

Thus,

$$0 < na_n \leq \frac{n}{n-m} r_m < \frac{n}{n-m} \varepsilon < 2\varepsilon,$$

which concludes the proof. □

2.5.22. Let $(a_n)_{n\geq 1}$ be a sequence of positive numbers such that the series $\sum_{n=1}^{\infty} a_n$ converges. Prove that there exists a nondecreasing divergent sequence $(b_n)_{n\geq 1}$ of positive integers such that the series $\sum_{n=1}^{\infty} a_n b_n$ also converges.

Solution. Since $\sum_{n=1}^{\infty} a_n$ converges, we deduce that for any $n \geq 1$, there exists a positive integer $N(n)$ depending on n such that

2.5 Qualitative Results

$$\sum_{j=N(n)}^{\infty} a_j < \frac{1}{n^3}.$$

To conclude the proof, it is enough to take $b_m = n$, provided $N(n) \leq m < N(n+1)$. □

"Prime numbers were invented to multiply them." Israel Gelfand attributes this famous quotation to the Russian physicist Lev Landau (1908–1968), who was awarded the 1962 Nobel Prize in Physics. Nonetheless, the additive properties of the prime numbers have been a fascinating subject for generations of mathematicians, and by now there is an ample supply of powerful and flexible methods to attack classical problems.

Problems concerning infinite sums of positive integers play an important role in number theory. For instance, the Hungarian mathematicians Paul Erdős (1913–1996) and Paul Turán (1910–1976) conjectured in 1936 that every infinite set $\{a_1, a_2, \ldots\}$ of positive integers contains arbitrarily long arithmetic progressions, provided that the series $\sum_{n=1}^{\infty} 1/a_n$ diverges. This conjecture is still open, and it is not even known whether such a set contains an arithmetic progression of length 3. A related theorem due to the Dutch mathematician Bartel Leendert van der Waerden (1903–1996) states that for any partition $\mathbb{N} = \bigcup_{j=1}^{n} S_j$ into finitely many subsets S_j, at least one of them contains arbitrarily long arithmetic progressions. A remarkable breakthrough in the field was achieved in 2004 by the British mathematician Ben Green and Fields Medalist Terence Tao [35], who proved a long-standing conjecture, namely, that the set of prime numbers contains arbitrarily long arithmetic progressions. For example, 7, 37, 67, 97, 127, 157 is an arithmetic progression of length 6 consisting only of primes. In 2004, Frind, Underwood, and Jobling found a progression of length 23 with first element of size $\approx 5.6 \times 10^{13}$. Notice that the Green–Tao theorem guarantees arbitrarily long arithmetic progressions of prime numbers, but infinitely long such progressions do not exist: $n + kd$ is not prime if $k = n$.

The following result establishes that the series of inverses of prime numbers diverges. We provide three different proofs to this property. It may be surprising to note that the sum of the reciprocals of the twin prime numbers (that is, pairs of primes that differ by 2, such as 11, 13 or 17, 19) is a convergent series, see [65, pp. 94–103].

2.5.23. Prove that the series

$$\sum_{p=\text{prime}} \frac{1}{p}$$

diverges.

Solution. We start with a simple proof due to Paul Erdős. Indeed, if $\sum 1/p$ converges, then we can choose b such that $\sum_{p>b} 1/p < \frac{1}{2}$. Take $a = 1$. Suppose $n \in M_x$, and write $n = k^2 m$, where m is square-free. Since $m = \prod_S p$, where S is some subset of P, m can assume at most $2^{|P|}$ values. Also $k \leq \sqrt{n} \leq \sqrt{x}$. Thus $|M_x| \leq 2^{|P|}\sqrt{x}$.

Now the number of positive integers $\leq x$ divisible by a fixed p does not exceed x/p. Thus $x - |M_x|$, the number of such integers divisible by some prime greater

than b, satisfies
$$x - |M_x| \le \sum_{p > b} \frac{x}{p} < \frac{x}{2}.$$

We see that
$$\frac{x}{2} < |M_x| \le 2^{|P|} \sqrt{x},$$

or $\sqrt{x} < 2^{|P|+1}$, which is clearly false for x sufficiently large.

The following very simple identity forms the basis for two more alternative proofs, namely,
$$\left(1 + \frac{1}{p}\right)\left(1 + \frac{1}{p^2} + \frac{1}{p^4} + \cdots + \frac{1}{p^{2k}}\right) = 1 + \frac{1}{p} + \frac{1}{p^2} + \cdots + \frac{1}{p^{2k+1}}.$$

Taking the product over P and letting $k \to \infty$ yields
$$\prod_P \left(1 + \frac{1}{p}\right) \sum_M \frac{1}{n^2} = \sum_M \frac{1}{n}. \tag{2.14}$$

Since $\sum 1/n^2$ converges and $\sum 1/n$ diverges, this means that
$$\text{for } a = 1, \ \prod_P \left(1 + \frac{1}{p}\right) \to \infty \quad \text{as } b \to \infty. \tag{2.15}$$

First alternative proof. Since for $C > 0$, $e^C = 1 + C + C^2/2! + \cdots > 1 + C$, we have
$$\prod_P \left(1 + \frac{1}{p}\right) < \prod_P e^{1/p} = \exp\left(\sum_P \frac{1}{p}\right).$$

This, with (2.15), shows that $\sum 1/p$ diverges.

Second alternative proof. By (2.15) $\prod_P (1 + 1/p) \to \infty$ as $b \to \infty$ for any fixed a, and so the same is true for $\sum_M 1/n$ by (2.14). If $\sum 1/p$ converges, we can choose a such that $\sum_P 1/p < \frac{1}{2}$ for all b, then choose b and x large enough such that $\sum_{M_x} 1/n > 2$. Since every n in M_x except 1 is of the form pn for $p \in P$ and $n \in M_x$, we have
$$\sum_P \frac{1}{p} \sum_{M_x} \frac{1}{n} \ge \sum_{M_x} \frac{1}{n} - 1.$$

Then
$$\frac{1}{2} > \sum_P \frac{1}{p} \ge 1 - \left(\sum_{M_x} \frac{1}{n}\right)^{-1} > 1 - \frac{1}{2},$$

a contradiction. $\square$

2.5.24. (i) Show that if $(a_n)_{n \ge 1}$ is a decreasing sequence of positive numbers then
$$\left(\sum_{i=1}^n a_i^2\right)^{1/2} \le \sum_{i=1}^n \frac{a_i}{\sqrt{i}}.$$

2.5 Qualitative Results

(ii) Show that there is a constant C such that if $(a_n)_{n\geq 1}$ is a decreasing sequence of positive numbers then

$$\sum_{m=1}^{\infty} \frac{1}{\sqrt{m}} \left(\sum_{i=m}^{\infty} a_i^2 \right)^{1/2} \leq C \sum_{i=1}^{\infty} a_i.$$

Solution. (i) We have

$$\left(\sum_{i=1}^{n} \frac{a_i}{\sqrt{i}} \right)^2 = \sum_{i,j}^{n} \frac{a_i a_j}{\sqrt{i}\sqrt{j}} \geq \sum_{i=1}^{n} \frac{a_i}{\sqrt{i}} \sum_{j=1}^{i} \frac{a_i}{\sqrt{j}}$$

$$\geq \sum_{i=1}^{n} \frac{a_i}{\sqrt{i}} i \frac{a_i}{\sqrt{i}} = \sum_{i=1}^{n} a_i^2.$$

(ii) Using (i) we obtain

$$\sum_{m=1}^{\infty} \frac{1}{\sqrt{m}} \left(\sum_{j=m}^{\infty} a_j^2 \right)^{1/2} \leq \sum_{m=1}^{\infty} \frac{1}{\sqrt{m}} \sum_{j=m}^{\infty} \frac{a_j}{\sqrt{j-m+1}}$$

$$= \sum_{j=1}^{\infty} a_j \sum_{m=1}^{j} \frac{1}{\sqrt{m}\sqrt{j-m+1}}.$$

Next, we observe that

$$\sum_{m=1}^{j} \frac{1}{\sqrt{m}\sqrt{j+1-m}} = 2 \sum_{m=1}^{j/2} \frac{1}{\sqrt{m}\sqrt{j+1-m}}$$

$$\leq 2 \frac{1}{\sqrt{j/2}} \sum_{m=1}^{j/2} \frac{1}{\sqrt{m}} \leq 2 \frac{1}{\sqrt{j/2}} \cdot 2\sqrt{j/2} = 4.$$

This completes the proof, with $C = 4$. □

Prove that the optimal constant is $C = \pi$.

2.5.25. Let $(a_n)_{n\geq 1}$ be a decreasing sequence of positive numbers such that the series $\sum_{n=1}^{\infty} a_n$ is convergent and $a_n \leq \sum_{i=n+1}^{\infty} a_i$ for all $n \geq 1$. Define $L = \sum_{n=1}^{\infty} a_n$. Prove that for every $0 \leq x \leq L$ there is a sequence $(\varepsilon_n)_{n\geq 1}$ with $\varepsilon_n \in \{0,1\}$ such that $x = \sum_{n=1}^{\infty} \varepsilon_n a_n$.

Miklós Schweitzer Competitions

Solution. Fix $0 \leq x \leq L$ and define the numbers ε_n inductively as follows:

$$\varepsilon_n = \begin{cases} 1 & \text{if } \sum_{i=1}^{n-1} \varepsilon_i a_i + a_n < x, \\ 0 & \text{if } \sum_{i=1}^{n-1} \varepsilon_i a_i \geq x. \end{cases}$$

For every n for which $\varepsilon_n = 0$ we have

$$0 \leq x - \sum_{i=1}^{\infty} \varepsilon_i a_i \leq x - \sum_{i=1}^{n-1} \varepsilon_i a_i \leq a_n.$$

We deduce that if there are infinitely many such n, then $x = \sum_{i=1}^{\infty} \varepsilon_i a_i$. If, however, there are only finitely many such n's, then for the largest one,

$$x - \sum_{i=1}^{n-1} \varepsilon_i a_i \leq a_n \leq \sum_{i=n+1}^{\infty} a_i = \sum_{i=n+1}^{\infty} \varepsilon_i a_i.$$

Therefore $x \leq \sum_{n=1}^{\infty} \varepsilon_n a_n$, so our claim follows in this case, too. □

2.5.26. Let $(a_n)_{n \geq 1}$ be a sequence of positive real numbers such that $\sum_{n=1}^{\infty} a_n$ converges.

(i) Prove that $\sum_{n=1}^{\infty} a_n^{\log n/(1+\log n)}$ also converges.
(ii) Prove or disprove: If $(b_n)_{n \geq 1}$ is an increasing sequence of positive real numbers tending to one, then $\sum_{n=1}^{\infty} a_n^{b_n}$ converges.

Christopher Hilar

Solution. (i) Define $I = \{n : a_n^{\log n/(1+\log n)} \leq e^2 a_n\}$ and $J = \mathbb{N} \setminus I$. If $n \in J$, then $a_n^{\log n} > (en)^2 a_n^{1+\log n}$, so $a_n < (en)^{-2}$. Therefore

$$\sum_{n=1}^{\infty} a_n^{\log n/(1+\log n)} \leq \sum_{n \in I} e^2 a_n + \sum_{n \in J} (en)^{-2} \leq e^2 \sum_{n=1}^{\infty} a_n + e^{-2} \sum_{n=1}^{\infty} n^{-2} < \infty.$$

(ii) We disprove the assertion. For $n \geq 3$, let

$$a_n = \frac{1}{n(\log n)^2}, \quad b_n = \frac{1}{1 + \frac{2\log\log n}{\log n}}.$$

Now $\sum_{n=1}^{\infty} a_n$ converges and $(b_n)_{n \geq 1}$ increases to 1, but $a_n^{b_n} = n^{-1}$, so $\sum_{n=1}^{\infty} a_n^{b_n}$ diverges. □

2.5.27. Let a_n, φ_n be positive constants with

$$\sum_{n=1}^{\infty} a_n \text{ convergent, and } \varphi_n = O\left(\frac{1}{\log n}\right).$$

Show that the series $\sum_{n=1}^{\infty} a_n^{1-\varphi_n}$ converges.

Jet Wimp

2.5 Qualitative Results

Solution. We prove the following equivalent statement: let $(b_n)_{n\geq 1}$ be a sequence of real numbers with $0 < b_n < 1$ and $1 - b_n = O(1/\log n)$. If $(a_n)_{n\geq 1}$ is a sequence of positive real numbers such that $\sum_{n=1}^{\infty} a_n$ converges, then $\sum_{n=1}^{\infty} a_n^{b_n}$ converges, too.

To prove this, suppose that $\sum_{n=1}^{\infty} a_n < \infty$, and choose N large enough that $a_n < 1$ and $b_n \geq 1 - K/\log n > 0$ for all $n \geq N$. Now $(1/n)^{(K/\log n)} = e^{-K}$, and

$$\sum_{n=N}^{\infty} a_n^{b_n} \leq e^{2K} \sum_{n=N}^{\infty} a_n^{1-K/\log n} \cdot \left(\frac{1}{n^2}\right)^{K/\log n}$$

$$\leq e^{2K} \sum_{n=N}^{\infty} \left[\left(1 - \frac{K}{\log n}\right) a_n + \frac{K}{\log n} \cdot \frac{1}{n^2}\right]$$

by the AM–GM inequality. The last series converges, and this concludes the proof. $\square$

2.5.28. Prove that if $(a_n)_{n\geq 1}$ is a sequence of positive numbers with $\sum_{n=1}^{\infty} a_n < \infty$, then for all $p \in (0,1)$,

$$\lim_{n\to\infty} n^{1-1/p} (a_1^p + \cdots + a_n^p)^{1/p} = 0.$$

Grahame Bennet

Solution. (Eugene A. Herman) Define the sequence $(x_n)_{n\geq 1}$ by $x_n = (1/n)^{1-p} (a_1^p + \cdots + a_n^p)$. Since

$$n^{1-1/p} (a_1^p + \cdots + a_n^p)^{1/p} = ((1/n)^{1-p} (a_1^p + \cdots + a_n^p))^{1/p} = x_n^{1/p},$$

it suffices to show that $x_n \to 0$ as $n \to \infty$. Using Hölder's inequality we obtain

$$\sum_{k=1}^{n} \left(\frac{1}{n}\right)^{1-p} a_k^p \leq \left(\sum_{k=1}^{n} \frac{1}{n}\right)^{1-p} \left(\sum_{k=1}^{n} a_k\right)^p.$$

Given any $\varepsilon > 0$ choose N such that $\sum_{n=N}^{\infty} a_n < (\varepsilon/2)^{1/p}$. For any n larger than both N and $((a_1^p + \cdots + a_N^p)2/\varepsilon)^{1/(1-p)}$, we have

$$x_n = \left(\frac{1}{n}\right)^{1-p} (a_1^p + \cdots + a_N^p) + \left(\frac{1}{n}\right)^{1-p} \sum_{k=N+1}^{n} a_k^p$$

$$< \frac{\varepsilon}{2} + \left(\sum_{k=N+1}^{n} \frac{1}{n}\right)^{1-p} \left(\sum_{k=N+1}^{n} a_k\right)^p = \frac{\varepsilon}{2} + \left(\frac{n-N}{n}\right)^{1-p} \left(\sum_{k=N+1}^{n} a_k\right)^p$$

$$< \frac{\varepsilon}{2} + \frac{\varepsilon}{2} = \varepsilon,$$

which concludes the proof. $\square$

2.5.29. Let $(a_n)_{n\geq 1}$ be a sequence of real numbers such that $\sum_{n=1}^{\infty} |a_n|$ is convergent and $\sum_{n=1}^{\infty} a_{kn} = 0$ for each positive integer k. Prove that $a_n = 0$ for all $n \geq 1$.

Solution. Let p_n be the nth prime greater than 1, and let R_n be the set of positive integers that are not divisible by any $p_1, \ldots, p_n$. Note that the first two members of R_n are 1 and p_{n+1}.

We first claim that
$$\sum_{j \in R_n} a_j = 0$$
for each n. From this it follows that $|a_1| \leq \sum_{p_{n+1}}^{\infty} |a_j|$ for all n, and therefore that $a_1 = 0$. A similar argument shows that for any k, $\sum_{j \in R_n} a_{kj} = 0$, and hence that $|a_k| = 0$.

For a bounded sequence $b = (b_n)_{n \geq 1}$, write $\langle a, b \rangle = \sum a_n b_n$, where $a = (a_n)_{n \geq 1}$.

Let c^k be the sequence with 1 in place nk ($n = 1, 2, \ldots$) and 0 elsewhere. The hypothesis states that $\langle a, c^k \rangle = 0$ for each k.

Fix an integer n, and for $1 \leq r \leq n$, let T_r be the set of products of r distinct primes chosen from $p_1, \ldots, p_n$. Consider the sequence b defined by
$$b = c^1 + \sum_{r=1}^{n} (-1)^r \sum_{k \in T_r} c^k.$$

If we show that b_j is 1 for $j \in R_n$ and 0 for other j, then our claim follows. If j is in R_n, then $c_j^k = 0$ for all k in $\bigcup_{r=1}^n T_r$, so $b_j = c_j^1 = 1$. Suppose now that j is not in R_n, and let $q_1, \ldots, q_m$ be the primes that divide j and are not greater than p_n. Let U_r be the set of products of r distinct primes chosen from $q_1, \ldots, q_m$. Then U_r has C_m^r members, and $c_j^k = (-1)^r$ for each k in U_r. These, together with c^1, are the only c^k that make a nonzero contribution to b_j, which is therefore equal to
$$1 + \sum_{r=1}^{m} (-1)^r C_m^r = (1-1)^m = 0.$$

This completes the proof. □

We say that a function $f : \mathbb{R} \to \mathbb{R}$ is *convergence-preserving* if for every convergent series $\sum_{n=1}^{\infty} a_n$, the series $\sum_{n=1}^{\infty} f(a_n)$ also converges. For example, all functions $f(x) = ax$ are convergence-preserving, where $a \in \mathbb{R}$ is an arbitrary constant. The following properties are also true:

(i) if f is convergence-preserving, then $f(0) = 0$;
(ii) if f and g are convergence-preserving, so are $f + g$, af, $f \circ g$, and $g(x) = f(ax)$, where a is a constant.

The following exercise establishes that every convergence-preserving function has at most linear growth in a neighborhood of the origin.

2.5.30. Let $f : \mathbb{R} \to \mathbb{R}$ be a convergence-preserving function. Prove that there exist a real number M and $\varepsilon > 0$ such that $f(x) < Mx$ for all $0 < x < \varepsilon$.

Solution. Arguing by contradiction, we deduce that for any positive integer n we can find $0 < x_n < 1/n^2$ such that $f(x_n) > nx_n$.

Let j_n be the least integer such that $j_n \geq 1/(n^2 x_n)$. Then $j_n x_n < 2/n^2$. Consider the series

2.5 Qualitative Results

$$x_1 + x_1 + \cdots + x_1 \ (j_1 \text{ times}) + x_2 + x_2 + \cdots + x_2 \ (j_2 \text{ times}) + \cdots.$$

This series converges by comparison with $\sum_{n=1} 2/n^2$.

When we apply f to each term of the series and sum the first n blocks, we obtain

$$\sum_{n=1}^{\infty} j_n f(x_n) \geq \sum_{n=1}^{\infty} n j_n x_n \geq \sum_{n=1}^{\infty} \frac{1}{n},$$

which diverges, contradicting the hypothesis that f is convergence-preserving.

As proved in [116], a function f is convergence-preserving if and only if there exists a constant $a \in \mathbb{R}$ such that $f(x) = ax$ in a neighborhood of the origin. This result guarantees the existence of many series having strange properties. For example, there is a sequence $(a_n)_{n \geq 1}$ for which $\sum_{n=1}^{\infty} a_n^3$ diverges but $\sum_{n=1}^{\infty} \tan a_n$ converges; otherwise, $(\arctan x)^3$ would be convergence-preserving. Similarly, there is a sequence $(b_n)_{n \geq 1}$ with $\sum_{n=1}^{\infty} b_n^3$ convergent and $\sum_{n=1}^{\infty} \tan b_n$ divergent; otherwise, $\tan \sqrt[3]{x}$ would be convergence-preserving. □

2.5.31. Let $\sum_{n=1}^{\infty} a_n$ be a convergent series with sum A and with positive terms that satisfy $a_{n+1} \leq a_n \leq 2a_{n+1}$ for all $n \geq 1$. Prove that any positive number $B < A$ is the sum either of a finite number of terms or of a subseries of the considered series.

Marian Tetiva, Problem 101, GMA 1(2001)

Solution. We first assume that $a_1 \leq B < A$. Define $A_n = \sum_{k=1}^{n} a_k$. Then

$$[a_1, A) = \bigcup_{n \geq 1} [A_n, A_{n+1});$$

hence there exists $s \geq 1$ for which $B \in [A_s, A_{s+1})$. If $B = A_s$, we are done. Otherwise, we have $B \in (A_s, A_{s+1})$, and thus

$$0 < B - A_s < a_{s+1}.$$

Since $a_1 \geq a_2 \geq \cdots$ and $\lim_{n \to \infty} a_n = 0$, we have

$$(0, a_{s+1}) = \bigcup_{n \geq s+1} [a_{n+1}, a_n);$$

hence there exists $p \in \mathbb{N}^*$ such that $B - A_s \in [a_{s+p+1}, a_{s+p})$. Again, we are done if $B - A_s = a_{s+p+1}$. If not, according to the hypothesis,

$$0 < B - A_s - a_{s+p+1} < a_{s+p} - a_{s+p+1} \leq a_{s+p+1}.$$

In turn, $B - A_s - a_{s+p+1}$ is in an interval of the form $[a_{s+p+q+1}, a_{s+p+q})$ for some positive integer q. Therefore we obtain

$$0 \leq B - A_s - a_{s+p+1} - a_{s+p+q+1} < a_{s+p+q} - a_{s+p+q+1} \leq a_{s+p+q+1}.$$

So we can assume in general that we have natural numbers $s < t_1 < \cdots < t_k$ such that
$$0 \leq B - A_s - a_{t_1} - \cdots - a_{t_k} < a_{t_k - 1} - a_{t_k} \leq a_{t_k}.$$
We are done if we have equality in the left-hand inequality; otherwise, $B - A_s - a_{t_1} - \cdots - a_{t_k}$ is in an interval $[a_{t_k+u+1}, a_{t_k+u})$ (for some natural u), and with $t_{k+1} = t_k + u + 1$ we have
$$0 \leq B - A_s - a_{t_1} - \cdots - a_{t_k} - a_{t_{k+1}} < a_{t_{k+1}-1} - a_{t_{k+1}} \leq a_{t_{k+1}}$$
and so on. This method yields a strictly increasing sequence of natural numbers $s < t_1 < t_2 < \cdots$ with the property that
$$0 \leq B - A_s - a_{t_1} - \cdots - a_{t_n} < a_{t_n - 1} - a_{t_n} \leq a_{t_n}, \forall n \in \mathbb{N}^*.$$
If any of the left-hand inequalities becomes an equality, the problem is solved, since obviously, B is the sum of a finite number of terms of the series $\sum_{n=1}^\infty a_n$. On the other hand, if all these inequalities are strict, because of $\lim_{n \to \infty} t_n = \infty$ we also have $\lim_{n \to \infty} a_{t_n} = 0$ and
$$B = \lim_{n \to \infty} (A_s + a_{t_1} + \cdots + a_{t_n}),$$
which proves the claim.

In the case $0 < B < a_1$, notice that $a_1 \leq A - B < A$. Indeed, $2A - 2a_1 = \sum_{n=1}^\infty 2a_{n+1} \geq \sum_{n=1}^\infty a_n = A$, whence $A - a_1 \geq a_1 > B$ (the other case is clear). In conformity with the part above, there is a sequence $j_1 < j_2 < \cdots$ of natural numbers such that $A - B = \sum_{n=1}^\infty a_{j_n}$, which easily implies $B = \sum_{n=1}^\infty a_{i_n}$, for $\{i_1, i_2, \ldots\} = \mathbb{N}^* \setminus \{j_1, j_2, \ldots\}$. □

2.6 Independent Study Problems

> Mediocrity knows nothing higher than itself, but talent instantly recognizes genius.
>
> Sir Arthur Conan Doyle (1859–1930)

2.6.1. Prove that the series
$$\sum_{n=1}^\infty \left(1 + \frac{1}{2} + \cdots + \frac{1}{n}\right)^2 n^{-2}$$
converges to $4.59987\ldots$. Does the value of this sum equal $17\pi^4/360$?

2.6.2. Determine the values of $p > 0$ for which the following series converge:

(i) $\sum_{n=2}^\infty \dfrac{(\ln n)^{-2p}}{n^p + p^{-(\ln n)^{1/p}}}$;

2.6 Independent Study Problems

(ii) $\sum_{n=1}^{\infty} \frac{(-1)^{\sqrt{n}}}{n^p}$;

(iii) $\sum_{n=1}^{\infty} \arccos^p \frac{v(n)}{v(n+1)}$,

where $v(n)$ is the number of digits in the decimal expression for n.

Hint. (i) Consider separately the cases $p > 1$ and $0 < p < 1$.

(ii) Use the relation $\arccos x \sim \sqrt{2(1-x)}$ as $x \to 1-0$.

2.6.3. Let $\sum_{n=1}^{\infty} a_n^2$ and $\sum_{n=1}^{\infty} b_n^2$ converge, where $a_n \geq 0$ and $b_n \geq 0$ for all $n \geq 1$. Prove that the series $\sum_{n=1}^{\infty} a_n b_n$ converges.

2.6.4. Study the convergence of the series $\sum_{n=1}^{\infty} a_n$, where

$$a_n = \left[\frac{(2n-1)!!}{(2n)!!}\right]^2.$$

R.E. Moore, Amer. Math. Monthly, Problem 2752

2.6.5. Let $\sum_{n=1}^{\infty} a_n$ be a convergent series of positive numbers. Prove that the series $\sum_{n=1}^{\infty} \sqrt{a_n}/n$ converges.

2.6.6. Evaluate

$$\sum_{n=1}^{\infty} \frac{\left(\frac{3-\sqrt{5}}{2}\right)^n}{n^3}.$$

H.F. Sandham, Amer. Math. Monthly, Problem 4293

2.6.7. Let $\sum_{n=1}^{\infty} a_n$ be a divergent series of positive numbers and set $s_n = a_1 + \cdots + a_n$.

(i) Prove that the series $\sum_{n=1}^{\infty} a_n/(1+a_n)$ diverges.

(ii) Prove that

$$\frac{a_{N+1}}{s_{N+1}} + \cdots + \frac{a_{N+k}}{s_{N+k}} \geq 1 - \frac{s_N}{s_{N+k}}$$

and deduce that the series $\sum_{n=1}^{\infty} a_n/s_n$ diverges.

(iii) Prove that

$$\frac{a_n}{s_n^2} \leq \frac{1}{s_{n-1}} - \frac{1}{s_n}$$

and deduce that the series $\sum_{n=1}^{\infty} a_n/s_n^2$ converges.

What can be said about the series $\sum_{n=1}^{\infty} a_n/(1+a_n^2)$, $\sum_{n=1}^{\infty} a_n/(1+na_n)$, and $\sum_{n=1}^{\infty} a_n/(1+n^2 a_n)$?

2.6.8. Let $\sum_{n=1}^{\infty} a_n$ be a divergent series of positive numbers and set $s_n = a_1 + \cdots + a_n$. Prove that the series $\sum_{n=1}^{\infty} a_n/s_n^{1+\varepsilon}$ converges for any $\varepsilon > 0$.

Hint. Compare the given series with the series $\sum_{n=2}^{\infty}(s_{n-1}^{-\varepsilon} - s_n^{-\varepsilon})$.

2.6.9. Show that the series $\sum_{n=1}^{\infty}(\log n)/n^2$ converges. If $\varepsilon > 0$, does

$$\sum_{n=1}^{\infty}\frac{(\log n)^3}{n^{1+\varepsilon}}$$

converge? Given a positive integer d, does $\sum_{n=1}^{\infty}(\log n)^d/n^s$ converge for $s > 1$?

2.6.10. Given the sequence $(a_n)_{n\geq 1}$ defined by $a_1 = 2$, $a_2 = 8$, $a_n = 4a_{n-1} - a_{n-2}$ ($n = 2, 3, \ldots$), show that

$$\sum_{n=1}^{\infty}\operatorname{arccot} a_n^2 = \frac{\pi}{12}.$$

N. Anning, Amer. Math. Monthly, Problem 3051

2.6.11. Given the sequence $(a_n)_{n\geq 1}$ defined by $a_1 = 1$, $a_2 = 3$, $a_n = 4a_{n-1} - a_{n-2}$ ($n = 2, 3, \ldots$), show that

$$\sum_{n=1}^{\infty}\operatorname{arccot} 2a_n^2 = \frac{\pi}{6}.$$

A.C. Aitken, Amer. Math. Monthly, Problem 3348

2.6.12. If $a, b > 1$ prove that

$$\sum_{n=1}^{\infty}\frac{1}{\sqrt{n^a n^b(n^a+n^b)}} \leq \frac{1}{\sqrt{2}}\zeta\left(\frac{3(a+b)}{4}\right) \leq \frac{1}{4}(\zeta(a)+\zeta(b)+\zeta(a+b)),$$

where ζ denotes the Riemann zeta function.

2.6.13. Suppose that the series with positive terms $\sum_{n=1}^{\infty}a_n$ converges. Put $r_n = a_n + a_{n+1} + a_{n+2} + \cdots$.

(i) Prove that

$$\frac{a_m}{r_m} + \cdots + \frac{a_n}{r_n} > 1 - \frac{r_n}{r_m}$$

if $m < n$ and deduce that the series $\sum_{n=1}^{\infty} a_n/r_n$ diverges.

(ii) Prove that

$$\frac{a_n}{\sqrt{r_n}} < 2(\sqrt{r_n} - \sqrt{r_{n-1}})$$

and deduce that the series $\sum_{n=1}^{\infty} a_n/\sqrt{r_n}$ converges.

2.6.14. Show that γ, Euler's constant, is given by

$$\gamma = \sum_{n=2}^{\infty}(-1)^n\frac{a_n}{n}, \quad \text{where } a_n = \sum_{k=1}^{\infty}\frac{1}{k^n}.$$

A.M. Glicksman, Amer. Math. Monthly, Problem 4045

2.6 Independent Study Problems

2.6.15. Find a sequence of real numbers $(a_n)_{n\geq 1}$ such that $\sum_{n=1}^{\infty} a_n$ converges, $\sum_{n=1}^{\infty} a_n^3$ diverges, $\sum_{n=1}^{\infty} a_n^5$ converges, and so on. More generally, let $\mathscr{C}$ be an arbitrary (finite or infinite) set of positive integers. Then there exists a sequence of real numbers $(a_n)_{n\geq 1}$ depending on $\mathscr{C}$ such that for $j = 1, 2, \ldots$, the series $\sum_{n=1}^{\infty} a_n^{2j-1}$ converges or diverges according as j does or does not belong to $\mathscr{C}$.

G. Pólya, Amer. Math. Monthly, Problem 4142

2.6.16. Let $(a_n)_{n\geq 1}$ be a sequence of real numbers and set

$$s_n = \sum_{k=1}^{n} a_k \quad \text{and} \quad \sigma_n = \sum_{k=1}^{n} \left(1 - \frac{k}{n+1}\right) a_k.$$

Assume that the series $\sum_{n=1}^{\infty} |s_n - \sigma_n|^k$ converges for any $k > 0$. Prove that the series $\sum_{n=1}^{\infty} a_n$ is convergent.

R. Bellman, Amer. Math. Monthly, Problem 4250

2.6.17. Find all values of α and β for which the series $\sum_{n=1}^{\infty} n^{\alpha} \sin n^{\beta}$ converges.

R.P. Boas Jr. and W.K. Hayman, Amer. Math. Monthly, Problem 4415

2.6.18. Let $(a_n)_{n\geq 1}$ be a sequence of positive numbers converging to 0. Prove the following properties:

(i) if $\sum_{n=1}^{\infty} a_n = +\infty$, then $\sum_{n=1}^{\infty} \min(a_n, 1/n) = +\infty$;
(ii) if $\sum_{n=1}^{\infty} a_n/n = +\infty$, then $\sum_{n=1}^{\infty} n^{-1} \min(a_n, 1/\ln n) = +\infty$.

2.6.19. Let $(a_n)_{n\geq 1}$ be a decreasing sequence of positive numbers converging to 0. Prove that the series $\sum_{n=1}^{\infty} a_n$ is convergent if and only if $a_n = o(1/n)$ and $\sum_{n=1}^{\infty} (a_n - a_{n+1}) n < +\infty$.

2.6.20. Let $(a_n)_{n\geq 1}$ be a decreasing sequence of positive numbers converging to 0. Prove that the series $\sum_{n=1}^{\infty} a_n/n$ is convergent if and only if $a_n = o(1/\ln n)$ and $\sum_{n=1}^{\infty} (a_n - a_{n+1}) \ln n < +\infty$.

2.6.21. We modify the harmonic series by taking the first term positive, the next two negative, the next three positive, etc. Show that this modified series is convergent.

E.P. Starke, Amer. Math. Monthly, Problem E824

2.6.22. Let $p > 1$ and assume that $(a_n)_{n\geq 1}$ is a decreasing sequence of positive numbers converging to 0. Prove that the series $\sum_{n=1}^{\infty} a_n^p$ is convergent if the series $\sum_{n=1}^{\infty} a_n^{p-1}/n^{1/p}$ converges. Is the monotonicity essential?

Hint. A counterexample is given by the sequence $(a_n)_{n\geq 1}$ defined by $a_n = 1$ for $n = 2^k$ and $a_n = 0$ otherwise.

2.6.23. Suppose that $a_n > 0$ and $\sum_{n=1}^{\infty} a_n < +\infty$. Prove that

(i) $\sum_{n=1}^{\infty} a_n^{(n-1)/n} < +\infty$;

(ii) $\sum_{n=1}^{\infty} \frac{a_n}{\ln(1+n)} \ln \frac{1}{a_n} < +\infty$.

2.6.24. Find the sum of the series

$$\sum_{n=1}^{\infty} \left\{ e - \left(1 + \frac{1}{n}\right)^n \right\}.$$

2.6.25. Suppose that $0 < a_n < 1$. Prove that if the series $\sum_{n=1}^{\infty} a_n/(\ln a_n)$ converges, then so does the series $\sum_{n=1}^{\infty} a_n/\ln(1+n)$. Is the converse true if $(a_n)_{n \geq 1}$ is a decreasing sequence of positive numbers converging to 0? Can the monotonicity be dropped?

2.6.26. Let $(a_n)_{n \geq 1}$ be a sequence of positive numbers such that the series $\sum_{n=1}^{\infty} a_n$ is convergent. For any $n \geq 1$, set $r_n = \sum_{k=n+1}^{\infty} a_k$.

(i) Prove that if the series $\sum_{n=1}^{\infty} r_n$ is convergent, then the series $\sum_{n=1}^{\infty} n a_n$ converges.

(ii) Prove that if the series $\sum_{n=1}^{\infty} n a_n$ is convergent, then the sequence $(n r_n)_{n \geq 1}$ converges to 0.

(iii) Deduce that the series $\sum_{n=1}^{\infty} n a_n$ is convergent if and only if the series $\sum_{n=1}^{\infty} r_n$ converges.

2.6.27. Let $(a_n)_{n \geq 1}$ and $(b_n)_{n \geq 1}$ be sequences of real numbers. Show that if both series $\sum_{n=1}^{\infty} b_n$ and $\sum_{n=1}^{\infty} |a_{n+1} - a_n|$ converge, then the series $\sum_{n=1}^{\infty} a_n b_n$ is convergent.

2.6.28. Let $(p_n)_{n \geq 1}$ be a decreasing sequence of positive numbers that converges to zero and let $(a_n)_{n \geq 1}$ be a sequence of real numbers. Show that if the series $\sum_{n=1}^{\infty} a_n$ converges, then

$$\frac{1}{p_n} \sum_{k=1}^{n} p_k a_k \to 0 \quad \text{as } n \to \infty. \tag{2.16}$$

Conversely, if $\sum_{n=1}^{\infty} a_n$ does not converge, show that there exists a decreasing sequence of positive numbers $(p_n)_{n \geq 1}$ that converges to zero such that relation (2.16) does not hold.

Chapter 3
Limits of Functions

> *No human investigation can be called real science if it cannot be demonstrated mathematically.*
> —Leonardo da Vinci (1452–1519)

Abstract. Some of the subjects treated so far in this volume have dealt with processes that can be carried out in a *finite* number of steps. These processes, however, are not adequate for the purposes of mathematics itself. That is why it is essential to manage *infinite* processes. The most important new idea introduced in this chapter is the notion of *limit of a function*. We have already met this idea in the case of sequences (that is, functions defined on *countable* sets) and we shall meet it over and over again in the next chapters. We present a sample of the various ways in which the notion of limit may appear in new situations, merely as a preview for deeper applications in the next chapters.

3.1 Main Definitions and Basic Results

> One says that a quantity is the *limit* of another quantity if the second approaches the first closer than any given quantity, however small....
>
> D'Alembert (1717–1783),
> *Encyclopédie*, tome neuvième, 1765, à Neufchastel

The concept of *limit* is fundamental in mathematical analysis. This notion describes the behavior of a certain system (described by a function) when the variable approaches a certain value. The first rigorous definition of the limit is due to Cauchy in his *Analyse Algébrique* [16], 1821. We describe in what follows the main definitions and properties related to the concept of limit of a function.

Let $I \subset \mathbb{R}$ be a set and let f be a real-valued function whose domain is I. Fix a point $x_0 \in \mathbb{R}$ that is an accumulation point of I. We say that f has *limit* $\ell \in \mathbb{R}$ at x_0, and we write

$$\lim_{x \to x_0} f(x) = \ell,$$

if for any $\varepsilon > 0$ there exists $\delta > 0$ (depending on ε) such that if $x \in I$, $x \neq x_0$, then $|f(x) - \ell| < \varepsilon$.

The notation "lim" is also due to Cauchy, who wrote in 1821: "When a variable quantity converges towards a fixed limit, it is often useful to indicate this limit by a specific notation, which we shall do by setting the abbreviation

$$\lim$$

in front of the variable in question...."

As mentioned by Pringsheim (*Enzyclopädie der Math. Wiss.*, 1899), "the concept of the *limit* of a function was probably first defined with sufficient rigor by Weierstrass."

This definition can be extended to the cases in which either $x_0 \in \overline{\mathbb{R}} := \mathbb{R} \cup \{-\infty, +\infty\}$ or $\ell \in \overline{\mathbb{R}}$. In the general framework, let $f : I \to \mathbb{R}$ and assume that $x_0 \in \overline{\mathbb{R}}$ is an accumulation point of I. We say that f has *limit* $\ell \in \overline{\mathbb{R}}$ as $x \to x_0$ if for every neighborhood V of ℓ there exists a neighborhood U of x_0 such that for every $x_0 \in U \cap I$, $x \neq x_0$, we have $f(x) \in V$. Roughly speaking, a function f has a limit ℓ at a point x_0 if the values of f at points x near x_0 can be made as close as we wish to ℓ by making x close to x_0.

If the values of x that approach x_0 are taken only to the right (resp., to the left) of x_0 we obtain the notions of *lateral* limits. In such cases we write $\lim_{x \to x_0+} f(x)$ or $\lim_{x \searrow x_0} f(x)$ (for the limit to the right), resp. $\lim_{x \to x_0-} f(x)$ or $\lim_{x \nearrow x_0} f(x)$ (for the limit to the left). Moreover, $\lim_{x \to x_0} f(x)$ exists if and only if $\lim_{x \nearrow x_0} f(x) = \lim_{x \searrow x_0} f(x)$. In such a case,

$$\lim_{x \to x_0} f(x) = \lim_{x \nearrow x_0} f(x) = \lim_{x \searrow x_0} f(x).$$

If the function f tends to a limit as $x \to x_0$, then this limit is unique. Thus, if $x_0, \ell, \ell' \in \overline{\mathbb{R}}$, and if $f(x) \to \ell$ and $f(x) \to \ell'$ as $x \to x_0$, then $\ell = \ell'$.

The existence or nonexistence of the limit of f as $x \to x_0$, as well as its value, depends uniquely on the values of f in a (small) neighborhood of x_0.

The behavior of some elementary functions around 0 or $\pm\infty$ is described by the following fundamental limits:

$$\lim_{x \to -\infty} a^x = \begin{cases} +\infty & \text{if } 0 < a < 1, \\ 0 & \text{if } a > 1, \end{cases} \qquad \lim_{x \to +\infty} a^x = \begin{cases} 0 & \text{if } 0 < a < 1, \\ +\infty & \text{if } a > 1, \end{cases}$$

$$\lim_{x \to 0} \log_a x = \begin{cases} +\infty & \text{if } 0 < a < 1, \\ -\infty & \text{if } a > 1, \end{cases} \qquad \lim_{x \to +\infty} \log_a x = \begin{cases} -\infty & \text{if } 0 < a < 1, \\ +\infty & \text{if } a > 1, \end{cases}$$

$$\lim_{x \to 0} x^a = \begin{cases} +\infty & \text{if } a < 0, \\ 0 & \text{if } a > 0, \end{cases} \qquad \lim_{x \to +\infty} x^a = \begin{cases} 0 & \text{if } a < 0, \\ +\infty & \text{if } a > 0. \end{cases}$$

The existence of the limit at some $x_0 \in \overline{\mathbb{R}}$ implies qualitative properties of the function in a neighborhood of x_0. Indeed, let $I \subset \mathbb{R}$ be an interval and assume

3.1 Main Definitions and Basic Results

that x_0 is an accumulation point of I and $f : I \to \mathbb{R}$ is a function. Then the following properties are true.

(i) *Local boundedness*: if $f(x) \to \ell \in \mathbb{R}$ as $x \to x_0 \in \overline{\mathbb{R}}$, then f is *locally bounded* in a neighborhood of x_0, that is, there exist $M > 0$ and a neighborhood U of x_0 such that $|f(x)| \leq M$ for any $x \in (U \setminus \{x_0\}) \cap I$.

(ii) *Constancy of sign*: if $f(x) \to \ell$ as $x \to x_0 \in \overline{\mathbb{R}}$, and $\ell > 0$ or $\ell = +\infty$ (resp., $\ell < 0$ or $\ell = -\infty$), then there exists a neighborhood U of x_0 such that

$$f(x) > 0 \text{ (resp. } f(x) < 0) \quad \text{for all } x \in (U \setminus \{x_0\}) \cap I.$$

We now state some basic rules for the calculus of limits. Let $f, g : I \to \mathbb{R}$ be two functions, x_0 an accumulation point of I, and assume that $f(x) \to \ell_1 \in \overline{\mathbb{R}}$ and $g(x) \to \ell_2 \in \overline{\mathbb{R}}$ as $x \to x_0$. Then:

(i) if $\ell_1 + \ell_2$ is defined, then $f(x) + g(x) \to \ell_1 + \ell_2$ as $x \to x_0$;
(ii) if $\ell_1 \ell_2$ is defined, then $f(x)g(x) \to \ell_1 \ell_2$ as $x \to x_0$;
(iii) if ℓ_1 / ℓ_2 is defined, then $f(x)/g(x) \to \ell_1 \ell_2$ as $x \to x_0$;
(iv) if $\ell_1 = 0$ and $f > 0$ in a neighborhood of x_0, then $f(x) \to +\infty$ as $x \to x_0$;
(v) if $\ell_1 = 0$ and $f < 0$ in a neighborhood of x_0, then $f(x) \to -\infty$ as $x \to x_0$.

When we operate on the extended real axis $\overline{\mathbb{R}} := \mathbb{R} \cup \{-\infty, +\infty\}$, expressions such as $+\infty + (-\infty)$, $-\infty + \infty$, and $\pm\infty / \pm\infty$ are not defined. Limits leading to such expressions are among the so-called *indeterminate forms*:

$$\frac{0}{0}, \quad \frac{\infty}{\infty}, \quad 0 \cdot \infty, \quad \infty - \infty, \quad 0^0, \quad 1^\infty, \quad \infty^0.$$

Two useful tools for computing limits of functions are the following.

Heine's Criterion. *Let $f : I \to \mathbb{R}$ be a function defined on an interval I and let x_0 be an accumulation point of I. Then $f(x) \to \ell$ as $x \to x_0$ if and only if $f(x_n) \to \ell$ as $n \to \infty$, for any sequence $(x_n)_{n \geq 1}$ in I converging to x_0.*

Squeezing and Comparison Test. *Let f, g, h be three functions defined on the interval I and let x_0 be an accumulation point of I. Assume that*

$$g(x) \leq f(x) \leq h(x) \quad \text{for all } x \in I.$$

If $g(x) \to \ell$ and $h(x) \to \ell$ as $x \to x_0$, then $f(x) \to \ell$ as $x \to x_0$.

In the calculation of limits it is often useful to operate by changing variables. This procedure ie described in what follows.

Change of Variable. *Let $f : I \to \mathbb{R}$ be a function defined on an interval I, let x_0 be an accumulation point of I, and assume that $f(x) \to \ell$ as $x \to x_0$. Let $g : J \to I$ be a function defined on an interval J such that $g(y) \to x_0$ as $y \to y_0$, where y_0 is an accumulation point of J. If $g(y) \neq x_0$ for all $y \neq y_0$ in a neighborhood of y_0, then $f(g(y)) \to \ell$ as $y \to y_0$.*

We conclude this preliminary section with an example of a function without a limit at any point. This is the *characteristic function* defined by

$$f(x) = \begin{cases} 1 & \text{if } x \text{ is rational,} \\ 0 & \text{if } x \text{ is irrational.} \end{cases}$$

Indeed, if $x_0 \in \mathbb{R}$ is arbitrary, let us consider two sequences $(a_n)_{n\geq 1} \subset \mathbb{Q}$ and $(b_n)_{n\geq 1} \subset \mathbb{R} \setminus \mathbb{Q}$ such that $a_n \to x_0$ and $b_n \to x_0$ as $n \to \infty$. Then $f(a_n) = 1$ and $f(b_n) = 0$ for all $n \geq 1$. Thus, by Heine's criterion, the limit $\lim_{x \to x_0} f(x)$ does not exist.

3.2 Computing Limits

> We have here, in fact, a *passage to the limit of unexampled audacity.*
>
> Felix Klein, 1908

We start with an elementary example that involves the indeterminate form 1^∞.

3.2.1. Assume that $\lim_{x \to 0} f(x) = 1$, $\lim_{x \to 0} g(x) = \infty$, $\lim_{x \to 0} g(x) \cdot (f(x) - 1) = c$. Prove that $\lim_{x \to 0} f(x)^{g(x)} = e^c$.

Solution. We have

$$f(x)^{g(x)} = (1 + (f(x) - 1))^{g(x)}.$$

Hence

$$\lim_{x \to 0} f(x)^{g(x)} = \lim_{x \to 0} (1 + (f(x) - 1))^{g(x)} = e^{\lim_{x \to 0} g(x)(1 + (f(x) - 1))} = e^c. \quad \square$$

If f has limit in $b := \lim_{x \to a} g(x)$, does there exist $\lim_{x \to a} (f \circ g)(x)$?

3.2.2. Establish whether the following assertion is true: let $f, g : \mathbb{R} \to \mathbb{R}$ be such that

$$\lim_{x \to a} g(x) = b \quad \text{and} \quad \lim_{x \to b} f(x) = c.$$

Then

$$\lim_{x \to a} f(g(x)) = c.$$

Solution. The assertion is false. Indeed, set

$$f(x) = g(x) = \begin{cases} 0 & \text{if } x \neq 0, \\ 1 & \text{if } x = 0. \end{cases}$$

Then $\lim_{x \to 0} g(x) = \lim_{x \to 0} f(x) = 0$, but $\lim_{x \to 0} f(g(x)) = 1$. $\quad \square$

3.2 Computing Limits

The following example shows how to use some assumptions for showing the existence of a limit. In some cases the value of the limit follows easily as soon as we know that it exists. In other cases it is more difficult to compute the limit.

3.2.3. Let $f : (0,\infty) \to \mathbb{R}$ be an arbitrary function satisfying the following hypotheses:

(i) $\lim_{x \to 0} x(f(x) - 1) = 0$;
(ii) $f(1) = 2$;
(iii) $(x+2)f(x+2) - 2(x+1)f(x+1) + xf(x) = 0$, for all $x \in (0,\infty)$.

(a) Prove that $\lim_{x \to 0} f(x)$ exists and, moreover, $\lim_{x \to 0} f(x) = \infty$.
(b) Assume that $\lim_{x \to \infty} f(x)$ exists. Compute this limit.

Solution. (a) By (i) it follows that $\lim_{x \to 0} xf(x) = 1$. In particular, this implies the existence of some $\alpha > 0$ such that $f > 0$ in $(0, \alpha)$. So, $\lim_{x \to 0} f(x)$ exists, and moreover, $\lim_{x \to 0} f(x) = \infty$.

(b) Let $g : (0,\infty) \to \mathbb{R}$ be the function defined by $g(x) = xf(x)$. Using (ii), we obtain

$$g(x+2) - 2g(x+1) + g(x) = 0, \quad \forall x \in (0,\infty).$$

In particular,
$$g(n+2) - 2g(n+1) + g(n) = 0, \quad \forall n \geq 1.$$

Set $c_n = g(n)$. Then $c_{n+2} - 2c_{n+1} + c_n = 0$, for all $n \in \mathbb{N}$. It follows that

$$\sum_{j=1}^{n}(c_{j+2} - 2c_{j+1} + c_j) = 0, \quad \forall n \in \mathbb{N},$$

that is, $c_{n+2} - c_{n+1} = c_2 - c_1$, for all $n \in \mathbb{N}$. This means that (c_n) is an arithmetic progression, and thus we obtain $c_{n+2} = n(c_2 - c_1) + c_2$. Hence

$$f(n+2) = \frac{n}{n+2}(2f(2) - f(1)) + \frac{2f(2)}{n+2}, \quad \forall n \in \mathbb{N}.$$

Since $\lim_{x \to \infty} f(x)$ exists, it follows that

$$\lim_{x \to \infty} f(x) = \lim_{n \to \infty} f(n+2) = (2f(2) - f(1)). \quad \square$$

3.2.4. Let $f : [0,\infty) \to \mathbb{R}$ be a function satisfying

$$f(x)e^{f(x)} = x, \quad \text{for all } x \in [0,\infty).$$

Prove that

(a) f is monotone.
(b) $\lim_{x \to \infty} f(x) = \infty$.
(c) $f(x)/\ln x$ tends to 1 as $x \to \infty$.

Solution. (a) By our hypothesis it follows that $f(x) = xe^{-f(x)}$, for any $x \geq 0$. We prove that f is increasing on $[0, \infty)$. Indeed, arguing by contradiction, it follows that there exist $0 \leq x_1 < x_2$ such that $f(x_1) \geq f(x_2)$. Hence $e^{-f(x_1)} \leq e^{-f(x_2)}$, that is, $x_1 e^{-f(x_1)} < x_2 e^{-f(x_2)} \Leftrightarrow f(x_1) < f(x_2)$, contradiction. This means that f is increasing.

(b) By (a), it is enough to show that f is not bounded above. Arguing by contradiction, there exists $M > 0$ such that $f(x) < M$, for all $x \geq 0$. Let $x_M = Me^M$. We have

$$f(x_M) < M \Longrightarrow e^{f(x_M)} < e^M \Longrightarrow x_M = f(x_M)e^{f(x_M)} < f(x_M)e^M \leq Me^M,$$

which contradicts the choice of x_M. It follows that f is not bounded above on $[0, \infty)$.

(c) Our hypothesis implies that $f(x) > 0$, for all $x > 0$. Thus

$$\frac{\ln x}{f(x)} = 1 + \frac{\ln f(x)}{f(x)}, \quad \forall x > 0.$$

It follows that

$$\lim_{x \to \infty} \frac{\ln x}{f(x)} = 1 + \lim_{x \to \infty} \frac{\ln f(x)}{f(x)}.$$

Since $\lim_{y \to \infty} \ln y / y = 0$, the above relation implies that $\lim_{x \to \infty} f(x) / \ln x = 1$. □

3.2.5. Let $f, g : \mathbb{R} \to \mathbb{R}$ be periodic functions of periods a and b such that $\lim_{x \to 0} f(x)/x = u \in \mathbb{R}$ and $\lim_{x \to 0} g(x)/x = v \in \mathbb{R} \setminus \{0\}$. Compute

$$\lim_{n \to \infty} \frac{f((3+\sqrt{7})^n a)}{g((2+\sqrt{2})^n b)}.$$

Solution. We have

$$\frac{f((3+\sqrt{7})^n a)}{g((2+\sqrt{2})^n b)} = \frac{f\left((3+\sqrt{7})^n a + (3-\sqrt{7})^n a - (3-\sqrt{7})^n a\right)}{g\left((2+\sqrt{2})^n b + (2-\sqrt{2})^n b - (2-\sqrt{2})^n b\right)}.$$

But $(3+\sqrt{7})^n + (3-\sqrt{7})^n \in \mathbb{Z}$ and $(2+\sqrt{2})^n + (2-\sqrt{2})^n \in \mathbb{Z}$. It follows that

$$\frac{f((3+\sqrt{7})^n a)}{g((2+\sqrt{2})^n b)} = \frac{f(-(3-\sqrt{7})^n a)}{g(-(2-\sqrt{2})^n b)}, \quad \forall n \geq 1.$$

Hence

$$\lim_{n \to \infty} \frac{f((3+\sqrt{7})^n a)}{g((2+\sqrt{2})^n b)}$$

$$= \frac{a}{b} \cdot \lim_{n \to \infty} \left\{ \frac{f(-(3-\sqrt{7})^n a)}{-(3-\sqrt{7})^n a} \cdot \frac{-(2-\sqrt{2})^n b}{g(-(2-\sqrt{2})^n b)} \cdot \frac{(3-\sqrt{7})^n}{(2-\sqrt{2})^n} \right\}.$$

3.2 Computing Limits

Using the above relations combined with the fact that the positive numbers $3-\sqrt{7}$, $2-\sqrt{2}$, and $3-\sqrt{7}/2-\sqrt{2}$ are less than 1, it follows that the required limit is 0. □

3.2.6. Let $f:(0,\infty)\to\mathbb{R}$ be a function satisfying

$$\lim_{x\to\infty}\frac{f(x)}{x^k}=a\in\overline{\mathbb{R}},\quad k\in\mathbb{N}^*\setminus\{1\},$$

and

$$\lim_{x\to\infty}\frac{f(x+1)-f(x)}{x^{k-1}}=b\in\overline{\mathbb{R}}.$$

Prove that $b=ka$.

Solution. Our assumption implies $\lim_{n\to\infty}f(n+1)-f(n)/n^{k-1}=b$. Define $u_n=f(n)$ and $v_n=n^k$. Hence

$$\lim_{n\to\infty}\frac{u_{n+1}-u_n}{v_{n+1}-v_n}=\lim_{n\to\infty}\frac{u_{n+1}-u_n}{n^{k-1}}\cdot\lim_{n\to\infty}\frac{n^{k-1}}{v_{n+1}-v_n}$$

$$=\lim_{n\to\infty}\frac{f(n+1)-f(n)}{n^{k-1}}\cdot\frac{n^{k-1}}{(n+1)^k-n^k}=\frac{a}{k}.$$

On the other hand, by the Stolz–Cesáro lemma,

$$\lim_{n\to\infty}\frac{f(n+1)-f(n)}{(n+1)^k-n^k}=\lim_{n\to\infty}f(n)n^k.$$

Combining these relations, we deduce that $b=ka$. □

3.2.7. Let f be a real-valued function on the open interval $(0,1)$ such that $\lim_{x\to 0}f(x)=0$ and $\lim_{x\to 0}[f(x)-f(x/2)]/x=0$.
Compute $\lim_{x\to 0}f(x)/x$.

Solution. Fix $\varepsilon>0$. Since $\lim_{x\to 0}[f(x)-f(x/2)]/x=0$, there exists $\delta>0$ such that

$$\left|f(x)-f\left(\frac{x}{2}\right)\right|<\varepsilon x\quad\text{for all }x\in(0,\delta).$$

Thus, for any positive integer n,

$$\left|f\left(\frac{x}{2^{n-1}}\right)-f\left(\frac{x}{2^n}\right)\right|<\varepsilon\frac{x}{2^{n-1}}.$$

Summing these inequalities for $n=1,\ldots,m$, we obtain

$$\left|f(x)-f\left(\frac{x}{2^n}\right)\right|\leq\left|f(x)-f\left(\frac{x}{2}\right)\right|+\cdots+\left|f\left(\frac{x}{2^{n-1}}\right)-f\left(\frac{x}{2^n}\right)\right|$$

$$\leq\varepsilon x\left(1+\frac{1}{2}+\cdots+\frac{1}{2^{n-1}}\right)=2\varepsilon x\left(1-\frac{1}{2^n}\right)<2\varepsilon x,$$

for all $x\in(0,\delta)$ and any positive integer n.

Next, we observe that since $\lim_{x\to 0} f(x) = 0$, then

$$\lim_{n\to\infty}\left|f(x) - f\left(\frac{x}{2^n}\right)\right| = |f(x)|, \quad \text{for any } x \in (0,\delta).$$

We conclude that $|f(x)| \leq 2\varepsilon x$ for any $x \in (0,\delta)$, which shows that $\lim_{x\to 0} f(x)/x = 0$. □

3.2.8. *Find all positive numbers a, b, c such that*

$$\lim_{x\to\infty}\left(\frac{a^{1/x}+b}{c}\right)^x$$

exists and is finite. In this case, determine the value of the limit.

Solution. The limit exists if and only if $c \geq b+1$. If $c = b+1$, the limit is $a^{1/c}$. If $c > b+1$, the limit is zero. First assume $c = b+1$. Then

$$\frac{a^{1/x}+b}{c} = 1 + \frac{a^{1/x}-1}{c} = 1 + \frac{k_x}{x},$$

where $k_x = x(a^{1/x}-1)/c$. We observe that $\lim_{x\to\infty} k_x = (\ln a)/c$ and

$$\left(1+\frac{k_x}{x}\right)^x = \left[\left(1+\frac{k_x}{x}\right)^{x/k_x}\right]^{k_x}. \tag{3.1}$$

Hence

$$\lim_{x\to\infty}\frac{k_x}{x} = 0 \quad\text{and}\quad \lim_{x\to\infty}\left(1+\frac{k_x}{x}\right)^{x/k_x} = e.$$

Taking limits in (3.1) we obtain

$$\lim_{x\to\infty}\left(\frac{a^{1/x}+b}{c}\right)^x = \lim_{x\to\infty}\left(1+\frac{k_x}{x}\right)^x = e^{(\ln a)/c} = a^{1/c}.$$

If $c > b+1$, then since $\lim_{x\to\infty} a^{1/x} = 1$, there is a positive number δ such that $(a^{1/x}+b)/c < 1-\delta$ for all x sufficiently large. The conclusion follows after observing that $\lim_{x\to\infty}(1-\delta)^x = 0$. □

If $c < b+1$, there is a positive number δ such that $(a^{1/x}+b)/c > 1+\delta$ for all x sufficiently large. Now, we use $\lim_{x\to\infty}(1+\delta)^x = +\infty$.

3.2.9. *Compute with elementary arguments*

$$\lim_{x\to 0}\frac{\sin x}{x}.$$

Solution. Let P_n denote the perimeter of a regular n-gon inscribed in a circle of radius R, then $P_n = 2nR\sin\pi/n$, and we know from plane geometry that

3.2 Computing Limits

$\lim_{n \to \infty} P_n = 2\pi R$. Hence $\lim_{n \to \infty} (n/\pi) \sin \pi/n = 1$, and if we let $\pi/n = x$, then $x \to 0$ as $n \to \infty$ and conversely. Thus, $\lim_{x \to 0} \sin x/x = 1$.

The function $\mathbb{R} \setminus \{0\} \ni x \longmapsto \sin x/x$ is sometimes called the *sinc* function (or *sinus cardinalis*). The property established in the above exercise shows that this function has a *removable singularity* at the origin. Thus, we can define

$$\operatorname{sinc}(x) = \begin{cases} \frac{\sin x}{x} & \text{if } x \in \mathbb{R} \setminus \{0\}, \\ 1 & \text{if } x = 0. \end{cases} \quad \square$$

This function is useful in information theory and digital signal processing.

3.2.10. Prove that

$$\lim_{x \to +\infty} \left(1 + \frac{1}{x}\right)^x = e;$$

$$\lim_{x \to -\infty} \left(1 + \frac{1}{x}\right)^x = e.$$

Solution. To compute the first limit, let n_x denote the unique integer such that $n_x \leq x < n_x + 1$, where $x > 1$ is a fixed real number. We have

$$\left(1 + \frac{1}{x}\right)^x \leq \left(1 + \frac{1}{n_x}\right)^{n_x+1} = \left(1 + \frac{1}{n_x}\right)^{n_x} \left(1 + \frac{1}{n_x}\right)$$

and

$$\left(1 + \frac{1}{x}\right)^x \geq \left(1 + \frac{1}{n_x+1}\right)^{n_x} = \left(1 + \frac{1}{n_x+1}\right)^{n_x+1} \cdot \frac{1}{1 + \frac{1}{n_x+1}}.$$

Taking $x \to +\infty$, we have $n_x \to \infty$, and both right-hand sides of the above inequalities converge to e, because $(1 + n^{-1})^n \to e$ as $n \to \infty$. Thus, by the squeezing and comparison tests, we conclude that $\lim_{x \to +\infty} (1 + x^{-1})^x = e$.

For the second limit, we put $y = -x$ and we have to prove that

$$\lim_{y \to +\infty} \left(1 - \frac{1}{y}\right)^{-y} = e.$$

But

$$\left(1 - \frac{1}{y}\right)^{-y} = \left(\frac{y}{y-1}\right)^y = \left(1 + \frac{1}{y-1}\right)^{y-1} \left(1 + \frac{1}{y-1}\right) \to e$$

as $y \to +\infty$, in virtue of the previous limit.

The following exercise implies the following asymptotic estimates:

$$\ln(1+x) = x + x^2 o(1) \quad \text{as } x \to 0,$$
$$e^x = 1 + x + x^2 o(1) \quad \text{as } x \to 0. \quad \square$$

3.2.11. Fix $a > 0$, $a \neq 1$. Prove that

$$\lim_{x \to 0} \frac{\log_a(1+x)}{x} = \log_a e;$$

$$\lim_{x \to 0} \frac{a^x - 1}{x} = \ln a.$$

Solution. For the first limit we observe that

$$\frac{\log_a(1+x)}{x} = \log_a(1+x)^{1/x},$$

and then we use $\lim_{x \to 0}(1+x)^{1/x} = e$.

To compute the second limit, we put $a^x - 1 = y$. Hence $x = \log_a(1+y)$ and $(a^x - 1)/x = y/\log_a(1+y)$. Since $y \to 0$ as $x \to 0$, this limit reduces to the previous one.

We argue in what follows that for any $a \in \mathbb{R}$,

$$(1+x)^a = 1 + ax + x^2 o(1) \quad \text{as } x \to 0. \quad \square$$

3.2.12. Let a be a real number. Prove that

$$\lim_{x \to 0} \frac{(1+x)^a - 1}{x} = a.$$

Solution. We have

$$(1+x)^a - 1 = e^{a \ln(1+x)} - 1 = e^{ax + x^2 o(1)} - 1 = e^{ax} e^{x^2 o(1)} - 1$$
$$= \left(1 + ax + x^2 o(1)\right)\left(1 + x^2 o(1)\right) - 1 = ax + x^2 o(1),$$

as $x \to 0$. Thus,

$$\frac{(1+x)^a - 1}{x} = a + o(1) \quad \text{as } x \to 0. \quad \square$$

3.3 Qualitative Results

> Read Euler: he is our master in everything.
>
> Pierre-Simon de Laplace (1749–1827)

3.3.1. Let $f : \mathbb{R} \to \mathbb{R}$ be such that there exists $M > 0$ satisfying

$$|f(x_1 + \cdots + x_n) - f(x_1) - \cdots - f(x_n)| \leq M, \quad \forall x_1, \ldots, x_n \in \mathbb{R}.$$

Prove that $f(x+y) = f(x) + f(y)$, for all $x, y \in \mathbb{R}$.

3.3 Qualitative Results

Solution. We argue by contradiction and assume that there exist $a,b \in \mathbb{R}$ such that $f(a+b) \neq f(a)+f(b)$. This means that there exists $\varepsilon \neq 0$ such that $f(a+b) = f(a)+f(b)+\varepsilon$. Hence $f(na+nb)-nf(a+b) = f(na+nb)-nf(a)-nf(b)-n\varepsilon$. By hypothesis we obtain

$$|f(na+nb)-nf(a+b)| \leq M.$$

Our hypothesis also yields

$$|f(na+nb)-nf(a)-nf(b)| \leq M,$$

which implies

$$\lim_{n\to\infty} |f(na+nb)-nf(a)-nf(b)-n\varepsilon| = +\infty,$$

a contradiction. □

3.3.2. Let $f,g: \mathbb{R} \to \mathbb{R}$ be two periodic functions satisfying

$$\lim_{x\to\infty}[f(x)-g(x)] = 0.$$

Prove that

(a) f and g have the same period;
(b) f and g are equal.

Solution. (a) Let $T_1 > 0$, resp. $T_2 > 0$, be the period of f, resp. g. Fix $x_0 \in \mathbb{R}$. Then, for any $n \in \mathbb{N}$,

$$f(x_0+nT_1+T_2)-g(x_0+nT_1+T_2) = f(x_0+T_2)-g(x_0+nT_1).$$

As $n\to\infty$, the left-hand side tends to 0. So

$$\lim_{n\to\infty} g(x_0+nT_1) = f(x_0+T_2).$$

On the other hand, $\lim_{n\to\infty}[g(x_0+nT_1)-f(x_0+nT_1)] = 0$, and by means of the periodicity of f,

$$\lim_{n\to\infty} g(x_0+nT_1) = f(x_0). \quad \square$$

It follows that $f(x_0+T_2) = f(x_0)$, which shows that f is T_2-periodic. Analogously, f is T_1-periodic.

(b) Let $T > 0$ be the common period. Then, for any $x_0 \in \mathbb{R}$,

$$f(x_0)-g(x_0) = \lim_{n\to\infty}[f(x_0+nT)-g(x_0+nT)] = 0.$$

3.3.3. Find all real numbers p and q such that

$$2^{px}+3^{qx} = p \cdot 2^{x+p-1}+q \cdot 3^{x+q-1},$$

for any real number x.

Solution. For $x=1$ we have $(p-1)\cdot 2^p + (q-1)\cdot 3^q = 0$. Since $2^p > 0$ and $3^q > 0$, it follows that only the following situations can occur:

CASE 1: $p = q = 1$. In this case the relation is fulfilled.

CASE 2: $p < 1 < q$. This implies $2^p < 2 < 3 < 3^q$, and after dividing by 3^{qx}, we obtain

$$\left(\frac{2^p}{3^q}\right)^x + 1 = p\cdot 2^{p-1}\left(\frac{2}{3^q}\right)^x + q\cdot 3^{q-1}\left(\frac{3}{3^q}\right)^x.$$

Since $\lim_{x\to\infty}(2^p/3^q)^x + 1 = 1$ and

$$\lim_{x\to\infty}\left(p\cdot 2^{p-1}\left(\frac{2}{3^q}\right)^x + q\cdot 3^{q-1}\left(\frac{3}{3^q}\right)^x\right) = 0,$$

we deduce that our assumption cannot be true in this case.

CASE 3: $p > 1 > q$. Our hypothesis implies

$$\left(\frac{2^p}{3}\right)^x + \left(\frac{3^q}{3}\right)^x = p\cdot 2^{p-1}\left(\frac{2}{3}\right)^x + q\cdot 3^{q-1}.$$

It follows that $\lim_{x\to\infty}(2^p/3)^x = q\cdot 3^{q-1}$. Since the limit on the left-hand side is finite, it follows that just the following situations can occur:

(i) $2^p/3 < 1$ and $q\cdot 3^{q-1} = 0$. We deduce that $q = 0$ and

$$2^{px} + 1 = p\cdot 2^{x+p-1}, \quad \forall x \in \mathbb{R}.$$

Dividing by 2^{px} and passing to the limit as $x \to \infty$ we obtain $1 = 0$, so this situation is excluded.

(ii) $2^p = 3$ and $q\cdot 3^{q-1} = 1$. Since $f(x) = x\cdot 3^{x-1}$ increases on $(0, \infty)$, it follows that f is one-to-one, so the equation $q\cdot 3^{q-1} = 1$ has the unique solution $q = 1$. Taking into account that $p > 1$, it follows that $\lim_{x\to\infty} 2^{(p-1)x} = \infty$. So, passing to the limit as $x \to \infty$, we obtain $0 = p\cdot 2^{p-1}$, contradiction. □

Consequently, the only possible case is $p = q = 1$.

3.3.4. Let $f: [0, \infty) \to \mathbb{R}$ be a function such that

(i) $f(x) + f(y) \leq f(x+y)$, for all $x, y \geq 0$;
(ii) there exists $M > 0$ such that $|f(x)| \leq Mx$, for any $x \geq 0$.

Prove that the following properties hold:

(a) if there exists a sequence (x_n) such that $\lim_{n\to\infty} x_n = +\infty$ and $\lim_{n\to\infty} f(x_n)/x_n = \ell$, then $f(x) \leq \ell x$, for all $x > 0$;
(b) there exists $\lim_{x\to\infty} f(x)/x$.

Solution. (a) Let (x_n) be a sequence of real numbers such that $\lim_{n\to\infty} x_n = +\infty$ and $\lim_{n\to\infty} f(x_n)/x_n = \ell$. For $y > 0$ arbitrarily chosen, there exist $k_n \in \mathbb{N}$ and

$z \in [0, y)$ such that $x_n = k_n y + z$. On the other hand, we obtain by induction that $f(nx) \geq n f(x)$, for all $n \in \mathbb{N}$ and $x \geq 0$. Hence

$$f(x_n) \geq f(k_n y) + f(z) \geq k_n f(y) + f(z),$$

that is,

$$\frac{f(x_n)}{x_n} \geq \frac{k_n f(y)}{k_n y + z} + \frac{f(z)}{k_n y + z}.$$

Passing to the limit as $n \to \infty$ we obtain $\ell \geq f(y)/y$, for all $y > 0$.

(b) Assume, by contradiction, that there are two sequences (x_n) and (y_n) such that $\lim_{n \to \infty} x_n = \lim_{n \to \infty} y_n = \infty$, $\lim_{n \to \infty} f(x_n)/x_n = \ell_1$, and $\lim_{n \to \infty} f(y_n)/y_n = \ell_2$, with $\ell_1 \neq \ell_2$. Then, by (a), $\ell_1 > f(y)/y$, for all $y > 0$. It follows that in the neighborhood $(\ell_1 + \ell_2/2, \ell_1 + \ell_2)$ of ℓ_2 there exists no element of the form $f(y_n)/y_n$, contradiction. This shows that $\ell_1 = \ell_2$, so $\lim_{x \to \infty} f(x)/x$ exists. □

3.3.5. (a) Construct a function $f : \mathbb{R} \to [0, \infty)$ such that any point $x \in \mathbb{Q}$ is a local strict minimum point of f.

(b) Construct a function $f : \mathbb{Q} \to [0, \infty)$ such that any point is a local strict minimum point and f is unbounded on any set of the form $I \cap \mathbb{Q}$, where I is a nondegenerate interval.

(c) Let $f : \mathbb{R} \to [0, \infty)$ be a function that is unbounded on any set of the form $I \cap \mathbb{Q}$, where I is a nondegenerate interval. Prove that f does not have the property stated in (a).

Romanian Mathematical Olympiad, 2004

Solution. (a) Define f by $f(x) = 1$ if $x \in \mathbb{R} \setminus \mathbb{Q}$ and $f(x) = 1 - 1/p$, provided that $x = n/p$, $(n, p) = 1$, $p > 0$. We observe that $x = n/p$ is a local strict minimum point because there exists a neighborhood of x not containing rational numbers of the form a/b with $(a, b) = 1$, $b > 0$ and $b \in \{1, 2, \ldots, p\}$.

(b) Consider the function $f(n/p) = p$.

(c) We argue by contradiction and assume that such a function exists. We construct a sequence of nondegenerate intervals as follows. Let $I = [a_0, b_0]$ be a nondegenerate interval, and for given $n \geq 1$, consider the intervals $I_k = [a_k, b_k]$, $k = 0, 1, \ldots, n-1$, such that $I_0 \supset I_1 \supset \cdots \supset I_{n-1}$ and, for any $0 \leq k \leq n-1$, $f(I_k) \subset [k, \infty)$. Fix $x_n \in \mathbb{Q} \cap I_{n-1}$ with $f(x_n) \geq n$ and consider a neighborhood $I_n = [a_n, b_n] \subset I_{n-1}$ of x_n such that for all $x \in I_n$, $f(x) \geq f(x_n) \geq n$. Since $(a_n)_{n \geq 0}$ is nondecreasing, $(b_n)_{n \geq 0}$ is nonincreasing, and $a_n < b_n$, there exists $c \in \cap_{n \geq 0} I_n$. Thus, for all positive integers n, $f(c) \geq n$. This contradiction concludes our proof. □

3.3.6. Let $f : (A, \infty) \to \mathbb{R}$ be a function satisfying $\lim_{x \to \infty} [f(x+1) - f(x)] = +\infty$ and such which f is bounded on every bounded interval contained in (A, ∞). Prove that $\lim_{x \to \infty} f(x)/x = +\infty$.

Solution. Arguing by contradiction, there exists a sequence $(x_n)_{n\geq 1}$ and a real number M (which can be supposed to be positive) such that

$$x_n \to \infty \text{ as } n \to \infty \text{ and } f(x_n) \leq Mx_n, \text{ for all integers } n \geq 1.$$

On the other hand, since $\lim_{x\to\infty}[f(x+1) - f(x)] = +\infty$, there exists $a > A$ such that $f(x+1) - f(x) \geq 2M$, for all $x \geq a$. For all $n \geq 1$, write $x_n = a + k_n + r_n$, where $k_n = [x_n - a]$. Thus, $r_n \in [0, 1)$, and by our hypothesis, $k_n \to \infty$ as $n \to \infty$. So, for all $n \geq 1$,

$$Mx_n - f(a) \geq f(x_n) - f(a)$$
$$= \sum_{j=0}^{k_n}[f(x_n - j + 1) - f(x_n - j)] + f(a + r_n) - f(a)$$
$$\geq 2Mk_n + f(a + r_n) - f(a)$$
$$> 2Mx_n - 2M(a+1) + \inf_{a \leq x < a+1} f(x) - f(a). \quad \square$$

Taking n sufficiently large, the above inequality yields a contradiction.

3.3.7. *Prove that there is no function $f : \mathbb{R} \to \mathbb{R}$ with $f(0) > 0$ such that*

$$f(x+y) \geq f(x) + yf(f(x)) \quad \text{for all } x, y \in \mathbb{R}.$$

Solution. Suppose that such a function exists. If $f(f(x)) \leq 0$ for all $x \in \mathbb{R}$, then f is a decreasing function in view of the inequalities $f(x+y) \geq f(x) + yf(f(x)) \geq f(x)$ for any $y \leq 0$. Since $f(0) > 0 \geq f(f(x))$, then $f(x) > 0$ for all x, which is a contradiction. Hence, there exists $z \in \mathbb{R}$ such that $f(f(z)) > 0$. Then the inequality $f(z+x) \geq f(z) + xf(f(z))$ yields $\lim_{x\to\infty} f(x) = +\infty$ and therefore $\lim_{x\to\infty} f(f(x)) = +\infty$. In particular, there exist $x, y > 0$ such that $f(x) \geq 0$, $f(f(x)) > 1$, $y \geq x + 1/f(f(x)) - 1$, and $f(f(x+y+1)) \geq 0$. Then $f(x+y) \geq f(x) + yf(f(x)) \geq x + y + 1$ and hence

$$f(f(x+y)) \geq f(x+y+1) + (f(x+y) - (x+y+1))f(f(x+y+1))$$
$$\geq f(x+y+1) \geq f(x+y) + f(f(x+y))$$
$$\geq f(x) + yf(f(x)) + f(f(x+y)) > f(f(x+y)). \quad \square$$

This contradiction completes the solution of the problem.

3.3.8. *For any monic polynomial $f \in Q[X]$ of even degree, prove that there exists a polynomial $g \in Q[X]$ such that $\lim_{|x|\to\infty}(\sqrt{f(x)} - g(x)) = 0$.*

Gabriel Dospinescu

Solution. If f has degree $2n$, define $g(x) = x^n + a_1x_{n-1} + \cdots + a_0$. Next, we compute the coefficients a_j inductively on j such that $f(x) - g^2(x)$ has degree at most $n - 1$. This is always possible, since we obtain a linear equation in a_j that

3.3 Qualitative Results

ensures that a_j is rational. To conclude, it is enough to observe that

$$\sqrt{f(x)} - g(x) = \frac{f(x) - g^2(x)}{\sqrt{f(x)} + g(x)}. \qquad \square$$

3.3.9. *Let $a_1, \ldots, a_p$ be real numbers and let $b_1, \ldots, b_p$ be distinct positive numbers with b_1 being the greatest of them, and let $\alpha \in \mathbb{R} \setminus \mathbb{N}$. If there exists an infinite set of real numbers $A \subset (-b_1, \infty)$ such that*

$$\sum_{i=1}^{p} a_i (x + b_i)^\alpha = 0 \quad \text{for all } x \in A,$$

prove that $a_1 = \cdots = a_p = 0$.

Marian Tetiva, Problem 122, GMA 4(2002)

Solution. The infinite set A has an accumulation point x_0, which lies in $[-b_1, \infty) \cup \{\infty\}$. We then can find a sequence $(x_n)_{n \geq 1}$ of elements from A (which can be assumed to be distinct and all different from x_0) which has the limit x_0. There are three possible cases.

(i) $x_0 = \infty$. Since

$$\sum_{i=1}^{p} a_i \left(1 + \frac{b_i}{x_n}\right)^\alpha = 0, \quad \forall n \in \mathbb{N}^*,$$

and $\lim_{n \to \infty} x_n = \infty$, we get (passing to the limit for $n \to \infty$ in the above equality) $\sum_{i=1}^{p} a_i = 0$. Using this and again the assumed equality we have

$$x_n \sum_{i=1}^{p} a_i \left[\left(1 + \frac{b_i}{x_n}\right)^\alpha - 1\right] = 0,$$

for all $n \geq 1$. Passing to the limit here, and using

$$\lim_{x \to \infty} x \left[\left(1 + \frac{b}{x}\right)^\alpha - 1\right] = \alpha b$$

(for some real number b), we get $\alpha \sum_{i=1}^{p} a_i b_i = 0$; thus $\sum_{i=1}^{p} a_i b_i = 0$ (because $\alpha \neq 0$). Now one can write

$$x_n^2 \sum_{i=1}^{p} a_i \left[\left(1 + \frac{b_i}{x_n}\right)^\alpha - 1 - \alpha \frac{b_i}{x_n}\right] = 0, \quad \forall n \in \mathbb{N}^*,$$

and in this new relation one can pass to the limit for $n \to \infty$, to obtain $\sum_{i=1}^{p} a_i b_i^2 = 0$ (after division by $\alpha(\alpha - 1)/2 \neq 0$), since

$$\lim_{x \to \infty} x^2 \left[\left(1 + \frac{b}{x}\right)^\alpha - 1 - \alpha \frac{b}{x}\right] = \frac{\alpha(\alpha - 1)}{2} b^2.$$

And so on. So clearly, we will obtain $\sum_{i=1}^{p} a_i b_i^k = 0$, for all natural numbers k, hence the conclusion $a_1 = \cdots = a_p = 0$ is plain. (Indeed, the equations $\sum_{i=1}^{p} b_i^k a_i = 0$ for $k = 0, 1, \ldots, p-1$ form a homogeneous system with nonzero determinant, the Vandermonde determinant of the distinct numbers $b_1, \ldots, b_p$.)

(ii) Let now $x_0 \in (-b_1, \infty)$. From the given conditions we infer
$$\sum_{i=1}^{p} a_i (x_n + b_i)^\alpha = 0, \quad \forall n \in \mathbb{N}^*,$$
which yields (by letting n tend to infinity)
$$\sum_{i=1}^{p} a_i (x_0 + b_i)^\alpha = 0.$$
Then we have
$$\sum_{i=1}^{p} a_i \frac{(x_n + b_i)^\alpha - (x_0 + b_i)^\alpha}{x_n - x_0} = 0, \quad \forall n \in \mathbb{N}^*,$$
and passing to the limit, we get $\alpha \sum_{i=1}^{p} a_i (x_0 + b_i)^{\alpha-1} = 0$, due to the limit
$$\lim_{x \to x_0} \frac{(x+b)^\alpha - (x_0+b)^\alpha}{x - x_0} = \alpha (x_0 + b)^{\alpha-1}.$$
But $\alpha \neq 0$; therefore $\sum_{i=1}^{p} a_i (x_0 + b_i)^{\alpha-1} = 0$, and so we get
$$\sum_{i=1}^{p} a_i \frac{(x_n + b_i)^\alpha - (x_0 + b_i)^\alpha - \alpha(x_0 + b_i)^{\alpha-1}(x_n - x_0)}{(x_n - x_0)^2} = 0, \quad \forall n \in \mathbb{N}^*.$$
Together with the limit
$$\lim_{x \to x_0} \frac{(x+b)^\alpha - (x_0+b)^\alpha - \alpha(x_0+b)^{\alpha-1}(x - x_0)}{(x - x_0)^2} = \frac{\alpha(\alpha - 1)}{2} (x_0 + b)^{\alpha-2},$$
this leads us to
$$\frac{\alpha(\alpha - 1)}{2} \sum_{i=1}^{p} a_i (x_0 + b_i)^{\alpha-2} = 0,$$
hence to $\sum_{i=1}^{p} a_i (x_0 + b_i)^{\alpha-2} = 0$ (as long as we know that $\alpha \neq 0$, $\alpha \neq 1$).

We will get, continuing this process, $\sum_{i=1}^{p} a_i (x_0 + b_i)^{\alpha-k} = 0$, for all natural numbers k. Considering the first p (for $k = 0, 1, \ldots, p - 1$) equations of this type as a homogeneous system with unknowns $a_1, \ldots, a_p$, we see that its determinant is

$$(x_0 + b_1)^\alpha \cdots (x_0 + b_p)^\alpha \begin{vmatrix} 1 & \cdots & 1 \\ \frac{1}{x_0+b_1} & \cdots & \frac{1}{x_0+b_p} \\ \cdots & \cdots & \cdots \\ \frac{1}{(x_0+b_1)^{p-1}} & \cdots & \frac{1}{(x_0+b_p)^{p-1}} \end{vmatrix} \neq 0,$$

3.3 Qualitative Results

because $1/(x_0+b_1), \ldots, 1/(x_0+b_p)$ are distinct; consequently the system has the sole solution $a_1 = \cdots = a_p = 0$.

(iii) The case $x_0 = -b_1$ remains, but this does not differ very much from the previous one. Indeed, we have now $x_0 + b_1 = 0$, and by the same procedure as above we can obtain

$$\sum_{i=2}^{p} a_i(x_0+b_i)^{\alpha-k} = 0, \quad \forall k \in \mathbb{N}.$$

Then, with the same reasoning, we get $a_2 = \cdots = a_p = 0$, which, together with the given condition, leads to $a_1(x+b_1)^\alpha = 0$ for all $x \in A$ and thus to $a_1 = 0$. The proof is now complete. □

We point out that the above statement does not remain true if α is a natural number. Indeed, in such a case we can choose nonzero numbers $a_1, \ldots, a_p$ such that $\sum_{i=1}^{p} a_i b_i^k = 0$ for $k = 0, 1, \ldots, p-2$ ($b_1, \ldots, b_p$ being given), and we obtain, with Newton's binomial, $\sum_{i=1}^{p} a_i(x+b_i)^\alpha = 0$, for all positive integers $\alpha \leq p-2$.

3.3.10. Let $(a_n)_{n \geq 1}$ be an increasing sequence of positive numbers. Prove that the series $\sum_{n=1}^{\infty} \arccos^2 a_n/a_{n+1}$ converges if and only if the sequence $(a_n)_{n \geq 1}$ is bounded.

Solution. If a_n/a_{n+1} does not converge to 1, then the series diverges. Let us now assume that $a_n/a_{n+1} \to 1$ as $n \to \infty$. The basic idea in the proof is that

$$\lim_{x \to 1-0} \frac{\arccos x}{\sqrt{2(1-x)}} = 1.$$

Hence, the series $\sum_{n=1}^{\infty} \arccos^2 a_n/a_{n+1}$ converges simultaneously with the series $\sum_{n=1}^{\infty} (1 - a_n/a_{n+1})$, and the latter converges simultaneously with the series

$$\sum_{n=1}^{\infty} \ln\left(\frac{a_n}{a_{n+1}}\right) = \lim_{n \to \infty} \ln\left(\frac{a_1}{a_n}\right). \quad \square$$

This shows that the given series converges if and only if the sequence $(a_n)_{n \geq 1}$ is bounded.

3.3.11. Let $f, g : \mathbb{R} \to \mathbb{R}$ be periodic functions such that $\lim_{x \to \infty} (f(x) - g(x)) = 0$. Prove that $f = g$.

Solution. Let $T, T' > 0$ be such that for all $x \in \mathbb{R}$,

$$f(x+T) = f(x) \quad \text{and} \quad g(x+T') = g(x).$$

Set $h = f - g$. Then, by hypothesis,

$$\lim_{n \to \infty} h(x+nT) = \lim_{n \to \infty} h(x+nT') = \lim_{n \to \infty} h(x+nT+nT'), \quad \text{for any } x \in \mathbb{R}.$$

But

$$h(x) = h(x+nT) - h(x+nT+nT') + h(x+nT') \quad \text{for any } x \in \mathbb{R} \text{ and all } = n \in \mathbb{N}.$$

Passing to the limit as $n \to \infty$ we deduce that $h(x) = 0$, hence $f = g$. □

3.3.12. *Let $f:(0,1)\to\mathbb{R}$ be a function satisfying $\lim_{x\to 0+} f(x)=0$ and such that there exists $0<\lambda<1$ such that $\lim_{x\to 0+}[f(x)-f(\lambda x)]/x=0$.*
Prove that $\lim_{x\to 0+} f(x)/x=0$.

Solution. Fix $\varepsilon>0$. By hypothesis, there exists $\delta\in(0,1)$ such that for all $x\in(0,\delta)$,
$$|f(x)-f(\lambda x)|\leq \varepsilon x.$$
Thus, for any $x\in(0,\delta)$ and all positive integers n,
$$|f(x)-f(\lambda^n x)|\leq |f(x)-f(\lambda x)|+|f(\lambda x)-f(\lambda^n x)|+\cdots+|f(\lambda^{n-1}x)-f(\lambda^n x)|$$
$$\leq \varepsilon\left(x+\lambda x+\cdots+\lambda^{n-1}x\right)\leq \frac{\varepsilon x}{1-\lambda}.$$

Hence, for any $x\in(0,\delta)$ and all positive integers n,
$$|f(x)-f(\lambda^n x)|\leq \frac{\varepsilon x}{1-\lambda}.$$

Passing to the limit as $n\to\infty$, we obtain, for all $x\in(0,\delta)$,
$$|f(x)|\leq \frac{\varepsilon x}{1-\lambda}.$$

Since $\varepsilon>0$ is arbitrary, we conclude that $\lim_{x\to 0+} f(x)/x=0$. □

3.3.13. *Let $f:(a,b)\to\mathbb{R}$ be a function such that $\lim_{x\to x_0} f(x)$ exists, for any $x_0\in[a,b]$. Prove that f is bounded if and only if for all $x_0\in[a,b]$, $\lim_{x\to x_0} f(x)$ is finite.*

Solution. Let us first assume that f is bounded. Then there exists $M>0$ such that for all $x\in(a,b)$, $|f(x)|\leq M$. Fix $x_0\in[a,b]$ and consider a sequence $(x_n)_{n\geq 1}$ in (a,b) such that $x_n\to x_0$ as $n\to\infty$. Then
$$|f(x_n)|\leq M\quad\text{and}\quad \lim_{n\to\infty} f(x_n)=\ell_{x_0}:=\lim_{x\to x_0} f(x).$$

In particular, $|\ell_{x_0}|\leq M$, which shows that $\ell_{x_0}=\lim_{x\to x_0} f(x)$ is finite.

Next, we assume that $\lim_{x\to x_0} f(x)$ exists and is finite for any $x_0\in[a,b]$, and we show that f is bounded. Arguing by contradiction, there is a sequence $(x_n)_{n\geq 1}$ in (a,b) such that $|f(x_n)|\to\infty$ as $n\to\infty$. Let $(y_n)_{n\geq 1}$ be a subsequence of $(x_n)_{n\geq 1}$ that has a limit, say y. Then $y\subset[a,b]$, and by hypothesis, $\ell_y-\lim_{x\to y} f(x)$ exists and is finite. Since
$$|\ell_y|=\lim_{n\to\infty}|f(y_n)|=\infty,$$
we have obtained a contradiction. This shows that f is bounded.

We say that a function $f:\mathbb{R}\to\mathbb{R}$ is *superlinear* if
$$\lim_{x\to\pm\infty}\frac{f(x)}{|x|}=+\infty. \qquad (3.2)$$

We will prove in this volume several qualitative properties of superlinear functions. We start with the following result. □

3.3.14. Let $f : \mathbb{R} \to \mathbb{R}$ be a superlinear function and define
$$g(y) = \max_{x \in \mathbb{R}} [xy - f(x)].$$

(i) Prove that g is well defined.
(ii) Show that g is superlinear.
(iii) Assume, additionally, that f is even. Prove that g is even.

Solution. (i) If $x_n \to +\infty$, then by our assumption (3.2),
$$x_n y - f(x_n) = x_n \left(y - \frac{f(x_n)}{x_n} \right) \to -\infty.$$

With the same argument, $x_n y - f(x_n) \to -\infty$ as $x_n \to -\infty$. Hence for each fixed y, the function $h(x) = xy - f(x)$ satisfies $\lim_{x \to \pm\infty} h(x) = -\infty$. This shows that g is well defined.

(ii) Fix $x \in \mathbb{R}$ and an arbitrary sequence $(y_n)_{n \geq 1}$ such that $y_n \to +\infty$ as $n \to \infty$. We have $g(y_n) \geq xy_n - f(x)$. Therefore
$$\frac{g(y_n)}{y_n} \geq x - \frac{f(x)}{y_n}.$$
We deduce that
$$\liminf_{n \to \infty} \frac{g(y_n)}{y_n} \geq x.$$

Since the sequence $(y_n)_{n \geq 1}$ diverging to $+\infty$ and $x \in \mathbb{R}$ are arbitrary, we conclude that $\lim_{y \to +\infty} g(y)/y = +\infty$. A similar argument shows that $\lim_{y \to -\infty} g(y)/y = +\infty$.

(iii) We have
$$g(-y) = \max_{x \in \mathbb{R}} [x(-y) - f(x)] = \max_{x \in \mathbb{R}} [(-x)(-y) - f(-x)],$$
$$\max_{x \in \mathbb{R}} [xy - f(x)] = g(y). \quad \square$$

3.4 Independent Study Problems

> Say what you know, do what you must, come what may.
> — Sofia Kovalevskaya (1850–1891)

3.4.1. Fix $a > 1$ and $p > 0$. Prove that
$$\lim_{x \to +\infty} \frac{x^p}{a^x} = 0, \quad \lim_{x \to +\infty} \frac{\ln x}{x^p} = 0, \quad \lim_{x \to 0} x^p \ln x = 0.$$

3.4.2. Let r be a positive number and define

$$f(r) = \min\{|r - \sqrt{m^2 + 2n^2}|;\ m, n \in \mathbb{Z}\}.$$

Determine whether there $\lim_{r\to\infty} f(r)$ exists, and in the affirmative case, whether the limit equals 0.

3.4.3. Let $a > 0$, $a \neq 1$. Compute

$$\lim_{x\to\infty} \left(\frac{a^x - 1}{x(a-1)}\right)^{1/x}.$$

Answer. The limit equals 1 if $0 < a < 1$, and a for $a > 1$.

3.4.4. Compute

$$\lim_{x\to\infty} x\left(\frac{1}{e} - \left(\frac{x}{x+1}\right)^x\right).$$

Answer. $-1/2e$.

3.4.5. Let $0 < u < \pi$ and $0 < v < \pi/2$. Prove that

$$\cos(u\sin v)\cos(u\cos v) - \cos u > 0.$$

Compute $\lim_{v\searrow 0} u(v)$, where $u(v)$ denotes the least positive u such that the above expression equals 0.

3.4.6. Let $n \geq 3$ be an integer. Prove that the equation $(\ln x)^n = x$ has at least two solutions r_n, s_n in $[1, \infty)$ such that $r_n \to e$ and $s_n \to \infty$ as $n \to \infty$.
Hint. Set $t = \ln x$ and study the equivalent equation $1/n = \ln t/t$.

3.4.7. Let c be a real number and $f : \mathbb{R} \to \mathbb{R}$ be such that $g(a) = \lim_{t\to\infty} f(at)t^{-c}$ exists and is finite for all $a > 1$. Prove that there exists $K \in \mathbb{R}$ such that $g(a) = Ka^c$.

3.4.8. Let $f : \mathbb{R} \to [0, \infty)$ be a function whose restriction to the interval $[0, 1]$ is bounded and such that for all $x, y \in \mathbb{R}$, $f(x+y) \leq f(x)f(y)$.
Prove that $\lim_{x\to\infty}(f(x))^{1/x}$ exists and is finite.

3.4.9. For any integer $n \geq 1$, define $\sin_n = \sin \circ \cdots \circ \sin$ (n times). Prove that

$$\lim_{x\to 0} \frac{\sin_n x}{x} = 1 \quad \text{for all } n \geq 1.$$

3.4.10. Let f be a positive strictly increasing function on $[1, +\infty)$, with $f(x) \to +\infty$ as $x \to +\infty$. Prove that the series $\sum_{n=1}^{\infty} 1/f(n)$ and $\sum_{n=1}^{\infty} n^{-2}f^{-1}(n)$ converge only simultaneously (f^{-1} is the function inverse to f).

3.4 Independent Study Problems

3.4.11. Find each limit if it exists:

(i) $\lim_{x \to \infty} x^2 (1 - \cos 1/x)$;
(ii) $\lim_{x \to 0} (1 - \sin x)^{1/x}$;
(iii) $\lim_{x \to 0+} x^x$;
(iv) $\lim_{x \to 0+} (\tan x)^{1/x}$.

3.4.12. Compute
$$\lim_{x \to 0} \frac{\sin \tan x - \tan \sin x}{\arcsin \arctan x - \arctan \arcsin x}.$$

V.I. Arnold

3.4.13. Let $f_1, f_2, \ldots, f_n : \mathbb{R} \to \mathbb{R}$ be periodic functions such that
$$\lim_{x \to \infty} (f_1(x) + f_2(x) + \cdots + f_n(x)) = 0.$$
Prove that $f_1 = f_2 = \cdots = f_n$.

Part II
Qualitative Properties of Continuous and Differentiable Functions

In Part II we study two basic notions of the infinitesimal calculus: the concept of *continuity* and that of *differentiability*. Continuity of a function f at some point x_0 may be seen as the glue that holds together the points on the graph of f, preventing them from scattering vertically at x_0. The notion of derivative corresponds to many intuitive concepts, including those of velocity, rate of change, or slope of a graph. In the next two chapters we are also concerned with several relevant qualitative results, including the intermediate value property, mean value theorems, and Taylor expansions.

Chapter 4
Continuity

> *Nature does nothing in vain, and more is in vain when less will serve; for Nature is pleased for simplicity, and affects not the pomp of superfluous causes.*
> —Sir Isaac Newton (1642–1727), *Principia*

Abstract. A crucial use of the idea of limit is to separate out a class of functions to study: exactly those whose value at a point coincides with the limit at that point. This is the class of continuous functions, and it contains many familiar and useful functions. In this chapter, we are interested in the special case of real-valued continuous functions on an interval in $\mathbb{R}$. The concepts we develop here will be reexamined and developed in later chapters.

4.1 The Concept of Continuity and Basic Properties

> Structures are the weapons of the mathematician.
> —Bourbaki

Intuitively, a function is continuous on an interval if its graph can be drawn "continuously," that is, without lifting the pen from the paper. Rigorously, a function f defined on an interval (a,b) is *continuous* at some point $c \in (a,b)$ if for each $\varepsilon > 0$, there exists $\delta > 0$ depending on both ε and c such that $|f(x) - f(c)| < \varepsilon$ whenever $|x - c| < \delta$. Equivalently, f is continuous at c if and only if it has left-hand and right-hand limits at c and they are equal to each other and to $f(c)$, that is,

$$\lim_{x \to c-} f(x) = \lim_{x \to c+} f(x) = f(c),$$

or

$$\lim_{x \to c} f(x) = f(c).$$

In other words, f is continuous at c if f *commutes* with "$\lim_{x \to c}$," namely

$$\lim_{x \to c} f(x) = f\left(\lim_{x \to c} x\right).$$

This definition is equivalent to the following sequential version: a function f defined on an interval (a,b) is *continuous* at $c \in (a,b)$ if $\lim_{n\to\infty} f(x_n) = f(c)$ for any sequence $(x_n)_{n\geq 1} \subset I$ such that $\lim_{n\to\infty} x_n = c$.

The set of all functions $f : I \to \mathbb{R}$ that are continuous on I is denoted by $C(I)$. To simplify the notation, we shall often write $C(a,b)$ and $C[a,b]$ instead $C((a,b))$ and $C([a,b])$, respectively.

Continuity is a *local property* (that is, it depends on any point), but it can be extended to a whole set. More precisely, if $f : I \to \mathbb{R}$ is a function and $A \subset I$, then we say that f is continuous on A if it is continuous at every point $a \in A$. In such a case we have obtained a *global property*.

If f and g are functions that are continuous on an interval I, then $f+g$, $f-g$, and fg are continuous on I, and f/g is continuous at all points of I where $g \neq 0$. If g is continuous on I and f is continuous on J, where g is an interval that contains the range of g, then $f \circ g$ is continuous on I.

Continuity may be also defined by means of *open* sets. More precisely, we have $f \in C(I)$ if for any open set $V \subset \mathbb{R}$, there exists an open set $U \subset \mathbb{R}$ such that $f^{-1}(V) = I \cap U$. In particular, if I is open, then f is continuous on I if the *inverse image* under f of every open set is also an open set. The same property can be stated in terms of *closed* sets.

The *direct image* of an open (resp, closed) set under a continuous function is not necessarily an open (resp., closed) set. However, the image of a compact (that is, closed and bounded) set under a continuous function is also compact. In particular, if f is a continuous function with compact domain K, then there exists a positive constant C such that $|f(x)| \leq C$ for all $x \in K$. A more precise property is stated in the following result, which is called "Hauptlehrsatz" (that is, *principal theorem*) in Weierstrass's lectures of 1861.

Weierstrass's Theorem. *Every real-valued continuous function on a closed and bounded interval attains its maximum and its minimum.*

Let f be a function with domain I. Let $a \in I$ and assume that f is discontinuous at a. Then there are two ways in which this discontinuity can occur:

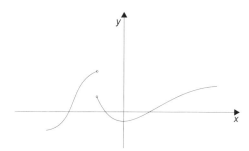

Fig. 4.1 Discontinuity of the first kind.

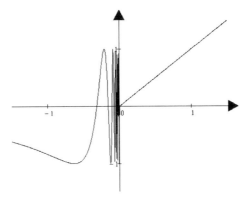

Fig. 4.2 Discontinuity of the second kind (limit does not exist).

(i) If $\lim_{x \to a-} f(x)$ and $\lim_{x \to a+} f(x)$ exist, but either do not equal each other or do not equal $f(a)$, then we say that f has a *discontinuity of the first kind* at the point a (see Figure 4.1).

(ii) If either $\lim_{x \to a-} f(x)$ does not exist or $\lim_{x \to a+} f(x)$ does not exist, then we say that f has a *discontinuity of the second kind* at the point a. We illustrate this notion in Figures 4.2 and 4.3, which correspond to the functions

$$f(x) = \begin{cases} \sin \frac{1}{x} & \text{if } x < 0, \\ x & \text{if } x \geq 0, \end{cases}$$

and

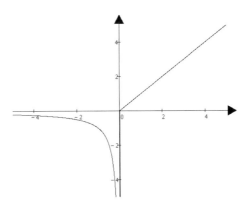

Fig. 4.3 Discontinuity of the second kind (infinite limit).

$$g(x) = \begin{cases} \frac{1}{x} & \text{if } x < 0, \\ x & \text{if } x \geq 0. \end{cases}$$

In the first case, $\sin(1/x)$ produces an infinity of oscillations in a neighborhood of the origin as x tends to zero. However, the function

$$h(x) = \begin{cases} x \sin \frac{1}{x} & \text{if } x \neq 0, \\ 0 & \text{if } x = 0, \end{cases}$$

is continuous (see Figure 4.4). As mentioned by Weierstrass (1874), "Here the oscillations close to the origin are less violent, due to the factor x, but they are still infinitely small."

Examples. Consider the functions

$$f_1(x) = \begin{cases} 1 & \text{if } x > 0, \\ 0 & \text{if } x = 0, \\ -1 & \text{if } x < 0, \end{cases}$$

$$f_2(x) = \begin{cases} \sin \frac{1}{x} & \text{if } x \neq 0, \\ 0 & \text{if } x = 0, \end{cases}$$

$$f_3(x) = \begin{cases} 1 & \text{if } x \text{ is rational}, \\ 0 & \text{if } x \text{ is irrational}. \end{cases}$$

Then f_1 has a discontinuity of the first kind at 0, while f_2 has a discontinuity of the second kind at the origin. The function f_3 has a discontinuity of the second kind at every point.

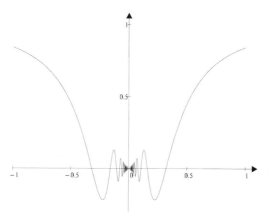

Fig. 4.4 Graph of the function $f(x) = \sin 1/x$, $x \neq 0$.

4.1 The Concept of Continuity and Basic Properties

Riemann (1854) found the following example of a function that is discontinuous in every interval:

$$f(x) = \sum_{n=1}^{\infty} \frac{B(nx)}{n^2}, \quad \text{with} \quad B(x) = \begin{cases} x - \langle x \rangle & \text{if } x \neq k/2, \\ 0 & \text{if } x = k/2, \end{cases}$$

where $k \in \mathbb{Z}$ and $\langle x \rangle$ denotes the nearest integer to x. This function is discontinuous at $x = 1/2, 1/4, 3/4, 1/6, 3/6, 5/6, \ldots$.

If f is a continuous, strictly monotone function with domain $[a,b]$, then its inverse f^{-1} exists and is continuous.

A refined type of continuity is now described. If f is a function with domain D, then f is called *uniformly continuous* on D if for any $\varepsilon > 0$, there exists $\delta > 0$ such that $|f(x) - f(y)| < \varepsilon$ whenever $x, y \in D$ and $|x - y| < \delta$. We observe that in this definition δ is "uniform" with respect to $x, y \in D$, in the sense that δ depends only on ε and is independent of x and y. The main property related to uniform continuity is that if f is a continuous function with compact domain K, then f is uniformly continuous on K.

A function $f : I \to \mathbb{R}$ defined on an interval I is said to have the *intermediate value property* if for all a and b in I with $a < b$ and for any number y between $f(a)$ and $f(b)$, there exists a number x in $[a,b]$ such that $f(x) = y$.

A basic property of continuous functions defined on an interval is that they have the intermediate value property. This theorem appears geometrically evident and was used by Euler and Gauss. Only Bolzano found that a "rein analytischer Beweis" was necessary to establish more rigor in analysis. It was widely believed by many mathematicians in the nineteenth century that the intermediate value property is equivalent to continuity. The French mathematician Gaston Darboux (1842–1917) proved in 1875 that this is not the case (see [20]). We will discuss in detail in the next chapter a celebrated result of Darboux that asserts that any derivative has the intermediate value property. Furthermore, Darboux gave examples of differentiable functions with discontinuous derivatives. In conclusion, for real-valued functions defined on intervals,

$$\text{Continuity} \implies \text{Intermediate Value Property,}$$

but the converse is **not** true.

Exercise. *Prove that the function $f : \mathbb{R} \to \mathbb{R}$ defined by*

$$f(x) = \begin{cases} \sin \dfrac{1}{x} & \text{if } x \neq 0, \\ a & \text{if } x = 0, \end{cases}$$

has the intermediate value property if and only if $-1 \leq a \leq 1$.

This result shows that a function that does not have the intermediate value property may be written as the sum of two functions that do have this property.

Indeed, if
$$f(x) = \begin{cases} 0 & \text{if } x \neq 0, \\ 1 & \text{if } x = 0, \end{cases}$$
then $f = f_1 + f_2$, where
$$f_1(x) = \begin{cases} \sin\frac{1}{x} & \text{if } x \neq 0, \\ 1 & \text{if } x = 0, \end{cases}$$
and
$$f(x) = \begin{cases} -\sin\frac{1}{x} & \text{if } x \neq 0, \\ 0 & \text{if } x = 0. \end{cases}$$

A deep result due to the Polish mathematician Wacław Sierpiński (1882–1969) asserts that if I is an interval of real numbers, then **any** function $f : I \to \mathbb{R}$ can be written as $f = f_1 + f_2$, where f_1 and f_2 have the intermediate value property.

A natural question related to the previous results is the following. If f and g are functions defined on $[a,b]$ such that f is continuous and g has the intermediate value property, does $f + g$ have the intermediate value property? Most persons would be inclined to answer "yes." Nevertheless, the correct answer is "no," and we refer to [80] for more details and a related (but complicated!) counterexample.

4.2 Elementary Problems

> An expert is a man who has made all the mistakes that can be made in a very narrow field.
>
> Niels Bohr (1885–1962)

4.2.1. Determine whether the following assertion is true: if $f : [0,1] \to \mathbb{R}$ is left-continuous, then f is bounded.

Solution. Set
$$f(x) = \begin{cases} \dfrac{1}{x} & \text{if } 0 < x \leq 1, \\ 0 & \text{if } x = 0. \end{cases}$$

This function is both left-continuous and unbounded. □

4.2.2. Find all continuous functions $f : \mathbb{R} \to \mathbb{R}$ such that for all $x \in \mathbb{R}$, $f(x) + f(2x) = 0$.

Solution. We first observe that $f(0) = 0$. Next, by recurrence, we deduce that for any positive integer n, $f(x) = (-1)^n f(x/2^n)$. Using the continuity of f at the origin we obtain that for all $x \in \mathbb{R}$,

4.2 Elementary Problems

$$f(x) = \lim_{n \to \infty} (-1)^n f\left(\frac{x}{2^n}\right) = f(0) = 0.$$

Conversely, the function $f \equiv 0$ satisfies the given relation, hence the unique function with this property is $f = 0$ everywhere. $\square$

4.2.3. Let $f : (0,1) \to \mathbb{R}$ be a continuous function such that $f(x)^2 = 1$ for all $x \in (0,1)$. Prove that either $f = 1$ or $f = -1$.

Solution. By hypothesis it follows that if $x \in (0,1)$, then either $f(x) = 1$ or $f(x) = -1$. We prove that if $f(x) = 1$ then $f \equiv 1$. Indeed, if not, there exists $y \in (0,1)$ such that $f(y) = -1$. Since f has the intermediate value property, there exists z between x and y such that $f(z) = 0$, which is a contradiction to $f(z)^2 = 1$. $\square$

4.2.4. Let $f : (0,1] \to [-1,1]$ be a function such that $\lim_{x \to 0+} f(x) = 0$. Prove that there are continuous functions f_1 and f_2 such that $f_1(0) = f_2(0) = 0$, $f_1 \leq f \leq f_2$, f_1 is nonincreasing, and f_2 is nondecreasing.

Solution. Fix $\varepsilon \in (0,1)$. Since $\lim_{x \to 0+} f(x) = 0$, there exists $\delta \in (0,1)$ such that $|f(x)| \leq \varepsilon$, provided $x \in (0, \delta]$. Thus, there is a decreasing sequence of real numbers $(\delta_n)_{n \geq 1} \subset (0,1)$ converging to 0 such that $|f(x)| \leq 1/n$ for all $x \in (0, \delta_{n-1}]$ and any $n \geq 2$.

Let $f_2 : [0,1] \to [-1,1]$ be a continuous function defined as follows: $f_2(0) = 0$, $f_2(1) = 1$, $f_2(\delta_n) = 1/n$ for all $n \geq 1$, and f_2 is linear on all intervals $[\delta_1, 1]$ and $[\delta_{n+1}, \delta_n]$, with $n \geq 1$. Set $f_1 = -f_2$. Then the functions f_1 and f_2 satisfy the required properties. $\square$

4.2.5. Let $f : \mathbb{R} \to \mathbb{R}$ be a continuous function such that

$$f\left(r + \frac{1}{n}\right) = f(r), \quad \text{for any rational number } r \text{ and positive integer } n.$$

Prove that f is constant.

Solution. We have

$$f\left(r + \frac{2}{n}\right) = f\left(r + \frac{1}{n}\right) = f(r),$$

hence

$$f\left(r + \frac{m}{n}\right) = f(r), \quad \text{for any rational number } r \text{ and all positive integers } m \text{ and } n.$$

Taking $r = 0$ and $r = -m/n$, we deduce that

$$f\left(\frac{m}{n}\right) = f(0) = f\left(-\frac{m}{n}\right). \quad \square$$

This shows that for all $r \in \mathbb{Q}$, $f(r) = f(0)$.

Fix $x_0 \in \mathbb{R} \setminus \mathbb{Q}$ arbitrarily and take $(r_n)_{n \geq 1} \subset \mathbb{Q}$ such that $r_n \to x_0$ as $n \to \infty$. Since f is continuous, we obtain $f(x_0) = \lim_{n \to \infty} f(r_n) = f(0)$, which concludes the proof.

4.2.6. Let $f(0) > 0$, $f(1) < 0$. Prove that $f(x_0) = 0$ for some x_0 under the assumption that there exists a continuous function g such that $f + g$ is nondecreasing.

A. Wilansky, Amer. Math. Monthly, Problem E 1336

Solution. Let A be the set of x such that $f(x) \geq 0$. Set $x_0 = \sup A$. Let $h = f + g$. Then, since h is nondecreasing,

$$h(x_0) \geq h(x) \geq g(x) \quad \text{for any } x \in A.$$

Hence, since g is continuous, $h(x_0) \geq g(x_0)$, and so $f(x_0) \geq 0$.

Next, $g(1) > h(1) \geq h(x_0) \geq g(x_0)$. Since g is continuous, there exists $t \geq x_0$ such that $g(t) = h(x_0)$. Then $h(t) \geq h(x_0) = g(t)$, so that $f(t) \geq 0$. By definition of x_0, $t = x_0$, so that $g(x_0) = h(x_0)$ and $f(x_0) = 0$. $\square$

By definition, every uniformly continuous function is continuous. The following property is due to E. Heine and establishes that the converse is true, provided the function is defined on a compact set.

4.2.7. Prove that any continuous function $f : [a,b] \to \mathbb{R}$ is uniformly continuous.

Solution. We give the original proof from 1872, which is due to Heine [45]. Arguing by contradiction, there exist $\varepsilon > 0$ and sequences $(x_n)_{n \geq 1}$ and $(y_n)_{n \geq 1}$ in $[a,b]$ such that for all $n \geq 1$, $|x_n - y_n| < 1/n$ and $|f(x_n) - f(y_n)| \geq \varepsilon$. By the Bolzano–Weierstrass theorem, passing to subsequences if necessary, we can assume that $x_n \to x$ and $y_n \to y$ as $n \to \infty$. Passing to the limit in $|x_n - y_n| < 1/n$, we deduce that $x = y$. On the other hand, since $|f(x_n) - f(y_n)| \geq \varepsilon$, we find as $n \to \infty$ that $0 \geq \varepsilon$, a contradiction.

Next, we give an alternative proof found in 1873 by Lüroth [72] based on Weierstrass's theorem. Fix $\varepsilon > 0$. For each $x \in [a,b]$ let $\delta(x) > 0$ be the length of the largest open interval I of center x such that $|f(y) - f(z)| < \varepsilon$ for $y, z \in I$ (see Figure 4.5). More precisely,

$$\delta(x) = \sup\{x > 0;\ |f(y) - f(z)| < \varepsilon \quad \text{for all } y, z \in [x - \delta/2, x + \delta/2]\},$$

where x, y, and z are supposed to lie in $[a,b]$.

By continuity of f at x, the set of all $\delta > 0$ in the definition of $\delta(x)$ is nonempty, for all $x \in [a,b]$. If $\delta(x) = \infty$ for some $x \in [a,b]$, then the estimate $|f(y) - f(z)| < \varepsilon$ holds without any restriction; hence any $\delta > 0$ satisfies the condition in the definition of uniform continuity.

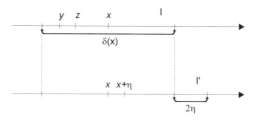

Fig. 4.5 Illustration of Lüroth's proof of the Weierstrass theorem.

If $\delta(x) < \infty$ for all $x \in [a,b]$, we move x to $x \pm \eta$. The new interval I' cannot be longer than $\delta(x) + 2|\eta|$; otherwise, I would be entirely in I' and could be extended, a contradiction. At the same time, I' cannot be smaller than $\delta(x) - 2|\eta|$. Thus, this $\delta(x)$ is a continuous function. By Weierstrass's theorem, there exists $x_0 \in [a,b]$ such that $\delta(x_0) \leq \delta(x)$ for all $x \in [a,b]$. This value $\delta(x_0)$ is positive by definition and can be used to satisfy the condition in the definition of uniform continuity.

A direct consequence of the above property is that any continuous periodic function $f : \mathbb{R} \to \mathbb{R}$ is uniformly continuous. □

4.2.8. Let $f, g : [a,b] \to (0,\infty)$ be continuous functions such that $g(x_0) < f(x)$ for all $x \in [a,b]$. Prove that there exists $\lambda > 1$ such that $f \geq \lambda g$ in $[a,b]$. Does the conclusion remains true if we replace $[a,b]$ with (a,b)?

Solution. We argue by contradiction. Thus, for any $n \geq 1$, there exists $x_n \in [a,b]$ such that $f(x_n) < (1 + 1/n)g(x_n)$. By the Bolzano–Weierstrass theorem, passing eventually to a subsequence, we can assume that $x_n \to x_0$. Moreover, since $[a,b]$ is compact, we have $x_0 \in [a,b]$. Passing to the limit and using the continuity of f and g, we deduce that $f(x_0) \leq g(x_0)$, a contradiction.

For the second part it is enough to consider the functions $f(x) = x$ and $g(x) = x^2$, $x \in (0,1)$. □

4.2.9. Let $f : [0,1] \to [0,1]$ be a continuous function such that $f \circ f = f$. Define $E_f = \{x \in [0,1]; f(x) = x\}$.

(i) Prove that $E_f \neq \emptyset$.
(ii) Show that E_f is an interval.
(iii) Find all functions with the above properties.

Solution. (i) We prove that $f([0,1]) \subset E_f$. Indeed, take $y = f(x) \in f([0,1])$, where $x \in [0,1]$. Then $f(f(x)) = f(x)$; hence $f(y) = y$, which shows that $y \in E_f$.

(ii) We show that $E_f \subset f([0,1])$. Since $f([0,1])$ is an interval, as the image of an interval by the continuous function f, it follows that E_f is an interval, too. Choose $x \in E_f$, that is, $x \in [0,1]$ and $f(x) = x$. This implies that $x \in f([0,1])$.
(iii) Continuous functions $f : [0,1] \to [0,1]$ satisfying $f \circ f = f$ are those functions such that $E_f = f([0,1])$. □

4.3 The Intermediate Value Property

> As for everything else, so for a
> mathematical theory: beauty can be
> perceived but not explained.
>
> Arthur Cayley (1821–1895)

4.3.1. Prove that if $f : [a,b] \to \mathbb{R}$ is one-to-one and has the intermediate value property, then f is strictly monotone.

Solution. We claim that if $u < v < w$ are arbitrary points in $[a,b]$, then $f(v)$ is between $f(u)$ and $f(w)$. Indeed, suppose $f(v)$ is outside this interval and, say, closer to $f(u)$. Since f has the intermediate value property, there exists ξ between u and w such that $f(u) = f(w)$, which contradicts the hypothesis that f is one-to-one.

Therefore, for $a < c < d < b$ the only possibilities are the following:

$$f(a) < f(c) < f(d) < f(b)$$

and

$$f(a) > f(c) > f(d) > f(b). \quad \square$$

All other configurations of the inequalities would contradict our above claim.

4.3.2. *Find all the functions $f : \mathbb{R} \to \mathbb{R}$ with the intermediate value property such that for some $n \geq 1$, $f^n(x) = -x$ for all x. (Here $f^2 = f \circ f$, etc.)*

M. Eşanu

Solution. Let f satisfy the iterative property in the statement of the problem. Then f is clearly a bijection. This, combined with the intermediate value property, says that f is monotone (see Exercise 4.2.1). Further,

$$f(-x) = f[f^n(x)] = f^n[f(x)] = -f(x),$$

so f is odd. In particular, then $f(0) = 0$. Then our hypothesis implies that f is decreasing and $xf(x) < 0$ for $x \neq 0$.

Now pick any $x_0 > 0$ and let $x_k = f(x_{k-1})$. Then

$$(-1)^k x_k > 0, \quad x_n = -x_0;$$

hence $(-1)^n x_n = (-1)^{n+1} x_0 > 0$, so n is an odd integer.

Assume $x_1 > -x_0$. Since f is decreasing and odd,

$$x_2 = f(x_1) < f(-x_0) = -f(x_0) = -x_1,$$

so we obtain inductively

$$(-1)^k x_k > (-1)^{k+1} x_{k+1},$$

which leads to the contradiction $x_0 > -x_n$.

Similarly, the assumption $x_1 < -x_0$ leads to the contradiction $x_n < -x_0$. Thus the only solution with the intermediate value property is $f(x) = -x$. $\quad \square$

4.3.3. *Let $f : \mathbb{R} \to \mathbb{R}$ be a continuous periodic function of period 1.*

(i) *Prove that the function is bounded above and below and that moreover, it achieves its bounds.*
(ii) *Show that there exists $x_0 \in \mathbb{R}$ such that $f(x_0 + \pi) = f(x_0)$.*

Solution. (i) Let f_1 be the restriction of f to $[0,2]$. By periodicity, the bounds of f and f_1 are the same. Moreover, f_1 is bounded and it achieves its bounds, as a continuous function defined on a compact set.

(ii) Let $x_M, x_m \in \mathbb{R}$ be such that $f(x_M) = \max_{x \in \mathbb{R}} f(x)$ and $f(x_m) = \min_{x \in \mathbb{R}} f(x)$. Let $g(x) = f(x+\pi) - f(x)$. It follows that

$$g(x_M) = f(x_M + \pi) - f(x_M) \leq 0 \quad \text{and} \quad g(x_m) = f(x_m + \pi) - f(x_m) \geq 0. \quad \square$$

The conclusion follows immediately, since g has the intermediate value property.

4.3.4. Let $f : (a,b) \to \mathbb{R}$ be a function such that for any $x \in (a,b)$ there exists a nondegenerate interval $[a_x, b_x]$ with $a < a_x \leq x \leq b_x < b$ such that f is constant in $[a_x, b_x]$.

(i) Prove that the image of f is at most countable.
(ii) Find all functions with the intermediate value property satisfying the above property.

Romanian Mathematical Olympiad, 2004

Solution. (i) Consider an element $f(x) \in \operatorname{Im} f$, where $x \in (a,b)$. If $r_x \in \mathbb{Q} \cap [a_x, b_x]$, then by hypothesis, $f(x) = f(r_x)$. It follows that $\operatorname{Im} f = \{f(r_x); \, x \in (a,b)\}$. Setting $\{r_x; \, x \in (a,b)\} = \{q_1, q_2, \ldots, q_n, \ldots\}$, we conclude that the set $\operatorname{Im} f = \{f(q_n)\}_{n \geq 1}$ is at most countable.

(ii) We prove that if f has the intermediate value property, then f should be constant. Indeed, arguing by contradiction, the function f has at least two values $\lambda < \mu$. It follows that $\operatorname{Im} f$ contains the interval $[\lambda, \mu]$, which is not possible, by (i). Hence f must be constant. $\quad \square$

4.3.5. Fix an integer $n \geq 1$.

(i) Prove that the equation

$$x^n + x^{n-1} + \cdots + x - 1 = 0$$

has a unique positive solution. Let a_n denote this solution. Show that $a_n \leq 1$.

(ii) Prove that the sequence $(a_n)_{n \geq 1}$ is decreasing.
(iii) Show that

$$a_n^{n+1} - 2a_n + 1 = 0$$

and deduce that $\lim_{n \to \infty} a_n = 1/2$.

(iv) Prove that

$$a_n = \frac{1}{2} + \frac{1}{4 \cdot 2^n} + o\left(\frac{1}{2^n}\right) \quad \text{as } n \to \infty.$$

Solution. (i) Set $f_n(x) = x^n + x^{n-1} + \cdots + x - 1$. Then f_n is continuous, $f_n(0) = -1$, and $f_n(1) = n - 1 \geq 0$. Thus, by the intermediate value property, there exists $a_n \in (0, 1]$ such that $f_n(a_n) = 0$. Since f_n is strictly increasing on $[0, \infty)$, it follows that the equation $x^n + x^{n-1} + \cdots + x - 1 = 0$ has a unique positive solution a_n.

(ii) We have $f_n(a_{n+1}) = -a_{n+1}^{n+1} < 0$. Since f_n increases in $[0, \infty)$ and $0 = f_n(a_n) > f_n(a_{n+1})$, we deduce that $a_n > a_{n+1}$.

(iii) We observe that
$$(x-1)(x^n + x^{n-1} + \cdots + x - 1) = x^{n+1} - 2x + 1.$$
Thus, $a_n^{n+1} - 2a_n + 1 = 0$. Therefore
$$0 < a_n - \frac{1}{2} = \frac{a_n^{n+1}}{2} < \frac{a_2^{n+1}}{2} \to 0 \quad \text{as } n \to \infty,$$
because $0 < a_2 < 1$. So, by the squeezing principle, we deduce that $\lim_{n \to \infty} a_n = 1/2$.

(iv) Set $b_n = a_n - 1/2$. Then $\ln(2b_n) = (n+1)\ln a_n$. But $\ln a_n = -\ln 2 + \ln(1 + 2b_n)$ and $\ln(1 + 2b_n) \sim 2b_n = o(1/n)$. We deduce that
$$\ln(2b_n) = -(n+1)\ln 2 + o(1) \quad \text{as } n \to \infty,$$
hence
$$a_n = \frac{1}{2} + \frac{1}{4 \cdot 2^n} + o\left(\frac{1}{2^n}\right) \quad \text{as } n \to \infty. \quad \square$$

4.3.6. Let $f : \mathbb{R} \to \mathbb{R}$ be a surjective continuous function that takes any value at most twice. Prove that f is strictly monotone.

Solution. Fix $a, b \in \mathbb{R}$ with $a < b$ and assume, without loss of generality, that $f(a) \leq f(b)$. Since f is continuous, $f([a,b])$ is a bounded interval, so there exist $x_0, x_1 \in [a, b]$ such that
$$f(x_0) \leq f(x) \leq f(x_1) \quad \text{for all } x \in [a, b].$$

We first claim that $f(x_0) < f(x_1)$. Indeed, if $f(x_0) = f(x_1)$, then f is constant in $[a, b]$, which contradicts our assumption that f takes any value at most twice. Thus, $f(x_0) < f(x_1)$.

Next, we prove that $f(x_0) \geq f(a)$ and $f(x_1) \leq f(b)$. We prove only the first inequality, the second one being left to the reader as an exercise with similar arguments. Arguing by contradiction, we assume that $f(x_0) < f(a)$. Then there exists $x_2 \in \mathbb{R}$ such that $f(x_2) < f(x_0)$. Then either $x_2 < a$ or $x_2 > b$. Let us assume that $x_2 < a$. Since f has the intermediate value property, it follows that f takes a certain value at least three times on the interval $[x_2, b]$, a contradiction. This shows that $f(x_0) \geq f(a)$.

It follows that for any real numbers $a < b$ and for all $x \in [a, b]$, $f(a) \leq f(x) \leq f(b)$. Thus, f is monotone. Moreover, since $f^{-1}\{y\}$ has at most two elements for all $y \in \mathbb{R}$, we conclude that f cannot take the same value at different points; hence f is strictly monotone. □

4.3.7. *Prove that there is no continuous function $f : \mathbb{R} \to \mathbb{R}$ that takes every value exactly twice.*

Solution. Suppose such a function exists. Consider two real numbers $a < b$ such that $f(a) = f(a')$. Since f takes every value exactly twice and f has the intermediate value property, it follows that either $f(x) < f(a)$ or $f(x) > f(a)$ for every $x \in (a, a')$. Assuming the first case, let $c \in [a, a']$ be such that $f(b) = \min_{x \in [a,a']} f(x)$.

We first assume that f attains its minimum in $[a, a']$ at a single point b. Since $f(b) < f(a)$, we have $b \in (a, a')$. By hypothesis, there exists $b' \in \mathbb{R}$ such that $f(b) = f(b')$. But $b' \notin [a, a']$, say $b' > a'$. Fix $y \in (f(b), f(a))$. Since $f(a) = f(a') > f(b) = f(b')$, there exist $x_1 \in (a, b)$, $x_2 \in (b, a')$, and $x_3 \in (a', b')$ such that $f(x_1) = f(x_2) = f(x_3) = y$, a contradiction.

Next, we assume that there exist $b_1 < b_2$ in $[a, a']$ such that $f(b_1) = f(b_2) = \min_{x \in [a,a']} f(x)$. Since $f(x) < f(a)$ for all $x \in (a, a')$, it follows that $b_1, b_2 \in (a, a')$. Fix $b \in (b_1, b_2)$ and set $y = f(b)$. We have $f(b_1) < y < f(a)$ and $f(b_2) < y < f(a')$; hence there exist $x_1 \in (a, b_1)$ and $x_2 \in (b_2, a')$ such that $f(x_1) = f(x_2) = y$. Since $f(b) = y$, we have obtained a contradiction. □

4.4 Types of Discontinuities

> When you have eliminated the impossible, whatever remains, however improbable, must be the truth.
>
> Sir Arthur Conan Doyle (1859–1930),
> *The Sign of Four*

We start with a basic property that asserts that monotone functions do not have discontinuity points of the second kind.

4.4.1. *Let I be a nondegenerate interval and assume that $f : I \to \mathbb{R}$ is monotone. Prove that if f has discontinuity points, then they are of the first kind, and the set of discontinuity points is at most countable.*

Solution. Assume that f is nondecreasing. Let $a \in I$ be such that $\lim_{x \to a-} f(x)$ makes sense. We claim that

$$\lim_{x \to a-} f(x) = \sup_{x < a} f(x) =: \ell.$$

Indeed, since $f(x) \le f(a)$ for all $x < a$, it follows that $\ell < +\infty$. Fix $\varepsilon > 0$. Thus, there exists $x_\varepsilon \in I$, $x_\varepsilon < a$, such that $\ell - \varepsilon < f(x_\varepsilon) \le \ell$. Since f is nondecreasing, we obtain

$$\ell - \varepsilon < f(x) \le \ell \quad \text{for all } x \in [x_\varepsilon, a),$$

which shows that $\ell = \lim_{x\to a-} f(x)$. We treat analogously the case of points $a \in I$ such that $\lim_{x\to a+} f(x)$ makes sense. If a is an interior point of I, then both one-sided limits make sense, and moreover, $\lim_{x\to a-} f(x) \leq \lim_{x\to a+} f(x)$.

Let $\mathscr{D}$ denote the set of discontinuity points of f. We associate to any $a \in \mathscr{D}$ the nondegenerate interval

$$I_a = \left(\lim_{x\to a-} f(x), \lim_{x\to a+} f(x) \right).$$

If one of these limits does not make sense, then we replace it by $f(a)$ in the definition of I_a. We observe that if $a, b \in \mathscr{D}$, then $I_a \cap I_b = \emptyset$, provided $a \neq b$. In such a way, to any $a \in \mathscr{D}$ we can associate $r_a \in I_a \cap \mathbb{Q}$; hence we define a one-to-one mapping $\mathscr{D} \ni a \longmapsto r_a \in \mathbb{Q}$. Since $\mathbb{Q}$ is countable, it follows that $\mathscr{D}$ is at most countable.

A deeper property established in 1929 by the Romanian mathematician Alexandru Froda [28] asserts that the set of discontinuity points of the first kind of **any** function $f : \mathbb{R} \to \mathbb{R}$ is at most countable. We refer to [82] for a detailed proof of this result.

Consider the Riemann function $f : \mathbb{R} \to \mathbb{R}$ defined by

$$f(x) = \begin{cases} 0 & \text{if } x = 0 \text{ or if } x \in \mathbb{R} \setminus \mathbb{Q}, \\ \frac{1}{n} & \text{if } x = \frac{m}{n}, m \in \mathbb{Z}, n \in \mathbb{N}^*, (m,n) = 1. \end{cases}$$

Then f is continuous on $(\mathbb{R} \setminus \mathbb{Q}) \cup \{0\}$ and discontinuous on $\mathbb{Q} \setminus \{0\}$. A natural question is whether there is a function $f : \mathbb{R} \to \mathbb{R}$ that is continuous at rational points and discontinuous at irrational points. The answer to this question is **no**. This is a direct consequence of the following general result, which is due to the Italian mathematician Vito Volterra (1860–1940), based on the observation that both $\mathbb{Q}$ and $\mathbb{R} \setminus \mathbb{Q}$ are dense subsets of $\mathbb{R}$.

In 1898, Osgood proved the following interesting property of the real axis: if $(U_n)_{n\geq 1}$ is a sequence of open and dense subsets in $\mathbb{R}$, then their intersection is dense in $\mathbb{R}$, too. In the more general framework of **complete** metric spaces this result is known as Baire's lemma, after Baire, who gave a proof in 1899. A nice application of this property was found in 1931 by Banach, who showed that "almost" continuous functions $f : [0,1] \to \mathbb{R}$ do not have a right derivative at **any** point. Indeed, let us observe that the space of all continuous functions on $[0,1]$ (denoted by $C[0,1]$) becomes a complete metric space with respect to the distance $d(f,g) = \max_{x \in [0,1]} |f(x) - g(x)|$, for all $f, g \in C[0,1]$. For any positive integer n, let F_n denote the set of all functions $f : [0,1] \to \mathbb{R}$ for which there exists some x (depending on f) such that for all $x \leq x' \leq x + 1/(n+1)$,

$$0 \leq x \leq 1 - \frac{1}{n+1} \quad \text{and} \quad |f(x') - f(x)| \leq n(x' - x).$$

Then the set F_n is closed in $C[0,1]$, so its complementary set U_n is open. Moreover, U_n is dense in $C[0,1]$. This follows from the fact that for any $f \in C[0,1]$ and for all $M > 0$ and $\varepsilon > 0$, there exists $g \in C[0,1]$ such that $d(f,g) < \varepsilon$, and the right derivative of g exists in any point, and moreover, its absolute value is at least M. For this purpose it is enough to choose a curve such as the "sawtooth," whose graph lies in a tape of width ε around the graph of f, and the number of "teeth" is sufficiently large that any segment of the graph has a slope that is either greater than M or less than $-M$. At this stage, by Baire's lemma, it follows that the set $U = \bigcap_{n \geq 1} U_n$ is dense in $C[0,1]$. Using now the definition of F_n, it follows that no function $f \in U$ can admit a finite right derivative at any point in $[0,1]$. □

We do not insist in this direction, but we will prove a property in the same direction as that of Osgood. More precisely, Volterra [113] proved in 1881 that if two real continuous functions defined on the real axis are continuous on dense subsets of $\mathbb{R}$, then the set of their common continuity points is dense in $\mathbb{R}$, too. The following result is a generalization of Volterra's theorem in the sense that it asserts, additionally, that the set of common continuity points is uncountable.

4.4.2. Consider the functions $f, g : \mathbb{R} \to \mathbb{R}$ and denote the set of continuity points of f (resp. g) by C_f (resp. C_g). Assume that both C_f and C_g are dense in $\mathbb{R}$. Prove that $C_f \cap C_g$ is an uncountable dense subset of $\mathbb{R}$.

Solution. We first observe that it is enough to prove that for any real a and for all $r > 0$,
$$\big((a-r, a+r) \cap C_f \cap C_g\big) \setminus C \neq \emptyset. \tag{4.1}$$

Indeed, by (4.1), $(a-r, a+r) \cap C_f \cap C_g \neq \emptyset$, so $C_f \cap C_g$ is a dense subset of $\mathbb{R}$. Relation (4.1) also implies that the set $C_f \cap C_g$ is uncountable. Indeed, if not, we choose $C = C_f \cap C_g$, so $\big((a-r, a+r) \cap C_f \cap C_g\big) \setminus C = \emptyset$, which contradicts relation (4.1).

For any fixed $a \in \mathbb{R}$ and $r > 0$, set $a_0 = a$ and $r_0 = r$. Define inductively the sequences $(a_n)_{n \geq 0} \subset E$ and $(r_n)_{n \geq 0} \subset \mathbb{R}_+$ as follows:

(i) $[a_{n+1} - r_{n+1}, a_{n+1} - r_{n+1}] \subset (a_n - r_n, a_n + r_n)$, for all positive integers n;
(ii) $c_n \notin (a_n - r_n, a_n + r_n)$, for all $n \geq 1$;
(iii) if n is odd, then $|f(x) - f(y)| < 1/n$, for all $x, y \in (a_n - r_n, a_n + r_n)$;
(iv) if $n \geq 2$ is even, then $|g(x) - g(y)| < 1/n$, for all $x, y \in (a_n - r_n, a_n + r_n)$.

These sequences are constructed as follows: assume that we have defined a_k and r_k, for all $k < n$. If n is odd, we choose $a_n \in (a_{n-1} - r_{n-1}, a_{n-1} + r_{n-1}) \cap C_f$. We observe that such an element exists, by the density of C_f in $\mathbb{R}$. Using now the continuity of f in a_n, there exists $\delta > 0$ such that $|f(x) - f(a_n)| < 1/2n$, provided $|x - a_n| < \delta$. Choosing now $r_n > 0$ such that $r_n < \min\{\delta, r/n, r_{n-1} - |a_n - a_{n-1}|\}$, we deduce that if $x, y \in (a_n - r_n, a_n + r_n)$, then

$$|f(x) - f(y)| \leq |f(x) - f(a_n)| + |f(a_n) - f(y)| < \frac{1}{n}.$$

We also observe that $[a_n - r_n, a_n + r_n] \subset (a_{n-1} - r_{n-1}, a_{n-1} + r_{n-1})$. Indeed, if $|x - a_n| \leq r_n$, then

$$|x - a_{n-1}| \leq |x - a_n| + |a_n - a_{n-1}| \leq r_n + |a_n - a_{n-1}| < r_{n-1}.$$

If n is even, we define similarly a_n and r_n, with the only observation that we replace f (resp. C_f) by g (resp. C_g).

By (i) and since (r_n) converges to zero, it follows by Cantor's principle that there exists $b \in \bigcap_{n \geq 0}[a_n - r_n, a_n + r_n]$. Moreover, (i) also implies that $b \in B(a_n, r_n)$, for all $n \geq 0$.

We prove in what follows that f and g are continuous in b, that is, $b \in C_f \cap C_g$. Fix $\varepsilon > 0$ and an odd integer N such that $1/N < \varepsilon$. By (iii), it follows that if $\delta < r_N - |b - a_N|$ then $|f(x) - f(b)| < 1/N < \varepsilon$, provided that $|x - b| < \delta$, so f is continuous in b. A similar argument shows that $b \in C_g$.

Since $b \in (a_n - r_n, a_n + r_n)$ and for all positive integers n, $c_n \notin (a_n - r_n, a_n + r_n)$, it follows that $b \notin C$, which implies relation (4.1) and concludes our proof. □

4.5 Fixed Points

> Descartes commanded the future from his study more than Napoleon from the throne.
>
> Oliver Wendell Holmes (1809–1894)

Part (i) of the next result is the simplest case of the Brouwer fixed-point theorem. In the general form, Brouwer's fixed-point theorem asserts that any continuous function with domain the closed unit ball B in $\mathbb{R}^N$ and range contained in B must have at least one fixed point. This deep result was proved in 1910 by the Dutch mathematician L.E.J. Brouwer (1881–1966). Properties stated in (i), resp. (ii), are illustrated in Figure 4.6, resp. Figure 4.7. The fixed-point property established in Exercise 4.4.1 (ii) is also known as the Knaster fixed-point theorem. A counterexample to (iii) is depicted in Figure 4.8.

4.5.1. Let $f : [a,b] \to [a,b]$ be an arbitrary function.

(i) Prove that if f is continuous, then f has a fixed point, that is, there exists $x_0 \in [a,b]$ such that $f(x_0) = x_0$.

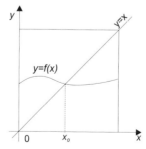

Fig. 4.6 Brouwer's fixed-point theorem (one-dimensional case).

4.5 Fixed Points

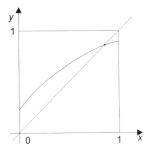

Fig. 4.7 Fixed-point theorem (case of increasing functions).

(ii) Show that the result remains true if f is nondecreasing.
(iii) Find a decreasing function $f : [a,b] \to [a,b]$ with no fixed points.

Solution. (i) Consider the continuous function $g : [a,b] \to \mathbb{R}$ defined by $g(x) = f(x) - x$. We have $g(a) = f(a) - a \geq 0$ and $g(b) = f(b) - b \leq 0$. Since $g(a)g(b) \leq 0$ and g is continuous, there exists $x_0 \in [a,b]$ such that $g(x_0) = 0$, that is, $f(x_0) = x_0$. We observe that in order to deduce that g vanishes, it is enough to show that g has the intermediate value property. In the proof, we have deduced this by using the continuity of g, as the difference of continuous functions. In the general case, by Sierpiński's theorem, it is **not** true that the difference of two functions having the intermediate value property is a function with the same property.

(ii) Set
$$A = \{a \leq x \leq b;\ f(x) \geq x\}$$
and $x_0 = \sup A$. The following situations may occur:

(a) $x_0 \in A$. By the definition of x_0 it follows that $f(x_0) \geq x_0$. If $f(x_0) = x_0$, then the proof is concluded. If not, we argue by contradiction and assume that $f(x_0) > x_0$. By the definition of x_0 we obtain $f(x) < x$, $\forall x > x_0$. On the other hand, for any $x_0 < x < f(x_0)$ we have $x > f(x) \geq f(x_0)$, contradiction, since

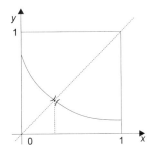

Fig. 4.8 A fixed-point theorem that is true for increasing functions but that fails in the decreasing case.

$x \in (x_0, f(x_0))$, that is, $x < f(x_0)$. It follows that the assumption $f(x_0) > x_0$ is false, so f has a fixed point.

(b) $x_0 \notin A$. We prove that in fact, it is impossible to have $x_0 \notin A$, so $x_0 \in A$, which reduces the problem to the case (a). If $x_0 \notin A$ then there exists a sequence $x_n \to x_0$, $x_n < x_0$, such that $x_n \in A$. Since f is increasing, it follows that $\lim_{n \to \infty} f(x_n) = x_0$. On the other hand, from $f(x_0) < x_0$ we deduce that there exists $x_n < x_0$ such that $f(x_n) > f(x_0)$, contradiction with the fact that f is increasing.

(iii) Consider the function

$$f(x) = \begin{cases} 1 - x & \text{if } 0 \leq x < \frac{1}{2}, \\ \frac{1}{2} - \frac{x}{2} & \text{if } \frac{1}{2} \leq x \leq 1. \end{cases}$$

A related counterexample is depicted in Figure 4.8.

We point out some simple facts regarding the hypotheses of the Brouwer fixed-point theorem on the real axis.

(i) While a fixed point in $[a,b]$ exists for a continuous function $f : [a,b] \to [a,b]$, it need **not** be unique. Indeed, any point $x \in [a,b]$ is a fixed point of the function $f : [a,b] \to [a,b]$ defined by $f(x) = x$.

(ii) The condition that f be defined on a closed subset of $\mathbb{R}$ is essential for the existence of a fixed point. For example, if $f : [0,1) \to \mathbb{R}$ is defined by $f(x) = (1+x)/2$, then f maps $[0,1)$ into itself, and f is continuous. However, f has **no** fixed point in $[0,1)$.

(iii) The condition that f be defined on a bounded subset of $\mathbb{R}$ is essential for the existence of a fixed point. For example, if $f : [1, \infty) \to \mathbb{R}$ is defined by $f(x) = x + x^{-1}$, then f maps $[1, \infty)$ into itself, f is continuous, but f has **no** fixed point in $[1, \infty)$.

(iv) The condition that f be defined on an interval in $\mathbb{R}$ is essential for the existence of a fixed point. For example, if $D = [-2, -1] \cup [1, 2]$ and $f : D \to \mathbb{R}$ is defined by $f(x) = -x$, then f maps D into itself, f is continuous, but f has **no** fixed point in D.

We have just observed that if $f : [a,b] \to [a,b]$ is a continuous function then f must have at least one fixed point, that is, a point $x \in [a,b]$ such that $f(x) = x$. A natural question in applications is to provide an algorithm for finding (or approximating) this point. One method of finding such a fixed point is by successive approximation. This technique is due to the French mathematician Emile Picard. More precisely, if $x_1 \in [a,b]$ is chosen arbitrarily, define $x_{n+1} = f(x_n)$, and the resulting sequence $(x_n)_{n \geq 1}$ is called the *sequence of successive approximations* of f (or a *Picard sequence* for the function f). If the sequence $(x_n)_{n \geq 1}$ converges to some x, then a direct argument based on the continuity of f shows that x is a fixed point of f. Indeed,

4.5 Fixed Points

$$f(x) = f\left(\lim_{n\to\infty} x_n\right) = f\left(\lim_{n\to\infty} x_{n-1}\right) = \lim_{n\to\infty} f(x_{n-1}) = \lim_{n\to\infty} x_n = x. \quad \square$$

The usual method of showing that the sequence $(x_n)_{n\geq 1}$ of successive approximations converges is to show that it satisfies the Cauchy convergence criterion: for every $\varepsilon > 0$ there is an integer N such that for all integers $j, k \geq N$, we have $|x_j - x_k| < \varepsilon$. The next exercise asserts that it is enough to set $j = k+1$ in the Cauchy criterion.

4.5.2. Let $f : [a,b] \to [a,b]$ be a continuous function. Let x_1 be a point in $[a,b]$ and let $(x_n)_{n\geq 1}$ denote the resulting sequence of successive approximations. Then the sequence $(x_n)_{n\geq 1}$ converges to a fixed point of f if and only if $\lim_{n\to\infty}(x_{n+1} - x_n) = 0$.

Solution. Clearly $\lim_{n\to\infty}(x_{n+1} - x_n) = 0$ if $(x_n)_{n\geq 1}$ converges to a fixed point. Suppose $\lim_{n\to\infty}(x_{n+1} - x_n) = 0$ and the sequence $(x_n)_{n\geq 1}$ does not converge. Since $[a,b]$ is compact, there exist two subsequences of $(x_n)_{n\geq 1}$ that converge to ξ_1 and ξ_2 respectively. We may assume $\xi_1 < \xi_2$. It suffices to show that $f(x) = x$ for all $x \in (\xi_1, \xi_2)$. Suppose this is not the case; hence there is some $x^* \in (\xi_1, \xi_2)$ such that $f(x^*) \neq x^*$. Then a $\delta > 0$ can be found such that $[x^* - \delta, x^* + \delta] \subset (\xi_1, \xi_2)$ and $f(\tilde{x}) \neq \tilde{x}$ whenever $\tilde{x} \in (x^* - \delta, x^* + \delta)$. Assume $\tilde{x} - f(\tilde{x}) > 0$ (the proof in the other case being analogous) and choose N such that $|f^n(x) - f^{n+1}(x)| < \delta$ for $n > N$. Since ξ_2 is a cluster point, there exists a positive integer $n > N$ such that $f^n(x) > x^*$. Let n_0 be the smallest such integer. Then, clearly,

$$f^{n_0-1}(x) < x^* < f^{n_0}(x),$$

and since $f^{n_0}(x) - f^{n_0-1}(x) < \delta$, we must have

$$f^{n_0-1}(x) - f^{n_0}(x) > 0, \text{ so that } f^{n_0}(x) < f^{n_0-1}(x) < x^*,$$

a contradiction. $\square$

4.5.3. Let $f : [0,1] \to [0,1]$ be a continuous function such that $f(0) = 0$, $f(1) = 1$. Define $f^n := f \circ f \circ \cdots \circ f$ (n times) and assume that there exists a positive integer m such that $f^m(x) = x$ for all $x \in [0,1]$. Prove that $f(x) = x$ for any $x \in [0,1]$.

Solution. Our hypothesis implies that f is one-to-one, so increasing (since f is continuous). Assume, by contradiction, that there exists $x \in (0,1)$ such that $f(x) > x$. Then, for any $n \in \mathbb{N}$, we have $f^n(x) > f^{n-1}(x) > \cdots > f(x) > x$. For $n = m$ we obtain a contradiction. A similar argument shows that the case $f(x) < x$ (for some x) is not possible. $\square$

4.5.4. Let a, b be real numbers, $a < b$, and consider a continuous function $f : [a,b] \to \mathbb{R}$.

(i) Prove that if $[a,b] \subset f([a,b])$ then f has a fixed point.
(ii) Assume that there exists a closed interval $I' \subset f([a,b])$. Prove that $I' = f(J)$, where J is a closed interval contained in $[a,b]$.

(iii) Assume that there exist n closed intervals $I_0, \ldots, I_{n-1}$ contained in $[a,b]$ such that for all $0 \leq k \leq n-2$, $I_{k+1} \subset f(I_k)$ and $I_0 \subset f(I_{n-1})$. Prove that f^n has a fixed point ($f^n = f \circ \cdots \circ f$).

École Normale Supérieure, Paris, 2003

Solution. (i) Define $f([a,b]) = [m,M]$ and let x_m, $x_M \in [a,b]$ be such that $f(x_m) = m$ and $f(x_M) = M$. Since $f(x_m) - x_m \leq 0$ and $f(x_M) - x_M \geq 0$, it follows by the intermediate value property that f has at least a fixed point.

(ii) Set $I' = [c,d]$ and consider u, $v \in I$ such that $f(u) = c$ and $f(v) = d$. Assume, without loss of generality, that $u \leq v$.

The set $A = \{x \in [u,v];\ f(x) = c\}$ is compact and nonempty, so there exists $\alpha = \max\{x;\ x \in A\}$ and, moreover, $\alpha \in A$. Similarly, the set $B = \{x \in [\alpha,v];\ f(x) = d\}$ has a minimum point β. Then $f(\alpha) = c$, $f(\beta) = d$ and for all $x \in (\alpha,\beta)$ we have $f(x) \neq c$ and $f(x) \neq d$. So, by the intermediate value property, $[c,d] \subset f((\alpha,\beta))$ and $f((\alpha,\beta))$ is an interval that contains neither c nor d. It follows that $I' = f(J)$, where $J = [\alpha,\beta]$.

(iii) Since $I_0 \subset f(I_{n-1})$, it follows by (ii) that there exists a closed interval $J_{n-1} \subset I_{n-1}$ such that $I_0 = f(J_{n-1})$. But $J_{n-1} \subset I_{n-1} \subset f(I_{n-2})$. So, by (ii), there exists a closed interval $J_{n-2} \subset I_{n-2}$ such that $J_{n-1} = f(J_{n-2})$. Thus, we obtain n closed intervals $J_0, \ldots, J_{n-1}$ such that

$$J_k \subset I_k, \quad \text{for all } 0 \leq k \leq n-1,$$

and

$$J_{k+1} = f(J_k), \quad \text{for all } 0 \leq k \leq n-2 \text{ and } I_0 = f(J_{n-1}).$$

Consequently, J_0 is included in the domain of the nth iterate f^n and $J_0 \subset I_0 = f^n(J_0)$. By (i) we deduce that f^n has a unique fixed point in J_0. □

We have already seen that if $f : [a,b] \to [a,b]$ is a continuous function, then f has at least one fixed point. One method of finding such a point is by successive approximation, that is, for a point x_0 in $[a,b]$, define $x_{n+1} = f(x_n)$, and the resulting sequence $(x_n)_{n \geq 0}$ is called the sequence of successive approximations of f. If the sequence $(x_n)_{n \geq 0}$ converges, then it converges to a fixed point of f.

The usual method of showing that the sequence $(x_n)_{n \geq 0}$ of successive approximations converges is to show that it satisfies the Cauchy convergence criterion. The next result establishes that this happens if and only if the difference of two consecutive terms in this iteration converges to zero. The American mathematician Felix Browder has called this condition asymptotic regularity.

4.5.5. *(i) Let $(x_n)_{n \geq 0}$ be a sequence of real numbers such that the sequence $(x_{n+1} - x_n)$ converges to zero. Prove that the set of cluster points of $(x_n)_{n \geq 0}$ is a closed interval in $\bar{\mathbb{R}}$, possibly degenerate.*

Barone's theorem, 1939

4.5 Fixed Points   159

(ii) Let $f : [a,b] \to [a,b]$ be a continuous function. Consider the sequence $(x_n)_{n\geq 0}$ defined by $x_0 \in [a,b]$ and, for any positive integer n, $x_n = f(x_{n-1})$. Prove that $(x_n)_{n\geq 0}$ converges if and only if $(x_{n+1} - x_n)$ converges to zero.

Hillam's fixed-point theorem, 1976

Solution. (i) Set $\ell_- := \liminf_{n\to\infty} x_n$, $\ell_+ := \liminf_{n\to\infty} x_n$ and choose $a \in (\ell_-, \ell_+)$. By the definition of ℓ_-, there exists $x_{n_1} < a$. Let n_2 be the least integer greater than n_1 such that $x_{n_2} > a$ (the existence of n_2 follows by the definition of ℓ_+). Thus, $x_{n_2-1} \leq a < x_{n_2}$. Since $\ell_- < a$, there exists a positive integer $n_3 > n_2$ such that $x_{n_3} < a$. Next, by the definition of ℓ_+, there exists an integer $N_4 > n_3$ such that $x_{N_4} > a$. If n_4 denotes the least integer with these properties, then $x_{n_4-1} \leq a < x_{n_4}$. In this manner we construct an increasing sequence of positive numbers $(n_{2k})_{k\geq 1}$ such that for all $k \geq 1$, $x_{n_{2k}-1} \leq a < x_{n_{2k}}$. Using the hypothesis we deduce that the sequences $(x_{n_{2k}-1})_{k\geq 1}$ and $(x_{n_{2k}})_{k\geq 1}$ converge to a, so a is a cluster point.

(ii) Assume that the sequence of successive approximations $(x_n)_{n\geq 0}$ satisfies $x_{n+1} - x_n \to 0$, as $n \to \infty$. With the same notation as above, assume that $\ell_- < \ell_+$. The proof of (i) combined with the continuity of f implies $a = f(a)$, for all $a \in (\ell_-, \ell_+)$. But this contradicts our assumption $\ell_- < \ell_+$. Indeed, choose $\ell_- < c < d < \ell_+$ and $0 < \varepsilon < (d-c)/3$. Since $x_{n+1} - x_n \to 0$, there exists N_ε such that for all $n \geq N_\varepsilon$, $-\varepsilon < x_{n+1} - x_n < \varepsilon$. Let $N_2 > N_1 > N_\varepsilon$ be such that $x_{N_1} < c < d < x_{N_2}$. Our choice of ε implies that there exists an integer $n \in (N_1, N_2)$ such that $a := x_n \in (c,d)$. Hence $x_{n+1} = f(a) = a$, $x_{n+2} = a$, and so on. Therefore $x_{N_2} = a$, contradiction. □

The reverse assertion is obvious.

The following result is a particular case of a fixed-point theorem due to Krasnoselski (see [61]).

4.5.6. Let $f : [a,b] \to [a,b]$ be a function satisfying $|f(x) - f(y)| \leq |x-y|$, for all $x, y \in [a,b]$. Define the sequence $(x_n)_{n\geq 1}$ by $x_1 \in [a,b]$ and, for all $n \geq 1$, $x_{n+1} = [x_n + f(x_n)]/2$. Prove that $(x_n)_{n\geq 1}$ converges to some fixed point of f.

Solution. We observe that it is enough to show that $(x_n)_{n\geq 1}$ converges. In this case, by the recurrence relation and the continuity of f, it follows that the limit of $(x_n)_{n\geq 1}$ is a fixed point of f. We argue by contradiction and denote by A the set of all limit points of $(x_n)_{n\geq 1}$, that is,

$$A := \{\ell \in [a,b]; \text{ there exists a subsequence}(x_{n_k})_{k\geq 1} \text{ of } (x_n)_{n\geq 1} \text{ such that } x_{n_k} \to \ell\}.$$

By our hypothesis and the compactness of $[a,b]$, we deduce that A contains at least two elements and is a closed set.

We split the proof into several steps.

(i) For any $\ell \in A$ we have $f(\ell) \neq \ell$. Indeed, assume that $\ell \in A$ and fix $\varepsilon > 0$ and $n_k \in \mathbb{N}$ such that $|x_{n_k} - \ell| \leq \varepsilon$. Then

$$|\ell - x_{n_k+1}| = \left| \frac{\ell + f(\ell)}{2} - \frac{x_{n_k} + f(x_{n_k})}{2} \right|$$

$$\leq \frac{|\ell - f(x_{n_k})|}{2} + \frac{|f(\ell) - f(x_{n_k})|}{2}$$

$$\leq |\ell - x_{n_k}| \leq \varepsilon$$

and so on. This shows that $|x_n - \ell| \leq \varepsilon$, for all $n \geq n_k$. Hence $(x_n)_{n \geq 1}$ converges to ℓ, contradiction.

(ii) There exists $\ell_0 \in A$ such that $f(\ell_0) > \ell_0$. Indeed, arguing by contradiction, set $\ell_- = \min_{\ell \in A} \ell$. Then $\ell_- \in A$ and $f(\ell_-) \leq \ell_-$. The variant $f(\ell_-) = \ell_-$ is excluded, by (i). But $f(\ell_-) < \ell_-$ implies that $[\ell_- + f(\ell_-)]/2 \in A$ and $[\ell_- + f(\ell_-)]/2 < \ell_-$, which contradicts the definition of ℓ_-.

(iii) There exists $\varepsilon > 0$ such that $|f(\ell) - \ell| \geq \varepsilon$, for all $\ell \in A$. For if not, let $\ell_n \in A$ be such that $|f(\ell_n) - \ell_n| < 1/n$, for all $n \geq 1$. This implies that any limit point of $(\ell_n)_{n \geq 1}$ (which lies in A, too) is a fixed point of f. This contradicts (i).

(iv) Conclusion. By (ii) and (iii), there exists a largest $\ell_+ \in A$ such that $f(\ell_+) > \ell_+$. Let $\ell' = [\ell_+ + f(\ell_+)]/2$ and observe that $f(\ell_+) > \ell' > \ell_+$ and $f(\ell') < \ell'$. By (iii), there exists a smallest $\ell'' \in A$ such that $\ell'' > \ell_+$ and $f(\ell'') < \ell''$. It follows that $\ell_+ < \ell'' < f(\ell_+)$. Next note that $f(\ell'') < \ell_+$; for if not, $\ell''' := [\ell'' + f(\ell'')]/2$ satisfies $\ell_+ < \ell''' < \ell''$ and, by definitions of ℓ_+ and ℓ'', it follows that $f(\ell''') = \ell'''$, contrary to (i). Thus $f(\ell'') < \ell_+ < \ell'' < f(\ell_+)$. It then follows that $|f(\ell'') - f(\ell_+)| > |\ell'' - \ell_+|$. This contradicts the hypothesis and concludes the proof. □

Remark. The iteration scheme described in the above Krasnoselski property does not apply to arbitrary continuous mappings of a closed interval into itself. Indeed, consider the function $f : [0,1] \to [0,1]$ defined by

$$f(x) = \begin{cases} \frac{3}{4}, & \text{if } 0 \leq x \leq \frac{1}{4}, \\ -3x + \frac{3}{2}, & \text{if } \frac{1}{4} < x < \frac{1}{2}, \\ 0, & \text{if } \frac{1}{2} \leq x \leq 1. \end{cases}$$

Then the sequence defined in the above statement is defined by $x_{2n} = 1/2$ and $x_{2n+1} = 1/4$, for any $n \geq 1$. So, $(x_n)_{n \geq 1}$ is a divergent sequence.

The contraction mapping theorem states that if $f : \mathbb{R} \to \mathbb{R}$ (or $f : [a, \infty) \to [a, \infty)$) is a map such that for some k in $(0, 1)$ and all x and y in $\mathbb{R}$, $|f(x) - f(y)| \leq k|x - y|$, then the iterates $f^n = f \circ \cdots \circ f$ (n terms) of f converge to a (unique) fixed point ξ of f. This theorem can be accompanied by an example to show that the inequality cannot be replaced by the weaker condition $|f(x) - f(y)| < |x - y|$. The most common example of this type is $f(x) = x + 1/x$ acting on $[1, \infty)$. Then $f(x) > x$, so that f has no fixed points. Also, for every x, the sequence x, $f(x)$, $f^2(x), \ldots$ is strictly increasing and so must converge in the space $[-\infty, +\infty]$. In fact, $f^n(x) \to +\infty$, for otherwise $f^n(x) \to a$ for some real a, and then $f(f^n(x)) \to f(a)$ (because f is continuous), so that $f(a) = a$, which is not so. Thus we define $f(+\infty)$

4.5 Fixed Points

to be $+\infty$ and deduce that this example is no longer a counterexample. The following property clarifies these ideas and provides an elementary, but interesting, adjunct to the contraction mapping theorem. We just point out that a mapping $f : \mathbb{R} \to \mathbb{R}$ satisfying $|f(x) - f(y)| < |x-y|$ for all $x \neq y$ is called a *contractive* function.

4.5.7. Suppose that $f : \mathbb{R} \to \mathbb{R}$ satisfies $|f(x) - f(y)| < |x-y|$ whenever $x \neq y$. Then there is some ξ in $[-\infty, +\infty]$ such that for any real x, $f^n(x) \to \xi$ as $n \to \infty$.

A.F. Beardon

Solution. We suppose first that f has a fixed point, say ξ, in $\mathbb{R}$. Then, from the contracting property of f, ξ is the only fixed point of f. We may assume that $\xi = 0$, and this implies that $|f(x)| < |x|$ for all nonzero x. Thus, for any x, the sequence $|f^n(x)|$ is decreasing, so converges to some nonnegative number $\mu(x)$. We want to show that $\mu(x) = 0$ for every x, so suppose now that x is such that $\mu(x) > 0$. Then f maps $\mu(x)$ and $-\mu(x)$ to points y_1 and y_2, say, where $y_j < |\mu(x)|$ for each j. Thus, since f is continuous, there are open neighborhoods of $\pm\mu(x)$ that are mapped by f into the open interval $I = (-\mu(x), \mu(x))$ that contains y_1 and y_2. This implies that for sufficiently large n, $f^n(x)$ lies in I, which contradicts the fact that $|f^n(x)| \geq \mu(x)$ for all n. Thus, for all x, $\mu(x) = 0$ and $f^n(x) \to 0$.

Now suppose that f has no fixed point in $\mathbb{R}$. Then the function $f(x) - x$ is continuous and nonzero in $\mathbb{R}$. By the intermediate value theorem, $f(x) > x$ for all x, or $f(x) < x$ for all x. We may assume that $f(x) > x$ for all x, since a similar argument holds in the other case. Now the sequence $f^n(x)$ is strictly increasing, hence converges to some ξ in $\mathbb{R} \cup \{+\infty\}$. Moreover, $\xi \notin \mathbb{R}$; otherwise, ξ would be a fixed point of f. Thus $f^n(x) \to +\infty$ for all x.

We say that a function $f : [a,b] \to \mathbb{R}$ satisfies the Lipschitz condition with constant $L > 0$ if for all x and y in $[a,b]$, $|f(x) - f(y)| \leq L|x-y|$. A function that satisfies a Lipschitz condition is clearly continuous. Geometrically, if $f : [a,b] \to \mathbb{R}$ satisfies the Lipschitz condition

$$|f(x) - f(y)| \leq L|x-y| \quad \text{for all } x, y \in [a,b],$$

then for any $x, y \in [a,b]$, $x \neq y$, the inequality

$$\left| \frac{f(x) - f(y)}{x - y} \right| \leq L$$

indicates that the *slope* of the chord joining the points $(x, f(x))$ and $(y, f(y))$ on the graph of f is bounded by L. □

Using the fact that the real line is totally ordered, the following more general theorem with much more elementary proof is possible.

4.5.8. Let $f : [a,b] \to [a,b]$ be a function that satisfies a Lipschitz condition with constant L. Let x_1 in $[a,b]$ be arbitrary and define $x_{n+1} = (1-\lambda)x_n + \lambda f(x_n)$, where

$\lambda = 1/(L+1)$. If $(x_n)_{n \geq 1}$ denotes the resulting sequence, then $(x_n)_{n \geq 1}$ converges monotonically to a point z in $[a,b]$ where $f(z) = z$.

Solution. Without loss of generality we can assume $f(x_n) \neq x_n$ for all n. Suppose $f(x_1) > x_1$ and let p be the first point greater than x_1 such that $f(p) = p$. Since $f(x_1) > x_1$ and $f(b) \leq b$, the continuity of f implies there is such a point.

Next, we prove the following claim. If $x_1 < x_2 < \cdots < x_n < p$ and $f(x_i) > x_i$ for $i = 1, 2, \ldots, n$, then $f(x_{n+1}) > x_{n+1}$ and $x_{n+1} < p$. Indeed, suppose $p < x_{n+1}$; then $x_n < p < x_{n+1}$; hence $0 < p - x_n < x_{n+1} - x_n = \lambda(f(x_n) - x_n)$. Therefore

$$0 < \frac{1}{\lambda}|x_n - p| = (L+1)|x_n - p| < |f(x_n) - x_n|$$
$$\leq |f(x_n) - f(p)| + |p - x_n|.$$

It follows that

$$L|x_n - p| < |f(x_n) - f(p)|,$$

which contradicts the fact that f is a Lipschitz function. Thus $x_{n+1} < p$ and $f(x_{n+1}) > x_{n+1}$ by the choice of p, and the claim is proved.

Using the induction hypothesis it follows that $x_n < x_{n+1} < p$ for all integers n. Since a bounded monotonic sequence converges, $(x_n)_{n \geq 1}$ converges to some point z. By the triangle inequality it follows that

$$|z - f(z)| \leq |z - x_n| + |x_n - f(x_n)| + |f(x_n) - f(z)|$$
$$= |z - x_n| + \frac{1}{\lambda}|x_{n+1} - x_n| + |f(x_n) - f(z)|.$$

Since the right-hand side tends to 0 as $n \to \infty$, we conclude that $f(z) = z$. If $f(x_1) < x_1$, a similar argument holds.

Applying a somewhat more sophisticated argument, one can allow λ to be any number less than $2/(L+1)$, but the resulting sequence $(x_n)_{n \geq 1}$ need not converge monotonically. The following example shows that this last results is best possible.

Let $f : [0,1] \to [0,1]$ be defined by

$$f(x) = \begin{cases} 1, & 0 \leq x < \frac{L-1}{2L}, \\ -Lx + \frac{1}{2}(L+1), & \frac{L-1}{2L} \leq x \leq \frac{L+1}{2L}, \\ 0, & \frac{L+1}{2L} < x \leq 1, \end{cases}$$

where $L > 1$ is arbitrary. Note that f satisfies a Lipschitz condition with constant L. Let $\lambda = 2/(L+1)$ and let $x_1 = (L-1)/2L$. Then $x_2 = (1-\lambda)x_1 + \lambda f(x_1) = (L+1)/2L$, $x_3 = (1-\lambda)x_2 + \lambda f(x_2) = (L-1)/2L$, etc. □

4.5.9. Let $f : [0,1] \to [0,1]$ be a function such that $|f(x) - f(y)| \leq |x - y|$ for all $x, y \in [0,1]$. Prove that the set of all fixed points of f is either a single point or an interval.

M.W. Botsko, Math. Magazine, Problem Q982

Solution. Let $F = \{x \in [0,1]; f(x) = x\}$. Since F is continuous, it follows that F is compact. Let a be the smallest number in F and b the largest number in F. It follows that $F \subset [a,b]$. Fix arbitrarily $x_0 \in [a,b]$. Since a is a fixed point of f, we have
$$f(x_0) - a \leq |f(x_0) - a| = |f(x_0) - f(a)| \leq x_0 - a.$$
Therefore, $f(x_0) \leq x_0$. Similarly,
$$b - f(x_0) \leq |b - f(x_0)| = |f(b) - f(x_0)| \leq b - x_0,$$
which shows that $f(x_0) \geq x_0$. It follows that $f(x_0) = x_0$, so x_0 is a fixed point of f. Thus, $F = [a,b]$. $\square$

4.6 Functional Equations and Inequalities

> ...the source of all great mathematics is the special case, the concrete example. It is frequent in mathematics that every instance of a concept of seemingly great generality is in essence the same as a small and concrete special case.
>
> Paul Halmos (1916–2006), *I Want to be a Mathematician*

Cauchy's functional equation is the equation
$$f(x+y) = f(x) + f(y) \quad \text{for all } x, y \in \mathbb{R}. \tag{4.2}$$

A function with this property is said to be *additive*.

A. Cauchy proved in 1821 in his celebrated *Cours d'Analyse de l'École Royale Polytechnique* that a function satisfying equation (4.2) is either continuous or totally discontinuous. Moreover, if f is continuous, then there is a real number a such that $f(x) = ax$ for any $x \in \mathbb{R}$. In 1875, G. Darboux showed [21] that the continuity hypothesis can be replaced by continuity at a single point, and five years later, he proved that it is enough to assume that f has constant sign in a certain interval $(0, \delta)$. Darboux also proved that if f satisfies equation (4.2) and is bounded in some interval, then f is continuous and of the form $f(x) = ax$. In 1905, G. Hamel proved [39] that there are noncontinuous solutions of the Cauchy functional equation using the axiom of choice and Hamel bases. Hamel also proved that if f is a solution of (4.2) that is not of the form $f(x) = ax$, then the graph of f has a point in every neighborhood of every point in the plane $\mathbb{R}^2$. The fifth of Hilbert's problems is a generalization of Cauchy's functional equation.

We are first concerned with equation (4.2) and other related functional equations.

4.6.1. Let $f : \mathbb{R} \to \mathbb{R}$ be a continuous function satisfying
$$f(x+y) = f(x) + f(y) \quad \text{for all } x, y \in \mathbb{R}.$$

Prove that there exists a real number a such that $f(x) = ax$ for all $x \in \mathbb{R}$.

Show that the same conclusion holds if the continuity assumption is replaced by that of monotonicity.

Solution. We observe that $f(0) = 0$. Set $a = f(1)$. By recurrence we deduce that $f(n) = na$ for any integer n. Fix $r \in \mathbb{Q}$, $r = m/n$ with $m \in \mathbb{Z}$ and $n \in \mathbb{N}^*$. Then
$$ma = f(m) = f\left(n \cdot \frac{m}{n}\right) = nf\left(\frac{m}{n}\right),$$
which implies $f(r) = ra$.

Fix $x \in \mathbb{R} \setminus \mathbb{Q}$ and take $(r_n)_{n \geq 1} \subset \mathbb{Q}$ such that $r_n \to x$ as $n \to \infty$. By continuity we deduce that
$$f(x) = \lim_{n \to \infty} f(r_n) = \lim_{n \to \infty} r_n a = ax.$$

Assuming now that f is monotone (say, nondecreasing) and $x \in \mathbb{R} \setminus \mathbb{Q}$, we consider two sequences of rational numbers $(p_n)_{n \geq 1}$ and $(q_n)_{n \geq 1}$ such that $p_n \nearrow x$ and $q_n \searrow x$ as $n \to \infty$. Then $f(p_n) \leq f(x) \leq f(q_n)$, that is, $p_n a \leq f(x) \leq q_n a$. Passing to the limit as $n \to \infty$, we conclude that $f(x) = ax$. $\square$

4.6.2. Let $f : \mathbb{R} \to \mathbb{R}$ be a continuous function satisfying
$$f(x+y) = f(x)f(y) \quad \text{for all } x, y \in \mathbb{R}.$$

Prove that either $f = 0$ or there exists a real number a such that $f(x) = e^{ax}$ for all $x \in \mathbb{R}$.

Solution. We first observe that f can take only positive values. Indeed, for any $x \in \mathbb{R}$,
$$f(x) = f\left(\frac{x}{2}\right) \cdot f\left(\frac{x}{2}\right) \geq 0.$$
Assuming that there is some $x_0 \in \mathbb{R}$ such that $f(x_0) = 0$, then
$$f(x) = f(x - x_0 + x_0) = f(x - x_0)f(x_0) = 0 \quad \text{for all } x \in \mathbb{R}.$$
If not, setting $g(x) = \ln f(x)$, we observe that for all $x, y \in \mathbb{R}$, $g(x+y) = g(x) + g(y)$. Thus, $g(x) = ax$ for some $a \in \mathbb{R}$, which implies that $f(x) = e^{ax}$.

With a similar argument one can show that if $f : (0, \infty) \to (0, \infty)$ is continuous and satisfies the functional equation
$$f(xy) = f(x)f(y) \quad \text{for all } x, y \in (0, \infty),$$
then $f(x) = x^a$, where $a \in \mathbb{R}$. $\square$

4.6.3. Let $f : \mathbb{R} \to \mathbb{R}$ be a function satisfying

4.6 Functional Equations and Inequalities

$$f(x+y) = f(x) + f(y) \quad \text{for all }, y \in \mathbb{R}.$$

If f is bounded on an interval $[a,b]$, then it is of the form $f(x) = ax$ for some real number a.

Solution. We show first that f is bounded on $[0, b-a]$. Suppose that for all $y \in [a,b]$, $|f(y)| \leq M$. If $x \in [0, b-a]$, then $f(x+a) \in [a,b]$, so that from $f(x) = f(x+a) - f(a)$, we deduce that

$$|f(x)| \leq M + |f(a)|.$$

Accordingly, if $b-a = c$ then f is bounded in $[0,c]$. Let $a = f(c)/c$ and set $g(x) = f(x) - ax$. Then g satisfies the same functional equation, that is, $g(x+y) = g(x) + g(y)$. We have $g(c) = f(c) - ac = 0$. It follows that g is periodic with period c and

$$g(x+c) = g(x) + g(c) = g(x).$$

Further, as the difference of two bounded functions on $[0,c]$, the function g is bounded on $[0,c]$. Hence, by periodicity, g is bounded on $\mathbb{R}$.

Suppose there is some $x_0 \in \mathbb{R}$ such that $g(x_0) \neq 0$. Then $g(nx_0) = ng(x_0)$. Thus, we can make $|ng(x_0)|$ as large as we wish by increasing n, which contradicts the boundedness of g. Therefore $g(x) = 0$, or $f(x) = ax$.

With a similar argument we can prove that the only solutions of

$$f(xy) = f(x) + f(y)$$

bounded on an interval $[1,a]$ are of the form $f(x) = k \log x$. Let $g(x) = f(x) - f(a)(\log x)/(\log a)$. Instead of periodicity, show that $g(ax) = g(x)$, which implies that g has the same bound in each interval $[1,a]$, $[a, a^2]$, ... But $g(x^n) = ng(x)$, which forces $g = 0$.

In particular, the above exercise implies that if $f : \mathbb{R} \to \mathbb{R}$ is continuous and satisfies

$$f(x+y) = f(x) + f(y) \quad \text{for all } x, y \in \mathbb{R},$$

then there exists $a \in \mathbb{R}$ such that $f(x) = ax$ for any $x \in \mathbb{R}$. In fact, the main conclusion remains true if f is additive and is continuous at a **single** point $p \in \mathbb{R}$. Indeed, in such a case, f is bounded in some neighborhood of p, and it suffices to apply the above conclusion. $\square$

4.6.4. Prove that there exists no function $f : (0, +\infty) \to (0, +\infty)$ such that $f^2(x) \geq f(x+y)(f(x) + y)$ for any $x, y > 0$.

International Mathematics Competition for University Students, 1999

Solution. Assume that such a function exists. The initial inequality can be written in the form

$$f(x) - f(x+y) \geq f(x) - \frac{f^2(x)}{f(x)+y} = \frac{f(x)y}{f(x)+y}.$$

Obviously, f is a decreasing function. Fix $x > 0$ and choose a positive integer n such that $nf(x+1) \geq 1$. For $k = 0, 1, \ldots, n-1$ we have

$$f\left(x + \frac{k}{n}\right) - f\left(x + \frac{k+1}{n}\right) \geq \frac{f\left(x + \frac{k}{n}\right)}{nf\left(x + \frac{k}{n}\right) + 1} \geq \frac{1}{2n}.$$

Summing these inequalities, we obtain

$$f(x+1) \leq f(x) - \frac{1}{2}.$$

Thus, $f(x+2m) \leq f(x) - m$ for all positive integers m. Taking $m \geq f(x)$, we get a contradiction to the condition $f(x) > 0$. $\square$

4.6.5. Find all functions $f : \mathbb{R}^+ \to \mathbb{R}^+$ such that for all $x, y \in \mathbb{R}^+$,

$$f(x)f(yf(x)) = f(x+y).$$

International Mathematics Competition for University Students, 1999

Solution. We first assume that $f(x) > 1$ for some $x \in \mathbb{R}^+$. Setting $y = x/f(x) - 1$, we obtain the contradiction $f(x) = 1$. Hence $f(x) \leq 1$ for each $x \in \mathbb{R}^+$, which implies that f is a decreasing function.

If $f(x) = 1$ for some $x \in \mathbb{R}^+$, then $f(x+y) = f(y)$ for each $y \in \mathbb{R}^+$, and by the monotonicity of f, it follows that $f \equiv 1$.

Let now $f(x) < 1$ for each $x \in \mathbb{R}^+$. Then f is strictly decreasing, in particular one-to-one. By the equalities

$$f(x)f(yf(x)) = f(x+y) = f(yf(x) + x + y(1 - f(x)))$$
$$= f(yf(x))f((x + y(1 - f(x)))f(yf(x))),$$

we deduce that $x = (x + y(1 - f(x)))f(yf(x))$. Setting $x = 1$, $z = xf(1)$, and $a = 1 - f(1)/f(1)$, we obtain $f(z) = 1/1 + az$.

Combining the two cases, we conclude that $f(x) = 1/1 + ax$ for each $x \in \mathbb{R}^+$, where $a \geq 0$. Conversely, a direct verification shows that the functions of this form satisfy the initial equality. $\square$

4.6.6. Let $f : [-1, 1] \to \mathbb{R}$ be a continuous function such that $f(2x^2 - 1) = 2xf(x)$, for all $x \in [-1, 1]$. Prove that f equals zero identically.

Solution. For any real number t that is not an integer multiple of π we define $g(t) = f(\cos t)/\sin t$. It follows that $g(t + \pi) = g(t)$. Moreover, by hypothesis,

$$g(2t) = \frac{f(2\cos^2 t - 1)}{\sin 2t} = \frac{2\cos t f(\cos t)}{\sin 2t} = g(t).$$

In particular,

$$g(1 + n\pi/2^k) = g(2^k + n\pi) = g(2^k) = g(1).$$

By the continuity of f we deduce that g is continuous on its domain of definition. But the set
$$\{1+n\pi/2^k;\ n,k \in \mathbb{Z}\}$$
is dense in $\mathbb{R}$. This implies that g should be constant on its domain. But g is odd, so $g(t) = 0$ for every t that is not an integer multiple of π. Therefore $f(x) = 0$, $\forall x \in (-1,1)$. Taking now $x = 0$ and $x = 1$ in the functional equation in the hypothesis, we deduce that $f(-1) = f(1) = 0$. □

4.6.7. *Let a and b be real numbers in $(0, 1/2)$ and $f : \mathbb{R} \to \mathbb{R}$ a continuous function such that $f(f(x)) = af(x) + bx$, for all x. Prove that there exists a real constant c such that $f(x) = cx$.*

Solution. We first observe that if $f(x) = f(y)$ then $x = y$, so f is one-to-one. So, by the continuity of f, we deduce that f is strictly monotone. Moreover, f cannot have a finite limit L as $x \to +\infty$. Indeed, in this case we have $f(f(x)) - af(x) = bx$, and the left-hand side is bounded, while the right-hand side is unbounded. Similarly, f cannot have a finite limit as $x \to -\infty$. Next, since f is monotone, we deduce that f is onto.

Let x_0 be arbitrarily chosen and $x_{n+1} = f(x_n)$, for all $n > 0$ and $x_{n-1} = f^{-1}(x_n)$ if $n < 0$. Let $r_1 = (a + \sqrt{a^2 + 4b})/2$ and $r_2 = (a - \sqrt{a^2 + 4b})/2$ be the roots of $x^2 - ax - b = 0$, so $r_1 > 0 > r_2$ and $|r_1| > |r_2|$. Thus, there exist $c_1, c_2 \in \mathbb{Z}$ such that $x_n = c_1 r_1^n + c_2 r_2^n$ for all $n \in \mathbb{Z}$.

Let us assume that f is increasing. If $c_1 \neq 0$, then x_n is dominated by r_2^n, provided that $n < 0$ is small enough. In this case we have $0 < x_n < x_{n+2}$, contradiction, since $f(x_n) > f(x_{n+2})$. It follows that $c_2 = 0$, so $x_0 = c_1$ and $x_1 = c_1 r_1$. Hence $f(x) = r_1 x$ for all x. Analogously, if f is decreasing then $f(x) = r_2 x$. □

4.6.8. *Let $g : \mathbb{R} \to \mathbb{R}$ be a continuous function such that $\lim_{x \to \infty} g(x) - x = \infty$ and such that the set $\{x;\ g(x) = x\}$ is finite and nonempty. Prove that if $f : \mathbb{R} \to \mathbb{R}$ is continuous and $f \circ g = f$, then f is constant.*

C. Joiţa, Amer. Math. Monthly, Problem 10818

Solution. (Robin Chapman) Let $x_1, \ldots, x_n$ with $x_1 < \cdots < x_n$ denote the values of x such that $g(x) = x$. Define intervals $I_0, \ldots, I_n$ as follows: $I_0 = (-\infty, x_1)$, $I_r = (x_r, x_{r+1})$ for $1 \leq r < n$, and $I_n = (x_n, \infty)$. The sign of $g(x) - x$ is constant on each I_r. Define $\varepsilon_r = (g(x) - x)/|g(x) - x|$ for $x \in I_r$.

Suppose that $f : \mathbb{R} \to \mathbb{R}$ is continuous and $f \circ g = f$. We first show that f takes the value $f(x_n)$ on all of I_n. Since $\lim_{x \to \infty}(g(x) - x) = \infty$, we have $\varepsilon_n = 1$ for large x. Choose $y_0 > x_n$. Since $g(x_n) = x_n$ and $g(y_0) > y_0$, the intermediate value property yields y_1 with $x_n < y_1 < y_0$ and $g(y_1) = y_0$. Repeating this argument yields a decreasing sequence $(y_m)_{m \geq 0}$ of elements of I_n with $g(y_{m+1}) = y_m$. Therefore, $f(y_m) = f(g(y_{m+1})) = f(y_{m+1})$. Consequently, $f(y_0) = f(y_m)$ for each m.

Since $(y_m)_{m \geq 0}$ is a decreasing sequence bounded below by x_n, it tends to a limit y_∞ with $y_\infty \geq x_n$. Continuity of g yields $g(y_\infty) = \lim_{m \to \infty} g(y_m) = \lim_{m \to \infty} y_{m-1} = y_\infty$. Therefore, $y_\infty = x_n$ and $f(y_\infty) = \lim_{m \to \infty} f(y_m) = f(y_0)$. It follows that $f(y) = f(x_n)$ for all $y \in I_n$.

We now show by descending induction on r that f has value $f(x_n)$ on all of I_r. Suppose that for some $r \in [0, n-1]$, we have $f(y) = f(x_n)$ for $y \in I_s$ with $r < s \le n$. By continuity, $f(y) = f(x_n)$ for $y \in [x_{r+1}, \infty)$. We divide this into cases according to whether ε_r is 1 or -1. Suppose first that $\varepsilon_r = 1$. For $y \in I_r$, consider the sequence $(y_m)_{m \ge 0}$ with $y_0 = y$ and $y_{m+1} = g(y_m)$. If all the y_m lie in I_r, then the sequence $(y_m)_{m \ge 0}$ is increasing and bounded above by x_{r+1}. It tends to a limit y_∞ with $y_\infty \le x_{r+1}$ and $g(y_\infty) = y_\infty$. Therefore, $y_\infty = x_{r+1}$. On the other hand, we also have $f(y_{m+1}) = f(g(y_m)) = f(y_m)$, so the sequence $(f(y_m))_{m \ge 0}$ is constant. Consequently, $f(y) = f(y_0) = f(y_\infty) = f(x_{r+1})$. Hence f is constant on I_r with value $f(x_n)$.

Suppose that $\varepsilon_r = -1$. It will be shown that $g(I_r) \supseteq I_r$. For $r > 0$, this is automatic from the intermediate value property. For $r = 0$ and $y \in I_0$, one has $g(y) < y < x_0 = g(x_0)$. By the intermediate value property, there exists $y' \in (y, x)$ with $g(y') = y$. We now mimic the argument used in the case $r = n$. Given $y \in I_0$, there is a sequence $(y_m)_{m \ge 0}$ with $y_0 = y$ and $g(y_{m+1}) = y_m$ for each m, since $g(y_0) < y_0$. By the intermediate value property, there exists y_1 such that $g(y_1) = y_0$. This sequence is increasing and tends to a limit y_∞ with $y_\infty \le x_{r+1}$. Again, $g(y_\infty) = y_\infty$. Thus, $y_\infty = x_{r+1}$. The sequence $(f(y_m))_{m \ge 0}$ is constant, and so $f(y) = f(y_0) = f(y_\infty) = f(x_{r+1}) = f(x_n)$. Again we conclude that f is constant on I_r. □

4.6.9. Prove that there is no continuous function $f : [0,1] \to \mathbb{R}$ such that

$$f(x) + f(x^2) = x \quad \text{for all } x \in [0,1].$$

Solution. Assume that there is a continuous function with this property. Thus, for any $n \ge 1$ and all $x \in [0, 1]$,

$$f(x) = x - f(x^2) = x - \left(x^2 - f(x^4)\right) = x - x^2 + \left(x^4 - f(x^8)\right) = \cdots$$
$$= x - x^2 + x^4 - \cdots + (-1)^n \left(x^{2^n} - f\left(x^{2^{n+1}}\right)\right).$$

Since $f(0) = 0$ and $\lim_{n \to \infty} x^{2^{n+1}} = 0$ for any $x \in (0, 1)$, it follows by the continuity of f that $\lim_{n \to \infty} f\left(x^{2^{n+1}}\right) = 0$, hence

$$f(x) = x - x^2 + x^4 - \cdots + (-1)^n x^{2^n} + \cdots,$$

for any $x \in (0, 1)$.

Fix $x_0 \in (0, 1)$ and define, for any $n \ge 1$, $x_n = \sqrt{x_{n-1}}$. Then $(x_n)_{n \ge 0}$ is an increasing sequence converging to 1. So, by the continuity of f, $\lim_{n \to \infty} f(x_n) = f(1) = 1/2$. Since $f(x) + f(x^2) = x$ for all $x \in [0, 1]$, we deduce that f is not constant; hence there exists $x_0 \in (0, 1)$ such that $f(x_0) \ne f(1) = 1/2$. Assume $f(x_0) = \alpha + 1/2$, with $\alpha > 0$.

Using now $f(x_n) = x_n - f(x_{n-1})$, we deduce that for all $n \ge 1$,

$$f(x_{2n+1}) = x_1 - x_2 + x_3 - x_4 + \cdots + x_{2n-1} - x_{2n} + x_{2n+1} - f(x_0).$$

But $x_{2k-1} - x_{2k} < 0$ for any $k \in \{1,\ldots,n\}$ and $x_{2n+1} < 1$. Therefore

$$f(x_{2n+1}) < 1 - f(x_0) = \frac{1}{2} - \alpha.$$

We deduce that

$$\frac{1}{2} = f(1) = \lim_{n \to \infty} f(x_{2n+1}) \leq \frac{1}{2} - \alpha,$$

a contradiction. This shows that there is no function with the required properties. □

4.7 Qualitative Properties of Continuous Functions

> I am interested in mathematics only as a creative art.
>
> Godfrey H. Hardy (1877–1947), *A Mathematician's Apology*

4.7.1. Let $f, g : [0,1] \to [0,1]$ be continuous functions such that

$$\max_{x \in [0,1]} f(x) = \max_{x \in [0,1]} g(x).$$

Prove that there exists $t_0 \in [0,1]$ such that $f^2(t_0) + 3f(t_0) = g^2(t_0) + 3g(t_0)$.

Solution. Let M be the common supremum of the functions f and g. Since f and g are continuous on $[0,1]$, there exist $\alpha, \beta \in [0,1]$ such that $f(\alpha) = g(\beta) = M$. The function $h := f - g$ satisfies $h(\alpha) = M - g(\alpha) \geq 0$, $h(\beta) = f(\beta) - M \leq 0$. Since h is continuous, there exists $t_0 \in [\alpha, \beta]$ such that $h(t_0) = 0$. It follows that $f(t_0) = g(t_0)$, so $f^2(t_0) + 3f(t_0) = g^2(t_0) + 3g(t_0)$. □

4.7.2. (i) *Prove that there exists no continuous and onto map $f : [0,1] \to (0,1)$.*

(ii) *Find a continuous and onto map $f : (0,1) \to [0,1]$.*
(iii) *Prove that a mapping as in (ii) cannot be bijective.*

Solution. (i) Assume that there exists $f : [0,1] \to (0,1)$ continuous and onto. Let $(x_n) \subset [0,1]$ be a sequence such that $0 < f(x_n) < 1/n$. By the Bolzano–Weierstrass theorem, we can assume that (x_n) converges to some $x \in [0,1]$. By continuity we find that $f(x) = 0$, contradiction. Consequently, such a function does not exist.

(ii) Take $f(x) = |\sin 2\pi x|$.
(iii) Let $g : (0,1) \to [0,1]$ be continuous and bijective. Let $x_0, x_1 \in (0,1)$ be such that $g(x_0) = 0$ and $g(x_1) = 1$. We can assume without loss of generality that $x_0 < x_1$ (if not, consider the mapping $1 - g$). Since g has the intermediate value

property, it follows that $g([x_0, x_1]) = [0,1]$. But $x_0, x_1 \in (0,1)$. So g is not one-to-one, a contradiction. □

4.7.3. *Prove that there exists no continuous function $f : \mathbb{R} \to \mathbb{R}$ such that*

$$f(x) \in \mathbb{Q} \iff f(x+1) \in \mathbb{R} \setminus \mathbb{Q}.$$

Solution. Consider the continuous function $g(x) = f(x+1) - f(x)$. By our hypothesis it follows that g has only irrational values, that is, g is constant. Let $a \in \mathbb{R} \setminus \mathbb{Q}$ be such that $g(x) = a$, for all $x \in \mathbb{R}$. This means that $f(x+2) - f(x) = 2a \in \mathbb{R} \setminus \mathbb{Q}, \forall x \in \mathbb{R}$. On the other hand, let $x_0 \in \mathbb{R}$ be such that $f(x_0) \in \mathbb{Q}$. We obtain that $f(x_0 + 2) \in \mathbb{Q}$, so $f(x_0 + 2) - f(x_0) \in \mathbb{Q}$. This contradiction shows that such a function cannot exist. □

4.7.4. *Let $f : [0, \infty) \to \mathbb{R}$ be a continuous function. Let A be the set of all real numbers a such that there exists a sequence $(x_n) \subset [0, \infty)$ satisfying $a = \lim_{n \to \infty} f(x_n)$.*

Prove that if A contains both a and b ($a < b$), then A contains the whole interval $[a, b]$.

Solution. Let $a < c < b$ and $x_n, y_n \in [0, \infty)$ be such that $a = \lim_{n \to \infty} f(x_n)$, $b = \lim_{n \to \infty} f(y_n)$. We can assume without loss of generality that $f(x_n) < c$ and $f(y_n) > c, \forall n \in \mathbb{N}$. Using the intermediate value property for f, we find some z_n between x_n and y_n such that $f(z_n) = c$. We conclude that $c \in A$. □

4.7.5. *Let $f : \mathbb{R} \to \mathbb{R}$ be a function having lateral limits at all points. Prove that the set of discontinuity points of f is at most countable.*

Solution. Let E be the set of discontinuity points of f and

$E_1 = \{x \in E; \ f(x-) = f(x+) < f(x)\}, \qquad E_2 = \{x \in E; \ f(x-) > f(x+)\},$
$E_3 = \{x \in E; \ f(x-) = f(x+) > f(x)\}, \qquad E_4 = \{x \in E; \ f(x-) < f(x+)\}.$

Obviously, $E = E_1 \cup E_2 \cup E_3 \cup E_4$.

For any $x \in E_1$, let $a_x \in \mathbb{Q}$ be such that $f(x-) < a_x < f(x+)$. Choose $b_x, c_x \in \mathbb{Q}$ such that $b_x < x < c_x$ and

$$b_x < t < c_x, \ x \neq t \text{ implies } f(t) < a_x.$$

The map $\varphi : E_1 \to \mathbb{Q} \times \mathbb{Q} \times \mathbb{Q}$ defined by $\varphi(x) = (a_x, b_x, c_x)$ is one-to-one, since $(a_x, b_x, c_x) = (a_y, b_y, c_y)$ implies $f(y) < a_x < f(y)$ for all $x \neq y$. This means that the set E_1 is at most countable.

If $x \in E_2$, let $a_x \in \mathbb{Q}$ be such that $f(x-) > a_x > f(x+)$ and choose $b_x, c_x \in \mathbb{Q}$ satisfying $b_x < x < c_x$ and

$$b_x < t < x \text{ implies } f(t) > a_x$$

and

$$t < c_x \text{ implies } f(t) < a_x. \quad \square$$

4.7 Qualitative Properties of Continuous Functions

With the same arguments as above, the mapping $E_2 \mapsto \mathbb{Q} \times \mathbb{Q} \times \mathbb{Q}$ is one-to-one, so E_2 is at most countable.

Similar arguments apply for E_3 and E_4. It follows that E is at most countable, as the finite union of at most countable sets.

4.7.6. Let $f : \mathbb{R} \to \mathbb{R}$ be a continuous function that maps open intervals into open intervals. Prove that f is monotone.

Solution. We argue by contradiction and assume that f is not monotone. Thus, without loss of generality, we can suppose that there exist $a < b < c$ such that $f(a) < f(b) > f(c)$. By Weierstrass's theorem, f achieves its maximum M on $[a,b]$ at a point that is different from a and b. It follows that the set $f((a,c))$ cannot be open because it contains M, but does not contain $M + \alpha$, for any $\alpha > 0$ less than some ε. So, necessarily, f is monotone. $\square$

4.7.7. Let $f : \mathbb{R} \to \mathbb{R}$ be a continuous function such that $|f(x) - f(y)| \geq |x - y|$, for all $x, y \in \mathbb{R}$. Prove that f is surjective.

Solution. By hypothesis it follows that f is one-to-one. Hence f is strictly monotone, as a continuous and one-to-one mapping. So, f transforms open intervals into open intervals, so the image of f is an open interval.

Let $y_n = f(x_n)$ be such that $y_n \to y \in \mathbb{R}$. The sequence (y_n) is thus a Cauchy sequence, so the same property holds for the sequence (x_n). Let $x = \lim_{n \to \infty} x_n$. By the continuity of f we obtain

$$f(x) = f(\lim_{n \to \infty} x_n) = \lim_{n \to \infty} f(x_n) = y. \quad \square$$

It follows that the image of f is a closed set. Since $f(\mathbb{R})$ is simultaneously open and closed, we deduce that $f(\mathbb{R}) = \mathbb{R}$, that is, f is surjective.

4.7.8. Let $f : [1,\infty) \to (0,\infty)$ be a continuous function. Assume that for every $a > 0$, the equation $f(x) = ax$ has at least one solution in the interval $[1,\infty)$.

(i) Prove that for every $a > 0$, the equation $f(x) = ax$ has infinitely many solutions.
(ii) Give an example of a strictly increasing continuous function f with these properties.

SEEMOUS 2008

Solution. (i) Suppose that one can find constants $a > 0$ and $b > 0$ such that $f(x) \neq ax$ for all $x \in [b,\infty)$. Since f is continuous, we obtain two possible cases:

(1) $f(x) > ax$ for $x \in [b,\infty)$. Define

$$c = \min_{x \in [1,b]} \frac{f(x)}{x} = \frac{f(x_0)}{x_0}.$$

Then, for every $x \in [1,\infty)$ one should have

$$f(x) > \frac{\min(a,c)}{2} x,$$

a contradiction.

(2) $f(x) < ax$ for $x \in [b, \infty)$. Define

$$C = \max_{x \in [1,b]} \frac{f(x)}{x} = \frac{f(x_0)}{x_0}.$$

Then, $f(x) < 2\max(a,C)x$ for every $x \in [1,\infty)$, and this is again a contradiction.

(ii) Choose a sequence $1 = x_1 < x_2 < \cdots < x_n < \cdots$ such that the sequence $(y_n)_{n \geq 1}$ defined by $y_n = 2^{n \cos n\pi} x_n$ is also increasing. Next define $f(x_n) = y_n$ and extend f linearly on each interval $[x_{n-1}, x_n]$, that is, $f(x) = a_n x + b_n$ for suitable a_n, b_n. In this way we obtain an increasing continuous function f, for which $\lim_{n \to \infty} f(x_{2n})/(2n) = \infty$ and $\lim_{n \to \infty} f(x_{2n-1})/(2n-1) = 0$. It now follows that the continuous function $f(x)/x$ takes every positive value on $[1, \infty)$. □

Let $f : I \to \mathbb{R}$ be an arbitrary function, where I is a bounded or unbounded interval. We say that f has a horizontal chord of length $\lambda > 0$ if there is a point x such that $x, x + \lambda \in [a, b]$ and $f(x) = f(x + \lambda)$.

Examples. (i) The function $f(x) = 1$ defined on the entire real line has horizontal chords of all lengths.

(ii) The function $f(x) = x^3 - x + 1$ has a horizontal chord of length 2 and two horizontal chords of length 1.

(iii) The function $f(x) = x^3$ has no horizontal chords.

The next exercise provides a class of functions having horizontal chords of any length.

4.7.9. Let $f : \mathbb{R} \to \mathbb{R}$ be a continuous and periodic function. Prove that f has horizontal chords of all lengths.

Solution. Let T be the period of f. Fix arbitrarily $\lambda > 0$. We must prove that there exists $\xi \in \mathbb{R}$ such that $f(\xi) = f(\xi + \lambda)$. To see this, consider the continuous function $g(x) = f(x + \lambda) - f(x)$ and choose $x_m, x_M \in [0, T]$ such that $f(x_m) = \min_{x \in [0,T]} f(x)$ and $f(x_M) = \max_{x \in [0,T]} f(x)$. Then $g(x_m) \geq 0$ and $g(x_M) \leq 0$. Thus, by the intermediate value property, there exists $\xi \in [0, T]$ such that $g(\xi) = 0$.

Notice that the above property is no longer true if we replace the continuity assumption with the intermediate value property. Indeed, consider the function $f : \mathbb{R} \to \mathbb{R}$ defined by

$$f(x) = \begin{cases} \cos\left(\frac{1}{\sin x}\right) + \frac{1}{2}(-1)^{[x/\pi]} & \text{if } x/\pi \notin \mathbb{Z}, \\ 0 & \text{if } x/\pi \in \mathbb{Z}, \end{cases}$$

where $[x]$ denotes the largest integer less than or equal to x. Then f is periodic with period 2π and has the intermediate value property. However, f has no chord of length π. □

4.7 Qualitative Properties of Continuous Functions

We prove in what follows that continuous periodic functions have particular chords of any length, not necessarily horizontal.

4.7.10. *Let $f : \mathbb{R} \to \mathbb{R}$ be a continuous and periodic function. Prove that f has a chord of any length, not necessarily horizontal, with midpoint on the graph of f.*

Solution. Let $T > 0$ be the period of f. Fix $\lambda > 0$ arbitrarily and consider the continuous function $g(x) = f(x+\lambda) - 2f(x) + f(x-\lambda)$. Let $x_m, x_M \in [0,T]$ be such that $f(x_m) = \min_{x \in [0,T]} f(x)$ and $f(x_M) = \max_{x \in [0,T]} f(x)$. Then $g(x_m) \geq 0$ and $g(x_M) \leq 0$. Thus, by the intermediate value property, there exists $\xi \in [0,T]$ such that $g(\xi) = 0$. This means that

$$f(\xi + \lambda) - f(\xi) = f(\xi) - f(\xi - \lambda),$$

which concludes the proof. □

4.7.11. *Let f be continuous and periodic with period T. Prove that f has two horizontal chords of any given length λ, with their left-hand points at different points of $[0,T)$.*

Solution. Consider the function $g(x) = f(x+\lambda) - f(x)$. Then $g(T) = g(0)$ and g changes sign between 0 and T. Thus, g vanishes at least twice in $[0,T)$, say at ξ and η. It follows that $f(\xi + \lambda) = f(\xi)$, $f(\eta + \lambda) = f(\eta)$ ($\xi \neq \eta$). This completes the proof. □

For functions that are continuous but not periodic, the situation can be quite different. A continuous function may have no horizontal chords, as for example, a strictly increasing function. However, the following *horizontal chord theorem* holds.

4.7.12. *Suppose that the continuous function $f : [0,1] \to \mathbb{R}$ has a horizontal chord of length λ. Prove that there exist horizontal chords of lengths λ/n, for any integer $n \geq 2$, but horizontal chords of any other length cannot exist.*

Solution. Without loss of generality we assume that $f(0) = f(1)$, hence $\lambda = 1$.

Fix an integer $n \geq 2$ and consider the function $g(x) = f(x + 1/n) - f(x)$, so that g has domain $[0, 1 - 1/n]$. We show that there exists ξ such that $g(\xi) = 0$, hence $f(\xi + 1/n) = f(\xi)$. This shows that f has a horizontal chord of length $1/n$, as required. Arguing by contradiction, we deduce that g is either positive for all $x \in [0, 1-1/n]$ or negative for all $x \in [0, 1-1/n]$. Otherwise, by the intermediate value property, we obtain that g vanishes at some ξ, as required. Assume $g > 0$ in $[0, 1-1/n]$. Taking $x = 0, 1/n, 2/n, \ldots, (n-1)/n$, we have

$$f(0) < f(1/n) < f(2/n) < \cdots < f((n-1)/n) < f(1) = f(0),$$

a contradiction.

We show in what follows that there exists a continuous function that has a horizontal chord of length 1, but none of length ρ for all $\rho \in (0, 1/2) \setminus \{1/n;\ n \geq 2\}$. The following example is due to Paul Lévy. Let $a \in (0,1)$ be fixed with $a \neq 1/n$, $n \geq 2$. Consider the function

$$f(x) = \sin^2(\pi x/a) - x \sin^2(\pi/a).$$

Since $f(0) = f(1) = 0$, $f(1) = 0$, we deduce that f has a horizontal chord of length 1. Next, if f has a horizontal chord of length a, then there is some α such that $f(\alpha) = f(\alpha + a)$, that is,

$$\sin^2(\pi\alpha/a) - \alpha\sin^2(\pi/a) = \sin^2(\pi(\alpha+a)/a) - (\alpha+a)\sin^2(\pi/a).$$

Thus, $a\sin^2(\pi/a) = 0$, hence $\pi/a = n\pi$, where n is a positive integer, which yields $a = 1/n$. This contradicts the choice of a.

The next exercise establishes an interesting property of continuous functions and is sometimes referred to as the Croft lemma. As a consequence, if $f : (0, \infty) \to \mathbb{R}$ is a continuous function such that $\lim_{n \to \infty} f(x/n) = 0$ for all $x > 0$, then $\lim_{x \to 0+} f(x) = 0$. □

4.7.13. Let $f : \mathbb{R} \to \mathbb{R}$ be a continuous function such that $\lim_{n \to \infty} f(n\delta) = 0$ for all $\delta > 0$. Prove that $\lim_{x \to \infty} f(x) = 0$.

Solution. Replacing f with $|f|$ if necessary, we can assume $f \geq 0$. Next, we claim that if $0 < a < b$, then $\cup_{n=1}^{\infty}(na, nb)$ contains an interval (x_0, ∞), for some $x_0 > 0$. Indeed, there is a positive integer n_0 such that $a/b < n/(n+1)$ for any $n \geq n_0$. This shows that the intervals (na, nb) and $((n+1)a, (n+1)b)$ have a common point, provided $n \geq n_0$. Since $nb \to \infty$ as $n \to \infty$, it follows that $\cup_{n=n_0}^{\infty}(na, nb) = (n_0 a, \infty)$.

Arguing by contradiction, we assume that $\lim_{x \to \infty} f(x)$ is not 0. Thus, there exist $\varepsilon > 0$ and a sequence $(x_n)_{n \geq 1}$ such that $x_n \geq n$ and $f(x_n) > \varepsilon$. Since f is continuous, there is an interval $I_n = [a_n, b_n]$ such that $x_n \in (a_n, b_n)$ and $f(x) \geq \varepsilon$ for all $x \in I_n$. We define inductively a sequence of bounded and closed intervals $(J_n)_{n \geq 1}$ as follows: $J_1 = I_1$; $J_2 \subset J_1$ and there are positive integers m_2 and n_2 such that $m_2 x \in I_{n_2}$ for all $x \in J_2$. More generally, for any $k \geq 2$, we assume that $J_k \subset J_{k-1}$ and there are positive integers m_k and n_k such that $m_k > m_{k-1}$, $n_k > n_{k-1}$, and $m_k x \in I_{n_k}$ for all $x \in J_k$. We remark that this construction is possible in virtue of the above claim.

Fix $x_0 \in \cap_{n=1}^{\infty} J_n$. Since $m_k x_0 \in I_{n_k}$ for all $k \geq 1$, it follows that $f(m_k x_0) \geq \varepsilon$ for any $k \geq 1$. On the other hand, $m_k \to \infty$ as $k \to \infty$. Thus, by hypothesis, $\lim_{k \to \infty} f(m_k x_0) = 0$, a contradiction. This concludes the proof. □

The following property is due to R.P. Boas [9] and is a source for new inequalities, as we will illustrate subsequently.

4.7.14. Let f be continuous with domain $0 \leq x < 1$ or $0 \leq x \leq 1$, $f(0) = 0$, $f(1) > 1$ (including the possibility that $f(1) = +\infty$). Let g be continuous with domain in the range of f, and $g(1) \leq 1$. Assume that $f(x)/x$ and $g(x)/x$ are strictly decreasing on their domains and $f(x) \neq x$ for $0 < x < 1$. Prove that $f(g(x)) \leq g(f(x))$ for $0 < x < 1$.

Solution. We observe first that since $f(x)/x$ decreases, then if $0 < y \leq 1$ we have $f(xy)/(xy) \leq f(x)/x$ for $0 < x < 1$ (or $0 < x \leq 1$), and consequently

$$f(xy) \leq yf(x) \text{ for all } 0 < x < 1, 0 < y \leq 1. \tag{4.3}$$

4.7 Qualitative Properties of Continuous Functions

By continuity and since $f(0) = 0$ and $f(1) > 1$, there exists $\varepsilon_1 \in (0,1)$ such that $f(\varepsilon_1) = 1$ and $f(x) > 1$ for $x \in [\varepsilon_1, 1)$. Since f increases strictly (because $f(x)/x$ increases), there is a unique ε_1 with the above property. Next, we determine ε_n ($n > 1$) recursively: $f(\varepsilon_{n+1}) = \varepsilon_n$ and $f(x) \geq \varepsilon_n$ for $x \in [\varepsilon_{n+1}, 1)$. Then $(\varepsilon_n)_{n \geq 1}$ is a decreasing sequence of positive numbers; hence it converges to some limit ε. If $\varepsilon > 0$, we would have $f(\varepsilon) = \lim_{n \to \infty} f(\varepsilon_n) = \varepsilon$, by continuity. Since we have assumed $f(x) \neq x$ for $x > 0$, we deduce that $\varepsilon = 0$.

Let $\varepsilon_0 = 1$. Since $\varepsilon_n \to 0$, each $x \in (0, 1)$ is in one of the intervals $[\varepsilon_{n+1}, \varepsilon_n]$, where $n \geq 0$. In this interval we have $f(x) \geq \varepsilon_n$, by the way in which the ε_n were constructed. Then relation (4.3) gives us

$$f(g(x)) = f(x \cdot g(x)/x) \leq f(x)g(x)/x \quad \text{for any } x \in [\varepsilon_{n+1}, \varepsilon_n]. \tag{4.4}$$

We observe that (4.3) is applicable because

$$g(x)/x \leq g(1)/1 \leq 1. \tag{4.5}$$

In the interval $(\varepsilon_1, 1)$ we have $f(x) > 1$ and so $g(1)/1 \leq g(f(x))/f(x)$ because $g(x)/x$ increases. If we substitute this into (4.5) and then (4.5) into (4.4), we obtain $f(g(x)) \leq g(f(x))$ for $x \in (\varepsilon_1, 1)$.

If $n \geq 1$, then since $g(x)/x$ increases, we can continue (4.4) to obtain

$$f(g(x)) \leq f(x)g(\varepsilon_n)/\varepsilon_n,$$

since $x \leq \varepsilon_n$. But $\varepsilon_n \leq f(x)$ and g increases, so that

$$f(g(x)) \leq f(x)g(f(x))/f(x) = g(f(x)),$$

for any $x \in [\varepsilon_{n+1}, \varepsilon_n)$. Since n is arbitrary, it follows that this equality is true for all $x \in (0, 1)$. □

We illustrate this property with some concrete examples.

EXAMPLE 1. Let

$$f(x) = -\log(1-x), \quad g(x) = x^a, \ a > 1.$$

Then

$$\log \frac{1}{1-x^a} \leq \left(\log \frac{1}{1-x}\right)^a \quad \text{for all } x \in (0,1).$$

Put $x = 1 - e^{-y}$, $y > 0$. We obtain the inequality

$$\log \frac{1}{1-(1-e^{-y})^a} \leq y^a \quad \text{for all } y > 0 \text{ and } a > 1.$$

EXAMPLE 2. Let

$$f(x) = \frac{2}{\pi} \tan \frac{\pi x}{2}, \quad g(x) = x^a, \ a > 1.$$

If we replace x by $2y/\pi$ we obtain

$$\tan \left[\left(\frac{2}{\pi}\right)^{a-1} y^a \right] \leq \left(\frac{2}{\pi}\right)^{a-1} (\tan y)^a.$$

In particular, if $a = 2$, we obtain

$$\tan \frac{2y^2}{\pi} \leq \frac{2}{\pi} \tan^2 y \quad \text{for all } y \in [0, \pi/2).$$

EXAMPLE 3. Let

$$f(x) = -\log(1-x), \quad g(x) = \frac{e^x - 1}{e - 1}.$$

We obtain the inequality

$$\log \frac{e-1}{e-e^x} \leq \frac{x}{(1-x)(e-1)} \quad \text{for all } x \in [0, 1).$$

4.7.15. Let $\omega : [0, \infty] \to \mathbb{R}$ be a continuous function such that $\lim_{x \to \infty} \omega(x)/x = 0$. Let $(c_n)_{n \geq 1}$ be a sequence of nonnegative real numbers such that $(c_n/n)_{n \geq 1}$ is bounded. Prove that $\lim_{n \to \infty} \omega(c_n)/n = 0$.

D. Andrica and M. Piticari

Solution. We distinguish the following situations.

CASE 1. There exists $\lim_{n \to \infty} c_n$ and it is $+\infty$. In that case we have

$$\left| \frac{\omega(c_n)}{n} \right| = \left| \frac{c_n}{n} \cdot \frac{\omega(c_n)}{c_n} \right| \leq M \left| \frac{\omega(c_n)}{c_n} \right|, \tag{4.6}$$

where $M = \sup \{c_n/n : n \geq 1\}$. Since $\lim_{x \to \infty} \omega(x)/x = 0$, from relation (4.6) it follows that $\lim_{n \to \infty} |\omega(c_n)/n| = 0$, hence the desired conclusion.

CASE 2. The sequence $(c_n)_{n \geq 1}$ is bounded. Consider $A > 0$ such that $c_n \leq A$ for all positive integers $n \geq 1$, and define $K = \sup_{x \in [0,A]} |\omega(x)|$. It is clear that $|\omega(c_n)/n| \leq K/n$ for all $n \geq 1$, that is, $\lim_{n \to \infty} \omega(c_n)/n = 0$.

CASE 3. The sequence $(c_n)_{n \geq 1}$ is unbounded and $\lim_{n \to \infty} c_n$ does not exist. For $\varepsilon > 0$ there exists $\delta > 0$ such that $|\omega(x)/x| < \varepsilon/M$ for any $x > \delta$, where $M = \sup \{c_n/n : n \geq 1\}$. Consider the sets

$$A_\delta = \{n \in \mathbb{N}^* : c_n \leq \delta\} \quad \text{and} \quad B_\delta = \{n \in \mathbb{N}^* : c_n > \delta\}.$$

If one of these sets is finite, then we immediately derive the desired result. Assume that A_δ and B_δ are both infinite. Since $\lim_{n \in A_\delta} \omega(c_n)/n = 0$, it follows that there exists $N_1(\varepsilon)$ with the property $|\omega(c_n)/n| < \varepsilon$ for all $n \in A_\delta$ with $n \geq N_1(\varepsilon)$. For $n \in B_\delta$, we have

$$\left|\frac{\omega(c_n)}{n}\right| = \left|\frac{\omega(c_n)}{c_n} \cdot \frac{c_n}{n}\right| < \frac{\varepsilon}{M} \cdot M = \varepsilon.$$

Therefore $\left|\frac{\omega(c_n)}{n}\right| < \varepsilon$ for any $n \geq N_1(\varepsilon)$. $\square$

4.8 Independent Study Problems

> The infinite! No other question has ever moved so profoundly the spirit of man.
>
> David Hilbert (1862–1943), *The World of Mathematics*

4.8.1. Find all continuous functions $f : \mathbb{R} \to \mathbb{R}$ such that

$$x_1 - x_2 \in \mathbb{Q} \iff f(x_1) - f(x_2) \in \mathbb{Q}.$$

Answer. $f(x) = ax + b$, with $a \in \mathbb{Q}$.

4.8.2. Let M be the set of all continuous functions $f : [0,1] \to \mathbb{R}$ satisfying $f(0) = f(1) = 0$.

Find the set of all positive numbers a such that for any $f \in M$ there exists a straight line of length a joining two points on the graph of f that is parallel to the Ox axis.

4.8.3. Let $C > 0$ be an arbitrary constant. Find all continuous functions $f : \mathbb{R} \to \mathbb{R}$ satisfying $f(x) = f(x^2 + C)$, for all $x \in \mathbb{R}$.

4.8.4. Let $f : \mathbb{R} \to \mathbb{R}$ be an arbitrary function satisfying $f(0) > 0$, $f(1) < 0$ and there exists a continuous function g such that $f + g$ is increasing. Prove that f vanishes at least once.

4.8.5. Let $f : [0,1] \to \mathbb{R}$ be a continuous function such that $f(0) \neq f(1)$ and f achieves each of its values finitely many times. Prove that at least one of the values of f is achieved by an odd number of values.

4.8.6. Let $f : \mathbb{R} \to \mathbb{R}$ be a continuous and decreasing function. Prove that

(a) f has a unique fixed point;
(b) the following alternative holds: either the set

$$\{x \in \mathbb{R}; \ (f \circ f)(x) = x\}$$

is infinite or it has an odd number of elements.

4.8.7. Let $f:\mathbb{R}\to\mathbb{R}$ be a continuous function such that its nth iterate has a unique fixed point x_0. Prove that $f(x_0) = x_0$.

4.8.8. Let $f:[a,b]\to[a,b]$ be a function such that
$$|f(x) - f(y)| \leq |x - y|, \quad \forall x, y \in [a, b].$$
Prove that the sequence (x_n) defined by $x_{n+1} = (x_n + f(x_n))/2$ converges to a fixed point of f.

4.8.9. Let $f, g : (a, b) \to \mathbb{R}$ be continuous functions such that at least one of these functions is monotone. Assume that there exists a sequence $(x_n) \subset (a, b)$ such that $f(x_n) = g(x_{n-1})$. Prove that the equation $f(x) = g(x)$ has at least one solution in (a, b).

4.8.10. Let (a_n) and (b_n) be sequences of real numbers converging to zero and such that $a_n \neq b_n$, for all n. Let $f : \mathbb{R} \to \mathbb{R}$ be a function such that for any $x \in (0, 1)$ there exists $N = N(x) \in \mathbb{N}$ satisfying $f(x + a_n) = f(x + b_n)$, for all $n \geq N$. If $0 < b_n < a_n$, can we assert that f is constant? The same question for $b_n = -a_n$.

4.8.11. Let $f_n : \mathbb{R} \to \mathbb{R}$ ($n \geq 1$) be functions satisfying the Lipschitz condition with constant L. Let $f : \mathbb{R} \to \mathbb{R}$ be a continuous function with the property that for all $x \in \mathbb{R}$ there exists an integer n such that $f(x) = f_n(x)$. Prove that f is also a Lipschitz function with the same constant L.
Hint. A function f satisfies the Lipschitz condition of constant 1 if and only if the mappings $x + f(x)$ and $x - f(x)$ are increasing.

4.8.12. Find all continuous functions $f : \mathbb{R} \to \mathbb{R}$ satisfying $f(2x + 1) = f(x)$, for all $x \in \mathbb{R}$.
Hint. Observe that if $x_0 = x$ and $x_{n+1} = (x_n - 1)/2$, then $\lim_{n\to\infty} x_n = -1$. This implies that any function with the required properties must be constant.

4.8.13. Let $f : \mathbb{R} \to \mathbb{R}$ be a continuous function at the origin. Solve the functional equation $f(x) = f(x/(1-x))$, for all $x \neq 1$.
Answer. The constant functions.

4.8.14. Find all continuous functions $f : \mathbb{R} \to \mathbb{R}$ such that for any geometric progression (x_n) the sequence $(f(x_n))$ is a geometric progression, too.

4.8.15. Find all continuous functions $f : \mathbb{R} \to \mathbb{R}$ such that
$$f(x) + f\left(\frac{x-1}{x}\right) = 1 + x, \quad \forall x \neq 0, 1.$$

Answer. $f(x) = (x^3 - x^2 - 1)/(2x(x-1))$.

4.8.16. Find all continuous and bijective functions $f : [0,1] \to [0,1]$ such that $f(2x - f(x)) = x$, for any $x \in [0, 1]$.
Answer. The identity map.

4.8 Independent Study Problems

4.8.17. Let $a \neq 0$ be a real number. Find all continuous functions $f : \mathbb{R} \to \mathbb{R}$ satisfying $f(2x - f(x)/a) = ax$, for all $x \in \mathbb{R}$.

4.8.18. Find all continuous functions $f : (0,1) \to \mathbb{R}$ such that

$$f\left(\frac{x-y}{\ln x - \ln y}\right) = \frac{f(x) + f(y)}{2}, \quad \forall x, y \in (0,1), \ x \neq y.$$

Answer. The constant functions.

4.8.19. Find all continuous functions $f : (-1,1) \to \mathbb{R}$ such that

$$f\left(\frac{x+y}{1+xy}\right) = f(x) + f(y), \quad \forall x, y \in (-1,1).$$

Hint. Make a change of variable to obtain the Cauchy functional equation.

4.8.20. Let k be a real number. Find all continuous functions $f : \mathbb{R} \to \mathbb{R}$ such that

$$f(x+y) = f(x) + f(y) + kxy, \quad \forall x, y \in \mathbb{R}.$$

Answer. $f(x) = kx^2/2 + cx$, where $c = f(1) - k/2$.

4.8.21. Let $n \geq 2$ be a positive integer. Find a function $f : \mathbb{R} \to \mathbb{R}$ such that $f(x) + f(2x) + \cdots + f(nx) = 0$ for all $x \in \mathbb{R}$ and $f(x) = 0$ if and only if $x = 0$.

4.8.22. Let $f : \mathbb{R} \to \mathbb{R}$ be a function satisfying the intermediate value property such that $f \circ f = 1_\mathbb{R}$. Prove that either the set of fixed points of f is $\mathbb{R}$ or it contains a unique element.

St. Alexe

4.8.23. Let $f : \mathbb{R} \to \mathbb{R}$ be a function such that f attains neither its maximum nor its minimum on any open interval $I \subset \mathbb{R}$.

(a) Prove that if f is continuous then f is monotone.
(b) Give examples of nonmonotone functions satisfying the above property.

C. Stoicescu and M. Chiriţă

4.8.24. Let $f_n : \mathbb{R} \to \mathbb{R}$, $n \in \mathbb{N}$, be functions satisfying, for all $x, y \in \mathbb{R}$ and any $n \in \mathbb{N}$, $|f_n(x) - f_n(y)| \leq |x - y|$. Assume that there exists a continuous function $f : [0,1] \to \mathbb{R}$ such that for all $x \in \mathbb{Q} \cap [0,1]$ we have $\lim_{n \to \infty} f_n(x) = f(x)$. Prove that this limit holds for any real $x \in \mathbb{R} \cap [0,1]$.

4.8.25. Let $f : [0,1] \to \mathbb{R}$ be a continuous function. Compute $\lim_{n \to \infty} n \int_0^1 x^n f(x) dx$.

4.8.26. Solve the simultaneous functional equations

$$f(x+y) = f(x) + \frac{f(y)g(x)}{1-f(x)f(y)},$$

$$g(x+y) = \frac{g(x)g(y)}{[1-f(x)f(y)]^2}.$$

J.L. Riley, Amer. Math. Monthly, Problem 2759

4.8.27. Let $f : [0,1] \to \mathbb{R}$ be a continuous function. Find all real numbers a for which $A_a(f) := \lim_{n\to\infty} n^a \int_0^1 e^{-nx} f(x) dx$ exists. If the limit $A_a(f)$ exists, compute it.

4.8.28. Let $f : \mathbb{R} \to \mathbb{R}$ be a continuous function such that for any real x and any positive integer n, $f(x+n^{-1}) > f(x-n^{-1})$. Prove that f is an increasing function.

4.8.29. Let $f : \mathbb{R} \to \mathbb{R}$ be a nonconstant function such that there exists a continuous strictly monotone function $g : \mathbb{R} \to \mathbb{R}$ such that $\lim_{t\to x} f(t) = g(f(x))$, for all $x \in \mathbb{R}$. Prove that there is a nondegenerate interval of fixed points of g.

G. Dospinescu

4.8.30. Let $I \subset \mathbb{R}$ be an interval and consider a continuous function $f : I \to \mathbb{R}$ such that for all $x \in I$ there exists a positive integer n_x with $f^{n_x}(x) = x$. [We have denoted by f^p the pth iterate of f, that is, $f^p = f \circ \cdots \circ f$ (p times)].

(a) Prove that if $I = [a, \infty)$ then f is the identity map.
(b) Prove that if f is not the identity map for $I = (a, \infty)$ then $\lim_{x\to\infty} f(x) = a$.

D. Schwartz

4.8.31. Prove that the function $f(x) = \cos x^2$ is not uniformly continuous on $\mathbb{R}$.

4.8.32. Let $f : \mathbb{R} \to \mathbb{R}$ be a uniformly continuous function. Prove that there exist positive constants a and b such that $|f(x)| \leq a|x| + b$ for all $x \in \mathbb{R}$.

4.8.33. Consider the function

$$f(x) = \begin{cases} x \sin \frac{1}{x} & \text{if } x > 0, \\ 0 & \text{if } x = 0. \end{cases}$$

Prove that $|f(x) - f(y)|/|x-y|^a$ is bounded for $x, y \in [0,1]$ if and only if $a \leq 1/2$.

J.H. Curtiss, Amer. Math. Monthly, Problem 3939

4.8.34. Let $f : (0, \infty) \to \mathbb{R}$ be a continuous function such that for all $\ell > 0$, $\lim_{x\to +\infty} (f(x+\ell) - f(x)) = 0$. Prove that for any $0 \leq a < b$,

$$\lim_{x\to +\infty} \sup_{\ell \in [a,b]} |f(x+\ell) - f(x)| = 0.$$

4.8 Independent Study Problems

4.8.35. Let $f : [0,1] \to \mathbb{R}$ be a continuous function. Prove that

$$\lim_{n \to \infty} \frac{\sum_{k=1}^n (-1)^k f(k/n)}{n} = 0.$$

4.8.36. (Rising Sun Lemma) Let $f : [a,b] \to \mathbb{R}$ be a continuous function and define the set

$$S = \{x \in [a,b]; \text{ there exists } y > x \text{ such that } f(y) > f(x)\}.$$

Prove that if $(c,d) \subset S$ but $c, d \notin S$, then $f(c) = f(d)$.

Chapter 5
Differentiability

> *Isaac Newton was not a pleasant man. His relations with other academics were notorious, with most of his later life spent embroiled in heated disputes.... A serious dispute arose with the German philosopher Gottfried Leibniz. Both Leibniz and Newton had independently developed a branch of mathematics called calculus, which underlies most of modern physics... Following the death of Leibniz, Newton is reported to have declared that he had taken great satisfaction in "breaking Leibniz's heart".*
> —Stephen Hawking, *A Brief History of Time*, 1988

Abstract. The concept of *derivative* is the main theme of differential calculus, one of the major discoveries in mathematics, and in science in general. Differentiation is the process of finding the best local linear approximation of a function. The idea of the derivative comes from the intuitive concepts of velocity or rate of change, which are thought of as instantaneous or infinitesimal versions of the basic difference quotient $(f(x) - f(x_0))/(x - x_0)$, where f is a real-valued function defined in a neighborhood of x_0. A geometric way to describe the notion of *derivative* is the *slope* of the tangent line at some particular point on the graph of a function. This means that at least locally (that is, in a small neighborhood of any point), the graph of a smooth function may be approximated with a straight line. Our goal in this chapter is to carry out this analysis by making the intuitive approach mathematically rigorous. Besides the basic properties, the chapter includes many variations of the mean value theorem as well as its extensions involving higher derivatives.

5.1 The Concept of Derivative and Basic Properties

> Data aequatione quotcunque fluentes quantitae involvente fluxiones invenire et vice versa. ["Given an equation which involves the derivatives of one or more functions, find the functions."]
> Sir Isaac Newton to G.W. Leibniz, October 24, 1676

The above quotation is the decipherment of an anagram in which Newton points out that differential equations are important because they express the laws of nature.[1] It is now fully recognized that the origins of the *differential calculus* go back

---

[1] Newton sent his letter to Henry Oldenburg, secretary of the Royal Society, and through him to G.W. Leibniz in Germany. Oldenburg was later imprisoned in the Tower of London for corresponding with foreigners.

to the works of Sir Isaac Newton (1643–1727) and Gottfried Wilhelm von Leibniz (1646–1716). However, when calculus was first being developed, there was a controversy as to who came up with the idea "first." It is thought that Newton had discovered several ideas related to calculus earlier than Leibniz. However, Leibniz was the first to publish. Today, both Leibniz and Newton are considered to have discovered calculus independently.

The concept of *derivative* has been applied in various domains and we mention in what follows only some of them:
– velocity and acceleration of a movement (Galileo 1638, Newton 1686);
– astronomy, verification of the law of gravitation (Newton, Kepler);
– calculation of the angles under which two curves intersect (Descartes);
– construction of telescopes (Galileo) and clocks (Huygens 1673);
– search for the maxima and minima of a function (Fermat 1638).

Let f be a real function with domain $I \subset \mathbb{R}$. If $x_0 \in I$ is an interior point of I, then the limit
$$\lim_{x \to x_0} \frac{f(x) - f(x_0)}{x - x_0},$$
when it exists, is called the *derivative* of f at x_0. In this case we say that f is *differentiable* at x_0 and the value of the above limit is called the *derivative* of f at x_0 and is denoted by $f'(x_0)$. If I is an open interval and if f is differentiable at every $x_0 \in I$, then we say that f is *differentiable on I*. We point out that the notation $f'(x_0)$ is a descendant of notation introduced by Newton, while Leibniz used the notation $(df/dx)(x_0)$. The nth-order derivative of f is denoted by $f^{(n)}$.

We notice that the *Newton quotient*
$$\frac{f(x) - f(x_0)}{x - x_0}$$

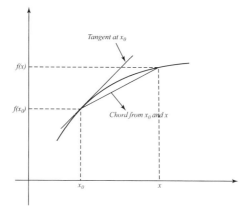

Fig. 5.1 Tangent and chord.

5.1 The Concept of Derivative and Basic Properties

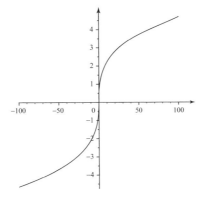

Fig. 5.2 Graph of the function $f(x) = \sqrt[3]{x}$.

measures the slope of the chord of the graph of f that connects the points $(x, f(x))$ and $(x_0, f(x_0))$ (see Figure 5.1).

The *one-sided derivatives* of f at x_0 (if they exist) are defined by

$$f'(x_0-) := \lim_{x \nearrow x_0} \frac{f(x) - f(x_0)}{x - x_0} \quad \text{and} \quad f'(x_0+) := \lim_{x \searrow x_0} \frac{f(x) - f(x_0)}{x - x_0}.$$

Moreover, f is differentiable at x_0 if and only if $f'(x_0-)$ and $f'(x_0+)$ exist, are finite, and $f'(x_0-) = f'(x_0+)$. The point x_0 is said to be an *angular point* if $f'(x_0-)$ and $f'(x_0+)$ are finite and $f'(x_0-) \neq f'(x_0+)$. We say that x_0 is a *cusp point* of f if $f'(x_0-) = \pm\infty$ and $f'(x_0+) = \mp\infty$ (see Figure 5.2), while we say that f has a *vertical tangent* at x_0 if $f'(x_0-) = f'(x_0+) = \pm\infty$ (see Figure 5.3).

The above definition implies that if f is differentiable at x_0, then f is continuous at x_0. The converse is not true, and the most usual counterexample in this sense is given by the mapping $f : \mathbb{R} \to \mathbb{R}$ defined by $f(x) = |x|$ (see Figure 5.4). Indeed, this function is continuous at the origin, but is not differentiable at $x_0 = 0$. Moreover,

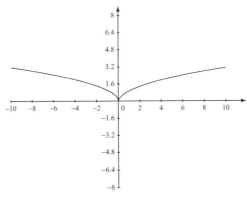

Fig. 5.3 Graph of the function $f(x) = \sqrt{|x|}$.

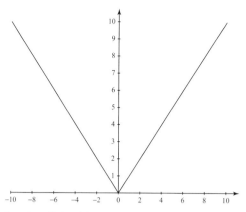

Fig. 5.4 Graph of the function $f(x) = |x|$.

$f'(0-) = -1$ and $f'(0+) = +1$. A function $f : \mathbb{R} \to \mathbb{R}$ that is *everywhere continuous* but *nowhere differentiable* is

$$f(x) = \sum_{n=1}^{\infty} 2^{-n} \sin^{n^2} x \quad \text{for all } x \in \mathbb{R}.$$

In fact, the first example of a continuous function that is nowhere differentiable is due to Weierstrass (1872), who shocked the mathematical world by showing that the continuous function

$$f(x) = \sum_{n=1}^{\infty} b^n \cos(a^n x) \quad (0 < b < 1)$$

is nowhere differentiable, provided $ab > 1 + 3\pi/2$.

The function

$$f(x) = \begin{cases} x, & \text{if } x \in \mathbb{Q}, \\ x^2 + x, & \text{if } x \in \mathbb{R} \setminus \mathbb{Q}, \end{cases}$$

is discontinuous for all $x \neq 0$, but it is differentiable at $x = 0$ and $f'(0) = 1$.

A related notion is the following. Let $I \subset \mathbb{R}$ be an interval. We say that a function $f : I \to \mathbb{R}$ has a *symmetric derivative* in $x_0 \in \text{Int}\, I$ if there exists

$$f'_s(x_0) : \lim_{t \to 0} \frac{f(x_0 + t) - f(x_0 - t)}{2t}.$$

If there exist $f'(x_0+)$ and $f'(x_0-)$, then there exists the symmetric derivative $f'_s(x_0)$ and, moreover,

$$f'_s(x_0) = \frac{f'(x_0+) + f'(x_0-)}{2}.$$

The converse is not true, as shown by the following counterexample:

$$f(x) = \begin{cases} x \sin \frac{1}{x}, & \text{if } x \neq 0, \\ 0, & \text{if } x = 0. \end{cases}$$

5.1 The Concept of Derivative and Basic Properties

We have already seen that Lipschitz maps are continuous. Of course, they do not need to be differentiable, as show by the absolute value map $f(x) = |x|$ ($x \in \mathbb{R}$), which satisfies

$$|f(x) - f(y)| \leq |x - y| \quad \text{for all } x, y \in \mathbb{R},$$

but which is not differentiable at the origin. The following question arises naturally: are there differentiable functions that are not Lipschitz? The answer is **yes**, and a function with such a property is

$$f(x) = \begin{cases} x^{3/2} \sin \frac{1}{x}, & \text{if } 0 < x \leq 1, \\ 0, & \text{if } x = 0. \end{cases}$$

In 1872, Weierstrass shocked the mathematical world by giving an example of a *function that is continuous at every point but whose derivative does not exist anywhere*. Consider the mapping $\varphi : \mathbb{R} \to \mathbb{R}$ defined by

$$\varphi(x) = \begin{cases} x - n, & \text{if } n \text{ is even and } n \leq x < n+1, \\ n+1 - x, & \text{if } n \text{ is odd and } n \leq x < n+1, \end{cases}$$

and set

$$f(x) := \sum_{n=1}^{\infty} \left(\frac{3}{4}\right)^n \varphi(4^n x).$$

Then f defined as above is continuous at **every** real x but differentiable at **no** real x. The celebrated French mathematician Henri Poincaré [*L'oeuvre mathématique de Weierstrass*, Acta Math., vol. 22 (1899)] mentioned about this striking example, "A hundred years ago such a function would have been considered an outrage on common sense."

The derivative of a function is directly relevant to finding its maxima and minima. We recall that if f is a real function defined on an interval I, then a point $x_0 \in I$ is called a *local minimum* for f if there exists $\delta > 0$ such that $(x_0 - \delta, x_0 + \delta) \subset I$ and $f(x) \geq f(x_0)$ for all $x \in (x_0 - \delta, x_0 + \delta)$. If x_0 is a local minimum of the function $-f$, we say that x_0 is a *local maximum* of f.

The following Fermat test is an important tool for proving inequalities, and since it holds even in higher dimensions, it also applies in critical-point theory, in order to prove the existence of weak solutions to partial differential equations. This theorem was discovered by the French mathematician Pierre de Fermat (1601–1665) and expresses a geometrically intuitive fact: if x_0 is an *interior extremal point* for f in I and if the graph of f has a tangent line at $(x_0, f(x_0))$, then such a line must be parallel to the x-axis.

Fermat's Theorem. *Let f be real-valued function defined on an interval I. If f has a local extremum at an interior point x_0 of I and if f is differentiable at x_0, then $f'(x_0) = 0$.*

A more precise answer can be formulated in terms of the second derivative. More exactly, if $f : I \to \mathbb{R}$ has a local maximum (resp., local minimum) at some interior point x_0 of I, then $f''(x_0) \leq 0$ (resp., $f''(x_0) \geq 0$).

Mean value theorems play a central role in analysis. The simplest form of the mean value theorem is stated in the next basic result, which is due to Michel Rolle (1652–1719).

Rolle's Theorem. *Let f be a continuous real-valued function defined on the closed interval $[a,b]$ that is differentiable on (a,b). If $f(a) = f(b)$, then there exists a point $c \in (a,b)$ such that $f'(c) = 0$.*

Geometrically, Rolle's theorem states that if $f(a) = f(b)$ then there is a point in the interval (a,b) at which the tangent line to the graph of f is parallel to the x-axis (see Figure 5.5). Another geometrical implication of Rolle's theorem is the following.

Polar Form of Rolle's Theorem. *Assume that f is a continuous real-valued function, nowhere vanishing in $[\theta_1, \theta_2]$, differentiable in (θ_1, θ_2), and such that $f(\theta_1) = f(\theta_2)$. Then there exists $\theta_0 \in (\theta_1, \theta_2)$ such that the tangent line to the graph $r = f(\theta)$ at $\theta = \theta_0$ is perpendicular to the radius vector at that point.*

Taking into account the geometric interpretation of Rolle's theorem, we expect that is would be possible to relate the slope of the chord connecting $(a, f(a))$ and $(b, f(b))$ with the value of the derivative at some interior point. In fact, this is the main content of a mean value theorem. The following is one of the basic results in elementary mathematical analysis and is known as the *mean value theorem for differential calculus* [Joseph-Louis Lagrange (1736–1813)]. Geometrically, this theorem states that there exists a suitable point $(c, f(c))$ on the graph of $f : [a,b] \to \mathbb{R}$ such that the tangent is parallel to the straight line through the points $(a, f(a))$ and $(b, f(b))$ (see Figure 5.6).

Lagrange's Mean Value Theorem. *Let f be a continuous real-valued function defined on the closed interval $[a,b]$ that is differentiable on (a,b). Then there exists a point $c \in (a,b)$ such that*

$$\frac{f(b) - f(a)}{b - a} = f'(c).$$

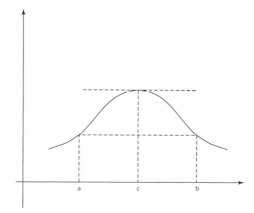

Fig. 5.5 Geometric illustration of Rolle's theorem.

5.1 The Concept of Derivative and Basic Properties

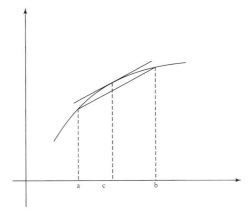

Fig. 5.6 Geometric illustration of Lagrange's theorem.

A simple statement that follows directly from the above result is that any differentiable function with **bounded** derivative is Lipschitz (compare with the comments at the beginning of this chapter!).

A useful consequence of the Lagrange mean value theorem shows that *computing the sign of the first derivative* of a differentiable function suffices to *localize maxima and minima of the function*. More precisely, if f is differentiable on an open interval I, then f is increasing on I if and only if $f'(x) \geq 0$ for all $x \in I$. This property is also called the *increasing function theorem*. Furthermore, if $f'(x) > 0$ for all $x \in I$, then f is strictly increasing in I. A direct consequence of this property is that if $f'(x) \leq g'(x)$ on $[a,b]$, then $f(x) - f(a) \leq g(x) - g(a)$ for all $x \in [a,b]$. This result is sometimes refereed as the *racetrack principle*: if one car goes faster than another, then it travels farther during any time interval.

The converse of the property

$$f' > 0 \text{ in } (a,b) \Longrightarrow f \text{ strictly increasing in } (a,b)$$

is **not** true. Indeed, the function

$$f(x) = \begin{cases} x(2 - \cos(\ln x) - \sin(\ln x)) & \text{if } x \in (0,1], \\ 0 & \text{if } x = 0, \end{cases}$$

is (strictly!) increasing on $[0,1]$, but there are infinitely many points $\xi \in (0,1)$ such that $f'(\xi) = 0$ (see Figure 5.7).

The definition of differentiability implies that if $f : (a,b) \to \mathbb{R}$ is differentiable at $x_0 \in (a,b)$ and $f'(x_0) > 0$, then there exists $\delta > 0$ such that

$$f(x) > f(x_0) \quad \text{for all } x \in (x_0, x_0 + \delta),$$
$$f(x) < f(x_0) \quad \text{for all } x \in (x_0 - \delta, x_0).$$

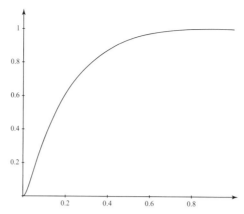

Fig. 5.7 A strictly increasing function with derivative vanishing infinitely many times.

The above statement does not imply that a function satisfying $f'(x_0) > 0$ is increasing in a neighborhood of x_0. As a counterexample, consider the function (see Figure 5.8)

$$f(x) = \begin{cases} x + x^2 \sin \frac{1}{x^2} & \text{if } x \neq 0, \\ 0 & \text{if } x = 0. \end{cases}$$

This function is differentiable everywhere and satisfies $f'(0) = 1$. However, for $x \neq 0$, the derivative

$$f'(x) = 1 + 2x \sin \frac{1}{x^2} - \frac{2}{x} \cos \frac{1}{x^2}$$

oscillates strongly near the origin (see Figure 5.9). Hence, even though the graph of f is contained between the parabolas $y_1 = x - x^2$ and $y_2 = x + x^2$, there are points with negative derivative arbitrarily close to the origin. If $f'(x) > 0$ for **all** $x \in (a,b)$,

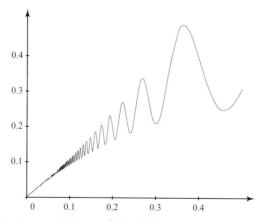

Fig. 5.8 Graph of the function $f(x) = x + x^2 \sin(1/x^2)$, $x \neq 0$.

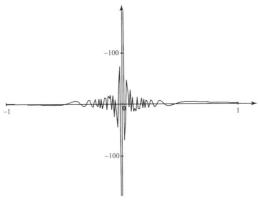

Fig. 5.9 Graph of the function $\mathbb{R} \setminus \{0\} \ni x \longmapsto 1 + 2x \sin \frac{1}{x^2} - \frac{2}{x} \cos \frac{1}{x^2}$.

then f is increasing. Thus, this counterexample is possible only because f is not continuously differentiable.

A much more elaborated notion of monotony is that of a *completely monotonic function*. This is a real-valued function defined on an interval I of the real line that satisfies $(-1)^n f^{(n)}(x) > 0$ for all $x \in I$ and $n \in \mathbb{N}^*$. Such functions do exist. An example is given by the mapping $f : (1, \infty) \to \mathbb{R}$ defined by $f(x) = \sqrt{x^2 - 1}$. An induction argument shows that for all $x > 1$, $f^{(n)}(x) > 0$ for odd n and $f^{(n)}(x) < 0$ for even $n > 0$.

Other immediate consequences of the mean value theorem establish that if f is a real-valued function that is continuous on $[a, b]$ and differentiable on (a, b), then the following hold:

(i) if $f'(x) \neq 0$ for all $x \in (a, b)$ then f is one-to-one on $[a, b]$;
(ii) if $f'(x) = 0$ for all $x \in (a, b)$ then f is constant on $[a, b]$.

The following example (*devil's staircase*) shows that the last result is not as trivial as it might appear. If $x \in [0, 1]$ has a representation in base 3 as, e.g., $x = 0.2022002101220\ldots$, then $f(x)$ is obtained in base 2 by converting all 2's preceding the first 1 to a 1 and deleting subsequent digits. In our case, $f(x) = 0.101100011$. In particular, $f(x) = 1/2$ on $[1/3, 2/3]$, $f(x) = 1/4$ on $[1/9, 2/9]$, $f(x) = 3/4$ on $[7/9, 8/9]$, etc. This function f is continuous and nondecreasing. Moreover, f is differentiable with derivative $f'(x) = 0$ on a set of measure

$$\frac{1}{3} + \frac{2}{9} + \frac{4}{27} + \frac{8}{81} + \cdots = 1,$$

hence almost everywhere. Nevertheless, $f(0) \neq f(1)$.

A useful consequence of the Lagrange mean value theorem is the following.

Corollary. *Let $f : [a, b) \to \mathbb{R}$ be a differentiable function that is continuous at $x = a$ and such that $\lim_{x \searrow a} f'(x)$ exists and is finite. Then f is differentiable at $x = a$ and $f'(a) = \lim_{x \searrow a} f'(x)$.*

A generalization of the Lagrange mean value theorem is stated in the next result, which is due to the celebrated French mathematician Augustin-Louis Cauchy (1789–1857).

Cauchy's Mean Value Theorem. *Let f and g be continuous real-valued functions defined on $[a,b]$ that are differentiable on (a,b) and such that $g(a) \neq g(b)$. Then there is a point $c \in (a,b)$ such that*

$$\frac{f(b)-f(a)}{g(b)-g(a)} = \frac{f'(c)}{g'(c)}.$$

A geometric interpretation of Cauchy's mean value theorem is that if $p(x) = (f(x), g(x))$ is a path in the plane, then under certain hypotheses, there must be an instant at which the velocity vector is parallel to the vector joining the endpoints of the path.

We point out that the mean value theorem is no longer valid if f takes values in $\mathbb{R}^N$, with $N \geq 3$. The mean value theorem is still valid in $\mathbb{R}^2$ provided one states it correctly. Specifically, say that two vectors $\mathbf{v}$ and $\mathbf{w}$ are parallel if there is some nontrivial linear relation $a\mathbf{v} + b\mathbf{w} = \mathbf{0}$, that is, if the span of $\mathbf{v}$ and $\mathbf{w}$ has dimension less than 2. Note that this is not an equivalence relation, since the zero vector is parallel to anything. Then for any C^1 function $f : [a,b] \to \mathbb{R}^2$ there is some $c \in (a,b)$ such that $f'(c)$ and $f(b) - f(a)$ are parallel. The result fails in $\mathbb{R}^3$ and, more generally, in any $\mathbb{R}^N$, where $N \geq 3$. Indeed, consider the function $f : [0, 2\pi] \to \mathbb{R}^3$ defined by $f(x) = (\cos x, \sin x, x)$. Then $f(2\pi) - f(0) = (0, 0, 2\pi)$ but $f'(x) = (-\sin x, \cos x, 1)$, for all $x \in [0, 2\pi]$.

In fact, the authentic sense of the mean value theorem is that of mean value **inequality**. For instance, if f is a real-valued function such that $m \leq f'(x) \leq M$ on the interval $[a,b]$, then

$$m(x-a) \leq f(x) - f(a) \leq M(x-a),$$

for all x in $[a,b]$. It is interesting to point out that before Lagrange and Cauchy, Ampère [34] saw the importance of the above mean value inequality and even used it as the defining property of the derivative.

The following result, discovered by Gaston Darboux (1842–1917), asserts that the derivative of a real-valued function defined on an interval has the intermediate value property.

Darboux's Theorem. *Let I be an open interval and let $f : I \to \mathbb{R}$ be a differentiable function. Then f' has the intermediate value property.*

If f' is continuous, then Darboux's theorem follows directly from the intermediate value property of continuous functions. But a derivative need **not** be continuous. Indeed, the function

$$f(x) = \begin{cases} x^2 \cdot \sin \frac{1}{x}, & \text{if } x \neq 0, \\ 0, & \text{if } x = 0, \end{cases}$$

is differentiable at $x = 0$ but $\lim_{x \to 0} f'(x)$ does not exist (see Figure 5.10).

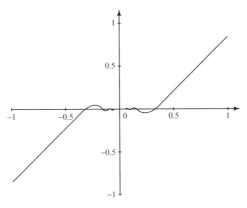

Fig. 5.10 A differentiable function whose derivative is not continuous at $x = 0$.

The standard proof of Darboux's theorem is based on the fact that a continuous function on a compact interval has a maximum. This proof relies on considering the auxiliary function $g(t) = f(t) - y_0 t$, where $a, b \in I$ are fixed and y_0 lies strictly between $f'(a)$ and $f'(b)$. Without loss of generality assume $a < b$ and $f'(a) > y_0 > f'(b)$. Then $g'(a) > 0 > g'(b)$, so a and b are not local maxima of g. Since g is continuous, it must therefore attain its maximum at an *interior* point x_0 of $[a, b]$. So, by Fermat's theorem, $g'(x_0) = 0$, and this concludes the proof. $\square$

The following new proof of Darboux's theorem is due to Olsen [86] and is based on an interesting approach that uses the Lagrange mean value theorem and the intermediate value theorem for continuous functions. We present this elegant proof in what follows. Assume that y_0 lies strictly between $f'(a)$ and $f'(b)$. Define the continuous functions $f_a, f_b : I \to \mathbb{R}$ by

$$f_a(t) = \begin{cases} f'(a) & \text{for } t = a, \\ \frac{f(t) - f(a)}{t - a} & \text{for } t \neq a, \end{cases}$$

and

$$f_b(t) = \begin{cases} f'(b) & \text{for } t = b, \\ \frac{f(t) - f(b)}{t - b} & \text{for } t \neq b. \end{cases}$$

Then $f_a(a) = f'(a)$, $f_a(b) = f_b(a)$, and $f_b(b) = f'(b)$. Thus, either y_0 lies between $f_a(a)$ and $f_a(b)$, or y_0 lies between $f_b(a)$ and $f_b(b)$. In the first case, by the continuity of f_a, there exists $s \in (a, b]$ such that

$$y_0 = f_a(s) = \frac{f(s) - f(a)}{s - a}.$$

The Lagrange mean value theorem implies that there exists $x_0 \in (a, s)$ such that

$$y_0 = \frac{f(s) - f(a)}{s - a} = f'(x_0).$$

In the second case, assuming that y_0 lies between $f_b(a)$ and $f_b(b)$, a similar argument based on the continuity of f_b shows that there exist $s \in [a,b)$ and $x_0 \in (s,b)$ such that $y_0 = [f(s) - f(b)]/(s-b) = f'(x_0)$. This completes the proof. □

The significance of Darboux's theorem is deep. Indeed, since f' always satisfies the intermediate value property (even when it is not continuous), its discontinuities are all of the second kind. Geometrically, Darboux's theorem means that although derivatives need not be continuous, they cannot suffer from jump discontinuities.

Another important result is Taylor's formula, which allows us to find approximate values of elementary functions. It was discovered by the British mathematician Brook Taylor (1685–1731). We first recall that if $f : (a,b) \to \mathbb{R}$ is n times differentiable at some point $x_0 \in (a,b)$, then the *Taylor polynomial* of f about x_0 of degree n is the polynomial P_n defined by

$$P_n(x) = \sum_{k=0}^{n} \frac{f^{(k)}(x_0)}{k!} (x-x_0)^k.$$

Often the Taylor polynomial with center $x_0 = 0$ is called the *Maclaurin polynomial* [*A Treatise of Fluxions*, Colin MacLaurin (1698–1746)].

Taylor's Formula with Lagrange's Reminder. *Let f be an arbitrary function in $C^{n+1}(a,b)$ and $x_0 \in (a,b)$. Then for all $x \in (a,b)$, there exists a point ξ in the open interval with endpoints x_0 and x such that*

$$f(x) - P_n(x) = \frac{f^{(n+1)}(\xi)}{(n+1)!} (x-x_0)^{n+1}.$$

Related to this result, in 1797, Lagrange wrote a treatise that allowed him (as he thought) to assert that any C^∞ function $f : (a,b) \to \mathbb{R}$ is equal to its Taylor series about $x_0 \in (a,b)$, that is, it can be expressed as

$$f(x) = \sum_{n=0}^{\infty} \frac{f^{(n)}(x_0)}{n!} (x-x_0)^n. \tag{5.1}$$

Lagrange's dream lasted some 25 years, until in 1823 Cauchy considered the function $f : \mathbb{R} \to \mathbb{R}$ defined by

$$f(x) = \begin{cases} e^{-1/x^2} & \text{if } x \neq 0, \\ 0 & \text{if } x = 0. \end{cases} \tag{5.2}$$

This function is continuous everywhere, but is extremely flat at the origin (see Figure 5.11), in the sense that $f^{(n)}(0) = 0$ for all n [Hint: $f^{(n)}(x) = p_n(x^{-1}) \cdot e^{-1/x^2}$, for all $x \neq 0$ and $n \in \mathbb{N}$, where p_n denotes a polynomial with integer coefficients]. Thus, the Maclaurin series for the function $f(x)$ is $0 + 0 + \cdots$ and obviously converges for all x. This shows that relation (5.1) is **wrong** for $x_0 = 0$ and $x \neq 0$. A function f satisfying relation (5.1) for **all** x in some neighborhood of x_0 is called *analytic* at x_0. This shows that the function f defined by (5.2) is **not** analytic at the

5.1 The Concept of Derivative and Basic Properties

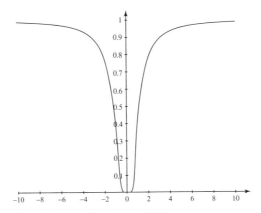

Fig. 5.11 Graph of the function defined by relation (5.2).

origin, even if its associated Taylor series converges. A related remarkable result of Emile Borel (1871–1956) states that if $(a_n)_{n\geq 0}$ is any real sequence, then there are infinitely many differentiable functions f such that $f^{(n)}(0) = a_n$ for all integer $n \geq 0$.

An important role in asymptotic analysis is played by the following notation, which is due to Landau:
(i) $f(x) = o(g(x))$ as $x \to x_0$ if $f(x)/g(x) \to 0$ as $x \to x_0$;
(ii) $f(x) = O(g(x))$ as $x \to x_0$ if $f(x)/g(x)$ is bounded in a neighborhood of x_0;
(iii) $f \sim g$ as $x \to x_0$ if $f(x)/g(x) \to 1$ as $x \to x_0$.

Using the above notation, we give in what follows the Taylor expansions of some elementary functions:

$$e^x = 1 + x + \frac{x^2}{2} + \frac{x^3}{3!} + \cdots + \frac{x^n}{n!} + o(x^n) \quad \text{as } x \to 0,$$

$$\ln(1+x) = x - \frac{x^2}{2} + \frac{x^3}{3} - \cdots + (-1)^n \frac{x^{n+1}}{n+1} + o(x^{n+1}) \quad \text{as } x \to 0,$$

$$\sin x = x - \frac{x^3}{3!} + \frac{x^5}{5!} - \cdots + (-1)^n \frac{x^{2n+1}}{(2n+1)!} + o(x^{2n+2}) \quad \text{as } x \to 0,$$

$$\cos x = 1 - \frac{x^2}{2!} + \frac{x^4}{4!} - \cdots + (-1)^n \frac{x^{2n}}{(2n)!} + o(x^{2n+1}) \quad \text{as } x \to 0,$$

$$\tan x = x + \frac{x^3}{3} + \frac{2x^5}{15} + \frac{17x^7}{315} + \frac{62x^9}{2835} + \frac{1382x^{11}}{155925} + \frac{21844x^{13}}{6081075} + o(x^{14}) \quad \text{as } x \to 0,$$

$$\arcsin x = x + \frac{1}{2}\frac{x^3}{3} + \frac{1 \cdot 3}{2 \cdot 4}\frac{x^5}{5} + \cdots + \frac{(2n-1)!!}{(2n)!!}\frac{x^{2n+1}}{2n+1} + o(x^{2n+2}) \quad \text{as } x \to 0,$$

$$\arccos x = \frac{\pi}{2} - x - \frac{1}{2}\frac{x^3}{3} - \frac{1 \cdot 3}{2 \cdot 4}\frac{x^5}{5} - \cdots - \frac{(2n-1)!!}{(2n)!!}\frac{x^{2n+1}}{2n+1} + o(x^{2n+2}) \quad \text{as } x \to 0,$$

$$\arctan x = x - \frac{x^3}{3} + \frac{x^5}{5} - \cdots + (-1)^n \frac{x^{2n+1}}{2n+1} + o(x^{2n+2}) \quad \text{as } x \to 0,$$

$$\sinh x = x + \frac{x^3}{3!} + \frac{x^5}{5!} + \cdots + \frac{x^{2n+1}}{(2n+1)!} + o(x^{2n+2}) \quad \text{as } x \to 0,$$

$$\cosh x = 1 + \frac{x^2}{2!} + \frac{x^4}{4!} + \cdots + \frac{x^{2n}}{(2n)!} + o(x^{2n+1}) \quad \text{as } x \to 0,$$

$$\tanh x = x - \frac{x^3}{3} + \frac{2x^5}{15} - \frac{17x^7}{315} + \frac{62x^9}{2835} - \frac{1382x^{11}}{155925} + \frac{21844x^{13}}{6081075} + o(x^{14}) \quad \text{as } x \to 0,$$

$$\frac{1}{1-x} = 1 + x + x^2 + x^3 + \cdots + x^n + o(x^n) \quad \text{as } x \to 0,$$

$$(1+x)^\alpha = 1 + C_\alpha^1 x + C_\alpha^2 x^2 + \cdots + C_\alpha^n x^n + o(x^n) \quad \text{as } x \to 0 \ (\alpha \in \mathbb{R}),$$

where the *binomial coefficients* are given by

$$C_\alpha^n := \frac{\alpha(\alpha-1)(\alpha-2)\cdots(\alpha-n+1)}{n!}.$$

The following theorem (whose "discrete" variant is the Stolz–Cesàro lemma for sequences) provides us a rule to compute limits of the indeterminate form $0/0$ or ∞/∞. This result is due to Guillaume François Antoine de l'Hôpital, Marquis de Sainte-Mesme et du Montellier Compte d'Autremonts, Seigneur d'Ouques et autre lieux (1661–1704), who published (anonymously) in 1691 the world's first textbook on calculus, based on Johann Bernoulli's notes.

L'Hôpital's Rule. *Let I be a nonempty open interval in $\mathbb{R}$, let $x_0 \in \mathbb{R} \cup \{-\infty, +\infty\}$ be one of its endpoints, and let $f, g : I \to \mathbb{R}$ be two differentiable functions, with g strictly monotone and $g'(x) \neq 0$ for all $x \in I$. Assume that either of the next two conditions is satisfied:*

(i) $f(x), g(x) \to 0$ as $x \to x_0$;
(ii) $g(x) \to \pm\infty$ as $x \to x_0$ (no assumptions on f).

Then the following implication holds:

$$\lim_{x \to x_0} \frac{f'(x)}{g'(x)} \text{ exists and equals } \ell \in \mathbb{R} \cup \{-\infty, +\infty\} \Rightarrow \lim_{x \to x_0} \frac{f(x)}{g(x)} \text{ exists and equals } \ell.$$

Some important comments related to l'Hôpital's rule are stated below.

(i) We can not assert that if $\lim_{x \to x_0} f(x)/g(x)$ exists, then $\lim_{x \to x_0} f'(x)/g'(x)$ exists, too. Indeed, consider the C^∞-functions

$$f(x) = \begin{cases} x \sin \frac{1}{x^4} e^{-1/x^2}, & \text{if } x \neq 0, \\ 0, & \text{if } x = 0, \end{cases} \quad g(x) = \begin{cases} e^{-1/x^2}, & \text{if } x \neq 0, \\ 0, & \text{if } x = 0. \end{cases}$$

Then $\lim_{x \to 0} f(x)/g(x) = 0$ but

$$\frac{f'(x)}{g'(x)} = -\frac{2}{x} \cos \frac{1}{x^4} + o(1) \quad \text{as } x \to 0,$$

which implies that $\lim_{x \to 0} f'(x)/g'(x) = 0$ does not exist.

(ii) We are **not** entitled to draw any conclusion about $\lim_{x\to x_0} f(x)/g(x)$ if $\lim_{x\to x_0} f'(x)/g'(x)$ does **not** exist. Strictly speaking, if g' has zeros in every neighborhood of x_0, then f'/g' is not defined around x_0 and we could assert that $\lim_{x\to x_0} f'(x)/g'(x)$ does not exist. Stolz [108] discovered a striking phenomenon related to the applicability of l'Hôpital's rule. More precisely, he showed that it is possible that f' and g' contain a common factor, that is, $f'(x) = A(x)f_1(x)$ and $g'(x) = A(x)g_1(x)$, where $A(x)$ does not approach a limit, but $\lim_{x\to x_0} f_1(x)/g_1(x)$ exists. This means that under these circumstances, $\lim_{x\to x_0} f'(x)/g'(x)$ may exist but $\lim_{x\to x_0} f(x)/g(x)$ not exist. For instance, let us consider the example given by Stolz:

$$f(x) = \frac{x}{2} + \frac{\sin 2x}{4} \quad \text{and} \quad g(x) = \left(\frac{x}{2} + \frac{\sin 2x}{4}\right) e^{\sin x}.$$

Hence

$$\lim_{x\to\infty} \frac{f'(x)}{g'(x)} = \lim_{x\to\infty} \frac{2\cos^2 x}{\cos x e^{\sin x}(x + 2\cos x + \sin x \cos x)}$$
$$= \lim_{x\to\infty} \frac{2\cos x}{e^{\sin x}(x + 2\cos x + \sin x \cos x)} = 0.$$

However, $\lim_{x\to\infty} f(x)/g(x)$ does not exist. This is due to the fact that g' has zeros in **every neighborhood** of ∞, and consequently, we are not entitled to apply l'Hôpital's rule.

(ii) Let f be differentiable on an interval containing $x = a$. Then

$$f'(a) = \lim_{x\to a} \frac{f(x) - f(a)}{x - a},$$

and (since f is continuous at $x = a$) both the numerator and the denominator of the difference quotient approach zero. L'Hôpital's rule then gives

$$f'(a) = \lim_{x\to a} f'(x), \tag{5.3}$$

a result that apparently shows that f' is continuous at $x = a$. However, taking the function

$$f(x) = \begin{cases} x^2 \sin \frac{1}{x} & \text{if } x \neq 0, \\ 0 & \text{if } x = 0, \end{cases}$$

we observe that not every differentiable function has a continuous derivative. The student should understand that because of the hypothesis of l'Hôpital's rule, relation (5.3) is true in the sense that "if $\lim_{x\to a} f'(x)$ exists, then $\lim_{x\to a} f'(x) = f'(a)$."

We conclude this section with the following useful result.

Differentiation Inverse Functions Theorem. *Suppose f is a bijective differentiable function on the interval $[a,b]$ such that $f'(x) \neq 0$ for all $x \in [a,b]$. Then f^{-1} exists and is differentiable on the range of f and, moreover, $(f^{-1})'[f(x)] = 1/f'(x)$ for all $x \in [a,b]$.*

5.2 Introductory Problems

> Nature not only suggests to us
> problems, she suggests their solution.
>
> Henri Poincaré (1854–1912)

The first problem in this section makes a point related to the notion of derivative of a real continuous function. More precisely, if f is continuous and if in the definition of the derivative of f at a point x_0 one looks only at **rational** displacements away from x_0, then the function is differentiable at $x = x_0$. Observe that a similar property does not hold for similar displacements of real functions in order to establish other properties, such as the continuity; consider, for example, the characteristic function $\mathbf{1}_\mathbb{Q} : \mathbb{R} \to \mathbb{R}$ defined by $\mathbf{1}_\mathbb{Q} = 1$ on $\mathbb{Q}$ and $\mathbf{1}_\mathbb{Q} = 0$ elsewhere. In this case, for any real x_0 we have $\lim_{h \to 0, h \in \mathbb{Q}} [\mathbf{1}_\mathbb{Q}(x_0 + h) - \mathbf{1}_\mathbb{Q}(x_0)] = 0$, but $\mathbf{1}_\mathbb{Q}$ is not continuous at x_0. We also point out the following two facts related to the result below: (i) the set $\mathbb{Q}$ of rational numbers can be replaced by any dense subset of $\mathbb{R}$; (ii) for the result stated in the next exercise it is not sufficient that f be continuous at x_0. Indeed, it suffices to consider the Dirichlet function (whose continuity points coincide with $(\mathbb{R} \setminus \mathbb{Q}) \cup \{0\}$), which is defined by

$$f(x) := \begin{cases} \frac{1}{q}, & \text{if } x = p/q, p \in \mathbb{Z}, q \in \mathbb{N}^*, p \text{ and } q \text{ are relatively prime,} \\ 0, & \text{if } x \in \mathbb{R} \setminus \mathbb{Q}. \end{cases}$$

5.2.1. *Let $f : \mathbb{R} \to \mathbb{R}$ be a continuous function such that for some $x_0 \in \mathbb{R}$, the limit $\lim_{h \to 0, h \in \mathbb{Q}} [f(x_0 + h) - f(x_0)]/h$ exists and is finite. Prove that f is differentiable in x_0.*

Solution. Denote $L := \lim_{h \to 0, h \in \mathbb{Q}} [f(x_0 + h) - f(x_0)]/h$. Fix $0 < \varepsilon < 1$. Then there is a $v > 0$ such that for all $h \in \mathbb{Q} \cap (-v, v)$ we have

$$\left| \frac{f(x_0 + h) - f(x_0)}{h} - L \right| < \frac{\varepsilon}{3}.$$

Suppose $\delta \neq 0$ and $|\delta| < v$. Then by density of rationals and continuity of f, there is an $h_\delta \in \mathbb{Q} \cap (-v, v)$ such that

$$|h_\delta - \delta| < \frac{\varepsilon \delta}{3(1 + |L|)} < \delta \quad \text{and} \quad |f(x_0 + \delta) - f(x_0 + h_\delta)| < \frac{\varepsilon \delta}{3}.$$

Then in the displayed inequality, the first term is at most $\varepsilon/3$ and the second term is at most

$$\left| \frac{h_\delta}{\delta} \right| \cdot \left| \frac{f(x_0 + h_\delta) - f(x_0)}{h_\delta} - L \right| + \left| \frac{h_\delta - \delta}{\delta} \right| \cdot |L| < \frac{\varepsilon}{3} + \frac{\varepsilon |L|}{3(1 + |L|)} < \frac{2\varepsilon}{3}.$$

5.2 Introductory Problems

Therefore for any $|\delta| < \nu$ we have

$$\left|\frac{f(x_0+\delta)-f(x_0)}{\delta} - L\right| < \varepsilon,$$

and since $0 < \varepsilon < 1$ was arbitrary, the conclusion follows. □

The definition of the derivative of a convenient function implies below an elementary inequality.

5.2.2. Let $f(x) = a_1 \sin x + a_2 \sin 2x + \cdots + a_n \sin nx$ and assume that $|f(x)| \leq |\sin x|$, for all $x \in \mathbb{R}$. Prove that $|a_1 + 2a_2 + \cdots + na_n| \leq 1$.

Solution. Since $f(0) = 0$ and $|f(x)| \leq |\sin x|$, we have

$$|f'(0)| = \lim_{x \to 0}\left|\frac{f(x)-f(0)}{x}\right| = \lim_{x \to 0}\left(\left|\frac{f(x)}{\sin x}\right| \cdot \left|\frac{\sin x}{x}\right|\right) = \lim_{x \to 0}\left|\frac{f(x)}{\sin x}\right| \leq 1.$$

This concludes the proof. □

The next exercise establishes a simple property of polynomials that are nonvanishing at some point.

5.2.3. Let $f(x)$ be a polynomial and a real number such that $f(a) \neq 0$. Show that there exists a polynomial with real coefficients $g(x)$ such that $p(a) = 1$, $p'(a) = 0$, and $p''(a) = 0$, where $p(x) = f(x)g(x)$.

Solution. Let $g(x) = a_0 + a_1(x-a) + \cdots + a_n(x-a)^n$. This polynomial satisfies our assumptions because

$$a_0 = \frac{1}{f(a)}, \quad a_1 = -\frac{f'(a)g(a)}{f(a)} = -\frac{f'(a)}{f(a)^2},$$

$$a_2 = -\frac{f''(a)g(a) + 2f'(a)g'(a)}{f(a)} = \frac{2f'(a)^2 - f''(a)f(a)}{f(a)^3}. \quad \square$$

Is the set of zeros of a differentiable function always finite? A sufficient condition is given in the following exercise.

5.2.4. Let $f : [0,1] \to \mathbb{R}$ be differentiable and such that there is no x with $f(x) = f'(x) = 0$. Show that the set $Z := \{x;\ f(x) = 0\}$ is finite.

Solution. Arguing by contradiction, there exists a sequence $(x_n)_{n \geq 1} \subset [0,1]$ such that $f(x_n) = 0$, for all $n \geq 1$. Since $[0,1]$ is a compact set, it follows that up to a subsequence, $x_n \to c \in [0,1]$ as $n \to \infty$. Then, by continuity, $f(c) = 0$. So, since $f(x_n) = 0$ for all n, we obtain

$$f'(c) = \lim_{x \to c}\frac{f(x)-f(c)}{x-c} = \lim_{n \to \infty}\frac{f(x_n)-f(c)}{x_n-c} = 0.$$

Hence $f(c) = f'(c) = 0$, a contradiction. □

Nonexistence results are quite frequent in mathematics. An elementary example is stated below.

5.2.5. *Does there exist a continuously differentiable function $f: \mathbb{R} \to (0, \infty)$ such that $f' = f \circ f$?*

International Mathematical Competition for University Students, 2002

Solution. Assume that there exists such a function. Then, by hypothesis, f is increasing. Thus, $f(f(x)) > f(0)$ for all $x \in \mathbb{R}$. So, $f(0)$ is a lower bound for $f'(x)$, and for all $x < 0$ we have $f(x) < f(0) + xf(0) = (1+x)f(0)$. Hence, if $x \leq -1$ then $f(x) \leq 0$, contradicting the property $f > 0$. Consequently, such a function cannot exist. □

We need only elementary knowledge of differentiability to solve the next system of functional equations.

5.2.6. *Find all differentiable functions f and g on $(0, \infty)$ such that $f'(x) = -g(x)/x$ and $g'(x) = -f(x)/x$, for all $x > 0$.*

B. Hogan, Math. Magazine, Problem 1005

Solution. We observe that

$$[x(f(x) + g(x))]' = xf'(x) + xg'(x) + f(x) + g(x) = 0.$$

So, there exists a real constant A such that $f(x) + g(x) = 2A/x$, for all $x > 0$. Similarly, we have

$$\left[\frac{f(x) - g(x)}{x}\right]' = \frac{xf'(x) - xg'(x) - f(x) + g(x)}{x^2} = 0,$$

which shows that $f(x) - g(x) = 2Bx$, for some real constant B. This yields

$$f(x) = \frac{A}{x} + Bx \quad \text{and} \quad g(x) = \frac{A}{x} - Bx,$$

for all $x > 0$, where $A, B \in \mathbb{R}$. □

How do we compute the second derivative of a function? A *symmetric* definition might sometimes be useful.

5.2.7. *Let $f: (a,b) \to \mathbb{R}$ be a function of class C^2. Prove that*

$$\lim_{h \to 0} \frac{f(x+h) - 2f(x) + f(x-h)}{h^2} = f''(x), \quad \text{for all } x \in (a,b).$$

Solution. Using Taylor's formula we have

$$f(x+h) - f(x) = f'(x)h + \frac{f''(t)}{2}h^2 \quad \text{for some } t \in (x, x+h), h > 0$$

and

$$f(x-h) - f(x) = -f'(x)h + \frac{f''(z)}{2}h^2 \quad \text{for some } z \in (x-h, x), h > 0.$$

Adding these equalities, dividing by h^2, and passing to the limit as $h \to 0$ ($h > 0$), our conclusion follows. Next, by the symmetry of the limit with respect to h we deduce that the same result holds if $h \to 0$, $h < 0$. □

An algebraic equation with two unknowns is solved with differentiability arguments.

5.2.8. Find all integers a and b such that $0 < a < b$ and $a^b = b^a$.

Solution. Let $f(x) = \ln x / x$. Then $a^b = b^a$ if and only if $f(a) = f(b)$. We have $f'(x) = (1 - \ln x)/x^2$, so f increases on $(0, e)$ and decreases on (e, ∞). So, in order to obtain $f(a) = f(b)$ we must have $0 < a < e$ (so, $a \in \{1, 2\}$) and $b > e$. For $a = 1$ no solutions exist, while for $a = 2$ we obtain $b = 4$. Since f is decreasing, this is the unique solution. □

Next, we give a characterization of the constants fulfilling a certain inequality. As a consequence, what number is bigger: π^e or e^π?

5.2.9. Show that a positive real number a satisfies

$$e^x > x^a, \quad \text{for all } x > 0,$$

if and only if $a < e$.

Solution. Let $f(x) = e^x/x^a$, $x > 0$. Since $f(x) \to \infty$ as $x \to 0$ and $x \to \infty$, it follows that f achieves its minimum. Let x_0 be the minimum point of f. So, by Fermat's theorem,

$$f'(x_0) = e^{x_0} x_0^{-a}(1 - a/x_0) = 0,$$

and the minimum occurs at $x_0 = a$ and is $f(a) = e^a/a^a$. This minimum is bigger than 1 if and only if $a < e$, and the conclusion follows.

An alternative argument may be given by using the ideas developed in the previous solution. Indeed, $e^x > x^a$ if and only if $1/a > g(x) := (\ln x)/x$. We have already seen that the maximum of g is $g(e) = 1/e$, so the inequality holds for all x if and only if $1/a > 1/e$, or $e > a$. □

Identities involving differentiable functions are important tools to deduce qualitative properties.

5.2.10. Let $f : [0, 1] \to \mathbb{R}$ be a C^2 function such that

$$f''(x) = e^x f(x), \quad \text{for all } x \in [0, 1].$$

(a) Prove that if $0 < x_0 < 1$, then f cannot have a positive local maximum value (a negative local minimum value) in x_0.
(b) We further assume that $f(0) = f(1) = 0$. Show that f vanishes identically.

Solution. (a) Assume that f has a local maximum point at x_0 and $f(x_0) > 0$. Then $f''(x_0) \leq 0$. But $f''(x_0) = e^{x_0} f(x_0) > 0$. This contradiction shows that f cannot have a positive local maximum point at x_0. A similar argument implies that f cannot have a negative local minimum point in the open interval $(0, 1)$.

(b) If there exists $x_0 \in (0,1)$ such that $f(x_0) \neq 0$, then by continuity arguments, f has a positive local maximum point or a negative local minimum point in $(0,1)$, which contradicts (a). □

The derivative (when it exists!) is not necessarily continuous. A sufficient condition is provided below.

5.2.11. Let f be a real-valued function. Show that if f' is defined at x_0 and if $\lim_{x \to x_0} f'(x)$ exists, then f' is continuous at x_0.

Solution. We have $f'(x_0) = \lim_{x \to x_0} [f(x) - f(x_0)]/(x - x_0)$. But by l'Hôpital's rule, the right-hand member equals $\lim_{x \to x_0} f'(x)$. This shows that f' is continuous at x_0. □

Taylor's formula is also an instrument for deducing quantitative information for derivatives.

5.2.12. Let f be a function that has a continuous third derivative on $[0,1]$. Assume that $f(0) = f'(0) = f''(0) = f'(1) = f''(1) = 0$ and $f(1) = 1$. Prove that there exists c in $[a,b]$ such that $f'''(c) \geq 24$.

M. Klamkin, Math. Magazine, Problem Q 913

Solution. Consider the Taylor series expansions about the points $x = 0$ and $x = 1$:

$$f(x) = f(0) + f'(0)x + \frac{f''(0)}{2}x^2 + \frac{f'''(\xi_x)}{6}x^3,$$

$$f(x) = f(1) + f'(1)(x-1) + \frac{f''(1)}{2}(x-1)^2 + \frac{f'''(\eta_x)}{6}(x-1)^3,$$

where $0 \leq \xi_x \leq x$ and $x \leq \eta_x \leq 1$. So, by hypothesis,

$$f(x) = \frac{f'''(\xi_x)}{6}x^3 \quad \text{and} \quad f(x) = 1 + \frac{f'''(\eta_x)}{6}(x-1)^3.$$

Setting $x = 1/2$, we find that there exist ξ and η such that $f'''(\xi) + f'''(\eta) = 48$. Thus at least one of $f'''(\xi)$ and $f'''(\eta)$ is greater than or equal to 24. □

Here is another problem about computing limits of functions, this time with differentiability arguments.

5.2.13. Let $f : (0,\infty) \to (0,\infty)$ be a differentiable function. Prove that

$$\lim_{\delta \to 0} \left(\frac{f(x + \delta x)}{f(x)} \right)^{1/\delta}$$

exists, is finite, and is not equal to 0, for any $x > 0$.

Solution. Fix $x > 0$ and $\delta > 0$. Since f has only positive values and "ln" is a continuous function,

we have

$$\ln \lim_{\delta \to 0} \left(\frac{f(x+\delta x)}{f(x)} \right)^{1/\delta} = \lim_{\delta \to 0} \ln \left(\frac{f(x+\delta x)}{f(x)} \right)^{1/\delta} = \lim_{\delta \to 0} \frac{\ln f(x+\delta x) - \ln f(x)}{\delta}$$
$$= \lim_{\delta \to 0} \frac{x(\ln f(x+\delta x) - \ln f(x))}{\delta x} = x(\ln f(x))' = \frac{xf'(x)}{f(x)}.$$

The conclusion now follows by passing to the exponential map on both sides. □

Uniqueness results are generally hard to prove. An important tool is related to the qualitative analysis of diffentiable functions.

5.2.14. (a) *Let $f: \mathbb{R} \to \mathbb{R}$ be a differentiable function. Assume that $f(x_0) = 0$ and $f'(x) > f(x)$ for all $x \in \mathbb{R}$. Prove that $f(x) > 0$ for all $x > x_0$.*

(b) *Let a be a positive number. Prove that the equation $ae^x = 1 + x + x^2/2$ has a unique real root.*

Solution. (a) Let $g(x) = e^{-x} f(x)$. Then $g'(x) = e^{-x}(f'(x) - f(x)) > 0$. Since g is increasing, we have $g(x) > g(x_0) = 0$, for all $x > x_0$.

(b) Let $f: \mathbb{R} \to \mathbb{R}$ be the function defined by $f(x) = ae^x - 1 - x - x^2/2$. We have

$$\lim_{x \to -\infty} f(x) = -\infty \quad \text{and} \quad \lim_{x \to \infty} f(x) = \infty.$$

It follows that f has at least one real root. On the other hand, for all $x \in \mathbb{R}$,

$$f'(x) = ae^x - 1 - x > ae^x - 1 - x - x^2/2 = f(x).$$

Applying (a), we deduce that f has no other real roots. □

Derivatives are often involved in extremum problems.

5.2.15. *Suppose f is a differentiable real function such that $f(x) + f'(x) \leq 1$ for all x, and $f(0) = 0$. What is the largest possible value of $f(1)$?*

Harvard–MIT Mathematics Tournament, 2002

Solution. Consider the auxiliary function $g(x) := e^x(f(x) - 1)$. It follows that $g'(x) = e^x(f(x) + f'(x) - 1) \leq 0$, for all $x \in \mathbb{R}$. In particular, this implies that $e(f(1) - 1) = g(1) \leq g(0) = -1$. Hence $f(1) \leq 1 - e^{-1}$. In order to find a function for which $f(1)$ achieves its largest possible value, we observe that g must be constant in $[0, 1]$, so $f(x) + f'(x) = 1$ for all $x \in [0, 1]$. Therefore, $f(x) = 1 - e^{-x}$. □

Sometimes there are not many functions satisfying even simple relations.

5.2.16. *Let $f: [0,1] \to \mathbb{R}$ be continuous in $[0,1]$ and differentiable in $(0,1)$ such that $f(0) = 0$ and $0 \leq f'(x) \leq 2f(x)$, for all $x \in (0,1)$. Prove that f vanishes identically.*

Solution. Let $g : [0,1] \to \mathbb{R}$ be the function defined by $g(x) = e^{-2x} f(x)$. Hence

$$g'(x) = e^{-2x}(f'(x) - 2f(x)) \leq 0,$$

so g is decreasing. Since $g(0) = 0$ and $g \geq 0$, we deduce that $g \equiv 0$ and the same conclusion holds for f. □

The sign of a function follows quite easily if one knows the behavior of the first derivative. What happens if the second derivative is involved, too?

5.2.17. Let $f : [0,1] \to \mathbb{R}$ be twice differentiable on $(0,1)$ such that $f(0) = f(1) = 0$ and $f'' + 2f' + f \geq 0$. Prove that $f \leq 0$ in $[0,1]$.

Solution. Define $g : [0,1] \to \mathbb{R}$ by $g(x) = e^x f(x)$. We have

$$g''(x) = e^x(f''(x) + 2f'(x) + f(x)) \geq 0,$$

so g is a convex function. It follows that the point $(x, g(x))$ lies under the straight line joining $(0, g(0))$ and $(1, g(1)) = (1, 0)$, for all $x \in (0, 1)$. So $g(x) \leq 0$, which implies $f \leq 0$. □

Next is another application to limits of functions.

5.2.18. Define $f : (0, \infty) \to \mathbb{R}$ by $f(x) = x \ln(1 + x^{-1})$ (see Figure 5.12).

(a) Prove that f is increasing.
(b) Compute the limits of f at 0 and at infinity.

Solution. (a) We have

$$f'(x) = \ln\left(1 + \frac{1}{x}\right) - \frac{1}{x+1}$$

and

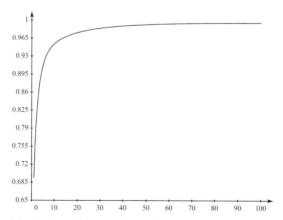

Fig. 5.12 Graph of the function $f(x) = x \ln(1 + x^{-1})$.

5.2 Introductory Problems

$$f''(x) = -\frac{1}{x(x+1)} + \frac{1}{(x+1)^2} = -\frac{1}{x(x+1)^2} < 0.$$

Since f' is decreasing and $\lim_{x\to\infty} f'(x) = 0$, we deduce that $f'(x) > 0$ for all $x > 0$.

(b) By L'Hôpital's rule we have

$$\lim_{x\to 0} f(x) = \lim_{x\to 0} \frac{\ln(x+1) - \ln x}{1/x} = \lim_{x\to 0} \frac{1/(x+1) - 1/x}{-1/x^2} = 0.$$

On the other hand,

$$\lim_{x\to\infty} \left(1 + \frac{1}{x}\right)^x = e,$$

so $\lim_{x\to\infty} f(x) = 1$. □

We provide below a functional equation that implies boundedness. The sign of g is essential! Why?

5.2.19. Let $f : \mathbb{R} \to \mathbb{R}$ be a twice differentiable function satisfying

$$f(x) + f''(x) = -xg(x)f'(x), \quad \text{for all } x \in \mathbb{R},$$

where $g(x) \geq 0$, for all $x \in \mathbb{R}$. Prove that f is bounded.

Solution. It is enough to show that f is bounded on $[0, \infty)$, because $f(-x)$ satisfies the same functional equation as $f(x)$. We have

$$\frac{d}{dx}\left((f(x))^2 + (f'(x))^2\right) = 2f'(x)(f(x) + f''(x)) = -2xg(x)(f'(x))^2 \leq 0.$$

We have

$$f^2(x) \leq f^2(0) + (f'(0))^2, \quad \text{for all } x \geq 0.$$

It follows that f is bounded on $[0, \infty)$. □

Is any critical point an extremum point? The sign of the second derivative is a useful criterion, but what happens if it also vanishes at that point?

5.2.20. Suppose that f''' is continuous on (a,b) and $x_0 \in (a,b)$. If $f'(x_0) = f''(x_0) = 0$ and $f'''(x_0) \neq 0$, show that f does not have a local extremum at x_0.

Solution. Without loss of generality, let us assume that $f'''(x_0) > 0$. Since f''' is continuous and $x_0 \in (a,b)$, there exists $\delta > 0$ such that $(x_0 - \delta, x_0 + \delta) \subset (a,b)$ and $f'''(x) > 0$ for all $x \in (x_0 - \delta, x_0 + \delta)$. This means that f'' is strictly increasing in $(x_0 - \delta, x_0 + \delta)$. Thus, since $f''(x_0) = 0$, we obtain that f' strictly decreases on $(x_0 - \delta, x_0)$ and f' is strictly increasing on $(x_0, x_0 + \delta)$. So, by our hypothesis $f'(x_0) = 0$, we obtain that f is strictly increasing in a neighborhood of x_0, which means that f does not have a local extremum at x_0. Similarly, if $f'''(x_0) < 0$, we obtain that f is strictly decreasing in some neighborhood of x_0.

A direct alternative argument is based on Taylor's formula without reminder:

$$f(x) = f(x_0) + \frac{1}{6} f'''(x_0)(x - x_0)^3 + o((x - x_0))^3. \quad □$$

We give below an existence result that is valid for all three times differentiable functions.

5.2.21. *Let $f : \mathbb{R} \to \mathbb{R}$ be a function of class C^3. Prove that there exists $a \in \mathbb{R}$ such that*
$$f(a) \cdot f'(a) \cdot f''(a) \cdot f'''(a) \geq 0.$$

Solution. If at least one of the numbers $f(a)$, $f'(a)$, $f''(a)$, and $f'''(a)$ is zero, for some point a, the proof is concluded. Let us now assume that any of the numbers $f(x)$, $f'(x)$, $f''(x)$, and $f'''(x)$ is either positive or negative, for all real x. Replacing if necessary $f(x)$ by $-f(x)$, we can assume that $f''(x) > 0$. At the same time, replacing $f(x)$ by $f(-x)$ if necessary, we can assume that $f'''(x) > 0$. [Indeed, these substitutions do not change the sign of the product $f(x)f'(x)f''(x)f'''(x)$.] From $f''(x) > 0$, it follows that f' is increasing, while $f'''(x) > 0$ implies that f' convex. Hence
$$f'(x+a) > f'(x) + af''(x), \quad \text{for all } a, x \in \mathbb{R}.$$

Taking here a sufficiently large (for x fixed), we deduce that $f'(x+a) > 0$. Since $f'(x)$ cannot change signs, it follows that $f'(x) > 0$, for all $x \in \mathbb{R}$. With similar arguments we deduce that $f'(x) > 0$ and $f''(x) > 0$ imply $f(x) > 0$, for all $x \in \mathbb{R}$. We conclude that $f(x)f'(x)f''(x)f'''(x) > 0$, for any $x \in \mathbb{R}$. □

First-order recurrent sequences are often treated by means of differentiable functions. Fixed-point arguments may be used to deduce the existence of the limit.

5.2.22. *Let $f : [1, \infty) \to \mathbb{R}$ be the function defined by*
$$f(x) = \frac{\sqrt{[x]} + \sqrt{\{x\}}}{\sqrt{x}},$$

where $[x]$ (resp., $\{x\}$) signifies the integer part (resp., the fractional part) of the real number x.

(a) *Find the smallest number z such that $f(x) \leq z$, for all $x \geq 1$.*
(b) *Fix $x_0 \geq 1$ and define the sequence (x_n) by $x_n = f(x_{n-1})$, for all $n \geq 1$. Prove that $\lim_{n \to \infty} x_n$ exists.*

Solution. (a) Let $x = y + z$, where $y = [x]$ and $z = \{x\}$. Then, by the arithmetic–geometric means value inequality, $f^2(x) \leq 1 + 2\sqrt{yz}/(y+z) \leq 2$. So $0 \leq f(x) \leq \sqrt{2}$, for all $x \geq 1$. Taking $y = 1$ we obtain
$$\lim_{x \nearrow 2} f^2(x) = \lim_{z \nearrow 1}\left(1 + \frac{2\sqrt{z}}{1+z}\right) = 2,$$

that is, $\sup\{f(x); x \geq 1\} = \sqrt{2}$.

(b) We first observe that $1 \leq x_n < 2$, for all $n \geq 1$. So, without loss of generality, we can assume that $1 \leq x_0 < 2$. If $x_n = 1$, then the sequence is constant and its limit equals 1. If not, then $f(x) = (1 + \sqrt{x-1})/\sqrt{x}$ in $(1,2)$. At this stage, the key point of the proof is to show that there exists a unique $v \in (1,2)$ such that $f(v) = v$, $f(x) > x$ if $1 < x < v$, and $f(x) < x$ for all $v < x < 2$. Let us assume for the moment that the result is true. An elementary computation by means of the derivative of f shows that f increases on $(1,2)$. So, for any $1 < x < v$ we have $x < f(x) < f(v) = v$. Consequently, for all $1 < x_0 < v$, the sequence $\{x_n\}$ of the nth iterates of f is bounded and increasing, so it converges. The same conclusion is valid if $v < x < 2$.

To conclude the proof, it remains to show that f has a unique fixed point v. Let $x = 1 + u$ with $u > 0$. Then $f(x) = x$ if and only if $1 + 2\sqrt{u} + u = 1 + 3u + 3u^2 + u^3$, so $u^5 + 6u^4 + 13u^3 + 12u^2 + 4u - 4 = 0$. Since the left-hand side is an increasing function with opposite sign values at 0 and 1, we deduce that the equation has a unique root u. Then it is enough to choose $v = 1 + u$. □

The next example is related to the sign of the Taylor polynomial of second order.

5.2.23. *Let $f : \mathbb{R} \to (0, \infty)$ be a twice differentiable function. Show that there exists $x_0 \in \mathbb{R}$ such that the Taylor polynomial of second order centered at x_0 is positive on the whole real axis.*
Note. *We recall that the Taylor polynomial of second order centered at x_0 is defined by $P_f(x_0) = f(x_0) + f'(x_0)(x - x_0) + [f''(x_0)]^2 (x - x_0)^2/2$.*

Solution. After completing the square of a binomial, we observe that x_0 has the required property if and only if either $f'(x_0) = f''(x_0) = 0$ or $f(x_0)f''(x_0) > [f'(x_0)]^2/2$. Let $g(x) = \sqrt{f(x)}$, so $g'(x) = f'(x)/(2g(x))$ and

$$g''(x) = \left(f(x)f''(x) - (f'(x))^2/2\right)/(2g^3(x)).$$

Assuming that there exists x_0 such that $g''(x_0) > 0$, then this is the good x_0. If not, then $g''(x) \leq 0$ for all x and the graph of g lies entirely under any tangent. Since g is positive, these tangent lines must be horizontal, so g and f are constant, and any point x_0 has the required property. □

A theorem of the German mathematician Hans Rademacher (1892–1969) asserts that Lipschitz functions are "almost" differentiable, in the following sense: If I is an open subset of $\mathbb{R}$ and $f : I \to \mathbb{R}$ is Lipschitz, then f is differentiable almost everywhere in I. If f is only continuous, a sufficient condition that the same property hold is provided in the next problem.

5.2.24. *Let $I \subset \mathbb{R}$ be an open interval and consider a continuous function $f : I \to \mathbb{R}$ satisfying, for all $x \in I$, $\lim_{y \to 0} y^{-1} [f(x+y) + f(x-y) - 2f(x)] = 0$. Prove that the set of points at which f is differentiable is a dense subset of I.*

Solution. Fix $a, b \in I$ with $a < b$. It is enough to show that there exists $x \in (a,b)$ such that f is differentiable at x. We split the proof into the following two steps.
STEP 1. If f has a local extremum in $x_0 \in (a,b)$ then f is differentiable in x_0 and $f'(x_0) = 0$. Indeed, since x_0 is a local extremum point of f, then for all $t > 0$

small enough, the quantities $t^{-1}[f(x_0+t)-f(x_0)]$ and $t^{-1}[f(x_0-t)-f(x_0)]$ have the same sign. Next, we use the identity

$$t^{-1}[f(x_0+t)+f(x_0-t)-2f(x_0)] = t^{-1}[f(x_0+t)-f(x_0)] + t^{-1}[f(x_0-t)-f(x_0)].$$

So, by hypothesis,

$$\lim_{t \searrow 0} \frac{f(x_0+t)-f(x_0)}{t} = \lim_{t \searrow 0} \frac{f(x_0-t)-f(x_0)}{t} = 0.$$

Thus, $f'(x_0) = 0$.

STEP 2. There exists a local extremum point $x_0 \in (a,b)$ of f. Indeed, consider the linear function $g(x) = cx + d$ such that $g(a) = f(a)$ and $g(b) = f(b)$. Then the function $h := f - g$ satisfies the same assumption as f, that is, for all $x \in (a,b)$,

$$\lim_{y \to 0} \frac{h(x+y) + h(x-y) - 2h(x)}{y} = 0.$$

Since h is continuous and $h(a) = h(b) = 0$, there exists a local extremum point $x_0 \in (a,b)$ of h. According to the first step, h is differentiable in x_0, and hence f has the same property. □

Functions with bounded first-order derivative are Lipschitz. Is the converse true?

5.2.25. Let $f : (0,1) \to \mathbb{R}$ be a differentiable function for which there exists a positive constant M such that for all $x, y \in (0,1)$, $|f(x) - f(y)| \le M|x-y|$. Prove that for any $x \in (0,1)$, $|f'(x)| \le M$.

Solution. Fix $x \in (0,1)$ and consider $t \in \mathbb{R} \setminus \{0\}$ such that $x + t \in (0,1)$. Thus, by hypothesis,

$$|f'(x)| \le \left| \frac{f(x+t)-f(x)}{t} \right| + \left| \frac{f(x+t)-f(x)}{t} - f'(x) \right|$$

$$\le M + \left| \frac{f(x+t)-f(x)}{t} - f'(x) \right|.$$

Passing now at the limit as $t \to 0$, we conclude the proof. □

We have seen that convex functions are **continuous** at **interior** points. What about the set of **differentiability** points of a convex function? The following problem shows that there are **many** such points. The complementary deep theorem due to Denjoy shows that the result is true for larger classes of functions.

5.2.26. Let $I \subset \mathbb{R}$ be an interval and assume that $f : I \to \mathbb{R}$ is a convex function. Prove that

(a) f has one-sided derivatives at any interior point x of I, and moreover, $f'(x-) \le f'(x+)$;
(b) the set of points in I where f is not differentiable is at most countable.

5.2 Introductory Problems

Solution. (a) Fix $x_0 \in \text{Int} I$. For convexity reasons, the function $g : I \setminus \{x_0\} \to \mathbb{R}$ defined by $g(x) = (f(x) - f(x_0))/(x - x_0)$ is nonincreasing and continuous. Hence there exists $f'(x_0-)$ and $f'(x_0+)$. Moreover, since for all $x, y \in I$ with $x < x_0 < y$ we have $g(x) \leq g(y)$, it follows that the lateral limits of f in x_0 are finite and $f'(x_0-) \leq f'(x_0+)$.

(b) Set $A := \{x \in \text{Int} I;\ f'(x-) < f'(x+)\}$. For any $x \in A$, define $I_x = (f'(x-), f'(x+))$. In order to prove that the set A is at most countable, it is enough to show that $I_x \cap I_y = \emptyset$, provided that $x, y \in A$ and $x \neq y$. Indeed, this implies that A is a disjoint union of open intervals, so A is at most countable.

Fix $x, y \in A$ with $x < y$. Then

$$f'(x+) \leq \frac{f(y) - f(x)}{y - x} \leq f'(y-).$$

By the definition of I_x we deduce that $I_x \cap I_y = \emptyset$, and the proof is concluded. $\square$

Remark. The French mathematician Arnaud Denjoy (1884–1974) proved the following result related to the set of points at which a function is differentiable.

Denjoy's Theorem. *Let $f : I \to \mathbb{R}$ be a function that admits one-sided derivatives at any point of $I \setminus A$, where A is at most countable. Then f admits a derivative at any point of I, excepting a set that is at most countable.*

Part (a) of the above exercise asserts that a convex function $f : I \to \mathbb{R}$ has one-sided derivatives at every interior point of I. So, according to Denjoy's theorem, we obtain the result established in (b).

The sign of the derivative ensures monotony properties of the function. Does the result remain true for the *symmetric derivative*? We recall that a function $f : I \to \mathbb{R}$ (where $I \subset \mathbb{R}$ an interval) has a symmetric derivative at $x_0 \in \text{Int} I$ if

$$f'_s(x_0) := \lim_{t \to 0} \frac{f(x_0 + t) - f(x_0 - t)}{2t}.$$

exists.

5.2.27. *Let $I \subset \mathbb{R}$ an interval. Assume that $f : I \to \mathbb{R}$ is continuous and has a symmetric derivative at every interior point of I.*

(a) *If the symmetric derivative of f is positive in $\text{Int} I$, prove that f is nondecreasing in $\text{Int} I$.*

(b) *If the symmetric derivative of f vanishes in $\text{Int} I$, prove that f is constant in $\text{Int} I$.*

A. Hincin, 1927

Solution. (a) Arguing by contradiction, let $x_1, x_2 \in \text{Int} I$ be such that $x_1 < x_2$ and $f(x_1) > f(x_2)$. Set $A_{x_1,x_2} := [f(x_1) + f(x_2)]/2$ and observe that $f(x_1) > A_{x_1,x_2}$. This enables us to define

$$c := \inf\{x \in [x_1, x_2];\ f(x) < A_{x,x_2}\}.$$

Then $f(c) = A_{c,x_2}$, and by the definition of c, there exists a sequence of positive numbers $(t_n)_{n \geq 1}$ such that for all positive integers n,

$$f(c - t_n) > \frac{f(c - t_n) + f(x_2)}{2}$$

and

$$f(c + t_n) < \frac{f(c + t_n) + f(x_2)}{2}.$$

Thus, for all $n \geq 1$,

$$\frac{f(c + t_n) - f(c - t_n)}{2t_n} < 0,$$

which implies that the symmetric derivative of f in c is nonpositive, a contradiction.

(b) We will argue by contradiction. Replacing $f(x)$ with $\pm f(\pm x)$ if necessary, we may assume there are $x_1, x_2 \in \text{Int}I$ such that $x_1 < x_2$ and $f(x_1) < f(x_2)$. Fix $0 < \varepsilon < [f(x_2) - f(x_1)]/(x_2 - x_1)$ and define $g(x) := f(x) - \varepsilon(x - x_0)$. Then, by hypothesis, the symmetric derivative of g is negative in $\text{Int}I$. So, by (a), g is nondecreasing in $\text{Int}I$. On the other hand, by the choice of ε, $g(x_1) > g(x_2)$. This contradiction shows that $f(x_1) = f(x_2)$. □

If $I \subset \mathbb{R}$ is an (unbounded!) interval, then a function $f : I \to \mathbb{R}$ is said to be *subadditive* if $f(x+y) \leq f(x) + f(y)$ for all $x, y \in I$. Are there any really useful subadditive functions? The answer is definitively **yes**, because the *distance function* given by the absolute value is a major representative of this class. What happens if differentiability properties are fulfilled?

5.2.28. Let $f : (0, \infty) \to \mathbb{R}$ be a differentiable function.

(a) *Prove that $f(x)/x$ is decreasing on $(0, \infty)$ if and only if $f'(x) < f(x)/x$ on $(0, \infty)$.*
(b) *Deduce that if $f'(x) < f(x)/x$ on $(0, \infty)$, then f is subadditive.*
(c) *Prove that if f is subadditive on $(0, \infty)$ and if $f(x) \leq -f(-x)$ for all $x > 0$, then f' is nonincreasing on $(0, \infty)$.*

Solution. (a) The proof is immediate on observing that

$$\left(\frac{f(x)}{x}\right)' = \frac{xf'(x) - f(x)}{x^2}.$$

(b) Follows from (a) and the observation that if $f(x)/x$ is nonincreasing, then f is subadditive.

5.2 Introductory Problems

(c) Fix $x, y \in (0, \infty)$. Thus, for any $t > 0$,

$$\frac{f(x+y+t) - f(x+y)}{t} \leq \frac{f(x+y+x+t-x) - f(x+y)}{t}$$
$$\leq \frac{f(x+y) + f(x+t) + f(-x) - f(x+y)}{t}$$
$$\leq \frac{f(x+t) - f(x)}{t}.$$

Taking now $t \searrow 0$ we obtain $f'(x+y) \leq f'(x)$. □

How much larger is $(|a|+|b|)^p$ than $|a|^p + |b|^p$? The ratio of course, should depend on p.

5.2.29. Define the function $f : [0,1] \to \mathbb{R}$ by $f(x) = (1+x)^p/1+x^p$, $p \geq 1$. Find the maximum and minimum values of f. Apply this result to prove the following inequalities:

$$|a|^p + |b|^p \leq (|a|+|b|)^p \leq 2^{p-1}(|a|^p + |b|^p), \quad \text{for all } a, b \in \mathbb{R}.$$

Solution. A straightforward computation shows that for all $x \in [0,1]$, $f'(x) \geq 0$. So, f is increasing in $[0,1]$ and the extreme values of f are 1 and 2^{p-1}. Next, without loss of generality, we can assume that $|a| \leq |b|$ and $b \neq 0$. Applying the inequalities $1 \leq f(x) \leq 2^{p-1}$ and replacing x by $|a|/|b|$, we deduce the conclusion. □

We prove below a very useful inequality.

5.2.30. (Young's Inequality). Let $p > 1$, $q = p/p - 1$, and consider the function defined by $f(x) = |a|^p/p + \frac{x^q}{q} - |a|x$ for all $x \geq 0$, $a \in \mathbb{R}$. Find the minimum of f on $[0, \infty)$ and deduce that

$$|ab| \leq \frac{|a|^p}{p} + |b|^q/q, \quad \text{for all } a, b \in \mathbb{R}.$$

Solution. The minimum of f is achieved in $x = |a|^{p-1}$ and equals 0. In particular, this implies $f(|b|) \geq 0$, which implies the conclusion. □

5.2.31. Let f be a $C^3(\mathbb{R})$ nonnegative function, $f(0) = f'(0) = 0$, $0 < f''(0)$. Let

$$g(x) = \left(\frac{\sqrt{f(x)}}{f'(x)}\right)'$$

for $x \neq 0$ and $g(0) = 0$. Show that g is bounded in some neighborhood of 0. Does the theorem hold for $f \in C^2(\mathbb{R})$?

International Mathematical Competition for University Students, 1997

Solution. The hypothesis implies that

$$f(x) = cx^2 + O(x^3), \ f'(x) = 2cx + O(x^2), \ f''(x) = 2c + O(x),$$

where $c = f''(0)/2$. Therefore

$$(f'(x))^2 = 4c^2x^2 + O(x^3),$$
$$2f(x)f''(x) = 4c^2x^2 + O(x^3),$$

and

$$2(f'(x))^2\sqrt{f(x)} = 2(4c^2x^2 + O(x^3))|x|\sqrt{c + O(x)}.$$

A direct computation shows that

$$g = \frac{(f')^2 - 2ff''}{2(f')^2\sqrt{f}}.$$

Thus, the function g is bounded in some neighborhood of 0 because

$$\frac{2(f'(x))^2\sqrt{f(x)}}{|x|^3} \to 8c^{5/2} \neq 0 \quad \text{as } x \to 0$$

and $f'(x)^2 - 2f(x)f''(x) = O(x^3)$ as $x \to 0$.

The theorem does not hold for some C^2 functions. Indeed, consider the C^2 function f defined by $f(x) = (x + |x|^{3/2})^2 = x^2 + 2x^2\sqrt{|x|} + |x|^3$. For $x > 0$ we have

$$g(x) = \frac{1}{2}\left(\frac{1}{1 + \frac{3}{2}\sqrt{x}}\right)' = -\frac{1}{2} \cdot \frac{1}{(1 + \frac{3}{2}\sqrt{x})^2} \cdot \frac{3}{4} \cdot \frac{1}{\sqrt{x}} \to -\infty \quad \text{as } x \to 0. \ \square$$

How can we find an *optimal* constant? Differentiability arguments are very powerful in such cases.

5.2.32. Two differentiable real functions f and g satisfy

$$\frac{f'(x)}{g'(x)} = e^{f(x) - g(x)}$$

for all x, and $f(0) = g(2003) = 1$. Find the largest constant c such that $f(2003) > c$ for all such functions f, g.

Harvard-MIT Mathematics Tournament, 2003

Solution. The main hypothesis may be rewritten, equivalently, $\left(e^{-f(x)}\right)' = \left(e^{-g(x)}\right)'$ for all $x \in \mathbb{R}$. Thus, $h(x) := e^{-f(x)} - e^{-g(x)} = \text{Const}$. Writing $h(0) = h(2003)$ we obtain, by hypothesis,

$$e^{-1} - e^{-g(0)} = e^{-f(2003)} - e^{-1}.$$

5.2 Introductory Problems

Thus, $e^{-f(2003)} = 2e^{-1} - e^{-g(0)} < 2e^{-1} = e^{-1+\ln 2}$. This implies that the largest constant c such that $f(2003) > c$ for all such functions f, g is $c = 1 - \ln 2$. □

5.2.33. Let p be a continuous function on the interval $[a,b]$, where $a < b$. Let $\lambda > 0$ be fixed. Show that the only function f satisfying

$$\begin{cases} f''(x) + p(x)f'(x) - \lambda f(x) = 0 & \text{for any } x \in (a,b), \\ f(a) = f(b) = 0, \end{cases}$$

is $f = 0$.

Solution. The equation implies that if f'' exists, then f'' must be continuous, so $f \in C^2[a,b]$. Assume that f is not identically zero on $[a,b]$. Then there is a point in $[a,b]$ where f is either strictly positive or strictly negative. Without loss of generality, we can assume that we have the first case. Then the maximal value of f on $[a,b]$ is positive, and it is attained at a point that is in the interior of $[a,b]$; let us call it x_0. Then $f'(x_0) = 0$ and $f''(x_0) \leq 0$. By hypothesis, $0 \geq f''(x_0) = \lambda f(x_0) > 0$. This contradiction proves our statement. □

5.2.34. Suppose that the differentiable functions $a, b, f, g : \mathbb{R} \to \mathbb{R}$ satisfy

$$f(x) \geq 0,\ f'(x) \geq 0,\ g(x) > 0,\ g'(x) > 0 \text{ for all } x \in \mathbb{R},$$
$$\lim_{x \to \infty} a(x) = A > 0,\ \lim_{x \to \infty} b(x) = B > 0,\ \lim_{x \to \infty} f(x) = \lim_{x \to \infty} g(x) = \infty,$$

and

$$\frac{f'(x)}{g'(x)} + a(x)\frac{f(x)}{g(x)} = b(x).$$

Prove that

$$\lim_{x \to \infty} \frac{f(x)}{g(x)} = \frac{B}{A+1}.$$

Solution. Let $0 < \varepsilon < A$ be an arbitrary real number. If x is sufficiently large, then $f(x) > 0$, $g(x) > 0$, $|a(x) - A| < \varepsilon$, $|b(x) - B| < \varepsilon$, and

$$B - \varepsilon < b(x) = \frac{f'(x)}{g'(x)} + a(x)\frac{f(x)}{g(x)} < \frac{f'(x)}{g'(x)} + (A+\varepsilon)\frac{f(x)}{g(x)}$$

$$< \frac{(A+\varepsilon)(A+1)}{A} \cdot \frac{f'(x)(g(x))^A + A \cdot f(x) \cdot (g(x))^{A-1} \cdot g'(x)}{(A+1) \cdot (g(x))^A \cdot g'(x)}$$

$$= \frac{(A+\varepsilon)(A+1)}{A} \cdot \frac{\left(f(x) \cdot (g(x))^A\right)'}{\left((g(x))^{A+1}\right)'}.$$

Hence

$$\frac{\left(f(x) \cdot (g(x))^A\right)'}{\left((g(x))^{A+1}\right)'} > \frac{A(B-\varepsilon)}{(A+\varepsilon)(A+1)}.$$

Similarly, for sufficiently large x,

$$\frac{\left(f(x)\cdot(g(x))^A\right)'}{\left((g(x))^{A+1}\right)'} < \frac{A(B+\varepsilon)}{(A-\varepsilon)(A+1)}.$$

As $\varepsilon \to 0$, we have

$$\lim_{x\to\infty} \frac{\left(f(x)\cdot(g(x))^A\right)'}{\left((g(x))^{A+1}\right)'} = \frac{B}{A+1}.$$

By l'Hôpital's rule we deduce that

$$\lim_{x\to\infty} \frac{f(x)}{g(x)} = \lim_{x\to\infty} \frac{f(x)\cdot(g(x))^A}{(g(x))^{A+1}} = \frac{B}{A+1}. \quad \square$$

5.2.35. Prove that there is no differentiable $f : (0,1) \to \mathbb{R}$ for which $\sup_{x\in E} |f'(x)|$ is finite, where E is a dense subset of the domain, and $|f|$ is a nowhere differentiable function.

P. Perfetti

Solution. Assume that a nonempty interval $(a,b) \subset (0,1)$ exists such that $f(x) \neq 0$ for all $x \in (a,b)$. Since f is differentiable on (a,b), thus continuous, f does not change sign on (a,b). Thus, either $|f| = f$ on all of (a,b) or $|f| = -f$ on (a,b). This shows that $|f|$ is differentiable on (a,b), which is a contradiction. Therefore, in every nonempty subinterval of $(0,1)$, f vanishes at least once. This shows that f is identically zero in $(0,1)$, hence $|f| \equiv 0$, a contradiction. There is therefore no differentiable function $f : (0,1) \to \mathbb{R}$ such that $|f|$ is nowhere differentiable. $\quad \square$

5.2.36. Let $f : (0,1) \to [0,\infty)$ be a function that is zero except at the distinct points a_n, $n \geq 1$. Let $b_n = f(a_n)$.

(i) Prove that if $\sum_{n=1}^{\infty} b_n < \infty$, then f is differentiable at least once in $(0,1)$.
(ii) Prove that for any sequence of nonnegative real numbers $(b_n)_{n\geq 1}$, with $\sum_{n=1}^{\infty} b_n = \infty$, there exists a sequence $(a_n)_{n\geq 1}$ such that the function f defined as above is nowhere differentiable.

Solution. (i) We first construct a sequence $(c_n)_{n\geq 1}$ of positive numbers such that $c_n \to \infty$ as $n\to\infty$ and $\sum_{n=1}^{\infty} c_n b_n < 1/2$. Let $B = \sum_{n=1}^{\infty} b_n$, and for each $k = 0, 1, \ldots$ denote by N_k the first positive integer for which

$$\sum_{n=N_k}^{\infty} b_n \leq \frac{B}{4^k}.$$

Now set $c_n = 2^k/5B$ for each n, $N_k \leq n < N_{k+1}$. Thus, $c_n \to \infty$ and

$$\sum_{n=1}^{\infty} c_n b_n = \sum_{k=0}^{\infty} \sum_{N_k \leq n < N_{k+1}} c_n b_n \leq \sum_{k=0}^{\infty} \frac{2^k}{5B} \sum_{n=N_k}^{\infty} b_n \leq \sum_{k=0}^{\infty} \frac{2^k}{5B} \cdot \frac{B}{4^k} = \frac{2}{5}.$$

Consider the intervals $I_n = (a_n - c_n b_n,\ a_n + c_n b_n)$. The sum of their lengths is $2\sum_{n\geq 1} c_n b_n < 1$; thus there exists a point $x_0 \in (0,1)$ that is not contained in any I_n. We show that f is differentiable at x_0 and $f'(x_0) = 0$. Since x_0 is outside of the intervals I_n, then $x_0 \neq a_n$ for any n and $f(x_0) = 0$. For arbitrary $x \in (0,1) \setminus \{x_0\}$, if $x = a_n$ for some n, then

$$\left| \frac{f(x) - f(x_0)}{x - x_0} \right| = \frac{f(a_n) - 0}{|a_n - x_0|} \leq \frac{b_n}{c_n b_n} = \frac{1}{c_n}.$$

Otherwise, $(f(x) - f(x_0))/(x - x_0) = 0$. Since $c_n \to \infty$, this implies that for arbitrary $\varepsilon > 0$ there are only finitely many $x \in (0,1) \setminus \{x_0\}$ for which

$$\left| \frac{f(x) - f(x_0)}{x - x_0} \right| < \varepsilon$$

does not hold. This shows that f is differentiable at x_0 and $f'(x_0) = 0$.

(ii) We remove the zero elements from the sequence $(b_n)_{n\geq 1}$. Since $f(x) = 0$ excepting a countable subset of $(0,1)$, we deduce that if f is differentiable at some point x_0, then $f(x_0)$ and $f'(x_0)$ must be 0.

Let $(\beta_n)_{n\geq 1}$ be a sequence satisfying $0 < \beta_n \leq b_n$, $b_n \to 0$ as $n \to \infty$, and $\sum_{n=1}^{\infty} \beta_n = \infty$.

Choose the numbers a_n ($n \geq 1$) such that the intervals $I_n = (a_n - \beta_n,\ a_n + \beta_n)$ cover each point of $(0,1)$ infinitely many times (this is possible, since the sum of lengths is $2\sum_{n=1}^{\infty} b_n = \infty$). Fix $x_0 \in (0,1)$ such that $f(x_0) = 0$. For any $\varepsilon > 0$, there exists $n \geq 1$ such that $\beta_n < \varepsilon$ and $x_0 \in I_n$. Therefore

$$\frac{|f(a_n) - f(x_0)|}{|a_n - x_0|} > \frac{b_n}{\beta_n} \geq 1;$$

hence f is not differentiable at x_0. $\square$

5.2.37. *Suppose f and g are nonconstant real-valued differentiable functions on $(-\infty, \infty)$. Furthermore, suppose $f(x+y) = f(x)f(y) - g(x)g(y)$ and $g(x+y) = f(x)g(y) - g(x)f(y)$, for all $x, y \in \mathbb{R}$.*
If $f'(0) = 0$, prove that $(f(x))^2 + (g(x))^2 = 1$ for all $x \in \mathbb{R}$.

Solution. Differentiating both sides of the first equation, we obtain

$$f'(x+y) = f'(x)f(y) - g'(x)g(y).$$

Letting $x = 0$, letting $g'(0) = k$, and writing f as a function of another variable t, we have $f'(y) = -g'(0)g(y)$, hence $f'(t) = -kg(t)$. Doing the same with the second equation, we deduce that $g'(t) = kf(t)$.

Let $h(x) := f(x)^2 + g(x)^2$. Differentiating with respect to x and substituting, we obtain

$$2f(x)f'(x) + 2g(x)g'(x) = h'(x),$$
$$2f(x) \cdot (-kg(x)) + 2g(x) \cdot (kf(x)) = h'(x),$$
$$0 = h'(x).$$

So, h is a constant function, say $h \equiv C$. Using the two given equations, we compute $h(x+y)$ and we obtain

$$\begin{aligned}h(x+y) &= f(x+y)^2 + g(x+y)^2 \\ &= (f(x)f(y) - g(x)g(y))^2 + (f(x)g(y) + g(x)f(y))^2 \\ &= \left(f(x)^2 + g(x)^2\right)\left(f(y)^2 + g(y)^2\right) \\ &= h(x) \cdot h(y).\end{aligned}$$

We therefore arrive at the equation $C = C^2$; hence $C = 1$ or $C = 0$. If $C = 0$, then $f(x)^2 + g(x)^2 = 0$. Thus, $f(x) = g(x) = 0$, which violates the condition that f and g are both nonconstant. If $C = 1$, however, no such problems occur, and we have shown that $h(x) = f(x)^2 + g(x)^2 = 1$. $\square$

A *bump function* is a function $f : \mathbb{R} \to \mathbb{R}$ that is both smooth (in the sense of having continuous derivatives of all orders) and compactly supported. The space of all bump functions on $\mathbb{R}$ is denoted by $C_0^\infty(\mathbb{R})$ or $C_c^\infty(\mathbb{R})$. The next exercise gives two examples of bump functions.

5.2.38. Let $a < b$ be real numbers. Prove that the functions

$$f(x) = \begin{cases} e^{-1/(x-a)(b-x)} & \text{if } x \in (a,b), \\ 0 & \text{if } x \in \mathbb{R} \setminus (a,b), \end{cases}$$

and

$$g(x) = \begin{cases} e^{-1/(x-a)} e^{-1/(b-x)} & \text{if } x \in (a,b), \\ 0 & \text{if } x \in \mathbb{R} \setminus (a,b), \end{cases}$$

are infinitely differentiable.

Solution. For $a < x < b$ we have

$$f'(x) = \frac{(b-x) - (x-a)}{(x-a)^2(x-b)^2} e^{-1/(x-a)(b-x)}.$$

An induction argument shows that there exists a sequence of polynomials $(P_n)_{n \geq 1}$ and a sequence of positive integers $(k_n)_{n \geq 1}$ such that for all $x \in (a,b)$,

$$f^{(n)}(x) = \frac{P_n(x)}{[(x-a)(b-x)]^{k_n}} e^{-1/(x-a)(b-x)}.$$

5.2 Introductory Problems

A straightforward computation shows that f is differentiable at $x = a$ and at $x = b$ and $f^{(n)}(a) = f^{(n)}(b) = 0$ for all $n \geq 1$. $\square$

Standard arguments apply to the function g.

In Exercise 4.5.1 we have established that all continuous functions $f : \mathbb{R} \to \mathbb{R}$ satisfying the Cauchy functional equation

$$f(x+y) = f(x) + f(y) \tag{5.4}$$

are the linear functions $f(x) = ax$, for any $a \in \mathbb{R}$. In Chapter 4 we have also established that the same property holds if f is assumed to be continuous at a single point. Taking $y = x$ in (5.4) we obtain

$$f(2x) = 2f(x). \tag{5.5}$$

The next property deals with the solutions of this functional equation, and it is due to Peter Lax [67], Abel Prize Laureate in 2005.

5.2.39. *Prove that every solution of (5.5) that is once differentiable at $x = 0$ is linear.*

Solution. Setting $x = 0$ in (5.5) shows that $f(0) = 0$. Applying (5.5) n times gives

$$f(x) = 2^n f\left(\frac{x}{2^n}\right). \tag{5.6}$$

Since f is differentiable at $x = 0$,

$$f(y) = my + \varepsilon y, \tag{5.7}$$

where $m = f'(0)$ and $\varepsilon = \varepsilon(y)$ tends to zero as y tends to zero.

Set $y = x/2^n$ into (5.7) and use (5.6):

$$f(x) = 2^n \left(\frac{x}{2^n} + \varepsilon \frac{x}{2^n}\right) = mx + \varepsilon x.$$

As $n \to \infty$, then $\varepsilon \to 0$, which yields $f(x) = mx$.

The condition that f be differentiable at $x = 0$ cannot be replaced by requiring mere Lipschitz continuity.

A continuous analogue of equation (5.5) is

$$\frac{1}{x} \int_0^x f(y) dy = f\left(\frac{x}{2}\right). \tag{5.8}$$

Clearly all functions of the form $f(x) = ax + b$ satisfy (5.8). P. Lax [67] proved that a solution f of (5.8) that is infinitely differentiable at $x = 0$ is of the form $f(x) = ax + b$. $\square$

5.3 The Main Theorems

> It is very simple to be happy, but it is very difficult to be simple.
>
> Rabindranath Tagore (1861–1941)

Continuous functions defined on $[a,b]$ with values in $[a,b]$ have at least a fixed point, by a celebrated result due to the Dutch mathematician L.E.J. Brouwer (1881–1966), which is obvious only on the real line! Is this fixed point unique? Elementary examples show the contrary. That is why we provide below a sufficient condition in this sense.

5.3.1. *Let $f:[a,b]\to[a,b]$ be a continuous function that is differentiable on (a,b) and satisfying $f'(x)\neq 1$, for all $x\in(a,b)$. Prove that f has a unique fixed point.*

Solution. Consider the function g defined by $g(x)=f(x)-x$. It remains to apply the intermediate value property. By the Brouwer fixed-point theorem, f has at least one fixed point. We argue by contradiction and assume that there exist $a\leq x_1 < x_2 \leq b$ such that $g(x_1)=g(x_2)=0$. So, by Rolle's mean value theorem, we can find $c\in(a,b)$ such that $g'(c)=0$, that is, $f'(c)=1$. This contradiction implies the uniqueness of the fixed point.

In the above solution we can avoid defining the function g by using the Lagrange mean value theorem instead of Rolle's theorem. □

The next result is related to the existence of a fixed point for a higher-order derivative.

5.3.2. *Let f be a real function with $n+1$ derivatives on $[a,b]$. Suppose $f^{(i)}(a)=f^{(i)}(b)=0$ for $i=0,1,\ldots,n$. Prove that there exists $\xi\in(a,b)$ such that $f^{(n+1)}(\xi)=f(\xi)$.*

G.Q. Zhang, Amer. Math. Monthly, Problem E 3214

Solution. We first assume that $n=0$ and consider the function $g(x)=e^{-x}f(x)$. By Rolle's theorem, there exists $\xi\in(a,b)$ such that $0=g'(\xi)=e^{-\xi}(f'(\xi)-f(\xi))$, that is, $f'(\xi)=f(\xi)$.

Next, we assume that $n\geq 1$ and define the function $h(x)=\sum_{k=0}^{n}f^{(k)}(x)$. Then $h(a)=h(b)=0$ and $h(x)-h'(x)=f(x)-f^{(n+1)}(x)$. Applying to the function h the result proven in the case $n=0$, we conclude the proof. □

Differentiability arguments may be used even if such a property is not assumed in the hypotheses.

5.3.3. *Let $f:[0,1]\to[0,1]$ be a nonconstant continuous function. Prove that there exist $x_1, x_2 \in [0,1]$, $x_1 \neq x_2$, such that $|f(x_1)-f(x_2)|=|x_1-x_2|^2$.*

R. Gologan

Solution. We first prove the following auxiliary result. □
Claim. *There exist $a,b \in (0,1]$ such that $|f(a)-f(b)|>|a-b|^2$.*

5.3 The Main Theorems

We argue by contradiction and assume that for all $x, y \in (0,1]$, $x \neq y$, we have

$$\left|\frac{f(y)-f(x)}{y-x}\right| \leq |y-x|.$$

Passing to the limit as $y \to x$ we deduce that f is differentiable in $(0,1]$ and, moreover, $f'=0$ on this interval. Thus, f is constant in $[0,1]$, and this contradiction proves our claim.

Next, we assume that $|f(0)-f(1)| < 1$ and consider the continuous function $g : [0,1] \to \mathbb{R}$ defined by

$$g(t) = |f((1-t)a+t) - f((1-t)b)| - |[(1-t)a+t] - (1-t)b|^2.$$

Using the claim, we deduce that $g(0) > 0$ and $g(1) < 0$. Thus, there exists $\lambda \in (0,1)$ such that $g(\lambda) = 0$. This implies our conclusion, provided that $x_1 = (1-\lambda)a + \lambda$ and $x_2 = (1-\lambda)b$.

The next result is an "almost" mean value theorem.

5.3.4. Let f be differentiable with f' continuous on $[a,b]$ such that there exists x_0 in $(a,b]$ with $f'(x_0) = 0$. Prove that there is some ξ in (a,b) for which

$$f'(\xi) = \frac{f(\xi)-f(a)}{b-a}.$$

Solution. Define the function $g : [a,b] \to \mathbb{R}$ by

$$g(x) = f'(x) - \frac{f(x)-f(a)}{b-a}.$$

We assume, without loss of generality, that $f(x_0) > f(a)$. Let $x_1 \in (a,x_0]$ be such that $f(x_1) = \max\{f(x); x \in [a,x_0]\}$. It follows that $f(x_1) - f(a) > 0 \geq (b-a)f'(x_1)$, which shows that $g(x_1) < 0$.

By the Lagrange mean value theorem, there exists $x_2 \in (a,x_1)$ such that $f'(x_2) = [f(x_1) - f(a)]/(x_1 - a)$. Therefore

$$g(x_2) = \frac{f(x_1)-f(a)}{x_1-a} - \frac{f(x_2)-f(a)}{b-a} \geq 0 > g(x_1).$$

So, since g is continuous, there exists $\xi \in [x_2, x_1)$ such that $g(\xi) = 0$. This concludes the proof. □

Here are more problems on proving inequalities using differentiable functions.

5.3.5. Prove that $a^b + b^a > 1$, for all $a,b > 0$.

D.S.Mitrinović

Solution. The inequality is obvious if at least one of the numbers a or b is greater than or equal to 1. So, it is enough to prove the inequality for $a,b \in (0,1)$. In this case we can write $a = 1-c$, $b = 1-d$, with $c,d \in (0,1)$. Consider the function

$f(x) = (1-cx)^d$, $x \in [0,1]$. Applying Lagrange's mean value theorem, there exists $x_0 \in (0,1)$ such that

$$(1-c)^d = 1 - \frac{cd}{(1-cx_0)^{1-d}} < 1 - cd.$$

Hence

$$a^b + b^a = \frac{1-c}{(1-c)^d} + \frac{1-d}{(1-d)^c} > \frac{1-c}{1-cd} + \frac{1-d}{1-cd} = \frac{(1-c)(1-d)}{1-cd} + 1 > 1. \quad \square$$

We always have to pay attention to the values of the exponents! An instructive example is provided in what follows (see Figure 5.13 for the graph of the associated function in a particular case corresponding to $p = 5/2$).

5.3.6. Define the function $f : (0,1) \to \mathbb{R}$ by $f(x) = (1+x^{1/p})^p + (1-x^{1/p})^p$.

(a) For any $p \geq 2$, prove the following inequalities for any $0 < x < 1$:
  (a_1) $f''(x) \leq 0$;
  (a_2) $f(x) \leq f(y) + (x-y)f'(y)$, for all $y \in (0,1)$;
  (a_3) $f(x) \leq 2^{p-1}(x+1)$;
(b) If $1 \leq p < 2$, show that the above inequalities are true with $\geq$ instead of $\leq$.

Solution. (a_1) We have

$$f''(x) = p^{-1}(1-p)x^{(1-2p)/p}\left[\left(1+x^{1/p}\right)^{p-2} - \left(1-x^{1/p}\right)^{p-2}\right] \leq 0,$$

for all $x \in (0,1)$.

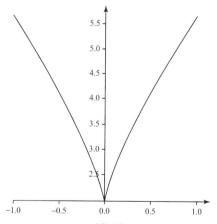

Fig. 5.13 Graph of the function $f(x) = (1+x^{2/5})^{5/2} + (1-x^{2/5})^{5/2}$.

5.3 The Main Theorems

(a_2) This property characterizes concave functions. Indeed, it is enough to use the fact that f' is nonincreasing, in conjunction with the Lagrange mean value theorem applied to f on the interval $[x,y] \subset (0,1)$.

(a_3) A straightforward computation shows that the function g defined by $g(x) = (x+1)^{-1}f(x)$ is increasing in $(0,1]$. This yields the conclusion. □

(b) Apply the same ideas as above. Exercise!

The mean value theorem and beyond! A first step to the **true** sense of this major result is the following exercise, which can, however, be viewed as a weaker version of the Lagrange mean value theorem. The nice part of our argument is that it avoids using the (somewhat subtle) fact that a continuous function attains its maximum on a closed interval.

5.3.7. *Let $f : [a,b] \to \mathbb{R}$ be a function that admits a derivative (not necessarily finite!) at any point of $[a,b]$. Prove that there exists $x_0 \in [a,b]$ such that*

$$\left| \frac{f(b) - f(a)}{b - a} \right| \leq |f'(x_0)|.$$

I. Halperin

Solution. For any integer $0 \leq n \leq 10$, set $c_n = a + \frac{(b-a)n}{10}$. We have

$$|f(b) - f(a)| \leq \sum_{n=0}^{9} |f(c_{n+1}) - f(c_n)|.$$

Thus, there exists an integer $0 \leq k \leq 9$ such that

$$\left| \frac{f(c_{k+1}) - f(c_k)}{c_{k+1} - c_k} \right| \geq \left| \frac{f(b) - f(a)}{b - a} \right|.$$

Let k_1 be the least integer with this property and define $a_1 = c_{k_1}$ and $b_1 = c_{k_1+1}$.

Next, we repeat the same arguments for the interval $I_1 := [a_1, b_1]$, and so on. Thus, we obtain a sequence of intervals $I_n := [a_n, b_n]$ such that $b_n - a_n = 10^{-n}(b-a)$ and

$$\left| \frac{f(b_n) - f(a_n)}{b_n - a_n} \right| \geq \left| \frac{f(b) - f(a)}{b - a} \right|. \tag{5.9}$$

Since I_n are closed intervals, $|I_n| = (b_n - a_n) \to 0$, and $I_1 \supset I_2 \supset \cdots$, it follows by the Cantor principle that there exists a unique element $x_0 \in \bigcap_{n \geq 1} I_n$. In particular, this shows that $x_0 = \lim_{n \to \infty} a_n = \lim_{n \to \infty} b_n$. We also observe that in relation (5.9) we can assume that either the a_n or b_n are replaced by x_0. This follows from (5.9) combined with the elementary identity

$$\frac{f(b_n) - f(a_n)}{b_n - a_n} = \frac{f(b_n) - f(x_0)}{b_n - x_0} \cdot \frac{b_n - x_0}{b_n - a_n} + \frac{f(x_0) - f(a_n)}{x_0 - a_n} \cdot \frac{x_0 - a_n}{b_n - a_n}.$$

In conclusion, we can assume that relation (5.9) holds for infinitely many indices n_k ($k \geq 1$) and, say, for a_n replaced by x_0. Taking $n_k \to \infty$ and using the definition of the derivative of f in x_0, we obtain the conclusion. □

Remark. Under the same assumptions, the above arguments apply to show that there exist $x_1, x_2 \in (a,b)$ satisfying

$$f'(x_1) \leq \frac{f(b) - f(a)}{b - a} \leq f'(x_2).$$

Actually, the Lagrange mean value theorem asserts that there exists $c \in (a,b)$ for which **both** of these inequalities hold simultaneously. Furthermore, a **general mean value theorem** holds for functions with values in **arbitrary** spaces endowed with a topology. However, the sense of this result is that it establishes a mean value inequality and not an equality, as in the one–dimensional case. The following property was discovered by the French mathematician Arnaud Denjoy in 1915, and it is usually known as the Denjoy–Bourbaki theorem, due especially to the elegance of the proof given in 1949 by the Bourbaki group of French mathematicians. We refer to [75] for a history of the celebrated Bourbaki group. We state in what follows this result.

Denjoy–Bourbaki Theorem. *Let E be a normed vector space and consider the continuous function $f : [a,b] \to E$. Let $\varphi : [a,b] \to \mathbb{R}$ be a continuous nondecreasing function. Assume that both f and φ admit a right derivative at every point of $[a,b) \setminus A$, where the set A is at most countable and, moreover, for all $x \in [a,b) \setminus A$, we have $\|f'(x+)\| \leq \varphi'(x+)$. Then $\|f(b) - f(a)\| \leq \varphi(b) - \varphi(a)$.*

An important consequence of the Denjoy–Bourbaki theorem is the following:

Corollary. *Let $f : [a,b] \to \mathbb{R}$ be a continuous function that admits a right derivative at every point of $[a,b) \setminus A$, where A is at most countable. Assume that for all $x \in [a,b) \setminus A$ we have $f'(x+) = 0$. Then f is constant on $[a,b]$.*

Are there converses of the mean value theorem?

5.3.8. (i) Let f be a twice differentiable function in $[a,b]$ such that $f''(x) > 0$ for all $x \in [a,b]$. Prove that for each $\xi \in (a,b)$, there exists $x_0 \in [a,b]$ such that either

$$f'(\xi) = \frac{f(b) - f(x_0)}{b - x_0} \quad or \quad f'(\xi) = \frac{f(a) - f(x_0)}{a - x_0}.$$

(ii) Let f be a twice differentiable function in $[a,b]$ such that $f'(x_0) = 0$, for some $x_0 \in (a,b)$. Prove that there exists $x_1 \in [a,b]$ such that either $f(a) = f(x_1)$ or $f(b) = f(x_1)$.

R.S. Luthar, Amer. Math. Monthly, Problem E 2057

Solution. (i) Define the functions

$$g(x) = \begin{cases} \frac{f(x) - f(a)}{x - a} & \text{if } x \in (a,b], \\ f'(a+) & \text{if } x = a, \end{cases} \qquad h(x) = \begin{cases} \frac{f(b) - f(x)}{b - x} & \text{if } x \in [a,b), \\ f'(b-) & \text{if } x = b. \end{cases}$$

Then g, h are continuous on $[a,b]$, and $g(b) = h(a)$. Since f' is increasing on $[a,b]$, we have $g(a) < f'(\xi) < h(b)$. If $g(a) < f'(\xi) < g(b)$, then by the continuity of g, there exists $x_0 \in (a,b)$ such that $f'(\xi) = g(x_0)$. Otherwise, we have $g(b) = h(a) \leq f'(\xi) < h(b)$. So, by the continuity of h, there exists $x_0 \in [a,b]$ such that $f'(\xi) = h(x_0)$. Finally, we remark that the case $x_0 = a$ cannot be excluded (see, e.g., $f(x) = x^2$, $[a,b] = [0,2]$, $\xi = 1$).

(ii) This follows from (i) if one replaces x_0 in (ii) by ξ, and replaces x_1 in (ii) by x_0. As above, the case $x_0 = a$ (or $x_0 = b$) cannot be excluded (see, e.g., $f(x) = x^2$, $[a,b] = [-1,1]$, $\xi = 0$). □

The following problem presents a simple differential inequality that does not follow with elementary arguments.

5.3.9. *Let $f : \mathbb{R} \to \mathbb{R}$ be a function of class C^3 such that f, f', f'', and f''' are all positive. Moreover, we assume that $f'''(x) \leq f(x)$ for all $x \in \mathbb{R}$. Prove that $f'(x) < 2f(x)$ for any $x \in \mathbb{R}$.*

Putnam Competition, 1999

Solution. We first observe that our hypotheses imply

$$\lim_{x \to -\infty} f'(x) = \lim_{x \to -\infty} f''(x) = 0.$$

Indeed, by Lagrange's mean value theorem, there exists $\xi_n \in (-n^2, -n)$ such that

$$f(-n^2) - f(-n) = (n - n^2) f'(\xi_n) < 0.$$

By our assumptions, the limits of f and f' at $-\infty$ exist, and moreover, they are nonnegative. Arguing by contradiction, let us suppose that $\lim_{x \to -\infty} f'(x) > 0$. Passing to the limit at $-\infty$ in the above relation we obtain $0 = -\infty$, contradiction. Hence $\lim_{x \to -\infty} f'(x) = 0$. A similar argument shows that $\lim_{x \to -\infty} f''(x) = 0$.

We shall apply the following elementary result several times in the proof. □

Lemma. *Let $f : \mathbb{R} \to \mathbb{R}$ be a differentiable function such that $\lim_{x \to -\infty} f(x) \geq 0$ and $f'(x) > 0$, for all $x \in \mathbb{R}$. Then $f(x) > 0$, for all $x \in \mathbb{R}$.*

Proof of the lemma. If there is some $x_0 \in \mathbb{R}$ such that $f(x_0) < 0$, then by our assumption that $f' > 0$, we deduce $f(x) < f(x_0)$, for all $x < x_0$. Therefore $\lim_{x \to -\infty} f(x) \leq f(x_0) < 0$, contradiction.

We continue the proof of our problem. By our assumption $f'''(x) \leq f(x)$ we obtain

$$f''(x) f'''(x) \leq f''(x) f(x) < f''(x) f(x) + f'^2(x), \quad \text{for all } x \in \mathbb{R}.$$

Using this inequality and applying the above lemma to the function $f(x) f'(x) - 1/2 (f''(x))^2$, we deduce that

$$\frac{1}{2}(f''(x))^2 < f(x) f'(x), \quad \text{for all } x \in \mathbb{R}. \tag{5.10}$$

On the other hand, by $f > 0$ and $f''' > 0$ it follows that

$$2f'(x)f''(x) < 2f'(x)f''(x) + 2f(x)f'''(x), \quad \text{for all } x \in \mathbb{R}.$$

Applying the lemma again, we obtain

$$f'^2(x) \leq 2f(x)f''(x), \quad \text{for all } x \in \mathbb{R}. \tag{5.11}$$

Combining relations (5.10) and (5.11), we obtain

$$\frac{1}{2}\left(\frac{f'^2(x)}{2f(x)}\right)^2 < \frac{1}{2}(f''(x))^2 < f(x)f'(x),$$

that is, $f'^3(x) < 8f^3(x)$, for all $x \in \mathbb{R}$. It follows that $f'(x) < 2f(x)$, for all $x \in \mathbb{R}$.

We remark that it is possible to prove a stronger inequality than that in our statement, more precisely, with a constant less than 2. Indeed, adding $1/2 f'(x)f''(x)$ to both sides of (5.10) and applying then the hypothesis $f'''(x) \leq f(x)$, we have

$$\frac{1}{2}[f'(x)f'''(x) + (f''(x))^2] < f(x)f'(x) + \frac{1}{2}f'(x)f'''(x) \leq \frac{3}{2}f(x)f'(x).$$

By the above lemma we deduce

$$\frac{1}{2}f'(x)f''(x) < \frac{3}{4}f^2(x), \quad \text{for all } x \in \mathbb{R}.$$

Multiplying here by $f'(x)$ and applying again the lemma, we deduce that

$$\frac{1}{6}f'^3(x) < \frac{1}{4}f^3(x), \quad \text{for all } x \in \mathbb{R}.$$

Hence $f'(x) < (3/2)^{1/3} f(x) < 2f(x)$.

We do not know at this stage the best constant satisfying the inequality in our statement. Using the fact that the function $f(x) = e^x$ satisfies our hypotheses, we can only assert that the best constant cannot be less than 1, so 1 is not a sharp bound. Let us define $f(x) = 1$ for $x \leq 0$ and $f(x) = 1/3 e^x + 2/3 e^{-x/2} \cos(\sqrt{3}x/2)$ for $x \geq 0$. Then f''' has a singularity at $x = 0$, but "smoothing" (this is a standard procedure in higher analysis, say, by means of "mollifiers") a little removes it without changing the basic result below. On $x \geq 0$, this f satisfies $f''' = f$, so it is easy to check that f and its first three derivatives are nonnegative and that $f \geq f'''$. However, one can compute that $f'(x)/f(x) = 1.01894\ldots$ at $x = 3.01674\ldots$. Since other related examples can be built, it is difficult to think that the constant $1.01894\ldots$ is sharp.

A qualitative property involving the symmetric derivative and the mean value theorem is stated below.

5.3.10. Let f be a twice differentiable function on $\mathbb{R}$ with f, f', f'' increasing. Fix numbers a and b with $a \leq b$. For each $x > 0$, define $\xi = \xi(x)$ so that

5.3 The Main Theorems

$$\frac{f(b+x) - f(a-x)}{b - a + 2x} = f'(\xi)$$

by the mean value theorem. Prove that $\xi(x)$ is an increasing function of x. Can the hypothesis "f'' increasing" be replaced with "f'' positive"?

M. McAsey and L.A. Rubel, Amer. Math. Monthly, Problem E 3033

Solution. The equation

$$\xi(x) = (f')^{-1}\left(\frac{f(b+x) - f(a-x)}{b - a + 2x}\right)$$

shows that ξ is differentiable. Differentiating the equation defining $f'(\xi)$ yields

$$\frac{f'(b+x) + f'(a-x) - 2f'(\xi)}{b - a + 2x} = f''(\xi) \cdot \xi'(x). \qquad (5.12)$$

Since f' is increasing, $f'' \geq 0$. Thus $\xi'(x) > 0$ would follow from (5.12), provided that

$$\frac{f'(b+x) + f'(a-x)}{2} > f'(\xi). \qquad (5.13)$$

But f'' increasing implies f' convex; hence (5.13) holds. However, this is not an obvious step, and we advise the reader to provide details for this statement.

The condition that f'' increases cannot be replaced with $f'' > 0$. Indeed, if $f(x) = \pi x/2 + x \arctan x - \ln(1+x^2)/2$, then f and f' are increasing, $f'' > 0$, but f'' decreases on $\mathbb{R}_+$. Setting $a = b = 0$, we find that the equation defining $\xi(x)$ becomes

$$\frac{\pi}{2} = \frac{\pi}{2} + \arctan \xi(x);$$

hence $\xi(x) \equiv 0$. □

Are there changes if in the Rolle mean value theorem the hypothesis $f(a) = f(b)$ refers to higher-order derivatives?

5.3.11. (a) Let $f : [a,b] \to \mathbb{R}$ be a differentiable function such that $f'(a) = f'(b)$. Prove that there exists $\xi \in (a,b)$ such that

$$f(\xi) - f(a) = (\xi - a) f'(\xi).$$

(b) Let $f : [a,b] \to \mathbb{R}$ be a twice differentiable function such that $f''(a) = f''(b)$. Prove that there exists $\xi \in (a,b)$ such that

$$f(\xi) - f(a) = (\xi - a) f'(\xi) - \frac{(\xi - a)^2}{2} f''(\xi).$$

T. Flett

Solution. (a) Consider the continuous function $g : [a,b] \to \mathbb{R}$ defined by

$$g(x) = \begin{cases} \frac{f(x)-f(a)}{x-a}, & \text{if } x \in (a,b], \\ f'(a), & \text{if } x = a. \end{cases}$$

If g achieves an extremum at an interior point, say $\xi \in (a,b)$, then by Fermat's theorem, $g'(\xi) = 0$ and we conclude the proof. □

Let us now assume the contrary, so the only extremum points of g are a and b. Without loss of generality we can suppose that for all $x \in [a,b]$ we have $g(a) \leq g(x) \leq g(b)$. The second inequality can be rewritten as

$$f(x) \leq f(a) + (x-a)g(b), \quad \text{for all } x \in [a,b].$$

This yields, for any $x \in [a,b)$,

$$\frac{f(b)-f(x)}{b-x} \geq \frac{f(b)-f(a)-(x-a)g(b)}{b-x} = \frac{f(b)-f(a)}{b-a}.$$

Taking $x \nearrow b$ we obtain $f'(b) \geq g(b)$. Using the hypothesis we obtain $f'(a) \geq g(b)$, so $g(a) \geq g(b)$. This implies that g is constant, that is, $g' = 0$ in (a,b). Thus, for all $\xi \in (a,b)$, $f(\xi) - f(a) = (\xi - a)f'(\xi)$.

(b) We have left the details to the reader, since the proof applies the same ideas as above. A key ingredient is a good choice of the auxiliary function g.

A "double" mean value theorem is provided below. The proof combines the standard Lagrange mean value theorem with the basic property that the derivative has the intermediate value theorem (the Darboux theorem).

5.3.12. Let a,b,c be real numbers with $a < b < c$ and consider a differentiable function $f : [a,c] \to \mathbb{R}$. Prove that there exist $\xi \in (a,b)$, $\eta \in (a,c)$, $\xi < \eta$, such that $f(a) - f(b) = (a-b)f'(\xi)$ and $f(a) - f(c) = (a-c)f'(\eta)$.

D. Voiculescu

Solution. Applying the Lagrange mean value theorem, we obtain $\xi \in (a,b)$ and $\zeta \in (b,c)$ such that $f(a) - f(b) = (a-b)f'(\xi)$ and $f(b) - f(c) = (b-c)f'(\zeta)$. Therefore

$$\frac{f(c)-f(a)}{c-a} = \frac{f(c)-f(b)}{c-b} \cdot \frac{c-b}{c-a} + \frac{f(b)-f(a)}{b-a} \cdot \frac{b-a}{c-a} = tf'(\zeta) + (1-t)f'(\xi),$$

where $t = (c-b)/(c-a)$. Since $t \in (0,1)$, it follows that $tf'(\zeta) + (1-t)f'(\xi)$ lies between $f'(\xi)$ and $f'(\zeta)$. So, by the intermediate value property, there exists $\eta \in (\xi, \zeta)$ such that $f'(\eta) = (f(c) - f(a))/(c-a)$. □

Given a differentiable function on an interval $[a,b]$, the next exercise provides lower and upper estimates of its derivative at a certain intermediate point c in terms of $a - c$ and $b - c$.

5.3 The Main Theorems

5.3.13. *Let a and b be real numbers with $a < b$ and consider a continuous function $f : [a,b] \to \mathbb{R}$ that is differentiable on (a,b). Prove that there exists $c \in (a,b)$ such that*

$$\frac{2}{a-c} < f'(c) < \frac{2}{b-c}.$$

José Luis Díaz-Barrero and Pantelimon George Popescu

Solution. We first assume that f' changes sign in (a,b). So, there exist x_1 and x_2 in (a,b) such that $f'(x_1) \geq 0$ and $f'(x_2) \leq 0$. Thus, by Darboux's theorem, there exists c between x_1 and x_2 such that $f'(c) = 0$, and the conclusion follows.

Next, we assume that f' does not change its sign in (a,b). Without loss of generality, we can suppose that $f' > 0$ in (a,b). Arguing by contradiction and observing that $a - x < 0$ for any $x \in (a,b)$, we deduce that

$$f'(x) \geq \frac{2}{b-x}, \quad \text{for all } x \in (a,b).$$

Define $x_n = b - (b-a)/2^n$, for any positive integer n. Then, combining the Lagrange mean value theorem with the above inequality, we obtain, for some $c_1 \in (a, x_1)$,

$$f(x_1) - f(a) = f\left(\frac{a+b}{2}\right) - f(a) = f'(c_1)\frac{b-a}{2} \geq \frac{2}{b-c_1} \cdot \frac{b-a}{2} \geq 1.$$

We construct a sequence $(c_n)_{n \geq 1}$ such that $c_{n+1} \in (x_n, x_{n+1})$ and

$$f(x_{n+1}) - f(x_n) = f'(c_{n+1})(x_{n+1} - x_n) \geq \frac{2}{b - c_{n+1}} \cdot \frac{b-a}{2^{n+1}} \geq 1.$$

Therefore $f(x_n) - f(a) \geq n$ for any positive integer n, which implies that f is unbounded. This contradiction concludes our proof. $\square$

The following result is a Rolle-type theorem for the second derivative on **unbounded** intervals.

5.3.14. *Let $f : I = (a, \infty) \to \mathbb{R}$ be a twice differentiable function such that $\lim_{x \searrow a} f(x)$ and $\lim_{x \to +\infty} f(x)$ exist. Assuming that these limits are finite and equal, prove that there exists $\xi \in I$ such that $f''(\xi) = 0$.*

Solution. Set $\ell := \lim_{x \searrow a} f(x) = \lim_{x \to +\infty} f(x)$. Replacing f by $f - \ell$, we can assume that $\ell = 0$.

Arguing by contradiction, let us assume that for all $x \in I$, $f''(x) \neq 0$. Then, by the Darboux theorem, f'' has a constant sign. Suppose that $f'' > 0$ on I. Then f' is increasing, so $\lim_{x \to +\infty} f'(x)$ exists. Since $\lim_{x \to +\infty} f(x) = 0$, it follows that $\lim_{x \to +\infty} f(x)/x = 0$. Applying l'Hôpital's rule, we deduce that $0 = \lim_{x \to +\infty} f(x)/x = \lim_{x \to +\infty} f'(x)$. Combining this with the fact that f' is increasing, we obtain $f' \leq 0$, so f is nonincreasing. Using $\lim_{x \searrow a} f(x) = \lim_{x \to +\infty} f(x) = 0$, we deduce that $f \equiv 0$, contradiction. $\square$

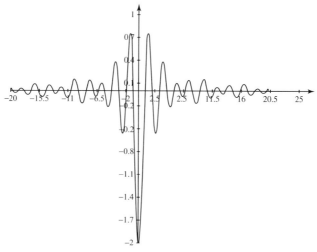

Fig. 5.14 Graph of the function $f(x) = \frac{\sin(2x)}{2x} - \frac{\sin(3x)}{3x}$.

We point out that the above result does not remain true if I is a bounded interval (give a counterexample!).

A sharp double inequality involving trigonometric functions is stated below.

5.3.15. Prove that

$$\frac{2}{\pi}\left(1 - \frac{a}{b}\right) \leq \sup_{x>0} \left|\frac{\sin(ax)}{ax} - \frac{\sin(bx)}{bx}\right| \leq 4\left(1 - \frac{a}{b}\right),$$

for all $0 < a < b$.

J. Rosenblatt, Amer. Math. Monthly, Problem 10604

Solution. Before starting the proof, we point out that the graph of the function $\mathbb{R} \ni x \longmapsto \sin(ax)/ax - \sin(bx)/bx$ in the particular case $a = 2$, $b = 3$ is depicted in Figure 5.14.

Fix $0 < a < b$ and set $\lambda = a/b$, $f(x) = \sin x/x$ ($x > 0$), and $\mu(\lambda, x) = [f(x) - f(\lambda x)]/(1 - \lambda)$. Since $|f(x)| < 1$ for all $x > 0$, it follows that

$$|\mu(\lambda, x)| < \frac{2}{1 - \lambda} \leq 4,$$

provided that $0 < \lambda \leq 1/2$. Next, by Lagrange's mean value theorem, there exists $\xi_x \in [\lambda x, x]$ such that $\mu(\lambda, x)/x = f'(\xi_x)$. Now $xf'(\xi_x) = (x/\xi_x)[\cos \xi_x - f(\xi_x)]$, so

$$|\mu(\lambda, x)| \leq 2\frac{x}{\xi_x} \leq \frac{2}{\lambda} \leq 4,$$

for all $1/2 \leq \lambda < 1$. This proves that $\sup_{x>0} |\mu(\lambda, x)| \leq 4$ for all $\lambda \in (0, 1)$.

5.3 The Main Theorems

To prove the lower bound, we observe that

$$\mu(\lambda,\pi) = \frac{\sin(\lambda\pi)}{\lambda(1-\lambda)\pi} = \mu(1-\lambda,\pi).$$

This means that it is enough to consider the case $\lambda \in (0, 1/2]$. Since f is decreasing on $(0, \pi/2]$, we obtain

$$\mu(\lambda,\pi) \geq \frac{1}{1-\lambda} \frac{\sin\pi/2}{\pi/2} \geq \frac{2}{\pi},$$

for all $\lambda \in (0, 1/2]$. □

What about the bounded increasing and convex functions? Do they have a common property?

5.3.16. Let $f : (a, \infty) \to \mathbb{R}$ be a nonconstant bounded twice differentiable function such that $f' \geq 0$ and $f'' \geq 0$ on (a, ∞). Prove that $\lim_{x \to \infty} f(x) = 0$.

Solution. By hypothesis, the functions f and f' are nondecreasing, so they have limits at $+\infty$. Set $\ell := \lim_{x \to +\infty} f(x)$. So, by l'Hôpital's rule, $0 = \lim_{x \to +\infty} f(x)/x = \lim_{x \to +\infty} f'(x)$. Thus, since $f' \geq 0$ is nondecreasing, it follows that $f' = 0$ on (a, ∞). Thus, f is a constant map, a contradiction. □

There are not many positive concave functions defined on the whole real axis! Is the last assumption really important?

5.3.17. Let $f : \mathbb{R} \to [0, \infty)$ be a twice differentiable function such that $f'' \leq 0$. Prove that f is constant.

Solution. Since $f'' \leq 0$, it follows that f' is nonincreasing, so f' has a limit at $-\infty$ (resp., at $+\infty$). Denote by ℓ_- (resp., by ℓ_+) this limit. Applying l'Hôpital's rule, we obtain $\ell_- = \lim_{x \to -\infty} f(x)/x = \lim_{x \to -\infty} f'(x)$ and $\ell_+ = \lim_{x \to +\infty} f(x)/x = \lim_{x \to +\infty} f'(x)$. Since $f \geq 0$, we deduce that $f(x)/x \leq 0$ if $x < 0$ and $f(x)/x \geq 0$ if $x > 0$. Hence $\ell_- \leq 0 \leq \ell_+$, which implies $\ell_- = \ell_+ = 0$. On the other hand, f' is monotone. Hence f is constant.

The following alternative proof is based on Taylor's formula. Fix $x \in \mathbb{R}$ arbitrarily. It is enough to show that $f'(x) = 0$. Indeed, for any real number t, there exists ξ between x and $x+t$ such that

$$f(x+t) = f(x) + tf'(x) + \frac{t^2}{2}f''(\xi).$$

By hypothesis, for any $x, t \in \mathbb{R}$,

$$0 \leq f(x+t) \leq f(x) + tf'(x).$$

Since t is arbitrary, it follows that $f'(x) = 0$. □

The next example illustrates how mean value arguments may be used to compute limits.

5.3.18. Let $f : (0,1) \to \mathbb{R}$ be differentiable on $(0,1)$ such that such that the functions f and $xf'(x)$ have limit at the origin. Prove that $xf'(x)$ tends to zero as x tends to zero.

Solution. Set $\ell := \lim_{x \searrow 0} xf'(x)$ and assume, by contradiction, that $\ell \neq 0$. Without loss of generality we can suppose that $\ell > 0$. Thus, there exists $\delta > 0$ such that for any $x \in (0, \delta)$, $f'(x) \geq \ell/(2x)$. On the other hand, by the mean value theorem, we deduce that

$$f\left(\frac{x}{2^j}\right) - f\left(\frac{x}{2^{j+1}}\right) \geq \frac{\ell}{4}, \quad \text{for any } 0 \leq j \leq n.$$

By summation we obtain

$$f(x) - f\left(\frac{x}{2^{n+1}}\right) \geq (n+1)\frac{\ell}{4},$$

which yields a contradiction as $n \to \infty$. $\square$

A refined asymptotic behavior is deduced below with differentiability arguments.

5.3.19. Define the sequence $(a_n)_{n \geq 2}$ by

$$a_n = \frac{(n^2+1)(n^2+2)\cdots(n^2+n)}{(n^2-1)(n^2-2)\cdots(n^2-n)}.$$

Prove that $\lim_{n \to \infty} a_n = e$ and find $\lim_{n \to \infty} n(a_n - e)$.

Z. Sasvári, Amer. Math. Monthly, Problem 10650

Solution. Define the function $f : (0,1) \to \mathbb{R}$ by $f(x) = \ln((1+x)/(1-x))$. Then, by Taylor's formula,

$$|f(x) - 2x| \leq \frac{2x^3}{(1-x^2)^2}, \quad \text{for all } x \in (0,1).$$

Since $\ln a_n = \sum_{k=1}^n f(k/n^2)$, it follows that

$$\left|\ln a_n - 1 - \frac{1}{n}\right| \leq \sum_{k=1}^n \left|f\left(\frac{k}{n^2}\right) - \frac{2k}{n^2}\right| \leq \frac{2}{n^6(1-1/n)^2} \sum_{k=1}^n k^3 = O\left(\frac{1}{n^2}\right),$$

as $n \to \infty$. Therefore

$$a_n = e\left(1 + \frac{1}{n} + O\left(\frac{1}{n^2}\right)\right),$$

which implies that $\lim_{n \to \infty} a_n = e$ and $\lim_{n \to \infty} n(a_n - e) = e$. $\square$

Independent Study. Find a sequence $(x_n)_{n \geq 1}$ such that $\lim_{n \to \infty} x_n (e - n(a_n - e))$ is finite.

5.3.20. Let f be twice continuously differentiable on $(0, +\infty)$ such that

5.3 The Main Theorems

$$\lim_{x \to 0+} f'(x) = -\infty \quad \text{and} \quad \lim_{x \to 0+} f''(x) = +\infty.$$

Show that

$$\lim_{x \to 0+} \frac{f(x)}{f'(x)} = 0.$$

Solution. Since f' tends to $-\infty$ and f'' tends to $+\infty$ as x tends to $0+$, there exists an interval $(0, r)$ such that $f'(x) < 0$ and $f''(x) > 0$ for all $x \in (0, r)$. Hence f is decreasing and f' is increasing on $(0, r)$. By the Lagrange mean value theorem, for every $0 < x < x_0 < r$ we obtain

$$f(x) - f(x_0) = f'(\xi)(x - x_0) > 0,$$

for some $\xi \in (x, x_0)$. Taking into account that f' is increasing, $f'(x) < f'(\xi) < 0$, we have

$$x - x_0 < \frac{f'(\xi)}{f'(x)}(x - x_0) = \frac{f(x) - f(x_0)}{f'(x)} < 0.$$

Taking limits as x tends to $0+$, we obtain

$$-x_0 \leq \liminf_{x \to 0+} \frac{f(x)}{f'(x)} \leq \limsup_{x \to 0+} \frac{f(x)}{f'(x)} \leq 0.$$

Since this happens for all $x_0 \in (0, r)$, we deduce that $\lim_{x \to 0+} f(x)/f'(x)$ exists and $\lim_{x \to 0+} f(x)/f'(x) = 0$. □

5.3.21. Let $f : \mathbb{R} \to \mathbb{R}$ be twice differentiable such that $f(0) = 2$, $f'(0) = -2$, and $f(1) = 1$. Prove that there exists $\xi \in (0, 1)$ such that $f(\xi) \cdot f'(\xi) + f''(\xi) = 0$.

International Mathematics Competition for University Students, 1997

Solution. Define the function

$$g(x) = \frac{1}{2}f^2(x) + f'(x).$$

Since $g(0) = 0$ and $f(x) \cdot f'(x) + f''(x) = g'(x)$, it is enough to prove that there exists a real number $0 < \eta \leq 1$ such that $g(\eta) = 0$.

We first assume that f has no zero. Set

$$h(x) = \frac{x}{2} - \frac{1}{f(x)}.$$

Because $h(0) = h(1) = -1/2$, there exists a real number $0 < \eta < 1$ such that $h'(\eta) = 0$. But $g = f^2 \cdot h'$, and we are done.

Next, we suppose that f has at least one zero. Since the set of the zeros of a function is closed, we may assume that z_1 is the smallest zero of f and z_2 is the largest one. Thus, $0 < z_1 \leq z_2 < 1$. The function f is positive on the intervals $[0, z_1)$ and $(z_2, 1]$. Thus, $f'(z_1) \leq 0$ and $f'(z_2) \geq 0$.

Then $g(z_1) = f'(z_1) \leq 0$ and $g(z_2) = f'(z_2) \geq 0$, and there exists $\eta \in [z_1, z_2]$ such that $g(\eta) = 0$. □

5.3.22. Let $f : (0, \infty) \to \mathbb{R}$ be a twice continuously differentiable function such that
$$|f''(x) + 2xf'(x) + (x^2 + 1)f(x)| \leq 1$$
for all x. Prove that $\lim_{x \to \infty} f(x) = 0$.

Solution. We first point out that l'Hôpital's rule is valid if the denominator converges to infinity, without any assumption on the numerator. Thus, by l'Hôpital's rule applied twice to the fraction $f(x)e^{x^2/2}/e^{x^2/2}$, we deduce that

$$\lim_{x \to \infty} f(x) = \lim_{x \to \infty} \frac{f(x)e^{x^2/2}}{e^{x^2/2}} = \lim_{x \to \infty} \frac{(f'(x) + xf(x))e^{x^2/2}}{xe^{x^2/2}}$$

$$= \lim_{x \to \infty} \frac{(f''(x) + 2xf'(x) + (x^2 + 1)f(x))e^{x^2/2}}{(x^2 + 1)e^{x^2/2}}$$

$$= \lim_{x \to \infty} \frac{f''(x) + 2xf'(x) + (x^2 + 1)f(x)}{(x^2 + 1)} = 0.$$

The following alternative solution uses just elementary knowledge of integral calculus. Indeed, we first set $g(x) = f'(x) + xf(x)$; hence
$$f''(x) + 2xf'(x) + (x^2 + 1)f(x) = g'(x) + xg(x).$$

We claim that if h is a continuously differentiable function such that $h'(x) + xh(x)$ is bounded, then $\lim_{x \to \infty} h(x) = 0$. Applying this lemma for $h = g$ then for $h = f$, the statement follows. In order to prove our claim, let M be an upper bound of $|h'(x) + xh(x)|$ and let $p(x) = h(x)e^{x^2/2}$. Then
$$|p'(x)| = |h'(x) + xh(x)|e^{x^2/2} \leq Me^{x^2/2}$$
and
$$|h(x)| = \left|\frac{p(x)}{e^{x^2/2}}\right| = \left|\frac{p(0) + \int_0^x p'(t)dt}{e^{x^2/2}}\right| \leq \frac{|p(0)| + M\int_0^x e^{t^2/2}dt}{e^{x^2/2}}.$$

Since $\lim_{x \to \infty} e^{x^2/2} = \infty$ and, by l'Hôpital's rule, $\lim_{x \to \infty} \int_0^x e^{t^2/2}dt/e^{x^2/2} = 0$, we conclude that $\lim_{x \to \infty} h(x) = 0$. □

5.3.23. Let $f(x)$ be a continuously differentiable real-valued function on $\mathbb{R}$. Show that if $f'(x)^2 + f^3(x) \to 0$ as $x \to +\infty$, then $f(x) \to 0$ as $x \to +\infty$.

5.3 The Main Theorems

Solution. There are three cases to be considered.

(i) Suppose $f(x)$ changes sign at x_n, $x_n \to +\infty$. Then $f(x)$ has a maximum or minimum at ξ_n, $x_n < \xi_n < x_{n+1}$, $f'(\xi_n) = 0$, $|f(x)| \leq |f(\xi_n)|$ for $x_n \leq x \leq x_{n+1}$, $|f(\xi_n)| \to 0$, $f(x) \to 0$.

(ii) $f(x)$ does not change sign for $x \geq u$, say $f(x) \geq 0$ for $x \geq u$. Then $f(x) \to 0$ for $x \to \infty$.

(iii) $f(x) \leq 0$ for $x \geq u$. Set $g = -f$, then $g'^2 - g^3 \to 0$, and since $(g'^2 - g^3)$ is arbitrarily small for sufficiently large x, $g(x)$ differs arbitrarily little, for sufficiently large x, from $h(x)$ for which $h'^2 - h^3 = 0$, or $(h' - h^{3/2})(h' + h^{3/2}) = 0$. If $h' + h^{3/2} \neq 0$ at some x, then $h' + h^{3/2} \neq 0$ on an interval I_1, so $h' - h^{3/2} = 0$ on I_1. If I is finite, then there is an abutting interval I_2 on which $h' + h^{3/2} = 0$. Thus, $h(x) = (-1/2x + c_1)^{-2}$ on I_1 and $h(x) = -(1/2x + c_2)^{-2}$ on I_2. Also $h'(x) = (-1/2x + c_1)^{-3}$ on I_1, $h'(x) = -(1/2x + c_2)^{-3}$ on I_2. At the point $x = a$ where I_1 and I_2 abut, we have $(-1/2a + c_1)^{-3} = -(1/2a + c_2)^{-3}$; hence $c_1 = -c_2$ and $h(x) = (1/2x + c)^{-2}$ on $I_1 \cup I_2$. It follows that $h(x) = (1/2x + c)^{-2}$ for all x; hence $h(x) \to 0$, $g(x) \to 0$, $f(x) \to 0$ as $x \to +\infty$. □

5.3.24. Let f be differentiable with f' continuous on $[a,b]$. Show that if there is a number c in $(a,b]$ such that $f'(c) = 0$, then we can find a number ξ in (a,b) such that
$$f'(\xi) = \frac{f(\xi) - f(a)}{b - a}.$$

S. Penner, Math. Magazine, Problem 987

Solution. For x in $[a,b]$, define
$$g(x) = f'(x) - \frac{f(x) - f(a)}{b - a}.$$

Assume first that $f(c) > f(a)$. Choose d in $(a,c]$ such that $f(d)$ is a maximum for f in $[a,c]$, and let e be a point in (a,d) such that $f'(e) = (f(d) - f(a))/(d - a)$. Then $g(e) > 0 > g(d)$, so, since g is continuous, there is a point ξ in (e,d) such that $g(\xi) = 0$. A similar argument takes care of the cases $f(c) < f(a)$ and $f(c) = f(a)$. □

5.3.25. Show that $\sum_{n=1}^{\infty} (-1)^{n-1} \sin(\log n)/n^\alpha$ converges if and only if $\alpha > 0$.

Solution. Set $f(t) = \sin(\log t)/t^\alpha$. We have
$$f'(t) = -\frac{\alpha}{t^{\alpha+1}} \sin(\log t) + \frac{\cos(\log t)}{t^{\alpha+1}}.$$

It follows that $|f'(t)| \leq (1 + \alpha)/t^{\alpha+1}$ for any $\alpha > 0$. Thus, by the mean value theorem, there exists $\theta \in (0, 1)$ such that
$$|f(n+1) - f(n)| = |f'(n + \theta)| \leq \frac{1 + \alpha}{n^{\alpha+1}}.$$

Since $\sum_{n=1}^{\infty}(1+\alpha)/n^{\alpha+1} < +\infty$ for $\alpha > 0$ and $f(n) \to 0$ as $n \to \infty$, we deduce that

$$\sum_{n=1}^{\infty}(-1)^{n-1}f(n) = \sum_{n=1}^{\infty}(f(2n-1) - f(2n))$$

converges.

Next, we prove that $\sin(\log n)/n^{\alpha}$ does not converge to 0 for $\alpha \leq 0$. It suffices to consider $\alpha = 0$. We show that $a_n = \sin(\log n)$ does not tend to zero. Assume the contrary. Then there exist a positive integer k_n and $\lambda_n \in [-1/2, 1/2]$ for $n > e^2$ such that $\log n/\pi = k_n + \lambda_n$. Then $|a_n| = \sin \pi |\lambda_n|$. Since $a_n \to 0$, we obtain $\lambda_n \to 0$. We have

$$k_{n+1} - k_n = \frac{\log(n+1) - \log n}{\pi} - (\lambda_{n+1} - \lambda_n) = \frac{1}{\pi}\log\left(1 + \frac{1}{n}\right) - (\lambda_{n+1} - \lambda_n).$$

Then $|k_{n+1} - k_n| < 1$ for all n big enough. Hence there exists n_0 such that $k_n = k_{n_0}$ for $n > n_0$. So $\log n/\pi = k_{n_0} + \lambda_n$ for all $n > n_0$. Since $\lambda_n \to 0$, we obtain a contradiction to $\log n \to \infty$ as $n \to \infty$. $\square$

The geometric interpretation of the following *straddle lemma* is that if the points u and v "straddle" z, then the slope of the chord between the points $(u, f(u))$ and $(v, f(v))$ on the graph of $f : [a,b] \to \mathbb{R}$ is close to the slope of the tangent line at $(z, f(z))$.

5.3.26. Let $f : [a,b] \to \mathbb{R}$ be differentiable at $z \in [a,b]$. Prove that for each $\varepsilon > 0$, there exists $\delta > 0$ such that

$$|f(v) - f(u) - f'(z)(v-u)| \leq \varepsilon(v-u)$$

whenever $u \leq z \leq v$ and $[u,v] \subseteq [a,b] \cap (z-\delta, z+\delta)$.

Solution. Since f is differentiable at z, there exists $\delta > 0$ such that

$$\left|\frac{f(x) - f(z)}{x-z} - f'(z)\right| < \varepsilon$$

for $0 < |x-z| < \delta$, $x \in [a,b]$. If $z = u$ or $z = v$, the conclusion is immediate, so suppose $u < z < v$. Then

$$|f(v) - f(u) - f'(z)(v-u)|$$
$$\leq |f(v) - f(z) - f'(z)(v-z)| + |f(z) - f(u) - f'(z)(z-u)|$$
$$< \varepsilon(v-z) + \varepsilon(z-u) = \varepsilon(v-u).$$

This concludes the proof. $\square$

5.4 The Maximum Principle

> Nature to Newton was an open book...
> In one person he combined the
> experimenter, the theorist, the
> mechanic, and, not least, the artist in
> exposition. Newton stands before us
> strong, certain, and alone.
>
> Albert Einstein (1879–1955)

The maximum principle is one of the simplest tools in nonlinear mathematical analysis. It is very efficient in the qualitative study of wide classes of ordinary differential equations, as well as in the treatment of partial differential equations.

The maximum principle can be described very easily in elementary terms. We give in what follows two different statements of this basic property.

Weak Maximum Principle. *Let $u : [a,b] \to \mathbb{R}$ be a twice differentiable convex function. Then u attains its maximum on $[a,b]$ either in a or in b. In particular, if $u(a) \leq 0$ and $u(b) \leq 0$, then $u \leq 0$ in $[a,b]$.*

Strong Maximum Principle. *Let $u : [a,b] \to \mathbb{R}$ be a twice differentiable convex function such that $u(a) = u(b) = 0$. Then the following alternative holds: either*

(i) $u \equiv 0$ in $[a,b]$

or

(ii) $u < 0$ in (a,b) *and, moreover,* $u'(a) < 0$, *and* $u'(b) > 0$.

The purpose of this section is to present some variants of the maximum principle and to illustrate the force of this method in the qualitative analysis of some ordinary differential equations or differential inequalities. We refer to the excellent monograph [94] for more details and examples on this subject.

5.4.1. (**The Maximum Principle**). *Let $f : [a,b] \to \mathbb{R}$ be a continuous function that is twice differentiable on (a,b) such that for all $x \in (a,b)$ we have $f''(x) = \alpha f(x)$, for some constant $\alpha > 0$. Prove that*

$$|f(x)| \leq \max\{|f(a)|, |f(b)|\}, \quad \text{for all } x \in [a,b].$$

Solution. The continuous function f is bounded, so it achieves the minimum and the maximum on the compact interval $[a,b]$. The idea of the proof is to show that f can have neither a positive maximum nor a negative minimum in (a,b). Fix $x_0 \in [a,b]$ and without loss of generality suppose $f(x_0) > 0$. Let $M = \max_{x \in [a,b]} f(x) \geq f(x_0) > 0$. If this maximum is attained at $x_M \in (a,b)$, then $f''(x_M) \leq 0$ and hence $M = f(x_M) \leq 0$, a contradiction. Thus the maximum is attained at either a or b and

$$f(x_0) \leq M = \max\{f(a), f(b)\} \leq \max\{|f(a)|, |f(b)|\}.$$

Since x_0 is arbitrary, the result follows. □

5.4.2. (The Maximum Principle) Let $f : (a,b) \to \mathbb{R}$ be a bounded function and let $u : (a,b) \to \mathbb{R}$ be a function satisfying

$$u''(x) + f(x)u'(x) \geq 0, \quad \text{for all } x \in (a,b).$$

Assume that $u(x) \leq M$, for all $x \in (a,b)$. Prove that if there exists a point c in (a,b) such that $u(c) = M$, then $u \equiv M$ în (a,b).

Solution. Assume by contradiction that $u(c) = M$ and there exists $d \in (a,b)$ such that $u(d) < M$. Let us suppose that $d > c$. Define the function $\varphi(x) = e^{\alpha(x-c)} - 1$, where $\alpha > 0$ is a constant to be determined. Observe that $\varphi(x) < 0$ for all $a < x < c$, $\varphi(x) > 0$ if $c < x < b$ and $\varphi(c) = 0$. We have

$$\Phi_\varphi(x) := \varphi''(x) + f(x)\varphi'(x) = \alpha[\alpha + f(x)]e^{\alpha(x-c)}.$$

Choose $\alpha > 0$ sufficiently large that $\Phi_\varphi(x) > 0$ for any $a < x < d$. In other words, we choose α such that $\alpha > -f(x)$, for all $x \in (a,b)$. This is possible, since f is bounded. Define $w(x) = u(x) + \varepsilon\varphi(x)$, where $\varepsilon > 0$ is chosen such that $\varepsilon < (M - u(d))/\varphi(d)$. Such a choice for ε is possible due to $u(d) < M$ and $\varphi(d) > 0$. Since $\varphi < 0$ in (a,c) we have $w(x) < M$, for all $a < x < c$. By the choice of ε we obtain

$$w(d) = u(d) + \varepsilon\varphi(d) < u(d) + M - u(d).$$

Hence $w(d) < M$. On the other hand,

$$w(c) = u(c) + \varepsilon\varphi(c) = M.$$

It follows that w has a maximum value greater than or equal to M, and this maximum is attained in an interior point of (a,d). But

$$\Phi_w(x) = \Phi_u(x) + \varepsilon\Phi_\varphi(x) > 0, \quad \text{for all } x \in (a,d).$$

This shows that there is no $x_0 \in (a,d)$ such that $w'(x_0) = 0$ and $w''(x_0) \leq 0$. Thus, w cannot achieve its maximum in (a,d), contradiction.

If $d < c$, then we use the auxiliary function $\varphi(x) = e^{-\alpha(x-c)} - 1$, with $\alpha > f(x)$ for all $x \in (a,c)$. □

5.4.3. Let $f : (a,b) \to \mathbb{R}$ be a function that is bounded on every closed interval contained in (a,b) and let u be a nonconstant function satisfying the differential inequality $u''(x) + f(x)u'(x) \geq 0$, for all $x \in (a,b)$. Moreover, we suppose that u has one-sided derivatives in a and in b.

(i) Assume that the maximum of u is achieved at $x = a$ and that f is bounded from below near $x = a$. Prove that $u'(a+) < 0$.
(ii) Assume that the maximum of u is achieved at $x = b$ and that f is bounded from above near $x = b$. Prove that $u'(b-) > 0$.

5.4 The Maximum Principle

Solution. (i) Assume that $u(a) = M$ and $u(x) \leq M$ for all $a \leq x \leq b$. Let $d \in (a,b)$ be such that $u(d) < M$. Consider the auxiliary function $\varphi(x) = e^{\alpha(x-a)} - 1$, with $\alpha > 0$. Choose $\alpha > -f(x)$, for all $a \leq x \leq d$. Hence $\Phi_\varphi(x) := \varphi''(x) + f(x)\varphi'(x) > 0$. Let $w(x) = u(x) + \varepsilon\varphi(x)$, where $0 < \varepsilon < (M - u(d))/\varphi(d)$. Since $\Phi_w(x) > 0$, it follows that the maximum of w in $[a,d]$ is achieved at one of the endpoints. Since $w(a) = M > w(d)$, the maximum is achieved for $x = a$. So, the right derivative in a cannot be positive, so $w'(a) = u'(a) + \varepsilon\varphi'(a) \leq 0$. But $\varphi'(a) = \alpha > 0$. Therefore $u'(a) < 0$.

(ii) Apply the same arguments as above. □

We give in what follows a variant of the maximum principle for **nonlinear** differential inequalities. More precisely, the nonlinearity is $u - u^3$, and this goes back to the Ginzburg–Landau theory arising in superconductivity.

5.4.4. (a) Fix $R > 0$ and let $u_i = u_i(r) : [R, \infty) \to [3^{-1/2}, \infty)$ $(i = 1, 2)$ be twice differentiable functions satisfying

$$-u_1'' - \frac{1}{r}u_1' + \frac{1}{r^2}u_1 - u_1(1 - u_1^2) \geq 0 \geq -u_2'' - \frac{1}{r}u_2' + \frac{1}{r^2}u_2 - u_2(1 - u_2^2).$$

We also assume that $u_1(R) \geq u_2(R)$ and $\limsup_{r \to \infty} (u_1 - u_2)(r) \geq 0$. Prove that $u_1 \geq u_2$ on $[R, \infty)$.

(b) Let $f = f(r)$ be a twice differentiable function such that $\liminf_{r \to \infty} f(r) = 1$ and

$$-f'' - \frac{1}{r}f' + \frac{1}{r^2}f = f(1 - f^2) \quad \text{in } (0, \infty).$$

Prove that

$$f(r) = 1 - \frac{1}{2r^2} - \frac{9}{8r^4} + O\left(\frac{1}{r^5}\right) \quad \text{as } r \to \infty.$$

Solution. (a) Set $u := u_1 - u_2$. Then

$$-u''(r) - \frac{1}{r}u'(r) + a(r)u(r) \geq 0 \quad \text{for all } r \in (R, \infty),$$

where $a(r) := u_1^2(r) + u_2^2(r) + u_1(r)u_2(r) + r^{-2} - 1$. Our hypotheses imply $a(r) > 0$ on (R, ∞), $u(R) \geq 0$, and $\limsup_{r \to \infty} u(r) \geq 0$. So, by the maximum principle, $u \geq 0$ on $[R, \infty)$, which concludes the proof.

(b) Define, for all $r > 0$, $\varphi(r) = 1 - (2r^2)^{-1} - 2^{-3}9r^{-4}$. For a positive constant C, set $u_1(r) = \varphi(r) + Cr^{-5}$ and $u_2(r) = \varphi(r) - Cr^{-5}$. Then $u_1(R) \geq u_2(R) \geq 3^{-1/2}$, provided that $R > 0$ is sufficiently large. We fix R with this property and choose $C > 0$ such that $u_1(R) \geq f(R) \geq f_2(R)$. Thus, the pairs (u_1, f) and (u_2, f) satisfy the hypotheses imposed at (a). Hence $u_1 \geq f \geq u_2$ in (R, ∞). In particular, this implies that for all $r > R$, $|\varphi(r) - f(r)| \leq Cr^{-5}$, and the conclusion follows. □

5.5 Differential Equations and Inequalities

> In mathematics you don't understand things. You just get used to them.
>
> John von Neumann (1903–1957)

The key role played by differential equations is synthesized as follows by the famous contemporary Russian mathematician Vladimir I. Arnold [2]: "Differential equations are one of the basic tools of mathematics. They were first considered systematically by Sir Isaac Newton (1642–1727), although problems leading to differential equations had in fact arisen earlier. Before Newton, however, only such geniuses as Christiaan Huygens (1629–1695), president of the French Academy of Sciences, and Isaac Barrow (1630–1677), a mathematician and theologian, who was Newton's teacher, could solve them. Today, thanks to Newton, many differential equations are solvable by college students and even school children."

We start with a differential inequality involving the second-order iterate of a differentiable function.

5.5.1. *Prove that there does not exist a positive continuously differentiable function f on $[0, \infty)$ such that $f'(x) \geq f(f(x))$, for all $x \geq 0$.*

Solution. Assume that f is such a function. Since $f'(x) \geq f(f(x)) > 0$, f is increasing, so that $f'(x) \geq f(f(x)) \geq f(f(0)) > 0$. This means that $f'(x)$ is bounded away from 0, and so $\lim_{x \to \infty} f(x) = \infty$. Hence $\lim_{x \to \infty} f(f(x)) = \infty$, and so $\lim_{x \to \infty} f'(x) = \infty$. Set $g(x) := f(x) - x - 1$. Then $g'(x)$ also approaches ∞, which implies that $g(x)$ approaches ∞. Thus there exists some x_0 such that $f(x_0) > x_0 + 1$. Next, applying the mean value theorem to the interval $[x_0, f(x_0)]$, we obtain a point $\xi \in (x_0, f(x_0))$ for which

$$f(f(x_0)) = f(x_0) + f'(\xi)(f(x_0) - x_0) > f'(\xi)(f(x_0) - x_0)$$
$$\geq f(f(\xi))(f(x_0) - x_0) > f(f(x_0))(f(x_0) - x_0) > f(f(x_0)),$$

which is a contradiction. □

The following inequality asserts that if a twice differentiable function $f : \mathbb{R} \to \mathbb{R}$ is bounded together with its second derivative, then the derivative of f is bounded, too. In 1932, the British mathematicians Godfrey Harold Hardy (1877–1947) and John Edensor Littlewood (1885–1977) [42] extended this inequality to larger classes of functions.

5.5.2. (**Landau's Inequality, [63]**). *Let $f : \mathbb{R} \to \mathbb{R}$ be a function of class C^2. Assume that both f and f'' are bounded and set*

$$M_0 = \sup_{x \in \mathbb{R}} |f(x)|, \quad M_2 = \sup_{x \in \mathbb{R}} |f''(x)|.$$

Prove that f' is bounded and, moreover,

5.5 Differential Equations and Inequalities

$$\sup_{x\in\mathbb{R}} |f'(x)| \leq 2\sqrt{M_0 M_2}.$$

Solution. We first observe that if $M_2 = 0$, then the only functions satisfying our hypotheses are the constant mappings. So, we can assume without loss of generality that $M_2 > 0$.

Let $x \in \mathbb{R}$ and fix arbitrarily $h > 0$. By Taylor's formula, there is some $t \in (x, x+2h)$ such that

$$f(x+2h) = f(x) + 2hf'(x) + 2h^2 f''(t),$$

that is,

$$f'(x) = \frac{f(x+2h) - f(x)}{2h} - hf''(t).$$

Taking the modulus, and applying our hypothesis we obtain

$$|f'(x)| \leq \frac{M_0}{h} + M_2 h \quad \text{for all } x \in \mathbb{R}.$$

Choosing now $h = (M_0/M_2)^{1/2}$, we obtain our conclusion. □

Remark. Applying Taylor's formula twice, namely between x and both $x \pm h$, we can obtain (exercise!) the better estimate

$$\sup_{x\in\mathbb{R}} |f'(x)| \leq \sqrt{2M_0 M_2}.$$

We also point out that the above Landau inequality establishes that

$$\|f'\|_{L^\infty(\mathbb{R})} \leq 2\sqrt{\|f\|_{L^\infty(\mathbb{R})} \cdot \|f''\|_{L^\infty(\mathbb{R})}},$$

where $\|\cdot\|_{L^\infty(\mathbb{R})}$ denotes the *norm* in the Banach space $L^\infty(\mathbb{R})$ and is defined by

$$\|u\|_{L^\infty(\mathbb{R})} := \sup_{x\in\mathbb{R}} |u(x)|.$$

The above L^∞-*inequality* was established in the so-called L^2 *framework* by Hardy and Littlewood [42] as follows:

$$\int_0^\infty |f'(x)|^2 dx \leq 2\sqrt{\int_0^\infty |f(x)|^2 dx} \sqrt{\int_0^\infty |f''(x)|^2 dx}.$$

Hardy and Littlewood's inequality has been given considerable space and even several proofs in the famous book [43], and has subsequently attracted further attention.

A variant of the above inequality for higher-order derivatives is stated in what follows. This result is due to the Russian mathematician Andrey Nikolaevich Kolmogorov (1903–1987), who is known for his major contributions in probability theory and topology. The Landau and Kolmogorov inequalities are extended in what follows.

5.5.3. (**Kolmogorov's Inequality**). Let $f : \mathbb{R} \to \mathbb{R}$ be a function of class C^3. Assume that both f and f''' are bounded and set

$$M_0 = \sup_{x \in \mathbb{R}} |f(x)|, \quad M_3 = \sup_{x \in \mathbb{R}} |f'''(x)|.$$

(a) Prove that f' is bounded and, moreover,

$$\sup_{x \in \mathbb{R}} |f'(x)| \leq \frac{1}{2} \left(9 M_0^2 M_3\right)^{1/3}.$$

(b) Is f'' bounded, too?

Solution. (a) Fix $x \in \mathbb{R}$ and $h \neq 0$. Applying Taylor's formula between x and $x+h$, we obtain, successively,

$$\left| f(x+h) - f(x) - h f'(x) - \frac{h^2}{2} f''(x) \right| \leq M_3 \frac{h^3}{6}$$

and

$$\left| f(x-h) - f(x) + h f'(x) - \frac{h^2}{2} f''(x) \right| \leq M_3 \frac{h^3}{6}.$$

Therefore

$$2h|f'(x)| = \left| \left(f(x-h) - f(x) + h f'(x) - \frac{h^2}{2} f''(x) \right) \right.$$
$$\left. - \left(f(x+h) - f(x) - h f'(x) - \frac{h^2}{2} f''(x) \right) + f(x+h) - f(x-h) \right|$$
$$\leq \left| f(x-h) - f(x) + h f'(x) - \frac{h^2}{2} f''(x) \right|$$
$$+ \left| f(x+h) - f(x) - h f'(x) - \frac{h^2}{2} f''(x) \right| + |f(x+h)| + |f(x-h)|$$
$$= \frac{M_3 h^3}{3} + 2 M_0.$$

Hence

$$|f'(x)| \leq \frac{M_0}{h} + \frac{M_3 h^2}{6} =: \psi(h).$$

But $\psi'(h) = -\frac{M_0}{h^2} + \frac{M_3 h}{3}$, which shows that ψ achieves its minimum for $h_0 = \left(3 M_0 M_3^{-1}\right)^{1/3}$. Since $\psi(h) = 2^{-1} \left(9 M_0^2 M_3\right)^{1/3}$, our conclusion follows.

(b) By hypothesis and (a), the functions f' and $f''' = (f')''$ are bounded. Applying Landau's inequality to these functions, we deduce that f'' is bounded. □

5.5 Differential Equations and Inequalities

5.5.4. (Landau–Kolmogorov Generalized Inequalities). Let $f : \mathbb{R} \to \mathbb{R}$ be a nonconstant function of class C^n such that both f and $f^{(n)}$ are bounded.

(a) Prove that $f^{(n-1)}$ is bounded.
(b) Deduce that all the derivatives $f^{(k)}$ are bounded, $1 \leq k \leq n-1$.
    For any integer $0 \leq k \leq n$, set $M_k = \sup_{x \in \mathbb{R}} |f^{(k)}(x)|$.
(c) Show that $M_k > 0$, for any integer $0 \leq k \leq n$.
(d) Using the functions $u_k = 2^{k-1} M_k M_{k-1}^{-1}$, $0 \leq k \leq n$, as needed, prove that

$$M_k \leq 2^{k(n-k)/2} M_0^{1-k/n} M_n^{k/n}.$$

Solution. (a) The Taylor formula implies

$$\left| \sum_{j=1}^{n-1} \frac{k^j}{j!} f^{(j)}(x) \right| \leq 2M_0 + \frac{M_n k^n}{n!}.$$

Hence

$$\left| \sum_{k=1}^{n-1} (-1)^{n-k-1} C_{n-1}^k \sum_{j=1}^{n-1} \frac{k^j}{j!} f^{(j)}(x) \right| \leq \sum_{k=1}^{n-1} C_{n-1}^k \left(2M_0 + \frac{M_n k^n}{n!} \right),$$

which implies

$$\left| \sum_{j=1}^{n-1} \frac{1}{j!} f^{(j)}(x) \sum_{k=1}^{n-1} (-1)^{n-1-k} C_{n-1}^k k^j \right| \leq \sum_{k=1}^{n-1} C_{n-1}^k \left(2M_0 + \frac{M_n k^n}{n!} \right).$$

Using the elementary formula

$$\sum_{k=1}^{m} (-1)^{m-k} C_m^k k^j = \begin{cases} 0 & \text{if } 1 \leq j \leq m-1, \\ m! & \text{if } j = m, \end{cases}$$

we observe that all the terms on the left-hand side of the above inequality are zero, excepting that corresponding to $j = n-1$. It follows that for all $x \in \mathbb{R}$ we have

$$|f^{(n-1)}(x)| \leq \sum_{k=1}^{n-1} C_{n-1}^k \left(2M_0 + \frac{M_n k^n}{n!} \right).$$

(b) We use a standard argument by induction.
(c) If $M_k = 0$ then f is a polynomial of degree at most $k-1$. Since f is bounded, it follows that f must be constant, which is excluded by our hypotheses. So, $M_k > 0$ for all $0 \leq k \leq n$.
(d) Applying Landau's inequality, we obtain $M_k \leq \sqrt{2 M_{k-1} M_{k+1}}$ for all $1 \leq k \leq n-1$. This inequality shows that $u_1 \leq u_2 \leq \cdots \leq u_n$, which yields

$$(u_1 u_2 \cdots u_k)^n \leq (u_1 u_2 \cdots u_n)^k.$$

Applying this inequality in our case, we obtain

$$2^{nk(k-1)/2}\frac{M_k^n}{M_0^n} \leq 2^{kn(n-1)/2}\frac{M_n^k}{M_0^k},$$

which concludes the proof. □

The function g in the next example behaves like at least like a superlinear power at $+\infty$. So, the next problem involves a second-order differential inequality with **superlinear** growth in the nonlinear term.

5.5.5. *Let $g : (0,+\infty) \to (0,+\infty)$ be a continuous function such that*

$$\lim_{x \to +\infty} \frac{g(x)}{x^{1+\alpha}} = +\infty, \tag{5.14}$$

for some $\alpha > 0$. Let $f : \mathbb{R} \to (0,+\infty)$ be a twice differentiable function. Assume that there exist $a > 0$ and $x_0 \in \mathbb{R}$ such that

$$f''(x) + f'(x) > ag(f(x)), \quad \text{for all } x \geq x_0. \tag{5.15}$$

Prove that $\lim_{x \to +\infty} f(x)$ exists, is finite and compute its value.

Vicenţiu Rădulescu, Amer. Math. Monthly, Problem 11024

Solution. If $x_1 > x_0$ is a critical point of f, then by (5.15), $f''(x_1) > 0$, so x_1 is a relative minimum point of f. This implies that $f'(x)$ does not change sign if x is sufficiently large. Consequently, we can assume that f is monotone on $(x_0, +\infty)$; hence $\ell := \lim_{x \to +\infty} f(x)$ exists.

The difficult part of the proof is to show that ℓ is finite. This will be deduced after applying in a decisive manner our superlinear growth assumption (5.14). Arguing by contradiction, let us assume that $\ell = +\infty$. In particular, it follows that f is monotone increasing on $(x_0, +\infty)$. Define the function

$$u(x) = e^{x/2} f(x), \quad x \geq x_0.$$

Then u is increasing, and for any $x \geq x_0$,

$$u''(x) = \frac{1}{4}u(x) + e^{x/2}\left(f''(x) + f'(x)\right) > \frac{1}{4}u(x) + ae^{x/2}g(f(x)). \tag{5.16}$$

Our hypothesis (5.14) and the assumption $\ell = +\infty$ yield some $x_1 > x_0$ such that

$$g(f(x)) \geq f^{1+\alpha}(x), \quad \text{for all } x \geq x_1. \tag{5.17}$$

So, by (5.16) and (5.17),

$$u''(x) > \frac{1}{4}u(x) + Cu(x)f^\alpha(x) > Cu(x)f^\alpha(x), \quad \text{for all } x \geq x_1, \tag{5.18}$$

5.5 Differential Equations and Inequalities

for some $C > 0$. In particular, since $\ell = +\infty$, there exists $x_2 > x_1$ such that

$$u''(x) > u(x), \quad \text{for all } x \geq x_2. \tag{5.19}$$

We claim a little more, namely that there exists $C_0 > 0$ such that

$$u''(x) > C_0 u^{1+\alpha/2}(x), \quad \text{for all } x \geq x_2. \tag{5.20}$$

Indeed, let us first choose $0 < \delta < \min\{e^{-x_2} u(x_2), e^{-x_2} u'(x_2)\}$. We prove that

$$u(x) > \delta e^x, \quad \text{for all } x \geq x_2. \tag{5.21}$$

For this purpose, consider the function $v(x) = u(x) - \delta e^x$. Arguing by contradiction and using $v(x_2) > 0$ and $v'(x_2) > 0$, we deduce the existence of a relative maximum point $x_3 > x_2$ of v. So, $v(x_3) > 0$, $v'(x_3) = 0$, and $v''(x_3) \leq 0$. Hence $\delta e^{x_3} = u'(x_3) < u(x_3)$. But by (5.19), $u''(x_3) > u(x_3)$, which yields $v''(x_3) > 0$, a contradiction. This concludes the proof of (5.21).

Returning to (5.18) and using (5.21), we obtain

$$u''(x) > C u^{1+\alpha/2}(x) u^{\alpha/2}(x) e^{-\alpha x/2} > C_0 u^{1+\alpha/2}(x), \quad \text{for all } x > x_2,$$

where $C_0 = C \delta^{\alpha/2}$. This proves our claim (5.20). So

$$u'(x) u''(x) > C_0 u^{1+\alpha/2}(x) u'(x), \quad \text{for all } x > x_2.$$

Hence

$$\left(\frac{1}{2} u'^2(x) - C_1 u^{2+\beta}(x) \right)' > 0, \quad \text{for all } x > x_2,$$

where $C_1 = 2C_0/(4+\alpha)$ and $\beta = \alpha/2 > 0$. Therefore

$$u'^2(x) \geq C_2 + C_3 u^{2+\beta}(x), \quad \text{for all } x > x_2,$$

for some positive constants C_2 and C_3. So, since u is unbounded, there exists $x_3 > x_2$ and $C_4 > 0$ such that

$$u'(x) \geq C_4 u^{1+\gamma}(x), \quad \text{for all } x > x_3,$$

where $\gamma = \beta/2 > 0$.

Applying the mean value theorem, we obtain

$$u^{-\gamma}(x_3) - u^{-\gamma}(x) = \gamma(x - x_3) u^{-\gamma-1}(\xi_x) u'(\xi_x) \geq C_4 \gamma (x - x_3), \quad \text{for all } x > x_3,$$

where $\xi_x \in (x_3, x)$. Taking $x \to +\infty$ in the above inequality, we obtain a contradiction, since the left-hand side converges to $u^{-\gamma}(x_3)$ (because $\ell = +\infty$), while the right-hand side diverges to $+\infty$. This contradiction shows that $\ell = \lim_{x \to +\infty} f(x)$ must be finite.

We prove in what follows that $\ell = 0$. Arguing by contradiction, let us assume that $\ell > 0$. We first observe that relation (5.15) yields, by integration,

$$f'(x) - f'(x_0) + f(x) - f(x_0) \geq a \int_{x_0}^{x} g(f(t)) dt. \tag{5.22}$$

Since ℓ is finite, it follows by (5.22) that $\lim_{x \to +\infty} f'(x) = +\infty$. But this contradicts the fact that $\lim_{x \to +\infty} f(x)$ is finite. □

Remark. The result stated in our problem does not remain true if g has linear growth at $+\infty$, so if (5.14) fails. Indeed, it is enough to choose $f(x) = e^x$ and g the identity map. We also remark that "ℓ is finite" does not follow if the growth hypothesis (5.14) is replaced by the weaker one $\lim_{x \to \infty} g(x)/x = +\infty$. Indeed, if $g(x) = x\ln(1+x)$ and $f(x) = e^{x^2}$, then $\ell = +\infty$.

The next problem is inspired by the following classical framework. Consider the **linear** differential equation $g'' + g = 0$ on $[0, \infty)$. All solutions of this equation are of the form $g(x) = C_1 \cos x + C_2 \sin x$, where C_1 and C_2 are real constants. In particular, this implies that there are **no** solutions that are positive on the **whole** semiaxis $[0, \infty)$. The purpose of this problem is to find a class of functions f such that $f(x)/x \neq \text{Const}$ for all $x > 0$ and the **nonlinear** differential equation $g'' + f \circ g = 0$ has no positive solution on the positive semiaxis.

5.5.6. *Find a class of positive continuous functions f defined on $(0, \infty)$ that are not a multiple of the identity map and such that there is no positive twice differentiable function g on $[0, \infty)$ satisfying $g'' + f \circ g = 0$.*

Solution. The equality $g'' + f \circ g = 0$ can be rewritten as

$$g' = h \quad \text{on } [0, \infty) \tag{5.23}$$

combined with

$$h' + f \circ g = 0 \quad \text{on } [0, \infty). \tag{5.24}$$

The following situations can occur.

CASE 1: there exists $x_0 \in [0, \infty)$ such that $h(x_0) < 0$. Thus, by (5.24), $h(x) < h(x_0)$ for all $x > x_0$. Then, by integration in (5.23), we obtain

$$g(x) < g(x_0) + h(x_0)(x - x_0), \quad \text{for all } x > x_0.$$

So, since $h(x_0) < 0$ and $g > 0$ in (x_0, ∞), the above relation yields a contradiction, for x sufficiently large.

CASE 2: $h(x_0) = 0$, for some $x_0 \geq 0$. Thus, by (5.24), it follows that h is decreasing in (x_0, ∞). In particular, we have $h < 0$ in (x_0, ∞). With the same arguments as in Case 1 we obtain again a contradiction. Consequently, Cases 1 and 2 can never occur.

CASE 3: $h > 0$ in $[0, \infty)$. In this situation, by (5.23), it follows that

$$g(x) > g(0) > 0, \quad \text{for all } x > 0. \tag{5.25}$$

We will assume that
$$\liminf_{x \to \infty} f(x) > 0. \tag{5.26}$$

So, by (5.25) and (5.26), there exists some $A > 0$ (sufficiently small, but **positive**) such that $f(g(x)) > A$ for all $x > 0$. Thus, by (5.24),

$$h(x) < h(0) - Ax, \quad \text{for all } x > 0,$$

a contradiction, since h is positive. In conclusion, the required sufficient condition is formulated in relation (5.26). We point out that this condition is not necessary. A necessary and sufficient condition is $\int_0^\infty f(t)dt = +\infty$. To see this, define $F(x) = \int_0^x f(t)dt$. Then multiplying by $2g'$ and integrating gives $(g')^2 + 2F \circ g = C$, where C is a constant. Rearranging gives

$$\frac{g'(t)}{\sqrt{C - 2F(g)}} = \pm 1.$$

Taking the positive case (that is, assuming that g is increasing) and integrating gives

$$\int_{g(0)}^{g(t)} \frac{du}{\sqrt{C - 2F(u)}} = t.$$

If $\int_0^\infty f(t)dt < \infty$, then F is bounded, and taking $C > 2\sup F(u)$ gives a solution (upon rearranging). On the other hand, if $\int_0^\infty f(t)dt = +\infty$, then F is unbounded, and hence there is some u_0 with $F(u_0) = C/2$. Then by Taylor's theorem (without remainder),

$$F(u) = \frac{C}{2} + f(u_0)(u - u_0) + o((u - u_0)^2) \quad \text{as } u \to u_0.$$

Hence we see that

$$\int_{g(0)}^{u_0} \frac{du}{\sqrt{C - 2F(u)}} = t_0 < \infty.$$

Thus, at $t = t_0$ we will have $g'(t_0) = 0$, and since $g''(x) < 0$, g will be decreasing for $t > t_0$. Hence for $t > t_0$ we have

$$\int_{g(t)}^{u_0} \frac{du}{\sqrt{C - 2F(u)}} = t - t_0.$$

Since

$$\int_0^{u_0} \frac{du}{\sqrt{C - 2F(u)}} = t_1 < \infty,$$

we have $g(t_0 + t_1) = 0$ and $g(t) < 0$ for $t > t_0 + t_1$. Thus there is no positive solution. □

We study in what follows a singular (at the origin) second-order differential equation of Ginzburg–Landau type.

5.5.7. Let $f : (0,1) \to \mathbb{R}$ be an arbitrary solution of the differential equation

$$-f''(x) - \frac{f'(x)}{x} + \frac{f(x)}{x^2} = f(x)\left(1 - f^2(x)\right) \quad \text{in } (0,1).$$

(a) Assume that there exists $x_0 \in (0,1)$ such that $f(x_0) \geq 1$ and $f'(x_0) \leq 0$. Prove that $\lim_{x \searrow 0} f(x) = +\infty$.
(b) Prove that the same conclusion holds if $f > 1$ and $f' > 0$ on a certain interval $(0, x_0)$.

Solution. We first observe that the function $g: (-\infty, 0) \to \mathbb{R}$ defined by $g(x) = f(e^x)$ satisfies

$$g''(x) = \left[1 + e^{2x}\left(g^2(x) - 1\right)\right]g(x), \quad \text{for all } x < 0. \tag{5.27}$$

The above differential equation implies the following "logarithmic convexity" property of g. Set

$$\omega = \{(x,t) \in \mathbb{R}^2;\ 1 + e^{2x}(t^2 - 1) \leq 0\}.$$

Then either g is convex and positive or g is concave and negative, excepting the case in which the graph of g is in ω.

(a) We first claim that $f \geq 1$ on the **whole** interval $(0, x_0)$ or, equivalently, $g \geq 1$ on $(-\infty, \ln x_0)$. Arguing by contradiction, there is a local maximum x_1 in this interval with $g(x_1) \geq 1$, $g'(x_1) = 0$, and $g''(x_1) \leq 0$, a contradiction. Hence it follows that $g'' \geq g \geq 1$ on $(-\infty, \ln x_0)$.

For any fixed $a < \ln x_0$ and $x < a$ we have

$$g'(a) - g'(x) = \int_x^a g''(t)dt \geq \int_x^a dt = a - x.$$

Hence $g'(x) \to -\infty$ as $x \to -\infty$. Using now $g(a) - g(x) = \int_x^a g'(t)dt$, we deduce that $g(x) \to +\infty$ as $x \to -\infty$, which implies $\lim_{x \searrow 0} f(x) = +\infty$.

(b) Since $g > 1$ in $(-\infty, \ln x_0)$, it follows that $(x, g(x)) \notin \omega$, for all $x \in (0, x_0)$. The logarithmic convexity of g implies that $g > 1$ and $g'' > 0$ in $(-\infty, \ln x_0)$. From now on, with an argument similar to that used in the proof of (a) we obtain the conclusion. □

We are now interested in the behavior of a function satisfying a nonlinear second-order differential equation.

5.5.8. *Let φ be a continuous positive function on the open interval (A, ∞), and assume that f is a C^2-function on (A, ∞) satisfying the differential equation*

$$f''(t) = (1 + \varphi(t)(f^2(t) - 1))f(t).$$

(a) *Given that there exists $a \in (A, \infty)$ such that $f(a) \geq 1$ and $f'(a) \geq 0$, prove that there is a positive constant K such that $f(x) \geq Ke^x$ whenever $x \geq a$.*
(b) *Given instead that there exists $a \in (A, \infty)$ such that $f'(a) < 0$ and $f(x) > 1$ if $x > a$, prove that there exists a positive constant K such that $f(x) \geq Ke^x$ whenever $x \geq a$.*
(c) *Given that f is bounded on (A, ∞) and that there exists $\alpha > 0$ such that $\varphi(x) = O(e^{-(1+\alpha)x})$ as $x \to \infty$, prove that $\lim_{x \to \infty} e^x f(x)$ exists and is finite.*

Vicențiu Rădulescu, Amer. Math. Monthly, Problem 11137

Solution. (a) We first claim that $f \geq 1$ in $[a, \infty)$. Indeed, assuming the contrary, it follows that f has a local maximum point $x_0 \geq a$ such that $f(x_0) \geq 1$,

5.5 Differential Equations and Inequalities

$f'(x_0) = 0$, and $f''(x_0) \leq 0$. Using now the differential equation satisfied by f, we get a contradiction.

In particular, the above claim shows that $f'' \geq f$ in $[a, \infty)$. Set $g := f' - f$. Then $g' + g \geq 0$ in $[a, \infty)$ and the function $h(x) := g(x)e^x$ satisfies $h' \geq 0$ in $[a, \infty)$. We deduce that for all $x \geq a$ we have $g(x) = f'(x) - f(x) \geq g(a)e^{a-x}$. Setting $v(x) := f(x)e^{-x}$, we obtain $v'(x) \geq g(a)e^{a-2x}$ on $[a, \infty)$. By integration on $[a, x]$, we obtain, for all $x \geq a$,

$$f(x) \geq f(a)e^{x-a} - \frac{g(a)}{2}e^{a-x} + \frac{g(a)}{2}e^{x-a}$$
$$= f(a)e^{x-a} + \frac{f(a) - f'(a)}{2}e^{a-x} + \frac{f'(a) - f(a)}{2}e^{x-a} \quad (5.28)$$
$$= \frac{f(a) + f'(a)}{2}e^{x-a} + \frac{f(a) - f'(a)}{2}e^{a-x}.$$

This shows that there exists a positive constant C such that $f(x) \geq Ce^x$, for any $x \geq a$.

(b) Our hypothesis implies $f'' > f$ in $[a, \infty)$. However, since $f(a) + f'(a)$ is not necessarily positive, estimate (5.28) does not conclude the proof, as above. For this purpose, using the fact that $f'' > 1$ in (a, ∞), we find some $x_0 > a$ such that $f'(x_0) > 0$. Since f is positive in $[x_0, \infty)$, we can repeat the arguments provided in (a), using x_0 instead of a in relation (5.28). Thus, we find $C > 0$ such that $f(x) \geq Ce^x$, for any $x \geq x_0$. Choosing eventually a smaller positive constant C, we deduce that the same conclusion holds in $[a, \infty)$.

(c) Make the change of variable $e^{-x} = t \in (0, e^{-A})$ and set $g(t) = f(x)$. Then g satisfies the differential equation

$$g''(t) + \frac{g'(t)}{t} - \frac{g(t)}{t^2} = \frac{\varphi(-\ln t)}{t^2} g(t)\left(g^2(t) - 1\right), \quad \text{for all } t \in (0, e^{-A}). \quad (5.29)$$

We observe that the above equation is equivalent to the first-order differential system

$$\begin{cases} g'(t) + \frac{g(t)}{t} = h(t), & t \in (0, e^{-A}), \\ h'(t) = \frac{\varphi(-\ln t)}{t^2} g(t)\left(g^2(t) - 1\right), & t \in (0, e^{-A}). \end{cases} \quad (5.30)$$

The growth assumption on φ can be written, equivalently, $\varphi(-\ln t) = O\left(t^{1+\alpha}\right)$ as $t \to 0$, where α is a positive number. This implies that the right-hand side of the second differential equation in (5.30) is integrable around the origin, and moreover,

$$h(t) = O(1) \quad \text{as } t \searrow 0. \quad (5.31)$$

On the other hand, since g is bounded around the origin, the first differential equation in (5.30) implies

$$g(t) = \frac{1}{t}\int_0^t rh(r)\,dr, \quad \text{for all } 0 < t < e^{-A}. \quad (5.32)$$

Relations (5.31) and (5.32) imply that $g(t) = O(t)$ as $t \searrow 0$. Since $tg'(t) + g(t) = th(t)$, we deduce that

$$g'(t) = O(1) \quad \text{as } t \searrow 0. \tag{5.33}$$

Let g and g_1 be two arbitrary solutions of (5.29). Then

$$\left\{t\left[g'(t)g_1(t) - g(t)g_1'(t)\right]\right\}' = \frac{\varphi(-\ln t)}{t} g(t)g_1(t)\left(g^2(t) - g_1^2(t)\right), \quad t \in (0, e^{-A}). \tag{5.34}$$

Relation (5.33) and the growth assumption on φ imply that the right-hand side of (5.34) is $O(t^{4+\alpha})$ as $t \searrow 0$. So, using again (5.34),

$$g'(t)g_1(t) - g(t)g_1'(t) = O(t^{4+\alpha}) \quad \text{as } t \searrow 0. \tag{5.35}$$

Next, we observe that we can choose g_1 such that $g_1(t) \sim t$ as $t \searrow 0$. Indeed, this follows from the fact that the initial value problem

$$\begin{cases} g''(t) + \frac{g'(t)}{t} - \frac{g(t)}{t^2} = \frac{\varphi(-\ln t)}{t^2} g(t)\left(g^2(t) - 1\right), \\ g(0) = 0, \ g'(0) = 1, \end{cases}$$

has a solution defined on some interval $(0, \delta)$. Thus, by (5.35), $\lim_{t \searrow 0} (g(t)/g_1(t))' = 0$. Hence, for any sequence $\{t_n\}_{n \geq 1}$ of positive numbers converging to 0, the sequence $\{g(t_n)/g_1(t_n)\}_{n \geq 1}$ is a Cauchy sequence. So, there exists $\lim_{t \searrow 0} g(t)/g_1(t) = \ell \in \mathbb{R}$. Since $g_1(t) \sim t$ as $t \searrow 0$, we deduce that $\lim_{t \searrow 0} g(t)/t = \ell$ or, equivalently, $\lim_{x \to \infty} e^x f(x) = \ell$. □

Remarks. (i) The conclusion stated in (c) does not remain true for general potentials φ (as in (a) or (b)). Indeed, the function $f(x) = x^{-1}$ satisfies the assumption $f(x) \in (-1, 1)$ for all $x \in (1, \infty)$ and is a solution of the differential equation $f'' = [1 + \varphi(f^2 - 1)]f$, provided that $\varphi(x) = (x^2 - 2)(x^2 - 1)^{-1}$. In this case, $\lim_{x \to \infty} e^x f(x)$ exists but is not finite.

(ii) Under the growth assumption on φ imposed in (c), our result shows that an arbitrary solution f of the differential equation $f'' = [1 + \varphi(f^2 - 1)]f$ in (A, ∞) satisfies the following alternative: either

(i) f is unbounded and, in this case, $f(x)$ tends to $+\infty$ as $x \to +\infty$ (at least like e^x, for general positive potentials φ)

or

(ii) f is bounded, and in this case, $f(x)$ tends to 0 as $x \to +\infty$ (at least like e^{-x}).

A uniqueness result related to the radial Ginzburg–Landau equation is the following.

5.5 Differential Equations and Inequalities

5.5.9. Assume that there exists a smooth positive function f on $(0,1)$ satisfying the differential equation

$$-f'' - \frac{f'}{r} + \frac{f}{r^2} = f(1-f^2) \quad \text{in } (0,1)$$

together with with boundary conditions $f(0) = 0$ and $f(1) = 1$. Prove that f is unique.

Solution. We give an argument of I. Shafrir [6] that is based on a method introduced by H. Brezis and L. Oswald[12]. Let f_1 and f_2 be two positive functions satisfying the hypotheses. Dividing the differential equation by f and subtracting the corresponding equations, we obtain

$$-\frac{f_1''}{f_1} + \frac{f_2''}{f_2} - \frac{1}{r}\left(\frac{f_1'}{f_1} - \frac{f_2'}{f_2}\right) = -(f_1^2 - f_2^2) \quad \text{in } (0,1).$$

Multiplying the above equality by $r(f_1^2 - f_2^2)$ and integrating over $(0,1)$ yields

$$\int_0^1 \left(f_1' - \frac{f_2}{f_1}f_2'\right)^2 r\,dr + \int_0^1 \left(f_2' - \frac{f_1}{f_2}f_1'\right)^2 r\,dr = -\int_0^1 (f_1^2 - f_2^2)^2 r\,dr.$$

Therefore $f_1 = f_2$ on $(0,1)$. □

The next property is very useful for proving uniqueness results.

5.5.10. Let $f, g : [a,b] \to \mathbb{R}$ be functions such that f is continuous, g is differentiable, and g vanishes at least once. Assume that there exists a real constant $\lambda \neq 0$ such that for all $x \in [a,b]$, $|f(x)g(x) + \lambda g'(x)| \leq |g(x)|$. Prove that $g = 0$ in $[a,b]$.

Solution. Arguing by contradiction, there exists $x_0 \in (a,b)$ such that $g(x_0) > 0$. Let $V = (c,d) \subset [a,b]$ be a neighborhood of x_0 such that $g > 0$ on V. Since g vanishes at least once on $[a,b]$, we can assume that $g(c) = 0$.

For any $x \in V$ we have, by hypothesis,

$$\left|\frac{g'(x)}{g(x)}\right| = \frac{1}{|\lambda|}\left|\left(f(x) + \lambda\frac{g'(x)}{g(x)}\right) - f(x)\right| \leq \frac{1}{|\lambda|}\left(1 + \sup_{x \in [a,b]} |f(x)|\right).$$

It follows that the mapping $V \ni x \longmapsto \ln g(x)$ is bounded. Observing that $g(x) \to -\infty$ as $x \searrow c$, we obtain the desired contradiction.

The above property implies the following classical result, which is very useful to establish the uniqueness of the solution in the theory of differential equations. □

Gronwall's lemma. Let $I \subset \mathbb{R}$ be an interval. Assume that $f : I \to \mathbb{R}$ is a differentiable function that does not have a constant sign on I and satisfying, for all $x \in I$, $|f'(x)| \leq C|f(x)|$, where C is a positive constant. Then $f = 0$ on I.

The following second-order differential equation with cubic nonlinearity has only bounded solutions.

5.5.11. *Consider the differential equation $x''(t) + a(t)x^3(t) = 0$ on $0 \leq t < \infty$, where $a(t)$ is continuously differentiable and $a(t) \geq \kappa > 0$. If $a'(t)$ has only finitely many changes of sign, prove that any solution $x(t)$ is bounded.*

Ph. Korman, Math. Magazine, Problem 1577

Solution. For any $t \geq 0$, define the function

$$E(t) = \frac{1}{2}\left(x'(t)\right)^2 + a(t)\frac{x^4(t)}{4}.$$

Using the differential equation, we obtain

$$E'(t) = a'(t)\frac{x^4(t)}{4}. \tag{5.36}$$

Assuming that $a'(t) \leq 0$ for all $t \in [t_1, t_2]$, then $E'(t) \leq 0$ on $[t_1, t_2]$. Thus, $E(t) \leq E(t_1)$ for all $t \in [t_1, t_2]$. If $a'(t) \geq 0$ on $[t_1, t_2]$, then $E'(t)/E(t) \leq a'(t)/a(t)$, for all $t_1 \leq t \leq t_2$. Hence

$$E(t) \leq \frac{E(t_1)}{a(t_1)}a(t) \leq \frac{a(t_2)}{a(t_1)}E(t_1), \quad \text{for all } t_1 \leq t \leq t_2. \tag{5.37}$$

This gives an idea about the growth rate of E. More precisely, $E(t)$ can increase by at most a factor of $a(t_2)/a(t_1)$ on $[t_1, t_2]$. Relation (5.37) combined with the definition of $E(t)$ yields

$$\frac{x^4(t)}{4} \leq \frac{E(t_1)}{a(t_1)}, \quad \text{for all } t_1 \leq t \leq t_2. \tag{5.38}$$

Let us now assume that $a'(t)$ changes sign at points $c_1, c_2, \ldots, c_n$. Since $E(t)$ is nonnegative and nonincreasing on any interval on which $a'(t) \leq 0$, and increases by at most a factor $a(c_{k+1})/a(c_k)$ on any interval $[c_k, c_{k+1}]$ on which $a'(t) \geq 0$, it follows that $E(t)$ remains bounded on $[0, c_n]$. We distinguish two cases:

(i) $a'(t) \leq 0$ on $[c_n, \infty)$. Then $E'(t) \leq 0$ for all $t \in [c_n, \infty)$. Thus, $E(t)$ is nonincreasing on $[c_n, \infty)$, so remains bounded. This implies that $x(t)$ is bounded on $[0, \infty)$.

(ii) $a'(t) \geq 0$ on (c_n, ∞). Thus, by (5.38), it follows that $x^4(t) \leq 4E(c_n)/a(c_n)$, for all $t \geq c_n$. Therefore $x(t)$ is bounded.

The next problem concerns the differential equation

$$f(x+h) - f(x-h) = 2hf'(x), \tag{5.39}$$

where h is a positive number (either fixed or variable) and $x \in \mathbb{R}$. This equation appears in the study of central forces and it is called the *gravity equation*.

5.5 Differential Equations and Inequalities

Archimedes observed that any quadratic polynomial satisfies (5.39), and he gave the following geometric interpretation of this fact. Select any two points A and B on a parabola and let C be the point of the parabola at which the tangent to the parabola is parallel to the chord AB. Then the line through C parallel to the axis of the parabola is midway between the lines through A and B parallel to the axis. □

We also observe that if $V(x)$ is a potential function associated with a radially symmetric central force and if a sphere of radius h attracts exterior particles as if all the mass of the sphere were at its center, then the function $f(x) := (xV(x))'$ satisfies the differential equation (5.39). The next result gives a sufficient condition to ensure that a solution of (5.39) has at most exponential growth.

5.5.12. *Let $f(x)$ satisfy (5.39) for all x and for two values of h, say for $h = a$ and $h = b$ with $0 < a < b$. Then there exist positive constants A and c such that $|f(x)| < Ae^{c|x|}$, for all real x.*

Sherman Stein

Solution. We assume for the sake of notational simplicity that $x \geq 0$ (a similar argument can be applied for $x < 0$). Since (5.39) is fulfilled for $h = a$ and $h = b$, we obtain

$$f(x+b) = f(x-b) + \frac{b}{a}[f(x+a) - f(x-a)].$$

Replacing x with $y - a$ in the above relation, we obtain

$$f(y+b-a) = f(y-b-a) + \frac{b}{a}[f(y) - f(y-2a)].$$

This identity relates $f(y+b-a)$ to values $f(x)$ for x no larger than y.

Set $M(x) := \max\{f(t); 0 \leq t \leq x\}$. It follows that for all $y \geq a+b$,

$$M(y+b-a) \leq M(y) + \frac{2b}{a}M(y).$$

Define $M := \max\{M(x); 0 \leq x \leq a+b\}$. Then, considering $M(x)$ over successive intervals of length $b-a$, we obtain

$$M(x) \leq A\left(1 + \frac{2b}{a}\right)^{x/(b-a)},$$

which concludes the proof. □

If f is twice differentiable on an interval I and $f'' = 0$ on I, then f is linear. Thus, $f \equiv 0$ in I, provided f has at least two zeros in I. This simple property is extended in the next exercise to a larger class of linear differential equations.

5.5.13. *Let p be a continuous real-valued function on $\mathbb{R}$ and let f be a solution of the differential equation*

$$f''(x) + p(x)f'(x) - f(x) = 0.$$

Prove that if f has more than one zero, then $f(x) \equiv 0$.

Solution. Let $a < b$ be two different zeros of f. Let x_M and x_m be two numbers in the interval $[a,b]$ for which $f(x_M)$ and $f(x_m)$ are the greatest and the smallest value of f in this interval respectively. Assume that f is not identically zero in the interval $[a,b]$. Then at least one of the numbers $f(x_M)$ and $f(x_m)$ is not 0.

If $f(x_M) \neq 0$ then $f(x_M) > 0$ and $a < x_M < b$. Moreover, since $(x_M, f(x_M))$ is a local maximum, then $f'(x_M) = 0$ and $f''(x_M) \leq 0$. But then

$$f''(x_M) + p(x_M)f'(x_M) - f(x_M) < 0,$$

which contradicts the definition of f.

If $f(x_m) \neq 0$ then $f(x_m) < 0$ and $a < x_m < b$. Moreover, since $(x_m, f(x_m))$ is a local minimum, $f'(x_m) = 0$ and $f''(x_m) \geq 0$. But then

$$f''(x_m) + p(x_m)f'(x_m) - f(x_m) > 0,$$

which contradicts the definition of f.

Thus f must be identically 0 in the interval $[a,b]$.

Let c be a number such that $a < c < b$. Then clearly $f(c) = f'(c) = 0$. Thus the problem follows from the theorem on the uniqueness of the solution of the second-order linear differential equation. □

5.6 Independent Study Problems

> All intelligent thoughts have already been thought; what is necessary is only to try to think them again.
>
> Johann Wolfgang von Goethe
> (1749–1832)

5.6.1. Let $f : (a,b) \to \mathbb{R}$ be a function having finite one-sided derivatives at all points in (a,b). Prove that f is differentiable in (a,b), excepting a set that is at most countable.

5.6.2. Let $f : [a,b] \to \mathbb{R}$ be a differentiable function in (a,b). Prove that there exists a nonconstant sequence of points $(a_n)_{n \geq 1} \subset (a,b)$ such that the sequence $(f(a_n))_{n \geq 1}$ is convergent.

5.6.3. Let $f : (a,b) \to \mathbb{R}$ be a differentiable function such that there exists $A > 0$ with $|f'(x)| \leq A$, for all $x \in (a,b)$. Prove that $\lim_{x \searrow a} f(x)$ and $\lim_{x \nearrow b} f(x)$ exist, and moreover,

$$\left| \lim_{x \searrow a} f(x) - f(c) \right| \leq A(c-a), \quad \left| \lim_{x \nearrow b} f(x) - f(c) \right| \leq A(b-c), \quad \text{for all } c \in (a,b).$$

5.6 Independent Study Problems

5.6.4. Let $f : \mathbb{R} \to \mathbb{R}$ be a differentiable function such that there exists $c > 0$ with
$$|f'(x)| \leq cf(x), \quad \text{for all } x \in \mathbb{R}.$$

(a) Establish the existence of some $\gamma \geq 0$ such that $|f(x)| \leq \gamma e^{c|x|}$, for all $x \in \mathbb{R}$.
(b) In addition, we assume that there exists $x_0 \in \mathbb{R}$ such that $f(x_0) = 0$. Prove that $f \equiv 0$.

5.6.5. Let $f : (a,b) \to \mathbb{R}$ be a differentiable function and let $x_0 \in (a,b)$ be such that f' is continuous at x_0. Let $(x_n)_{n \geq 1}$ and $(y_n)_{n \geq 1}$ be sequences in (a,b) such that $x_n \neq y_n$ for all $n \geq 1$ and $\lim_{n \to \infty} x_n = \lim_{n \to \infty} y_n = x_0$. Prove that
$$\lim_{n \to \infty} \frac{f(x_n) - f(y_n)}{x_n - y_n} = f'(x_0).$$

5.6.6. (Clarkson's inequalities). Using perhaps the above result, show that for all $a, b \in \mathbb{R}$, $p \geq 2$, $1/p + 1/q = 1$, the following inequalities hold:

(a) $2(|a|^p + |b|^p) \leq |a+b|^p + |a-b|^p \leq 2^{p-1}(|a|^p + |b|^p)$;
(b) $|a+b|^p + |a-b|^p \leq 2(|a|^q + |b|^q)^{p/q}$;
(c) $2(|a|^p + |b|^p)^{q/p} \leq |a+b|^q + |a-b|^q$.

If $1 \leq p < 2$, prove that the above inequalities hold with $\geq$ instead of $\leq$.

5.6.7. Let $f : \mathbb{R} \to [0, \infty)$ be a function of class C^1, periodic of period 1. Prove or disprove that
$$\lim_{c \to \infty} \left(\frac{f(x)}{1 + cf(x)} \right)' = 0.$$

Hint. The periodic continuous functions f and f' are bounded.

5.6.8. Let $f : (0, \infty) \to (0, \infty)$ be a function of class C^2 such that $f' \leq 0$ and f'' is bounded. Prove that $\lim_{x \to \infty} f'(x) = 0$.

5.6.9. For any integer $n \geq 0$ define the polynomial $f_n(x)$ by $f_0(x) = 1$, $f_n(0) = 0$ for $n \geq 1$ and
$$f'_{n+1}(x) = (n+1)f_n(x+1), \quad \text{for all } n \geq 0.$$
Find the decomposition into prime factors of $f_{100}(1)$.

5.6.10. Assume that the function $f : \mathbb{R} \to \mathbb{R}$ is twice differentiable and satisfies
$$f''(x) - 2f'(x) + f(x) = 2e^x, \quad \text{for all } x \in \mathbb{R}.$$

(a) If $f(x) > 0$ for all x, is it true that $f'(x) > 0$ for every x?
(b) If $f'(x) > 0$ for all x, is it true that $f(x) > 0$ for every x?

5.6.11. A frequent error is related to the derivative rule for the product of two functions, and it concerns the "formula" $(fg)' = f'g'$. If $f(x) = e^{x^2}$, establish whether there exists an open interval (a,b) and a differentiable function $g:(a,b)\to\mathbb{R}$, $g\neq 0$, such that the above "formula" is true for all $x \in (a,b)$.

5.6.12. Find all functions $f:\mathbb{R}\to\mathbb{R}$ of class C^1 such that

$$f^2(x) = \int_0^x [f^2(t) + f'^2(t)]dt + 2004, \quad \text{for all } x \in \mathbb{R}.$$

5.6.13. Let $f:\mathbb{R}\to\mathbb{R}$ be an infinitely differentiable function such that

$$f\left(\frac{1}{n}\right) = \frac{n^2}{n^2+1}, \quad n = 1, 2, \ldots.$$

Find $f^{(k)}(0)$, for every $k = 1, 2, \ldots$.

5.6.14. Let $p(x)$ be a nontrivial polynomial of degree less than 2004 that has no nonconstant factor in common with the polynomial $x^3 - x$. Let

$$\frac{d^{2004}}{dx^{2004}}\left(\frac{p(x)}{x^3-x}\right) = \frac{f(x)}{g(x)},$$

where $f(x)$ and $g(x)$ are polynomials. Find the least possible degree of $f(x)$.

5.6.15. Let $f:\mathbb{R}\to(0,\infty)$ be a differentiable function such that $f'(x) > f(x)$, for all $x \in \mathbb{R}$. Find the values of k for which there exists N such that $f(x) > e^{kx}$, for any $x > N$.

5.6.16. Let $f(x) = (x-x_1)\cdots(x-x_n)$, $x_i \neq x_j$ for $i \neq j$, $g(x) = x^{n-1} + a_{n-2}x^{n-2} + \cdots + a_0$. Prove that $\sum_{j=1}^n g(x_j)/f'(x_j) = 1$.
Hint. Decompose g/f, multiply by x, and pass to the limit as $x\to\infty$.

5.6.17. Let $a_1 < a_2 < \cdots < a_n$ be the roots of a polynomial of degree n and let $b_1 \leq \cdots \leq b_{n-1}$ be the roots of its derivative. Compute $\sum_{i,j}(b_i - a_j)^{-1}$.
Answer. The sum is 0 if $a_i \neq a_j$ for all $i \neq j$, and $+\infty$ elsewhere.

5.6.18. Suppose a and b are real numbers, $b > 0$, and f is defined on $[-1,1]$ by

$$f(x) = \begin{cases} x^a \sin(x^{-b}) & \text{if } x \neq 0, \\ 0 & \text{if } x = 0. \end{cases}$$

Prove the following statements:

(i) f is continuous if and only if $a > 0$;
(ii) $f'(0)$ exists if and only if $a > 1$;
(iii) f' is bounded if and only if $a \geq 1+b$;
(iv) f' is continuous if and only if $a > 1+b$;
(v) $f''(0)$ exists if and only if $a > 2+b$;

5.6 Independent Study Problems

(vi) f'' is bounded if and only if $a \geq 2 + 2b$;
(vii) f'' is continuous if and only if $a > 2 + 2b$.

5.6.19. Let $I \subset \mathbb{R}$ be an open interval and let $f, g : I \to \mathbb{R}$ be differentiable functions. Prove that between two consecutive zeros of f there is at least one zero of $f' + fg'$.
Hint. Consider the function fe^g.

5.6.20. Let $f : \mathbb{R} \to \mathbb{R}$ be a twice differentiable function in $\mathbb{R} \setminus \{0\}$ satisfying $f'(x) < 0 < f''(x)$ if $x < 0$ and $f'(x) > 0 > f''(x)$ for all $x > 0$. Prove that f is not differentiable at $x = 0$.

5.6.21. Given a twice differentiable function $f : (a,b) \to \mathbb{R}$ and $x \in (a,b)$, do there exist $a < x_1 < x < x_2 < b$ such that $f(x_2) - f(x_1) = f'(x)(x_2 - x_1)$?
The same question if we assume that $f'' \neq 0$ in (a,b).
Hint. No, provided that f' has a strict local extremum in x; yes, if f' is monotone.

5.6.22. Let $I \subset \mathbb{R}$ be an interval and $f : I \to \mathbb{R}$. Prove that the following conditions are equivalent:

(a) $\left| \det \begin{pmatrix} f(u) & f(v) & f(w) \\ u & v & w \\ 1 & 1 & 1 \end{pmatrix} \right| \leq |(u-v)(v-w)(w-u)|$, for all $u, v, w \in I$.

(b) f is differentiable and f' satisfies the Lipschitz condition with constant 2.

5.6.23. Let $f : [a,b] \to \mathbb{R}$ be a differentiable function with first derivative continuous and decreasing and satisfying $f'(b) > 0$. Define the sequence (b_n) by

$$b_0 = b, \quad b_n = f^{-1}(f(b_{n-1}) - f'(b_{n-1})(b_{n-1} - a)), \quad \text{for all } n \geq 1.$$

Prove that $f(b_n)$ converges to $f(a)$.

5.6.24. Assume that the function $f : [a,b] \to \mathbb{R}$ satisfies $f(a) = f(b) = 0$, $f(x) > 0$, for all $x \in (a,b)$ and $f + f'' > 0$. Prove that $b - a \geq \pi$.
Hint. Compare f with a function of the form $c \sin(x - a)$, for c sufficiently small.

5.6.25. Prove that if $f \in C^2(\mathbb{R})$ satisfies for all $x \in \mathbb{R}$ the relations $f^2(x) \leq 1$ and $[f'(x)]^2 + [f''(x)]^2 \leq 1$, then $f^2(x) + [f'(x)]^2 \leq 1$, for any $x \in \mathbb{R}$.

5.6.26. (**Hardy**) Let $f : [0, \infty) \to \mathbb{R}$ be a function satisfying $\lim_{x \to \infty} [f(x) + f'(x)] = 0$. Prove that $\lim_{x \to \infty} f(x) = 0$.
Hint. Apply Cauchy's mean value theorem to the functions $e^x f(x)$ and e^x.

5.6.27. Let $f : [0, \infty) \to \mathbb{R}$ be a function of class C^2 such that

$$\lim_{x \to \infty} [f(x) + f'(x) + f''(x)] = C.$$

Prove that $\lim_{x \to \infty} f(x) = C$.

5.6.28. (**Hardy–Littlewood**) Let $f : [0, \infty) \to \mathbb{R}$ be a twice differentiable function such that $\lim_{x \to \infty} f(x) = 0$ and f'' is bounded. Prove that $\lim_{x \to \infty} f'(x) = 0$.

5.6.29. Let $f \in C^2[a,b]$ be a function that is three times differentiable in (a,b). Prove that there exists $c \in (a,b)$ such that

$$\begin{vmatrix} f(b) & b^2 & b & 1 \\ f(a) & a^2 & a & 1 \\ f'(a) & 2a & 1 & 0 \\ f''(a) & 2 & 0 & 0 \end{vmatrix} = -f'''(c) \frac{(b-a)^3}{3}.$$

5.6.30. Let $f : \mathbb{R} \to \mathbb{R}$ be a function such that f^2 and f^3 are differentiable. Is f differentiable?

Answer. No. Consider the function f defined by $f(x) = x^{2/3} \sin(1/x)$ for $x \neq 0$ and $f(0) = 0$. Then $f(x)^2 = O(x^{4/3}) = o(x)$ as $x \to 0$, so f^2 is differentiable at 0 with derivative 0 and similarly for f^3. Note that f^2 and f^3 are not continuously differentiable at 0.

5.6.31. Assume that the function $u : (0,1) \to \mathbb{R}$ satisfies

$$u''(x) + e^{u(x)} = -x, \quad \text{for all } x \in (0,1).$$

Prove that u does not achieve its minimum in $(0,1)$.

5.6.32. Assume that the function $u : (0,1) \to \mathbb{R}$ satisfies

$$u''(x) - 2\cos(u'(x)) = 1, \quad \text{for all } x \in (0,1).$$

Prove that u does not have any local maximum point in $(0,1)$.

5.6.33. Assume that the function $u : (0,1) \to \mathbb{R}$ satisfies $u(0) = u(1) = 0$ and

$$u''(x) + e^x u'(x) = -1, \quad \text{for all } x \in (0,1).$$

(i) Prove that u does not achieve its minimum in $(0,1)$.
(ii) Show that $u'(0) > 0$ and $u'(1) < 0$.

5.6.34. Assume that the function $u : \mathbb{R} \to \mathbb{R}$ satisfies

$$u''(x) + e^{u(x)} = e, \quad \text{for all } x \in \mathbb{R}.$$

Prove that u cannot have a minimum value greater than 1 or a maximum value less than 1.

5.6.35. Prove the maximum principle (stated in problem 5.3.2) using the auxiliary function $\varphi(x) = (x-a)^\alpha - (c-a)^\alpha$ instead of $\varphi(x) = e^{\alpha(x-c)} - 1$.

5.6.36. Let $f, g : (a,b) \to \mathbb{R}$ be bounded functions on every closed subinterval of (a,b) and assume that $f \leq 0$ in (a,b). Suppose that u satisfies the differential inequality $u''(x) + f(x)u'(x) + g(x)u(x) \geq 0$, for any $x \in (a,b)$. In addition, we assume that u achieves a maximum value $M \geq 0$ at an interior point of (a,b). Show that $u \equiv M$ in (a,b).

Hint. Apply the same ideas as in the proof of problem 5.3.2.

5.6.37. Let $f, g : (a,b) \to \mathbb{R}$ be bounded functions on every closed subinterval of (a,b) and suppose that $f \leq 0$ in (a,b). Assume that u satisfies the differential inequality $u''(x) + f(x)u'(x) + g(x)u(x) \geq 0$, for any $x \in (a,b)$. In addition, we assume that u is nonconstant, has a nonnegative maximum in a, and that the function $f(x) - (x-a)g(x)$ is bounded below near $x = a$.

Prove that $u'(a+) < 0$.

If u has a nonnegative maximum at b and the function $f(x) - (b-x)g(x)$ is bounded above near $x = b$, prove that $u'(b-) > 0$.

Hint. Apply the same ideas as in the solution of problem 5.3.3.

5.6.38. Do there exist nonlinear functions $f : \mathbb{R} \to \mathbb{R}$ of class C^1 such that $f(x)$ is rational for every rational number x and $f(x)$ is irrational for every irrational number x?

Hint. Consider the function $f(x) = x/(1+|x|)$.

5.6.39. Let $f : \mathbb{R} \to \mathbb{R}$ be a differentiable function such that $A := \{x \in \mathbb{R};\ f'(x) > 0\}$ is dense in $\mathbb{R}$. Is f an increasing function?

5.6.40. Prove that there does not exist any function $f : [0,2] \to \mathbb{R}$ such that $f(0) = 0$ and, for all $x \in [0,2]$, $f'(x) - f^4(x) = 1$.

5.6.41. Compute
$$\lim_{x \to 0} \frac{\sin(\tan x) - \tan(\sin x)}{\arcsin(\arctan x) - \arctan(\arcsin x)}.$$

5.6.42. (**Laguerre's theorem**) Let $a_1 < \cdots < a_n$ be the roots of a polynomial of degree n and let $b_1 < \cdots < b_{n-1}$ be the roots of its derivative. Prove that b_j does not belong to the interval containing a_j and obtained by dividing $[a_j, a_{j+1}]$ into n equal intervals.

Hint. The roots of the derivative satisfy $(x-a_1)^{-1} + \cdots + (x-a_n)^{-1} = 0$.

5.6.43. Using perhaps Taylor's formula, compute the following limits:
$$\lim_{x \to 0} \frac{\cos x - e^{-x^2/2}}{x^4}; \quad \lim_{x \to 0} \frac{1}{x}\left(\frac{1}{x} - \frac{1}{\tan x}\right); \quad \lim_{x \to \infty}\left[x - x^2 \ln\left(1 + \frac{1}{x}\right)\right].$$

5.6.44. Let $f : \mathbb{R} \to \mathbb{R}$ be a function that is n times differentiable. Assume that there exists a positive constant L such that for all real numbers x and y, $\left|f^{(n)}(x) - f^{(n)}(y)\right| \leq L|x-y|$. Prove that if f is bounded, then all derivatives $f', f'', \ldots, f^{(n)}$ are bounded.

S. Rădulescu, I. Savu

5.6.45. Suppose $f : \mathbb{R} \to \mathbb{R}$ is a C^2 function such that for every $a, b \in \mathbb{R}$ with $a < b$, there exist a unique $\xi \in (a,b)$ such that $f'(\xi) = [f(b) - f(a)]/(b-a)$. Prove that f has no inflection points.

5.6.46. Let $f : [a,b] \to \mathbb{R}$ be a continuous function satisfying $f(a) = f(b)$. Prove that there are $c, d \in [a,b]$ such that $d - c = (b-a)/2$ and $f(c) = f(d)$. Deduce that for every $ep > 0$, there exist $x, y \in [a,b]$ such that $0 < y - x < \varepsilon$ and $f(x) = f(y)$.

5.6.47. (Subexponential Function) Let a be a real number and let f be a positive function defined on (a, ∞) [resp., on $(-\infty, a)$]. We say that f is subexponential if $f(x) = o(e^{\varepsilon x})$, as $x \to \infty$ [resp., $f(x) = o(-e^{\varepsilon x})$, as $x \to -\infty$], for all $\varepsilon > 0$.

(i) Show that if $\varepsilon > 0$ and if $f : (0, \infty) \to (0, \infty)$ satisfies $\lim_{x \to \infty} f(x) = \infty$ and $f'(x) = o(f(x))$ as $x \to \infty$, then $f(x)e^{-\varepsilon x}$ is decreasing in a neighborhood of $+\infty$. Deduce that f is subexponential.

(ii) Let $\langle x \rangle := \sqrt{1+x^2}$. Prove that $\exp(\langle x \rangle^\alpha)$ is subexponential for $\alpha < 1$.

(iii) Give an example of a nontrivial bounded function $f \in C^\infty(\mathbb{R})$ such that $f'(x) = o(f(x))$ as $|x| \to \infty$.

5.6.48. (Schwartzian Derivative) Let $f : I \to \mathbb{R}$ and assume that $f'''(x)$ exists and $f'(x) \neq 0$ for all $x \in I$. Define the Schwartzian derivative of f at x by

$$\mathscr{D}f(x) := \frac{f'''(x)}{f'(x)} - \frac{3}{2}\left[\frac{f''(x)}{f'(x)}\right]^2.$$

(i) Show that $\mathscr{D}(f \circ g) = (\mathscr{D}f \circ g) \cdot (g')^2 + \mathscr{D}g$.

(ii) Prove that if $f(x) = (ax+b)/(cx+d)$, then $\mathscr{D}f = 0$.

(iii) Show that $\mathscr{D}g = \mathscr{D}h$ if and only if $h = (ag+b)/(cg+d)$, where $ad - bc = 1$.

(iv) Prove that if $fg = 1$, then $\mathscr{D}f = \mathscr{D}g$.

5.6.49. Let $f : \mathbb{R} \to \mathbb{R}$ be differentiable and superlinear (in the sense of relation (3.2)). Prove that the max in the definition of

$$g(y) = \max_{x \in \mathbb{R}} [xy - f(x)]$$

is attained at a point x satisfying $f(x) = y$. Conclude that f' is surjective.

The following result (see Figure 5.15) describes Newton's method for computing the solution of wide classes of nonlinear equations.

5.6.50. Suppose f is twice differentiable on $[a,b]$, $f(a) < 0$, $f(b) > 0$, $f'(x) \geq \delta > 0$, and $0 \leq f''(x) \leq M$ for all $x \in [a,b]$. Let ξ be the unique point in (a,b) such that $f(\xi) = 0$. Choose $x_1 \in (\xi, b)$ and define the sequence $(x_n)_{n \geq 1}$ by

$$x_{n+1} = x_n - \frac{f(x_n)}{f'(x_n)} \quad \text{for all } n \geq 1.$$

(i) Prove that $x_{n+1} < x_n$ and that $\lim_{n \to \infty} x_n = \xi$.

5.6 Independent Study Problems

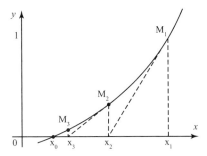

Fig. 5.15 Newton's sequence approaching the solution x_0 of an equation $f(x) = 0$.

(ii) Use Taylor's formula to show that

$$x_{n+1} - \xi = \frac{f''(x_n)}{2f'(t_n)}(x_n - \xi)^2,$$

for some $t_n \in (\xi, x_n)$.

(iii) If $A = M/(2\delta)$, deduce that

$$0 \leq x_{n+1} - \xi \leq \frac{1}{A}[A(x_1 - \xi)]^{2^n}.$$

(iv) Show that Newton's method amounts to finding a fixed point of the function g defined by

$$g(x) = x - \frac{f(x)}{f'(x)}.$$

How does $g'(x)$ behave for x near ξ?

Part III
Applications to Convex Functions and Optimization

The notion of convex function was introduced in the first part of the twentieth century, though it was implicitly used earlier by Gibbs and Maxwell in order to describe relationships between thermodynamic variables. In the next two chapters we are concerned with various properties of convex functions, one of the reasons being that among nonlinear functions, convex functions are in some sense the closest to linear.

Chapter 6
Convex Functions

> *Inequality is the cause of all local movements.*
> —Leonardo da Vinci (1452–1519)

Abstract. The convex functions form a special class of functions, defined on convex subsets of the real line (that is, intervals), and having a simple geometric property. Their importance in various fields of analysis is steadily growing. We are also concerned with two interesting themes, "convexity and continuity" and "convexity and differentiability."

6.1 Main Definitions and Basic Results

> All analysts spend half their time hunting through the literature for inequalities which they want to use and cannot prove.
>
> Harald Bohr (1887–1951)

Let $I \subset \mathbb{R}$ be an interval. A function $f : I \subset \mathbb{R}$ is said to be *convex* if for every $x, y \in I$ and any real number $\lambda \in [0, 1]$ (see Figure 6.1),

$$f(\lambda x + (1 - \lambda)y) \leq \lambda f(x) + (1 - \lambda)f(y).$$

If the above inequality is strict whenever $x \neq y$ and $\lambda \in (0, 1)$, then the function f is said to be *strictly convex*. The function f is said to be *concave* (resp., *strictly concave*) if $-f$ is *convex* (resp., *strictly convex*). Functions that are simultaneously convex and concave are said to be *affine*.

If $f : I \subset \mathbb{R}$ is twice differentiable, a useful test to establish the convexity of f is related to the sign of the second derivative of f. More precisely, if $f'' \geq 0$ (resp., if $f'' > 0$) on I, then f is convex (resp., strictly convex) on I.

Examples. (i) The function $f : [0, \infty) \to \mathbb{R}$ (where $n \in \mathbb{N}$) is convex, if $n \geq 2$ and it is affine if $n = 0$ or if $n = 1$.

(ii) The function $f(x) = e^x$ ($x \in \mathbb{R}$) is strictly convex, while the function $g(x) = \log x$, where $x \in (0, \infty)$, is strictly concave.

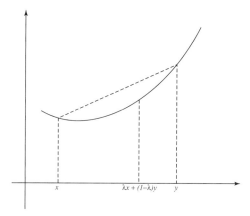

Fig. 6.1 Geometric illustration of a convex function.

A direct consequence (based on a recurrence argument) of the definition of a convex function is given in the next result, due to Johan Jensen (1859–1925), which has many applications in establishing various inequalities. As mentioned by Hardy, Littlewood, and Pólya [43], "this theorem is so fundamental that we propose to give a number of proofs, of varying degrees of simplicity and generality."

Theorem (Discrete Jensen Inequality). *Let f be a convex function defined on an interval $I \subset \mathbb{R}$. Assume that $x_1, x_2, \ldots, x_n \in I$ and let $\lambda_1, \lambda_2, \ldots, \lambda_n$ be nonnegative real numbers such that $\sum_{i=1}^{n} \lambda_i = 1$. Then*

$$f\left(\sum_{i=1}^{n} \lambda_i x_i\right) \leq \sum_{i=1}^{n} \lambda_i f(x_i). \tag{6.1}$$

Moreover, if f is strictly convex and equality holds in the above inequality for some $\lambda_1, \lambda_2, \ldots, \lambda_n$, all different from zero, then $x_1 = x_2 = \cdots = x_n$.

Examples. (i) **Arithmetic–Geometric Means Inequality.** Taking the convex function $f(x) = e^x$, inequality (6.1) yields

$$e^{\sum_{k=1}^{n} \lambda_k x_k} \leq \sum_{k=1}^{n} \lambda_k e^{x_k},$$

for all real numbers x_k ($1 \leq k \leq n$) and for any choice of nonnegative numbers λ_k ($1 \leq k \leq n$) with $\sum_{k=1}^{n} = 1$. Setting $a_k = e^{x_k}$, we obtain

$$a_1^{\lambda_1} t a_2^{\lambda_2} \cdots a_n^{\lambda_n} \leq \sum_{k=1}^{n} \lambda_k a_k, \tag{6.2}$$

for all $a_k > 0$ and $\lambda_k \geq 0$ with $\sum_{k=1}^{n} \lambda_k = 1$. Inequality (6.2) is sometimes referred to as the *generalized means inequality*.

Choosing $\lambda_1 = \cdots = \lambda_n = 1/n$ in (6.2), we obtain the *arithmetic–geometric means inequality*

$$(a_1 a_2 \cdots a_n)^{1/n} \leq \frac{1}{n} \sum_{k=1}^{n} a_k,$$

for any $a_1, a_2, \ldots, a_n \in \mathbb{R}_+$, with equality if and only if $a_1 = \cdots = a_n$.

(ii) **Young's Inequality**. In (6.2) we take $n = 2$, $a_1 = A$, $a_2 = B$, $\lambda_1 = 1/p$, $\lambda_2 = 1/q$, with $p, q > 0$ and $1/p + 1/q = 1$. Thus, we obtain

$$A^{1/p} B^{1/q} \leq \frac{A}{p} + \frac{B}{q},$$

which implies the following *Young's inequality* (William Young, 1863–1942):

$$ab \leq \varepsilon \frac{a^p}{p} + \frac{1}{\varepsilon} \frac{b^q}{q},$$

for all positive numbers a, b, and ε, where $1/p + 1/q = 1$.

The following version of Jensen's inequality is useful in combinatorics: let $f : [0, \infty) \to \mathbb{R}$ be a convex function and consider positive integers x_k ($1 \leq k \leq n$) such that $\sum_{k=1}^{n} x_k = s$. Then

$$\sum_{k=1}^{n} f(x_k) \geq r f(k+1) + (n-r) f(k),$$

where $s = nk + r$ and $0 \leq r < n$. In other words, if the sum of the integers is given, then the sum of functions is the least when the integers are closest to each other.

6.2 Basic Properties of Convex Functions and Applications

> Logic is the hygiene the mathematician practices to keep his ideas healthy and strong.
> —Hermann Weyl (1885–1955)

The set epi f defined in the next exercise is called the *epigraph* of the function f and is the set of points lying on or above the graph of f. This set plays an important role in optimization theory and is usually referred in economics as an *upper contour set*.

6.2.1. *Let f be a real function on the interval I. Prove that the following statements are equivalent:*

(i) *f is convex;*

(ii) for any $x, y, z \in I$ with $x < y < z$,

$$\frac{f(y) - f(x)}{y - x} \leq \frac{f(z) - f(x)}{z - x} \leq \frac{f(z) - f(y)}{z - y};$$

(iii) for any $a \in I$, the mapping

$$I \setminus \{a\} \ni t \longmapsto \frac{f(t) - f(a)}{t - a}$$

is nondecreasing;

(iv) for any $x, y, z \in I$ with $x < y < z$,

$$\begin{vmatrix} 1 & 1 & 1 \\ x & y & z \\ f(x) & f(y) & f(z) \end{vmatrix}$$

(v) the set $epi\, f = \{(x,y) \in \mathbb{R}^2;\ x \in I \text{ and } f(x) \leq y\}$ is convex.

Solution. $(i) \implies (ii)$ Fix $x, y, z \in I$ with $x < y < z$. There exists $\lambda \in (0,1)$ such that $y = \lambda x + (1 - \lambda) z$. Then

$$\lambda = \frac{y - z}{x - z} \quad \text{and} \quad 1 - \lambda = \frac{x - y}{x - z}.$$

Thus, since f is convex,

$$f(y) \leq \frac{y - z}{x - z} f(x) + \frac{x - y}{x - z} f(z).$$

Therefore

$$f(y) - f(x) \leq \frac{y - x}{x - z} f(x) + \frac{x - y}{x - z} f(z),$$

which implies

$$\frac{f(y) - f(x)}{y - x} \leq \frac{f(z) - f(x)}{z - x}.$$

The second inequality in (ii) is proved with a similar argument, and we leave the details to the reader.

$(ii) \implies (iii)$ Fix $t_1, t_2 \in I \setminus \{a\}$ with $t_1 < t_2$.
If $t_1 < t_2 < a$, we apply (ii) with $(x,y,z) = (t_1, t_2, a)$.
If $t_1 < a < t_2$, we apply (ii) with $(x,y,z) = (t_1, a, t_2)$.
If $a < t_1 < t_2$, we apply (ii) with $(x,y,z) = (a, t_1, t_2)$.
$(iii) \implies (iv)$ Assume that $x, y, z \in I$ and $x < y < z$.

6.2 Basic Properties of Convex Functions and Applications

Then

$$\begin{vmatrix} 1 & 1 & 1 \\ x & y & z \\ f(x) & f(y) & f(z) \end{vmatrix} = \begin{vmatrix} 1 & 0 & 0 \\ x & y-x & z-x \\ f(x) & f(y)-f(x) & f(z)-f(x) \end{vmatrix}$$

$$= (y-x)(z-x)\left(\frac{f(z)-f(x)}{z-x} - \frac{f(y)-f(x)}{y-x} \right) \geq 0.$$

$(iv) \Longrightarrow (i)$ Fix $x, y \in I$ and take $\lambda \in [0,1]$. Thus, by (iv),

$$D := \begin{vmatrix} 1 & 1 & 1 \\ x & \lambda x + (1-\lambda)y & y \\ f(x) & f(\lambda x + (1-\lambda)y) & f(y) \end{vmatrix} \geq 0.$$

But

$$D = \begin{vmatrix} 1 & 0 & 1 \\ x & 0 & y \\ f(x) & f(\lambda x + (1-\lambda)y) - \lambda f(x) - (1-\lambda)f(y) & f(y) \end{vmatrix}.$$

This implies $f(\lambda x + (1-\lambda)y) - \lambda f(x) - (1-\lambda)f(y) \leq 0$; hence f is convex.

$(i) \Longrightarrow (v)$ Fix $(x_1, y_1), (x_2, y_2) \in \mathbb{R}^2$ such that $x_1, x_2 \in I$ and $f(x_1) \leq y_1, f(x_2) \leq y_2$. We prove that if $\lambda \in [0,1]$ then $(\lambda x_1 + (1-\lambda)x_2, f(\lambda x_1 + (1-\lambda)x_2)) \in \mathrm{epi}\, f$. We first observe that since I is an interval and $x_1, x_2 \in I$, then $\lambda x_1 + (1-\lambda)x_2 \in I$. Next, since f is convex, we have

$$f(\lambda x_1 + (1-\lambda)x_2)) \leq \lambda f(x_1) + (1-\lambda)f(x_2),$$

which concludes the proof.

$(v) \Longrightarrow (i)$ Fix $x, y \in I$ and $\lambda \in [0,1]$. Since $\mathrm{epi}\, f$ is a convex set and $(x, f(x)) \in \mathrm{epi}\, f, (y, f(y)) \in \mathrm{epi}\, f$, we deduce that $(\lambda x + (1-\lambda)y, \lambda f(x) + (1-\lambda)f(y)) \in \mathrm{epi}\, f$. Therefore $f(\lambda x + (1-\lambda)y) \leq \lambda f(x) + (1-\lambda)f(y)$.

The property stated in Exercise 6.1.1 (ii) is also known as the *three chords lemma* (see Figure 6.2). In particular, this result establishes that if f is a real-valued function on the interval I, then for any $a, x, y \in I$ with $a < x < y$,

$$\frac{f(x) - f(a)}{x - a} \leq \frac{f(y) - f(a)}{y - a}.$$

Geometrically, this property asserts that the slope of a chord increases if the left end $(a, f(a))$ of the chord is fixed and the right end is moved to the right. □

6.2.2. Let $x_1, x_2, \ldots, x_n$ be positive real numbers. Show that

$$(x_1 x_2 \cdots x_n)^{\frac{x_1 + x_2 + \cdots + x_n}{n}} \leq x_1^{x_1} x_2^{x_2} \cdots x_n^{x_n}.$$

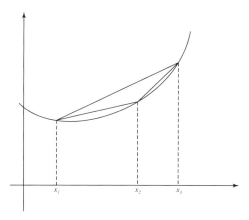

Fig. 6.2 A function is convex if and only if its sequential secants have increasing slopes.

Solution. The function $f(x) = x\ln x$, $x > 0$, is convex. Indeed, $f'(x) = 1 + \ln x$, $f''(x) = x^{-1} > 0$. So, by Jensen's inequality we obtain

$$\ln\left(\frac{x_1+\cdots+x_n}{n}\right)^{\frac{x_1+\cdots+x_n}{n}} \leq \ln\left(x_1^{x_1}\cdots x_n^{x_n}\right)^{\frac{1}{n}},$$

that is,

$$\left(\frac{x_1+\cdots+x_n}{n}\right)^{x_1+\cdots+x_n} \leq x_1^{x_1}\cdots x_n^{x_n}.$$

Applying now the mean value inequality, we obtain

$$(x_1\cdots x_n)^{\frac{x_1+\cdots+x_n}{n}} \leq \left(\frac{x_1+\cdots+x_n}{n}\right)^{x_1+\cdots+x_n} \leq x_1^{x_1}\cdots x_n^{x_n}. \quad \square$$

6.2.3. *Let I be an arbitrary interval and consider the function $f : I \to (0, \infty)$. Assume that the function $e^{cx} f(x)$ is convex in I, for any real number c. Prove that the function $\log f(x)$ is convex in I.*

Solution. Let $t \in [0, 1]$ and fix $x, y \in I$, $x \neq y$. Define

$$c = \frac{\log f(y) - \log f(x)}{x - y}.$$

By our hypothesis that the function $e^{cx} f(x)$ is convex we have

$$e^{c(tx+(1-t)y)} f(tx+(1-t)y) \leq te^{cx} f(x) + (1-t)e^{cy} f(y).$$

Hence

$$f(tx+(1-t)y) \leq te^{c(x-y)(1-t)} f(x) + (1-t)e^{-c(x-y)t} f(y)$$
$$= te^{(\log f(y)-\log f(x))(1-t)} f(x) + (1-t)e^{(\log f(x)-\log f(y))t} f(y).$$

6.2 Basic Properties of Convex Functions and Applications

$$= t \left(\frac{f(x)}{f(y)}\right)^{t-1} f(x) + (1-t) \left(\frac{f(x)}{f(y)}\right)^{t} f(y) = f(x)^{t} f(y)^{1-t}.$$

Passing to the logarithm, we deduce that the function $\log f$ is convex. □

6.2.4. Let $f : \mathbb{R} \to \mathbb{R}$ be a continuous function satisfying the mean value inequality

$$f(x) \leq \frac{1}{2h} \int_{x-h}^{x+h} f(y) dy, \quad \forall x \in \mathbb{R}, \ \forall h > 0.$$

Prove that the following properties hold:

(a) The maximum of f on any closed interval is achieved at one of the endpoints of the interval.
(b) The function f is convex.

Solution. (a) Let $[a,b]$ be an arbitrary closed interval and assume, by contradiction, that f achieves its maximum neither at a nor at b. Since f is continuous on the compact set $[a,b]$, there exists $c \in (a,b)$ such that $f(c) = \max_{x \in [a,b]} f(x)$. Using again the continuity of f, there exists $a < a_0 < c < b_0$ such that $f(x) < f(c)$, for all $x \in [a, a_0] \cup [b_0, b]$. Let $h = \min\{c - a, b - c\}$. We assume, without loss of generality, that $h = c - a$. By hypothesis and since $f(x) < f(c)$ for all $x \in [a, a_0] \subset [a, c+h]$ and $f \leq f(c)$ in $[a, c+h]$, it follows that

$$f(c) \leq \frac{1}{2h} \int_{a}^{c+h} f(t) dt < f(c).$$

This contradiction shows that f achieves its maximum either at a or at b.

(b) Let $a, b \in \mathbb{R}$, $a < b$. Define the mapping

$$L(x) = \frac{(x-a)f(b) - (x-b)f(a)}{b-a}$$

and consider the function $G(x) = f(x) - L(x)$. A simple computation shows that G satisfies the same mean value inequality as in the hypothesis. So, G achieves its maximum either at a or at b. But $G(a) = G(b) = 0$, so $f(x) \leq L(x)$ for all $x \in [a,b]$. Applying this result for $x = (1-t)a + tb \in [a,b]$ (where $t \in [0,1]$), we deduce that f is convex.

The *power mean* of two positive numbers a and b is defined by

$$M_p(a,b) := \left(\frac{a^p + b^p}{2}\right)^{1/p},$$

for any real number $p \neq 0$. We observe that $M_1(a,b)$ is the arithmetic mean of a and b, while $M_{-1}(a,b)$ coincides with the harmonic mean of a and b. Next, since $M_p(a,b) \longrightarrow \sqrt{ab}$ as $p \to 0$, we can adopt the convention that $M_0(a,b)$ is the geometric mean of a and b. We also point that for $p = 1/3$ one obtains the Lorentz mean

of a and b. The purpose of the next result is to prove a link between some power means $M_p(a,b)$ and the *logarithmic mean* of two distinct positive numbers a and b, which is defined by
$$L(a,b) := = \frac{a-b}{\ln a - \ln b}.$$ □

6.2.5. If $p \geq 1/3$ then $L(a,b) < M_p(a,b)$, for any distinct positive numbers a and b.

Solution. For any $x \geq 1$, consider the function
$$f(x) = \frac{3}{8}\ln x - \frac{x^3 - 1}{(x+1)^3}.$$

Then
$$f'(x) = \frac{3}{8} \frac{(x-1)^4}{x(x+1)^4}.$$

Hence $f''(x) > 0$ for all $x > 1$. Since $f(1) = 0$, we deduce that $f(x) > 0$ for $x \in (1, \infty)$. Now let us assume that $0 < b < a$ and apply $f(x) > 0$ for $x = a^{1/3}b^{-1/3}$. We thus obtain
$$\left(\frac{\sqrt[3]{a} + \sqrt[3]{b}}{2}\right)^3 > \frac{a-b}{\ln a - \ln b}.$$ □

6.2.6. Let $a_1, a_2, \ldots, a_n$ and $b_1, b_2, \ldots, a_n$ be nonnegative real numbers. Show that
$$\sqrt[n]{a_1 a_2 \cdots a_n} + \sqrt[n]{b_1 b_2 \cdots b_n} \leq \sqrt[n]{(a_1 + b_1)(a_2 + b_2)\cdots(a_n + b_n)}.$$

Putnam Competition, 2003

Solution. If $a_k = 0$, for some $1 \leq k \leq n$, then the inequality is trivial. Assuming that $a_k > 0$ for all k, we divide by $\sqrt[n]{a_1 a_2 \cdots a_n}$ and set $x_k = b_k/a_k$, for all $1 \leq k \leq n$. Thus, we obtain the equivalent inequality
$$1 + \sqrt[n]{x_1 x_2 \cdots x_n} \leq \sqrt[n]{(1+x_1)(1+x_2)\cdots(1+x_n)}.$$

Putting $x_k = e^{t_k}$ and taking logarithms of both sides of this inequality, we see that this is equivalent to proving that
$$\ln\left(1 + e^{(t_1 + \cdots + t_n)/n}\right) \leq \frac{1}{n}\sum_{k=1}^{n} \log(1 + e^{t_k}),$$

since "ln" is an increasing function. Setting $f(t) = \ln(1+e^t)$, we obtain $f''(t) = e^{2t}/(1+e^t)^2 \geq 0$; hence f is convex. The conclusion now follows by Jensen's inequality. □

6.2.7. Let $b_1, b_2, \ldots, b_n$ be an arbitrary permutation of the positive numbers $a_1, a_2, \ldots, a_n$. Prove that

(i) $a_1^{a_1} a_2^{a_2} \cdots a_n^{a_n} \geq b_1^{a_1} b_2^{a_2} \cdots b_n^{a_n}$;

(ii) $a_1^{1/b_1} a_2^{1/b_2} \cdots a_n^{1/b_n} \geq b_1^{1/b_1} b_2^{1/b_2} \cdots b_n^{1/b_n}$.

6.2 Basic Properties of Convex Functions and Applications

Solution. We first observe that

$$x - 1 \geq \ln x, \quad \text{for all } x > 0, \tag{6.3}$$

with equality if and only if $x = 1$. Indeed, taking $f(x) = \ln x - x + 1$, then $f'(x) = 1/x - 1$ and $f''(x) = -1/x^2$. It follows that $f(x)$ has an absolute maximum at $x = 1$, because $f'(x)$ vanishes if and only if $x = 1$ and $f'' < 0$.

(i) Taking $x = b_i/a_i$ in (6.3) we obtain, for all $1 \leq i \leq n$,

$$\frac{b_i}{a_i} - 1 \geq \ln \frac{b_i}{a_i}.$$

Multiplying by a_i and adding, we obtain

$$\sum_{i=1}^{n} b_i - \sum_{i=1}^{n} a_i \geq \sum_{i=1}^{n} \ln \left(\frac{b_i}{a_i}\right)^{a_i}.$$

Since $\sum_{i=1}^{n} b_i = \sum_{i=1}^{n} a_i$, it follows that

$$0 \geq \ln \left(\frac{b_1}{a_1}\right)^{a_1} \left(\frac{b_2}{a_2}\right)^{a_2} \cdots \left(\frac{b_n}{a_n}\right)^{a_n},$$

that is,

$$1 \geq \left(\frac{b_1}{a_1}\right)^{a_1} \left(\frac{b_2}{a_2}\right)^{a_2} \cdots \left(\frac{b_n}{a_n}\right)^{a_n}.$$

We also observe that equality occurs in the above inequality if and only if $a_i/b_i = 1$ for all $i = 1, 2, \ldots, n$.

(ii) Multiplying (6.3) by $1/b_i$ yields

$$\frac{1}{a_i} - \frac{1}{b_i} \geq \ln \left(\frac{b_i}{a_i}\right)^{1/b_i}.$$

Hence

$$\sum_{i=1}^{n} \frac{1}{a_i} - \sum_{i=1}^{n} \frac{1}{b_i} \geq \sum_{i=1}^{n} \ln \left(\frac{b_i}{a_i}\right)^{1/b_i}.$$

Since $\sum_{i=1}^{n} 1/a_i = \sum_{i=1}^{n} 1/b_i$, we obtain $0 \geq \sum_{i=1}^{n} \ln (b_i/a_i)^{1/b_i}$, which can be rewritten as

$$1 \geq \left(\frac{b_1}{a_1}\right)^{1/b_1} \left(\frac{b_2}{a_2}\right)^{1/b_2} \cdots \left(\frac{b_n}{a_n}\right)^{1/b_n},$$

which implies the conclusion. □

6.2.8. Find the greatest positive constant C such that

$$\sqrt{\frac{x}{y+z}} + \sqrt{\frac{y}{z+x}} + \sqrt{\frac{z}{x+y}} > C,$$

for all positive numbers x, y, and z.

Solution. We first observe that the inequality is homogeneous, that is, it remains unchanged if we replace (x,y,z) by $(\lambda x, \lambda y, \lambda z)$, for any $\lambda > 0$. This observation enables us to assume, without loss of generality, that $x+y+z = 1$. In this case we have

$$\sqrt{\frac{x}{y+z}} = \sqrt{\frac{x}{1-x}} \geq 2x,$$

with equality if and only if $x = 1/2$. Writing the similar inequalities corresponding to y and z, we deduce that

$$\sqrt{\frac{x}{y+z}} + \sqrt{\frac{y}{z+x}} + \sqrt{\frac{z}{x+y}} \geq 2(x+y+z) = 2.$$

However, in the above inequality, equality cannot hold. Why?

Next, it remains to show that 2 is, indeed, the best constant. For this purpose we fix $\varepsilon > 0$ small enough and take $x = y = 1/2 - \varepsilon$ and $z = 2\varepsilon$. Then $x+y+z = 1$ and

$$\sqrt{\frac{x}{y+z}} + \sqrt{\frac{y}{z+x}} + \sqrt{\frac{z}{x+y}} = 2\sqrt{\frac{1-2\varepsilon}{1+2\varepsilon}} + \sqrt{\frac{2\varepsilon}{1-2\varepsilon}} < 2 + \sqrt[4]{\varepsilon},$$

for any $\varepsilon > 0$ sufficiently small. This justifies that 2 is the greatest positive constant that can be chosen in our inequality. $\square$

6.2.9. *(a) Prove that if A, B, and C are the angles of a triangle then*

$$\frac{1}{\sin\frac{A}{2}} + \frac{1}{\sin\frac{B}{2}} + \frac{1}{\sin\frac{C}{2}} \geq 6.$$

(b) Let ABC be a triangle and let I be its incenter. Prove that at least one of the segments IA, IB, IC is greater than or equal to the diameter of the incircle of ABC.

Magkos Athanasios, Mathematical Reflections, Problem J34

Solution. (a) Consider the function $f : (0, \pi) \to \mathbb{R}$ defined by $f(x) = 1/(\sin x/2)$. Then f is convex, because

$$f''(x) = \frac{1+\cos^2 x}{\sin^3 x} > 0 \quad \text{for all } x \in (0, \pi).$$

Applying Jensen's inequality, we deduce that

$$\frac{1}{\sin\frac{A}{2}} + \frac{1}{\sin\frac{B}{2}} + \frac{1}{\sin\frac{C}{2}} \geq 3\left(\frac{1}{\sin\frac{\frac{A}{2}+\frac{B}{2}+\frac{C}{2}}{3}}\right) = \frac{3}{\sin\frac{\pi}{6}} = 6.$$

(b) Let r be the inradius of the triangle ABC. Arguing by contradiction, we assume that $IA < 2r$, $IB < 2r$, and $IC < 2r$. Hence

$$IA + IB + IC < 6r.$$

On the other hand,
$$IA = \frac{r}{\sin\frac{A}{2}}, \quad IB = \frac{r}{\sin\frac{B}{2}}, \quad IC = \frac{r}{\sin\frac{C}{2}}.$$

We deduce that
$$r\left(\frac{1}{\sin\frac{\frac{A}{2}+\frac{B}{2}+\frac{C}{2}}{3}}\right) < 6r,$$

which contradicts (a). This contradiction concludes the proof. □

Alternative proof of (b). We argue again by contradiction and assume that $IA < 2r$, $IB < 2r$, and $IC < 2r$. Since $AI \sin A/2 = BI \sin B/2 = CI \sin C/2 = r$, we deduce that
$$\sin\frac{A}{2} + \sin\frac{B}{2} + \sin\frac{C}{2} > \frac{1}{2} + \frac{1}{2} + \frac{1}{2} = \frac{3}{2}.$$

This inequality contradicts
$$\sin\frac{A}{2} + \sin\frac{B}{2} + \sin\frac{C}{2} \leq 3\sin\frac{A+B+C}{6} = \frac{3}{2},$$

which is a direct consequence of Jensen's inequality corresponding to the concave function $g(x) = \sin x/2$, $x \in (0, \pi)$.

6.3 Convexity versus Continuity and Differentiability

> If I feel unhappy, I do mathematics to become happy. If I am happy, I do mathematics to keep happy.
>
> Alfréd Rényi (1921–1970)

6.3.1. *Let I be an arbitrary interval and consider a convex function $f : I \to \mathbb{R}$. Prove that f is continuous at any interior point of I.*

Solution. Fix $x_0 \in \text{Int}\, I$ and let $\delta_0 > 0$ be such that $(x_0 - \delta_0, x_0 + \delta_0) \subset I$. Define, for all $x \in (x_0 - \delta_0, x_0 + \delta_0) \setminus \{x_0\}$, $g(x) = (f(x) - f(x_0))/(x - x_0)$. Since f is convex, the function g is increasing both on $(x_0 - \delta_0, x_0)$ and on $(x_0, x_0 + \delta_0)$. Thus, $f'(x_0 - 0)$ exists. It is important to observe that $f'(x_0 - 0)$ is finite. For this purpose, it is enough to show that for all $x_0 - \delta_0 < x < x_0 < y < x_0 + \delta_0$ we have $g(x) \leq g(y)$. Indeed, we can write $x_0 = \lambda x + (1 - \lambda)y$, where $\lambda = (y - x_0)/(y - x)$. Then, by the convexity of f,
$$\lambda \frac{f(x_0) - f(x)}{x_0 - x} \leq (1 - \lambda) \frac{f(y) - f(x_0)}{y - x_0}.$$

Using now the expression of λ, we obtain $g(x) \leq g(y)$.

By the definition of $f'(x_0 - 0)$ we deduce that for fixed $0 < \varepsilon < 1$, there exists $\delta < \delta_0$ such that for all $x \in (x_0 - \delta, x_0)$,

$$\left| \frac{f(x) - f(x_0)}{x - x_0} - f'(x_0 - 0) \right| < \varepsilon.$$

It follows that there exists $\eta > 0$ such that for all $x \in (x_0 - \eta, x_0)$, we have $|f(x) - f(x_0)| < 2\varepsilon$. Hence f is left-continuous at x_0, and a similar argument applies for the right continuity of f at x_0. □

Remark. The above result implies that if $f : [a,b] \to \mathbb{R}$ is convex then f is bounded. We point out that this result is no longer available if the domain of f is not compact. For instance, the function $f : (0,1] \to \mathbb{R}$ defined by $f(x) = x^{-1}$ is convex but it is not bounded above on $(0,1]$.

Let f be a real-valued function defined on an interval I. We recall that f is said to be *Lipschitz on* I if there is a positive constant M (which is called a *Lipschitz constant*) such that for every $x, y \in I$,

$$|f(x) - f(y)| \leq M|x - y|.$$

The function f is called *locally Lipschitz* if for every compact set $K \subset I$ there exists a constant M_K depending on K such that for every $x, y \in K$,

$$|f(x) - f(y)| \leq M_K |x - y|.$$

Obviously, any locally Lipschitz function is continuous.

The purpose of the next exercise is to show that convex functions are not only continuous at interior points of their domain, but moreover, are locally Lipschitz. Thus, by a celebrated theorem of Rademacher, convex functions are almost everywhere differentiable. More precisely, we will establish in Exercise 6.2.6 that the set of points where a convex function is not differentiable is at most countable.

6.3.2. *Let I be an arbitrary interval and consider a convex function $f : I \to \mathbb{R}$. Prove that f is locally Lipschitz.*

Solution. Let K be any compact subset of I. We can choose points $a < b < c < d$ in I such that for any $x, y \in K$ with $x < y$ we have $a < b < x < y < c < d$. Using Exercise 6.1.1 (ii), we obtain

$$\frac{f(b) - f(a)}{b - a} \leq \frac{f(y) - f(x)}{y - x} \leq \frac{f(d) - f(c)}{d - c}.$$

To conclude the proof, it is sufficient to choose

$$M_K = \max \left\{ \left| \frac{f(b) - f(a)}{b - a} \right|, \left| \frac{f(d) - f(c)}{d - c} \right| \right\}.$$

6.3 Convexity versus Continuity and Differentiability

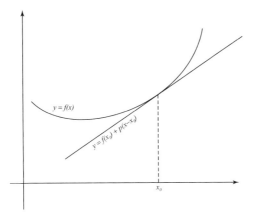

Fig. 6.3 A function is convex if and only if for each point on its graph there is a line through this point that lies below the graph.

A convex function f is characterized by having a support line at each point (see Figure 6.3), that is,
$$f(x) - f(x_0) \geq p(x - x_0),$$
for all x and x_0, where p depends on x_0. In fact, $p = f'(x_0)$ when $f'(x_0)$ exists, and p is any number between $f'(x_0-)$ and $f'(x_0+)$ in the countable set where these are different. We prove in what follows this property in the framework of convex differentiable functions. □

6.3.3. Let I be an arbitrary interval and let $f : I \to \mathbb{R}$ be a differentiable function. Prove the following properties:

(i) f is convex if and only if
$$f(x) \geq f(a) + f'(a)(x-a) \quad \text{for any } x, a \in I.$$

(ii) f is strictly convex if and only if
$$f(x) > f(a) + f'(a)(x-a) \quad \text{for any } x, a \in I, x \neq a.$$

Solution. (i) We first assume that f is convex. Fix $x, a \in I$ with $x \neq a$ and assume that $\lambda \in (0, 1]$. Since f is convex, we obtain
$$f(a + \lambda(x-a)) = f((1-\lambda)a + \lambda x)$$
$$\leq (1-\lambda)f(a) + \lambda f(x) = f(a) + \lambda(f(x) - f(a)).$$

Therefore
$$\frac{f(a + \lambda(x-a)) - f(a)}{\lambda(x-a)} \cdot (x-a) \leq f(x) - f(a).$$

Passing to the limit as $\lambda \searrow 0$, we obtain $f'(a)(x-a) \leq f(x) - f(a)$.

Next, we fix $x, y \in I$ and $\lambda \in [0, 1]$. Thus, by hypothesis,

$$f(x) \geq f((1-\lambda)x + \lambda y) + f'((1-\lambda)x + \lambda y) \cdot \lambda(x - y)$$

and

$$f(y) \geq f((1-\lambda)x + \lambda y) - f'((1-\lambda)x + \lambda y) \cdot (1-\lambda)(x - y).$$

Multiplying the first inequality by $(1 - \lambda)$ and the second inequality by λ and summing the new inequalities, we obtain

$$(1 - \lambda)f(x) + \lambda f(y) \geq f((1-\lambda)x + \lambda y),$$

which shows that f is convex.

(ii) For necessity, with the same arguments as in (i), we obtain

$$f(x) - f(a) > \frac{f(a + \lambda(x - a))}{\lambda}.$$

But by (i),

$$f(a + \lambda(x - a)) - f(a) \geq f'(a)(x - a).$$

We conclude that $f(x) > f(a) + f'(a)(x - a)$.

The sufficiency part is argued with the same proof as above.

The following useful test implies that the functions $\mathbb{R} \ni x \longmapsto e^x$, $(0, \pi/2) \ni x \longmapsto \tan x$, and $(0, \infty) \ni x \longmapsto x^a$ ($a > 1$) are strictly convex, while the mappings $(0, \infty) \ni x \longmapsto \log_a x$ ($a > 1$), $(0, \pi) \ni x \longmapsto \sin x$, and $(-\pi/2, \pi/2) \ni x \longmapsto \cos x$ are strictly concave. □

6.3.4. Let I be an arbitrary interval and let $f : I \to \mathbb{R}$ be a twice differentiable function such that $f''(x) \geq 0$ (resp., $f''(x) > 0$) for any $x \in I$. Prove that f is a convex (resp., strictly convex) function.

Solution. We apply Taylor's formula. Thus, for any $x, a \in I$, there exists ξ between x and a such that

$$f(x) = f(a) + \frac{x - a}{1!} f'(a) + \frac{(x - a)^2}{2!} f''(\xi) \geq f(a) + (x - a)f'(a).$$

At this stage it is enough to apply Exercise 6.2.2. □

6.3.5. Let $f : I \to \mathbb{R}$ be a convex function. Prove that f has finite one-sided derivatives at any interior point of I. Moreover, if x and y, with $x \leq y$, are interior points of I, then

$$f'(x-) \leq f'(x+) \leq f'(y-) \leq f'(y+).$$

Solution. Let a be an interior point of I and choose $x_1, x_2, y \in I$ such that $x_1 < x_2 < a < y$. Then, by Exercise 6.1.1 (ii), we have

$$\frac{f(x_1) - f(a)}{x_1 - a} \leq \frac{f(x_2) - f(a)}{x_2 - a} \leq \frac{f(y) - f(a)}{y - a}. \qquad (6.4)$$

6.3 Convexity versus Continuity and Differentiability

Hence, $f'(a-) = \lim_{x \to a-}[f(x) - f(a)]/(x-a)$ exists, and moreover, this limit is at most $[f(y) - f(a)]/(y-a)$; hence $f'(a-)$ is finite. A similar argument shows that $f'(a+)$ exists, is finite, and $f'(a-) \leq f'(a+)$. The last assertion follows directly from (6.4). □

In particular, since one-sided derivatives exist at interior points, we find again that a convex function $f : I \to \mathbb{R}$ is continuous at interior points of I. Exercise 6.2.4 also shows that if discontinuity points exist, then necessarily they are of the first kind.

A convex function does not have to be smooth, but it is differentiable in massive sets. The original result in this sense is due to Stanislav Mazur (1933) and implies that in a finite-dimensional space, a convex function is differentiable almost everywhere (see [36]).

6.3.6. *Let $f : I \to \mathbb{R}$ be a convex function. Prove that the set of points where f is not differentiable is at most countable.*

Solution. Let $\mathscr{S}$ denote the set of points $x \in I$ such that f is not differentiable at x. We associate to any point $x \in \mathscr{S}$ that is interior to I the open and nonempty interval $I_x = (f'(x-), f'(x+))$. According to Exercise 6.2.4. we have $I_x \cap I_y = \emptyset$, provided $x \neq y$. Next, for any $x \in \mathscr{S}$, we take $r_x \in I_x \cap \mathbb{Q}$. The one-to-one mapping $\mathscr{S} \ni x \longmapsto r_x \in \mathbb{Q}$ shows that the set $\mathscr{S}$ is at most countable. □

The following exercise provides an important test for locating the extremum points of a real function of class C^2.

6.3.7. *Let f be a twice differentiable convex function on an open interval I such that f'' is continuous. Assume that there exists $a \in I$ such that $f'(a) = 0$ and $f''(a) > 0$. Prove that a is a strict local minimum point of f.*

Solution. According to Taylor's formula, we have

$$f(x) = f(a) + f'(a)(x-a) + \frac{f''(a)}{2}(x-a)^2 + o\left((x-a)^2\right)$$
$$= f(a) + \frac{f''(a)}{2}(x-a)^2 + o\left((x-a)^2\right),$$

as $x \to a$. Thus, there exists $\delta > 0$ such that for any $x \in I \cap (a-\delta, a+\delta)$,

$$f(x) - f(a) \geq \frac{f''(a)}{4}(x-a)^2;$$

hence a is a strict local minimum point of f. □

The following exercise provides a general convexity property of convex C^1 functions satisfying a certain class of functional equations.

6.3.8. *Let I be an interval and assume that $f \in C^1(I)$ satisfies $f = \varphi \circ f'$, where $\varphi : \mathbb{R} \to \mathbb{R}$ is a continuous function. Prove that f is either convex or concave.*

Solution. The function f is convex (resp., concave) if and only if f' is nondecreasing (resp., nonincreasing). Assume that f' is neither nondecreasing nor nonincreasing on I. Hence there exist $a, b, c \in I$ with $a < b < c$ and $\max\{f'(a), f'(c)\} < f'(b)$ or $\min\{f'(a), f'(c)\} > f'(b)$.

Let $\max\{f'(a), f'(c)\} < f'(b)$. The second case can be treated analogously. Let $t \in (a,c)$ be such that $f'(t) = \max\{f'(s); \ s \in [a,c]\}$. We can choose $z \in [\max\{f'(a), f'(c)\}, f'(t))$ such that $0 \leq z < f'(t)$ or $z < f'(t) \leq 0$. For

$$\alpha = \max\{s \in [a,t]; \ f'(s) = z\} \quad \text{and} \quad \beta = \min\{s \in [t,c]; \ f'(s) = z\}$$

we have $a \leq \alpha < t < \beta \leq c$, $f'(\alpha) = f'(\beta) = z$, and $f'(s) > z$ for any $s \in (\alpha, \beta)$. We obtain $f(\alpha) = \varphi(z) = f(\beta)$. This yields a contradiction, since only one of the following situations can occur:

(i) if $0 \leq z < f'(t)$ then $f'(s) > 0$ for all $s \in (\alpha, \beta)$, which implies $f(\alpha) < f(\beta)$;
(ii) if $z < f'(t) \leq 0$ then $f'(s) \leq 0$ for all $s \in (\alpha, \beta)$ and $f'(\alpha) < 0$, which implies $f(\alpha) > f(\beta)$. □

6.4 Qualitative Results

> Vous avez trouvé par de long ennuis
> Ce que Newton trouva sans sortir de
> chez lui.
>
> Voltaire (1694–1778), Written to La Condamine after his measurement of the equator

6.4.1. Let x be a real number such that $0 < x < \pi/4$. Prove that

$$(\sin x)^{\sin x} < (\cos x)^{\cos x}.$$

W.W. Chao, Amer. Math. Monthly, Problem 10261

Solution. We find a positive lower bound on $\ln(\cos x) - \tan x \ln(\sin x)$ for $0 < x < \pi/4$. For all real numbers a, b, and λ with $a > 0$, $b > 0$, and $\lambda \in (0,1)$, the strict concavity of the natural logarithm gives

$$\ln(\lambda a + (1-\lambda)b) > \lambda \ln a + (1-\lambda)\ln b.$$

Let $x \in \mathbb{R}$ with $0 < x < \pi/4$ and define $a = \sin x$, $b = \sin x + \cos x$, $\lambda = \tan x$. The above inequality yields

$$\ln(\cos x) > \tan x \ln(\sin x) + (1 - \tan x)\ln(\sin x + \cos x).$$

The last term is positive, since $\tan x < 1$ and $\sin x + \cos x = \sqrt{2}\cos(\pi/4 - x) > 1$. □

6.4 Qualitative Results

6.4.2. Compare $\tan(\sin x)$ and $\sin(\tan x)$ for all $x \in (0, \pi/2)$.

Solution. Let $f(x) = \tan(\sin x) - \sin(\tan x)$. Then

$$f'(x) = \frac{\cos x}{\cos^2(\sin x)} - \frac{\cos(\tan x)}{\cos^2 x}$$
$$= \frac{\cos^3 x - \cos(\tan x) \cdot \cos^2(\sin x)}{\cos^2 x \cdot \cos^2(\tan x)}.$$

Let $0 < x < \arctan \frac{\pi}{2}$. It follows from the concavity of cosine on $(0, \frac{\pi}{2})$ that

$$\sqrt[3]{\cos(\tan x) \cdot \cos^2(\sin x)} < \frac{1}{3}[\cos(\tan x) + 2\cos(\sin x)]$$
$$\leq \cos\left[\frac{\tan x + 2\sin x}{3}\right] < \cos x.$$

The last inequality follows from

$$\left[\frac{\tan x + 2\sin x}{3}\right]' = \frac{1}{3}\left[\frac{1}{\cos^2 x} + 2\cos x\right] \geq \sqrt[3]{\frac{1}{\cos^2 x} \cdot \cos x \cdot \cos x} = 1.$$

This proves that $\cos^3 x - \cos(\tan x) \cdot \cos^2(\sin x) > 0$, so $f'(x) > 0$. Thus, f increases on the interval $[0, \arctan \pi/2]$. To end the proof it is enough to notice that

$$\tan\left[\sin\left(\arctan\frac{\pi}{2}\right)\right] = \tan\frac{\pi/2}{\sqrt{1+\pi^2/4}} > \tan\frac{\pi}{4} = 1.$$

This implies that if $x \in [\arctan\frac{\pi}{2}, \frac{\pi}{2}]$ then $\tan(\sin x) > 1$ and therefore $f(x) > 0$. $\square$

6.4.3. Find all positive real numbers x and y such that the expression

$$f(x,y) = \frac{x^4}{y^4} + \frac{y^4}{x^4} - \frac{x^2}{y^2} - \frac{y^2}{x^2} + \frac{x}{y} + \frac{y}{x}$$

is minimum and find the value of this minimum.

Solution. Define $\frac{x}{y} + \frac{y}{x} = t \geq 2$. We have

$$f(x,y) = g(t) = (t^2-2)^2 - 2 - t^2 + 2 + t = t^4 - 5t^2 + t + 4, \quad t \geq 2.$$

Therefore $g'(t) = 2t(2t^2 - 5) + 1 > 0$, so g is increasing in $[2, \infty)$. It follows that $g(t) \geq g(2) = 2$. In conclusion, the minimum of f is achieved for $x = y$ and $\min f(x,y) = 2$. $\square$

6.4.4. Let $f : [0,1] \to \mathbb{R}$ be a continuous function such that $f(0) = 0$. Prove that there exists a concave continuous function $g : [0,1] \to \mathbb{R}$ such that $g(0) = 0$ and $g(x) \geq f(x)$, for all $x \in [0,1]$.

Solution. Let $M = \max_{x \in [0,1]} |f(x)|$. Define the sequence (x_n) such that $x_0 = 1$ and $0 < x_n < 3^{-1} x_{n-1}$ satisfies $|f(x)| < M/2^n$, for all $0 < x < x_n$. On the other hand, for any $x \in (0,1]$ there exists a unique $n \geq 0$ such that $x = \lambda_x + (1-\lambda_x)x_n$, for some $\lambda_x \in [0,1]$. Define the continuous function $g : [0,1] \to [0,1]$ by $g(0) = 0$ and

$$g(x) = \lambda_x \frac{M}{2^n} + (1-\lambda_x) \frac{M}{2^{n-1}}, \quad \forall x \in (0,1].$$

We have $g \geq f$ and

$$\frac{g(x_n) - g(x_{n+1})}{x_n - x_{n+1}} = \frac{M/2^n}{x_n - x_{n+1}} > \frac{M/2^{n-1}}{x_n - x_{n+1}} = \frac{g(x_{n-1}) - g(x_n)}{x_{n-1} - x_n},$$

so g is concave. □

6.4.5. Find the minimum value of the function

$$f(x) = \frac{(x+1/x)^6 - (x^6 + 1/x^6) - 2}{(x+1/x)^3 + (x^3 + 1/x^3)}, \quad x > 0.$$

Solution. We have

$$\frac{(x+1/x)^6 - (x^6 + 1/x^6) - 2}{(x+1/x)^3 + (x^3 + 1/x^3)} = (x+1/x)^3 - (x^3 + 1/x^3) = 3(x+1/x).$$

Applying the mean value inequality, we deduce that the minimum of f is 6 and it is achieved for $x = 1$. □

6.4.6. Let $f : [0,1] \to \mathbb{R}$ be a concave increasing function such that $f(0) = 0$ and $f(1) = 1$. Show that $f(x)f^{-1}(x) \leq x^2$, for all $0 \leq x \leq 1$.

Solution. Let $0 < a \leq b \leq 1$. Taking $t = (b-a)/b$ and using the definition of a concave function, we obtain

$$f(a) \geq \frac{af(b) + (b-a)f(0)}{b} = \frac{af(b)}{b}.$$

If $a = x$ and $b = 1$, this inequality implies $f(x) \geq x$, so $x = f^{-1}(f(x)) \geq f^{-1}(x)$ (since f^{-1} is increasing). For $b = x \neq 0$ and $a = f^{-1}(x)$ we obtain

$$x = f(f^{-1}(x)) \geq \frac{f^{-1}(x)f(x)}{x}.$$

Hence $f(x)f^{-1}(x) \leq x^2$, for all $x \in [0,1]$. □

6.4.7. Let a, b, u and v be nonnegative numbers such that $a^5 + b^5 \leq 1$ and $u^5 + v^5 \leq 1$. Prove that

$$a^2 u^3 + b^2 v^3 \leq 1.$$

6.4 Qualitative Results

Solution. By the arithmetic–geometric means inequality we obtain

$$\frac{2a^5 + 3u^5}{5} = \frac{a^5 + a^5 + u^5 + u^5 + u^5}{5} \geq (a^{10} u^{15})^{1/5} = a^2 u^3. \quad \square$$

Writing the similar inequality for b and v and adding these relations, we obtain the assented inequality.

6.4.8. Let $x, y \in (0, \sqrt{\pi/2})$ with $x \neq y$. Prove that

$$\ln^2 \frac{1 + \sin(xy)}{1 - \sin(xy)} < \ln \frac{1 + \sin(x^2)}{1 - \sin(x^2)} \cdot \ln \frac{1 + \sin(y^2)}{1 - \sin(y^2)}.$$

Solution. For $t \in (-\infty, \ln(\pi/2))$, define the function
$f(t) = \ln(\ln 1 + \sin e^t / 1 - \sin e^t)$. We show that f is strictly convex. Indeed,

$$f''(t) = \frac{2e^t}{e^{2f(t)} \cos^2 e^t} \left[(\cos e^t + e^t \sin e^t) \ln \left(\frac{1 + \sin e^t}{1 - \sin e^t} \right) - 2e^t \right].$$

Define

$$g(u) = (\cos u + u \sin u) \ln \left(\frac{1 + \sin u}{1 - \sin u} \right) - 2u.$$

Then $g(0) = 0$ and

$$g'(u) = u \cos u \ln \left(\frac{1 + \sin u}{1 - \sin u} \right) + 2u \tan u > 0,$$

for any $0 < u < \pi/2$. Hence $g > 0$ in $(0, \pi/2)$. It follows that $f''(t) > 0$, so f is strictly convex in $(-\infty, \ln(\pi/2))$. Therefore

$$\ln^2 \frac{1 + \sin(xy)}{1 - \sin(xy)} = e^{2f(\ln x + \ln y)} < e^{f(2 \ln x) + f(2 \ln y)}$$

$$= \ln \frac{1 + \sin(x^2)}{1 - \sin(x^2)} \cdot \ln \frac{1 + \sin(y^2)}{1 - \sin(y^2)},$$

for any $x, y \in (0, \sqrt{\pi/2})$ with $x \neq y$.

Hilbert's double series theorem asserts the following. Assume $p > 1$, $p' = p/(p-1)$ and consider $A := \sum_{n=1}^{\infty} a_n^p$, $B := \sum_{k=1}^{\infty} b_k^{p'}$, where $(a_n)_{n \geq 1}$ and $(b_n)_{n \geq 1}$ are sequences of nonnegative numbers. Then

$$\sum_{n=1}^{\infty} \sum_{k=1}^{\infty} \frac{a_n b_k}{n + k} < \frac{\pi}{\sin(\pi/p)} A^{1/p} B^{1/p'}$$

with equality if and only if either $A = 0$ or $B = 0$. We prove below the following reverse Hilbert-type inequality. $\square$

6.4.9. Show that
$$\left(\sum_{i=1}^n \frac{a_i}{i}\right)^2 \le \sum_{i=1}^n \sum_{j=1}^n \frac{a_i a_j}{i+j-1}.$$

V.R. Rao Uppuluri, Amer. Math. Monthly, Problem 1682

Solution. For any $x > 0$ define
$$f(x) = x^{-1}\left(\sum_{i=1}^n \frac{a_i x^i}{i}\right)^2 - \sum_{i=1}^n \sum_{j=1}^n \frac{a_i a_j}{i+j-1}.$$

Then $f(0+) = 0$ and $f'(x) = -\left[\sum_{i=1}^n (i-1)a_i x^{i-1}/i\right]^2 \le 0$ for $x > 0$ with equality if and only if $a_2 = a_3 = \cdots = a_n = 0$. Thus, $f(1) \le f(0+)$ with equality if and only if $a_2 = a_3 = \cdots = a_n = 0$. □

6.4.10. Prove that
$$\sum_{k=1}^n \left(\frac{1}{k}\right)^{1/p} < \frac{p}{p-1} n^{1-1/p},$$
for all integers $n \ge 1$ and any real number $p > 1$.

Solution. We argue by induction. For $n = 1$ we obtain $1 < p/(p-1)$, obvious. We assume that the inequality holds for n and prove it for $n+1$. So, it is enough to show that
$$\frac{p}{p-1} n^{1-1/p} + \frac{1}{(n+1)^{1/p}} < \frac{p}{p-1}(n+1)^{1-1/p},$$
or, equivalently,
$$\frac{p}{p-1}\left[(n+1)^{1-1/p} - n^{1-1/p}\right] > \frac{1}{(n+1)^{1/p}}. \qquad (6.5)$$

In order to prove this inequality, we consider the function $f : [1, \infty) \to (0, \infty)$ defined by $f(x) = x^{1-1/p}$, $x > 1$. By the Lagrange mean value theorem, there exists $n < \xi < n+1$ such that
$$(n+1)^{1-1/p} - n^{1-1/p} = \frac{p-1}{p} \cdot \frac{1}{\xi^{1/p}}.$$

Since $\xi < n+1$, the above inequality implies (6.5).

We point out that the conclusion of the above exercise follows easily after observing that $\sum_{k=1}^n k^{-1/p}$ represents the lower Darboux sum of $f(x) = x^{-1/p}$ on $[0,n]$, while $p(p-1)^{-1} n^{1-1/p} = \int_0^n x^{-1/p} dx$. □

6.4.11. Let a, b, c be three positive real numbers. Prove that among the three numbers $a - ab$, $b - bc$, $c - ca$ there is one that is at most $1/4$. Prove that if $a + b + c = 1$, then among the numbers $a - ab$, $b - bc$, $c - ca$ there is one that is at least $2/9$.

Solution. We first observe that $a(1-a)b(1-b)c(1-c) \le 1/4^3$. This implies that at least one of $a - ab$, $b - bc$, $c - ca$ is less than or equal than $1/4$. Let us now

6.4 Qualitative Results

suppose that $a+b+c=1$. Assume, by contradiction, that all of $a-ab, b-bc, c-ca$ are less than $2/9$. Thus, by addition, $ab+bc+ca > 1/3$. But on the other hand,

$$ab+bc+ca \leq \frac{(a+b+c)^2}{3} = \frac{1}{3}.$$

This contradiction concludes the proof. □

6.4.12. Let $a_1, \ldots, a_n$ be real numbers. Prove that

$$\sqrt[n]{\prod_{k=1}^{n} a_k} \leq \ln\left(1 + \sqrt[n]{\prod_{k=1}^{n}(e^{a_k} - 1)}\right) \leq \frac{\sum_{k=1}^{n} a_k}{n}.$$

Mihály Bencze, Problem 1754, Mathematics Magazine

Solution. Set $A = \left(\sum_{k=1}^{n} a_k\right)/n$ and $G = \left(\prod_{k=1}^{n} a_k\right)^{1/n}$. In order to prove the first inequality, we have to show that

$$e^G - 1 \leq \sqrt[n]{\prod_{k=1}^{n}(e^{a_k} - 1)}.$$

Since the a_j's are positive numbers, we can write $a_j = e^{x_j}$ for all $1 \leq j \leq n$, where $x_j \in \mathbb{R}$. Define the function $f(x) = \ln\left(e^{e^x} - 1\right)$, for $x \in \mathbb{R}$. A straightforward computation shows that $f''(x) > 0$ for any $x \in \mathbb{R}$. Thus, f is a convex function, and hence by Jensen's inequality,

$$\ln(e^G - 1) \leq \frac{\sum_{k=1}^{n} \ln\left(e^{e^{x_k}} - 1\right)}{n},$$

which implies the first inequality.

The second inequality can be rewritten as

$$\sqrt[n]{\prod_{k=1}^{n}(e^{a_k} - 1)} \leq e^A - 1.$$

Consider the function $g(x) = \ln(1 - e^{-x})$, for $x > 0$. Then g is concave, since $g''(x) = -e^{-x}(1 - e^{-x})^{-2} < 0$ on $(0, \infty)$. Thus, Jensen's inequality yields

$$\frac{\sum_{k=1}^{n} \ln(1 - e^{-a_k})}{n} \leq \ln(1 - e^{-A}),$$

which concludes the proof. □

Analytic tools are often useful to justify that a certain constant is really the **best** for which a related property is fulfilled. We give in what follows such an example.

6.4.13. Let $n \geq 3$ be an integer. Assume that $x_1, \ldots, x_n$ are arbitrary positive numbers satisfying
$$(x_1 + \cdots + x_n)\left(\frac{1}{x_1} + \cdots + \frac{1}{x_n}\right) < C_n.$$
Find the largest constant C_n such that for any $1 \leq i < j < k \leq n$, there is a triangle with side lengths x_i, x_j, and x_k.

Solution. For symmetry reasons we can assume, without loss of generality, that $x_1 \leq x_2 \leq \cdots \leq x_n$. This shows that **any** three numbers x_i, x_j, and x_k are the side lengths of a triangle if $x_1 + x_2 > x_n$.

We first consider the case $n = 3$. Then
$$(x_1 + x_2 + x_3)\left(\frac{1}{x_1} + \frac{1}{x_2} + \frac{1}{x_3}\right) \geq (x_1 + x_2 + x_3)\left(\frac{4}{x_1 + x_2} + \frac{1}{x_3}\right)$$
$$\geq 10 + \left(4 - \frac{x_1 + x_2}{3}\right)\left(\frac{x_3}{x_1 + x_2} - 1\right) \geq 10.$$

This shows that if we take $C_3 = 10$, then the hypothesis is not fulfilled if x_1, x_2, and x_3 are not the sides of a triangle.

Next, we argue that the best constant is $C_n = (n - 3 + \sqrt{10})^2$. Our reasoning in the general case relies heavily on the particular case $n = 3$. Indeed, for an arbitrary integer $n \geq 3$, the Cauchy–Schwarz inequality yields
$$A := [(x_1 + x_2 + x_n) + x_3 + \cdots + x_{n-1}]\left[\left(\frac{1}{x_1} + \frac{1}{x_2} + \frac{1}{x_n}\right) + \frac{1}{x_3} + \cdots + \frac{1}{x_{n-1}}\right]$$
$$\geq \left[\sqrt{(x_1 + x_2 + x_n)\left(\frac{1}{x_1} + \frac{1}{x_2} + \frac{1}{x_n}\right)} + n - 3\right]^2.$$

We point out that in the above inequality we have applied the result found for $n = 3$. This shows that if we assume $x_n \geq x_1 + x_2$, then $A \geq (\sqrt{10} + n - 3)^2$. In conclusion, if
$$(x_1 + \cdots + x_n)\left(\frac{1}{x_1} + \cdots + \frac{1}{x_n}\right) < (n - 3 + \sqrt{10})^2,$$
then $x_1 + x_2 > x_n$, and hence there is a triangle with side lengths x_i, x_j, and x_k, for any $1 \leq i < j < k \leq n$. It remains to argue that $C_n = (n - 3 + \sqrt{10})^2$ is indeed the largest constant with this property. For this purpose we fix arbitrarily $\varepsilon > 0$ and take
$$x_1 = x_2 = 1, \quad x_3 = \cdots = x_{n-1} = 2\sqrt{\frac{2}{5}}, \quad x_n = 2 + \varepsilon.$$

Then $x_1 + x_2 < x_n$, and a straightforward computation (give the details!) shows that $A(\varepsilon) \to C_n$ as $\varepsilon \to 0$. This shows that C_n is the best constant with the given property. □

6.5 Independent Study Problems

> Chance favors only the prepared mind.
> ――――――――――――――――――――
> Louis Pasteur (1822–1895)

6.5.1. Let $f : (0,\infty) \to \mathbb{R}$ be an arbitrary function. Prove that if one of the functions $xf(x)$ and $f(1/x)$ is convex, then the other function is also convex.

6.5.2. Prove the following necessary and sufficient condition for a function $f : (0,1) \to \mathbb{R}$ to be convex: if $\alpha \in \mathbb{R}$ and $[a,b] \subset (0,1)$, then the function $f(x) + \alpha x$ achieves its maximum on $[a,b]$ at one of the endpoints of the interval.

6.5.3. Let $f : [0,\infty) \to \mathbb{R}$ be a decreasing function such that the function $xf(x)$ is increasing. Prove that
$$|f(|x|)x - f(|y|)y| \leq f(0)\,|x-y|, \quad \forall x,y \in \mathbb{R}.$$

6.5.4. Let a and b be positive numbers. Find (in terms of a and b) the greatest number c such that
$$a^x b^{1-x} \leq a\frac{\sinh x}{\sinh u} + b\frac{\sinh u(1-x)}{\sinh u},$$
for all u satisfying $0 < |u| \leq |c|$ and for any $x \in (0,1)$.
(Remark: $\sinh u = (e^u - e^{-u})/2$.)

6.5.5. Let a, b be positive numbers and $n \geq 1$ an integer. Prove that
$$\prod_{k=1}^{n}(a^k + b^k)^2 \geq (a^{n+1} + b^{n+1})^2.$$

Hint. Take into account the factors indexed by k and $n-k+1$.

6.5.6. Let $n \geq 1$ be an integer. Prove that
$$\frac{\prod_{j=1}^{n} x_j}{\left(\sum_{j=1}^{n} x_j\right)^n} \leq \frac{\prod_{j=1}^{n}(1-x_j)}{\left(\sum_{j=1}^{n}(1-x_j)\right)^n},$$
for all $x_j \in (0, 1/2]$, $j = 1, \ldots, n$.
Hint. Consider the function $f(x) = \ln(1/x - 1)$ and apply Jensen's inequality.

6.5.7. For any positive numbers $x_1, \ldots, x_n$ define the function
$$R(x_1, \ldots, x_n) = \frac{(x_1 + \cdots + x_n)^{x_1 + \cdots + x_n}}{x_1^{x_1} \cdots x_n^{x_n}}.$$

Prove that $R(x_1, \ldots, x_n) R(y_1, \ldots, y_n) \leq R(x_1 + y_1, \ldots, x_n + y_n)$.
Hint. Use the fact that the log function is concave and continue with an induction argument.

6.5.8. Prove that if $0 < x_j < \pi$ and $x = (x_1 + \cdots + x_n)/n$, then

$$\prod_{j=1}^{n} \frac{\sin x_j}{x_j} \leq \left(\frac{\sin x}{x}\right)^n.$$

Hint. The function $\ln(\sin x/x)$ is concave.

6.5.9. Let $a_0 = 0$, $a_n = e^{a_{n-1}}$. Prove that for all real numbers $t_1, \ldots, t_n$ and for any positive integer n, the following inequality holds:

$$\sum_{j=1}^{n}(1-t_j)e^{\sum_{k=1}^{j} t_k} \leq a_n$$

if and only if $t_n = a_0$, $t_{n-1} = a_1, \ldots, t_1 = a_{n-1}$.

Hint. Prove by induction that $e^{t_1((1-t_1)+a_{n-1})} \leq e^{a_{n-1}} = a_n$.

6.5.10. Prove that if $n \geq 2$ and $x_j > 0$ for $j = 1, \ldots, n$, then $x_1^{x_2} + x_2^{x_3} + \cdots + x_{n-1}^{x_n} + x_n^{x_1} \geq 1$.

Hint. We can assume that $0 < x_j < 1$, $j = 1, \ldots, n$, and that $x_1 = \min x_j$. Then $x_1^{x_2} + x_2^{x_3} \geq x_3^{x_2} + x_2^{x_3} \geq 1$. We point out that the above inequality cannot be improved. Indeed, for $n = 4$ and taking $x_1 = r^{-r^r}$, $x_2 = r^{-r}$, $x_3 = 1/r$, $x_4 = 1$, we have $3/r + 1 \geq 1$ and it is enough to take $r \to \infty$. A slight generalization of the above result is the following:

$$x_1^{x_2} + x_2^{x_3} + \cdots + x_{n-1}^{x_n} + x_n^{x_1} > 1 + (n-2)\min\{x_1^{x_2}, \ldots, x_n^{x_1}\}.$$

6.5.11. Let $p_1, \ldots, p_n$ and $q_1, \ldots, q_n$ be positive numbers such that

$$\sum_{k=1}^{n} p_k = \sum_{k=1}^{n} q_k.$$

Prove the following inequality arising in information theory:

$$\sum_{k=1}^{n} p_k \ln p_k \geq \sum_{k=1}^{n} p_k \ln q_k.$$

Hint. Use the inequality $t \ln t \geq t - 1$.

6.5.12. Let $f : \mathbb{R} \to [0, \infty)$ be a continuous function at the origin that is concave in $(0,1)$ and periodic of period 1. Prove that $f(nx) \leq nf(x)$, for all $x \in \mathbb{R}$ and any $n \in \mathbb{N}$.

Hint. Observe that $f(x)/x$ is decreasing and $f(x)/(1-x)$ is increasing on $(0,1)$. The result contained in this exercise is a generalization of the inequality $|\sin nx| \leq n|\sin x|$.

6.5.13. The functions $f, g : \mathbb{R} \to \mathbb{R}$ are said to be *similar* if there exists a bijection h such that $f = h^{-1} \circ g \circ h$. Are the functions $\sin$ and $\cos$ similar? The same question for the functions x^2 and $x^2 + ax + b$.

Hint. The functions sin and cos are not similar. The quadratic functions are similar if $4b = a^2 - 2a$.

6.5.14. Prove that the set of all real numbers x satisfying $\sum_{k=1}^{n} k/(x-k) \leq 1$ is a finite union of intervals and find the sum of the lengths of these intervals.

6.5.15. Assume that f is a continuous real function defined on (a,b) such that

$$f\left(\frac{x+y}{2}\right) \leq \frac{f(x)+f(y)}{2} \quad \text{for all } x, y \in (a,b).$$

Prove that f is convex.

6.5.16. Let I be an interval and assume that $f : I \to \mathbb{R}$ is a convex function. Prove that either f is monotone or there exists $c \in I$ such that f is nonincreasing on $I \cap (-\infty, c]$ and is nondecreasing on $I \cap [c, \infty)$.

6.5.17. Let f and g be positive nonincreasing convex functions on (a,b). Prove that the function fg is convex in the interval (a,b).

6.5.18. Prove that if f is a convex function on $[0,1]$, then the function $g(x) = f(x) + f(1-x)$ is decreasing on $[0, 1/2]$.

6.5.19. Prove that if a function f is convex and strictly monotone, then f^{-1} is either convex or concave.

6.5.20. Suppose that a function f is convex on $\mathbb{R}$ and

$$\lim_{x \to +\infty} f(x)/x = \lim_{x \to -\infty} f(x)/x = 0.$$

Prove that f is constant.

6.5.21. Prove that is f is a convex function on $(0, \infty)$, then the function $g(x) = f(x+a) - f(x)$ is increasing, where $a > 0$.

The following inequality is due to the Romanian mathematician Tiberiu Popoviciu (1906–1975).

6.5.22. Let I be an interval and assume that $f : I \to \mathbb{R}$ is a convex function. Prove that for any $x, y, z \in I$,

$$\frac{f(x)+f(y)+f(z)}{3} + f\left(\frac{x+y+z}{3}\right)$$
$$\geq \frac{2}{3}\left[f\left(\frac{x+y}{2}\right) + f\left(\frac{y+z}{2}\right) + f\left(\frac{z+x}{2}\right)\right].$$

6.5.23. Let $f : \mathbb{R} \to \mathbb{R}$ be differentiable, strictly convex, and superlinear (in the sense of relation (3.2)). Prove that the function $g : \mathbb{R} \to \mathbb{R}$ defined by

$$g(y) = \max_{x \in \mathbb{R}}\,[xy - f(x)]$$

is differentiable, strictly convex, and superlinear, and $g' = (f')^{-1}$.

Chapter 7
Inequalities and Extremum Problems

> *To myself I am only a child playing on the beach, while vast oceans of truth lie undiscovered before me.*
> —Sir Isaac Newton (1642–1727)

Abstract. This chapter is not intended to be a compendium of the most important inequalities. Instead, we provide an introduction to a selection of inequalities. The inequalities that we consider have a common theme; they relate to subjects arising in linear or nonlinear analysis, as well as to extremum problems.

7.1 Basic Tools

> It is not certain that everything is uncertain.
> ——————————
> Blaise Pascal (1623–1662), *Pensées*

The main ingredients in the proofs of the exercises contained in this chapter are the Lagrange mean value theorem and the increasing function theorem. We recall that this last result establishes the following property: if f is differentiable on an open interval I, then f is nondecreasing on I if and only if $f'(x) \geq 0$ for all $x \in I$. If $f'(x) > 0$ for all $x \in I$, then f is increasing in I. Some immediate consequences of the increasing function theorem are stated in the following properties. Assume that $a < b$ and f is differentiable on $[a,b]$.

(i) If $f'(x) \leq 0$ for all $x \in [a,b]$, then f is nonincreasing on $[a,b]$.
(ii) If $f'(x) = 0$ for all $x \in [a,b]$, then f is constant on $[a,b]$.
(iii) If $m \leq f'(x) \leq M$ on $[a,b]$, then

$$m(x-a) \leq f(x) - f(a) \leq M(x-a) \quad \text{for all } x \in [a,b].$$

(iv) If $f'(x) \leq g'(x)$ for all $x \in [a,b]$, then $f(x) - f(a) \leq g(x) - g(a)$ on $[a,b]$ (racetrack principle).

We conclude this introductory section with an elementary proof of the increasing function theorem that depends only on the nested intervals theorem: if $a_n \leq a_{n+1} \leq$

$b_{n+1} \leq b_n$ for all $n \geq 1$ and $\lim_{n \to \infty}(b_n - a_n) = 0$, then there is a number c such that $\lim_{n \to \infty} a_n = \lim_{n \to \infty} b_n = c$. The proof of the increasing function theorem given here does not require the continuity of f'. We start with the following simple observation: if slope $(a,b) = m$ and c is between a and b, then one of slope (a,c) and slope (c,b) is greater than or equal to m and one is less than or equal to m. Here, slope (a,b) denotes the *slope* of a given function f, which is defined by

$$\text{slope}(a,b) = \frac{f(b) - f(a)}{b - a}.$$

Suppose now that $f'(x) \geq 0$. Arguing by contradiction, there exist $a_1 < b_1$ with $f(a_1) > f(b_1)$. Let $m = $ slope $(a_1, b_1) < 0$. By repeated bisection and the above observation, we can find a nested sequence of intervals $[a_n, b_n]$ with slope $(a_n, b_n) \leq m$ and $\lim_{n \to \infty}(b_n - a_n) = 0$. Let $c = \lim_{n \to \infty} a_n = \lim_{n \to \infty} b_n$. Since $f'(c) \geq 0$ and $m < 0$, we deduce that for all x sufficiently near c, slope $(x, c) > m$. Thus for all large enough n, slope $(a_n, c) > m$ and slope $(c, b_n) > m$, which contradicts our initial observation and the fact that by construction, slope $(a_n, b_n) \leq m$. If $a_n = c$ or $b_n = c$, the contradiction is immediate.

7.2 Elementary Examples

> Truth is much too complicated to allow anything but approximations.
>
> John von Neumann (1903–1957)

7.2.1. Show that the sequence $(n + 1/n)^{n+1/2}$, $n = 1, 2, 3, \ldots$, is decreasing.

Solution. It is enough to show that the function $f(x) = (x + 1/x)^{x+1/2}$ is decreasing on $(0, \infty)$. We first observe that

$$f(x) = e^{\ln(1+x^{-1})^{x+1/2}} = e^{(x+1/2)\ln(1+x^{-1})}.$$

Hence

$$f'(x) = e^{(x+1/2)\ln(1+x^{-1})} \left[\left(\frac{-x^{-2}}{1+\frac{1}{x}}\right) \cdot \left(x + \frac{1}{2}\right) + \ln\left(1 + \frac{1}{x}\right) \right]$$

$$= e^{(x+1/2)\ln(1+x^{-1})} \left[\frac{-x^{-1} - \frac{1}{2}x^{-2}}{1+\frac{1}{x}} + \ln\left(1 + \frac{1}{x}\right) \right]$$

$$= e^{(x+1/2)\ln(1+x^{-1})} \left[\frac{(-2x-1)}{2x(x+1)} + \ln\left(1 + \frac{1}{x}\right) \right].$$

7.2 Elementary Examples

Since $e^{(x+1/2)\ln(1+1/x)} > 0$, we look at the other factor. Let

$$g(x) = \left[\frac{(-2x-1)}{2x(x+1)} + \ln\left(1+\frac{1}{x}\right)\right].$$

We obtain

$$g'(x) = \frac{1}{2(x^2+x)^2} > 0.$$

It follows that g is increasing on $(0,\infty)$. But

$$\lim_{x\to\infty} g(x) = \lim_{x\to\infty}\left[\frac{(-2x-1)}{2x(x+1)} + \ln\left(1+\frac{1}{x}\right)\right] = 0.$$

So as x gets big, $g(x)$ increases to 0. The only way that can happen is if $g(x)$ is negative on $(0,\infty)$. Thus $f'(x) = e^{(x+1/2)\ln(1+1/x)}g(x)$ is negative on $(0,\infty)$, and so $f(x)$ is decreasing on $(0,\infty)$. In particular, the sequence $(1+1/n)^{n+1/2}, n=1,2,\ldots,$ is decreasing. □

7.2.2. For any integer $n \geq 1$, let

$$a_n = b_n e^{-1/(12n)} \quad \text{and} \quad b_n = (n!e^n)n^{-n-1/2}.$$

Prove that each interval (a_n, b_n) contains the interval (a_{n+1}, b_{n+1}) as a subinterval.

Solution. For any $x \in (-1,1)$ we have

$$\ln\frac{1+x}{1-x} = 2x\left(1 + \frac{x^2}{3} + \frac{x^4}{5} + \cdots\right).$$

Setting $x = (2n+1)^{-1}$, we obtain

$$\ln\frac{n+1}{n} = \frac{2}{2n+1}\left[1 + \frac{1}{3(2n+1)^2} + \frac{1}{5(2n+1)^4} + \cdots\right].$$

Therefore

$$\left(n+\frac{1}{2}\right)\ln\left(1+\frac{1}{n}\right) = 1 + \frac{1}{3(2n+1)^2} + \frac{1}{5(2n+1)^4} + \cdots,$$

which is larger than 1, but less than

$$1 + \frac{1}{3}\left[\frac{1}{(2n+1)^2} + \frac{1}{(2n+1)^4} + \cdots\right] = 1 + \frac{1}{12n(n+1)}.$$

It follows that

$$e < \left(1+\frac{1}{n}\right)^{n+1/2} < e^{1+1/[12n(n+1)]}.$$

We have
$$\frac{b_n}{b_{n+1}} = \frac{(1+1/n)^{n+1/2}}{e}.$$

Hence
$$1 < \frac{b_n}{b_{n+1}} < e^{1/[12n(n+1)]} = \frac{e^{1/(12n)}}{e^{1/[12(n+1)]}}.$$

Thus, $b_n > b_{n+1}$ and
$$b_n e^{-1/(12n)} < b_{n+1} e^{-1/[12(n+1)]},$$

which shows that $a_n < a_{n+1}$. □

7.2.3. Show that the sequence $(a_n)_{n\geq 1}$ defined by
$$a_n = \left(1 + \frac{1}{n}\right)^{n+p}$$

is decreasing if and only if $p \geq 1/2$.

Solution. We have observed in the solution of Exercise 7.2.2 that
$$\ln a_n = \frac{2(n+p)}{2n+1}\left[1 + \frac{1}{3(2n+1)^2} + \frac{1}{5(2n+1)^4} + \cdots\right]$$
$$= 1 + \frac{p - \frac{1}{2}}{n + \frac{1}{2}}\left[1 + \frac{1}{3(2n+1)^2} + \frac{1}{5(2n+1)^4} + \cdots\right].$$

This implies that if $p \geq 1/2$, then the sequence $(a_n)_{n\geq 1}$ is decreasing. The above expansion also shows that
$$\ln a_n = \frac{\frac{1}{2} - p}{n + \frac{1}{2}} + \frac{1}{12n^2} + O(n^{-3}) \quad \text{as } n \to \infty.$$

Therefore
$$\ln a_{n+1} - \ln a_n = \frac{\frac{1}{2} - p}{\left(n + \frac{1}{2}\right)\left(n + \frac{3}{2}\right)} + O(n^{-3}) \quad \text{as } n \to \infty.$$

Thus, a_n increases for n large enough, provided $p < 1/2$. If $p \leq 0$, this is true already for all $n \geq 1$, as can be seen by expanding $(1+1/n)^n$ by means of Newton's binomial. □

7.2.4. Show that the sequence $(a_n)_{n\geq 1}$ defined by
$$a_n = \left(1 + \frac{1}{n}\right)^n \left(1 + \frac{x}{n}\right)$$

is decreasing if and only if $x \geq 1/2$.

7.2 Elementary Examples

Solution. We have

$$a_n = \left(1 + \frac{1}{n}\right)^{n+1/2} \frac{1 + \frac{x}{n}}{\left(1 + \frac{1}{n}\right)^{1/2}}.$$

Thus, by Exercise 7.2.3, the first factor on the right-hand side is decreasing. Since the square of the second factor is

$$1 + \frac{2x-1}{n+1} + \frac{x^2}{n(n+1)},$$

we conclude that the condition $x \geq 1/2$ is sufficient. On the other hand,

$$\ln a_n = 2n \left[\frac{1}{2n+1} + \frac{1}{3(2n+1)^3} + \frac{1}{5(2n+1)^5} + \cdots\right]$$
$$+ 2\left[\frac{x}{2n+x} + \frac{1}{3}\left(\frac{x}{2n+x}\right)^3 + \frac{1}{5}\left(\frac{x}{2n+x}\right)^5 + \cdots\right]$$
$$= \frac{2n}{2n+1} + \frac{2x}{2n+x} + \frac{1}{12n^2} + O(n^{-3}).$$

Since

$$\ln a_n - \ln a_{n+1} = \frac{2x-1}{2x^2} + O(n^{-3}) \quad \text{as } n \to \infty,$$

we deduce that the condition $x \geq 1/2$ is also necessary. $\square$

7.2.5. Show that for any positive integer n,

$$\frac{e}{2n+2} < e - \left(1 + \frac{1}{n}\right)^n < \frac{e}{2n+1}.$$

Solution. The first inequality can be rewritten as

$$\left(1 + \frac{1}{n}\right)^{n+1} < e\left(1 + \frac{1}{2n}\right). \tag{7.1}$$

For $x \in (0, 1/n]$, consider the function

$$f(x) = x + x\ln\left(1 + \frac{x}{2}\right) - (1+x)\ln(1+x).$$

Then $f(0) = 0$ and

$$f'(x) = \frac{x}{x+2} - \ln\frac{1+x}{1+\frac{x}{2}} > \frac{x}{x+2} - \frac{1+x}{1+\frac{x}{2}} + 1 = 0.$$

Thus, $f(x) > 0$ for all $x \in (0, 1/n]$, which implies relation (7.1).

The second inequality is equivalent to

$$e < \left(1+\frac{1}{n}\right)^n \left(1+\frac{1}{2n}\right),$$

which is a consequence of Exercise 7.2.4. □

7.3 Jensen, Young, Hölder, Minkowski, and Beyond

> In questions of science the authority of a thousand is not worth the humble reasoning of a single individual.
>
> Galileo Galilei (1564–1642)

7.3.1. Assume that $1 \le p < \infty$ and a, b are positive numbers. Then

(i) $\inf\limits_{t>0} \left[\frac{1}{p} t^{1/p-1} a + \left(1-\frac{1}{p}\right) t^{1/p} b\right] = a^{1/p} b^{1-1/p}.$

(ii) $\inf\limits_{0<t<1} \left[t^{1-p} a^p + (1-t)^{1-p} b^p\right] = (a+b)^p.$

Solution. We give two different proofs of this property, cf. Maligranda [74]. The first one relies on basic properties of differentiable functions, while the second proof is based on convexity arguments.

First proof. (i) Let, for $t > 0$, the function f be defined by

$$f(t) = \frac{1}{p} t^{1/p-1} a + \left(1 - \frac{1}{p}\right) t^{1/p} b.$$

Then the derivative f' satisfies

$$f'(t) = \frac{1}{p}\left(\frac{1}{p}-1\right) t^{1/p-2} a + \left(1-\frac{1}{p}\right) \frac{1}{p} t^{1/p-1} b = \frac{1}{p}\left(\frac{1}{p}-1\right) t^{1/p-2} (a - tb),$$

and so f' is negative for $t < t_0 = a/b$, zero for $t = t_0$, and positive for $t > t_0$. Hence, f has its minimum at the point $t_0 = a/b$, and this minimum is equal to

$$f(t_0) = f\left(\frac{a}{b}\right) = \frac{1}{p}\left(\frac{a}{b}\right)^{1/p-1} a + \left(1-\frac{1}{p}\right)\left(\frac{a}{b}\right)^{1/p} b = a^{1/p} b^{1-1/p}.$$

(ii) Let, for $0 < t < 1$, the function g be defined by

$$g(t) = t^{1-p} a^p + (1-t)^{1-p} b^p.$$

7.3 Jensen, Young, Hölder, Minkowski, and Beyond

Then the derivative g' satisfies the equation

$$g'(t) = (1-p)t^{-p}a^p - (1-p)(1-t)^{-p}b^p = 0$$

only when $t = t_1 = a/(a+b)$. Since

$$g''(t_1) = (1-p)(-p)t_1^{-p-1}a^p - (1-p)(-p)(1-t_1)^{-p-1}b^p > 0,$$

it follows that g has its local minimum at $t_1 = a/(a+b)$, which is equal to

$$g(t_1) = g\left(\frac{a}{a+b}\right) = \left(\frac{a}{a+b}\right)^{1-p} a^p + \left(1 - \frac{a}{a+b}\right)^{1-p} b^p$$

$$= \left(\frac{a}{a+b}\right)^{1-p} a^p + \left(\frac{b}{a+b}\right)^{1-p} b^p = (a+b)^p.$$

This local minimum of the function g is equal to its global minimum because g is continuous on $(0,1)$ and $\lim_{t \to 0^+} g(t) = \lim_{t \to 1^-} g(t) = +\infty$.

Second proof. (i) The function $\varphi(u) = \exp(u)$ is convex on $\mathbb{R}$. Thus

$$a^{1/p}b^{1-1/p} = \left[t^{1/p-1}a\right]^{1/p} \left[t^{1/p}b\right]^{1-1/p}$$

$$= \exp\left[\frac{1}{p}\ln(t^{1/p-1}a) + \left(1 - \frac{1}{p}\right)\ln(t^{1/p}b)\right]$$

$$\leq \frac{1}{p}\exp\left[\ln(t^{1/p-1}a)\right] + \left(1 - \frac{1}{p}\right)\exp\left[\ln(t^{1/p}b)\right]$$

$$= \frac{1}{p}t^{1/p-1}a + \left(1 - \frac{1}{p}\right)t^{1/p}b$$

for every $t > 0$. For $t = a/b$ we have equality.

(ii) The function $\psi(u) = u^p$ for $p > 1$ is convex on $[0, \infty)$. Therefore,

$$(a+b)^p = \left[t\frac{a}{t} + (1-t)\frac{b}{1-t}\right]^p$$

$$\leq t\left(\frac{a}{t}\right)^p + (1-t)\left(\frac{b}{1-t}\right)^p = t^{1-p}a^p + (1-t)^{1-p}b^p$$

for every $0 < t < 1$. For $t = a/(a+b)$ we have equality.

If $p \in (0,1)$ and we change in the equalities (i) and (ii) the infimum into supremum, then the results remain true. We also point out that the second proof of (i) gives an alternative proof of the Young inequality, as well as a different proof of the AM–GM inequality

$$a^{1/p}b^{1-1/p} \leq \frac{1}{p}a + \left(1 - \frac{1}{p}\right)b.$$

The classical Hölder inequality states that if f and g are continuous functions on the interval $[a,b]$ and if p and q are positive numbers such that $p^{-1} + q^{-1} = 1$, then

$$\int_a^b |f(x)g(x)|dx \leq \left(\int_a^b |f(x)|^p dx\right)^{1/p} \left(\int_a^b |g(x)|^q dx\right)^{1/q}.$$

Equivalently, if f and g are continuous on $[a,b]$, then

$$\int_a^b |f(x)|^{1/p}|g(x)|^{1/q}dx \leq \left(\int_a^b |f(x)|dx\right)^{1/p} \left(\int_a^b |g(x)|dx\right)^{1/q}.$$

According to Exercise 7.3.1, the inequality

$$a^{1/p}b^{1/q} \leq \frac{1}{p}t^{1/p-1}a + \left(1 - \frac{1}{p}\right)t^{1/p}b$$

holds for all $t > 0$. It follows that

$$\int_a^b |f(x)|^{1/p}|g(x)|^{1/q}dx \leq \int_a^b \left[\frac{1}{p}t^{1/p-1}|f(x)| + \left(1 - \frac{1}{p}\right)t^{1/p}|g(x)|\right] dx$$

$$= \frac{1}{p}t^{1/p-1}\int_a^b |f(x)|dx + \left(1 - \frac{1}{p}\right)t^{1/p}\int_a^b |g(x)|dx.$$

Taking the infimum for $t \in (0,\infty)$ and using Exercise 7.3.1 (i), we obtain Hölder's inequality.

The Minkowski inequality states that if f and g are continuous functions on the interval $[a,b]$ and if $p \geq 1$, then

$$\left(\int_a^b |f(x) + g(x)|^p dx\right)^{1/p} \leq \left(\int_a^b |f(x)|^p dx\right)^{1/p} + \left(\int_a^b |g(x)|^p dx\right)^{1/p}. \quad (7.2)$$

In order to prove (7.2), we apply Exercise 7.3.1 (ii). Thus, by the inequality

$$(a+b)^p \leq t^{1-p}a^p + (1-t)^{1-p}b^p$$

we deduce that for all $t \in (0,1)$,

$$\int_a^b |f(x) + g(x)|^p dx \leq \int_a^b [|f(x)| + |g(x)|]^p dx$$

$$\leq \int_a^b \left[t^{1-p}|f(x)|^p + (1-t)^{1-p}|g(x)|^p\right] dx$$

$$= t^{1-p}\int_a^b |f(x)|^p dx + (1-t)^{1-p}\int_a^b |g(x)|^p dx.$$

Taking the infimum for $t \in (0,1)$ and using Exercise 7.3.1 (ii), we obtain (7.2).

7.3 Jensen, Young, Hölder, Minkowski, and Beyond

We recall that Young's inequality asserts that for all $p > 1$ and $x, y > 0$,

$$xy \leq \frac{x^p}{p} + \frac{y^{p'}}{p'},$$

where p' stands for the conjugate exponent of p, that is, $1/p + 1/p' = 1$. A simple proof of Young's inequality is described in what follows. Set $x^p = e^a$ and $y^{p'} = e^b$. Let $1/p = t$ and observe that $0 \leq t \leq 1$. We have to show that

$$e^{(1-t)a+tb} \leq (1-t)e^a + te^b.$$

Let $f(t) = (1-t)e^a + te^b - e^{(1-t)a+tb}$. Then $f''(t) = -(a-b)^2 e^{(1-t)a+tb} \leq 0$. Thus, f is concave and $f(0) = f(1) = 0$. This forces $f(t) \leq 0$ for any $t \in [0,1]$. □

The next exercise is a refinement of Young's inequality.

7.3.2. Let $1 < p \leq 2$ and let p' be its conjugate exponent. Then for all $x, y \geq 0$,

$$\frac{1}{p'}(x^{p/2} - y^{p'/2})^2 \leq \frac{x^p}{p} + \frac{y^{p'}}{p'} - xy \leq \frac{1}{p}(x^{p/2} - y^{p'/2})^2.$$

Solution. If $p = 2 = p'$ the result is trivial, so assume $1 < p < 2$. We prove the first inequality because the second can be obtained via an essentially identical argument by interchanging the roles of p and p' and of x and y. If either $x = 0$ or $y = 0$, the first inequality is obviously true. Fix $1 < p < 2$, $x > 0$, and suppose $y > 0$. Expanding the square and simplifying, we see that it is enough to check the following inequality:

$$f(y) := \frac{2-p}{p} x^p + \frac{2}{p'} x^{p/2} y^{p'/2} - xy \geq 0.$$

Now $y = x^{p-1}$ is the unique solution of $f'(y) = 0$. Since $f'' > 0$, $f(x^{p-1}) = 0$ is the global minimum of f.

Jensen's inequality asserts that if $f : (a,b) \to \mathbb{R}$ is a convex function, then

$$f(\lambda_1 x_1 + \lambda_2 x_2 + \cdots + \lambda_n x_n) \leq \lambda_1 f(x_1) + \lambda_2 f(x_2) + \cdots + \lambda_n f(x_n),$$

for any integer $n \geq 2$, where $x_1, x_2, \ldots, x_n \in (a,b)$ are arbitrary points, and $\lambda_i \geq 0$ ($i = 1, \ldots, n$) are constrained by $\lambda_1 + \cdots + \lambda_n = 1$.

We give in what follows a direct proof of Jensen's inequality, whose main idea relies on the property that any convex function has a support line, in the sense that through each point of its graph, a line can be drawn such that the graph of the function does not lie below the line. We first observe that if $\varphi(x) = ax + b$, where a and b are real numbers with $b \neq 0$, then

$$\varphi\left(\sum_{i=1}^n \lambda_i x_i\right) = \sum_{i=1}^n \lambda_i \varphi(x_i) \quad \text{if and only if} \quad \sum_{i=1}^n \lambda_i = 1. \tag{7.3}$$

Indeed, relation (7.3) reduces to

$$a\sum_{i=1}^{n}\lambda_i x_i + b = a\sum_{i=1}^{n}\lambda_i x_i + \left(\sum_{i=1}^{n}\lambda_i\right)b,$$

which is true if and only if $\sum_{i=1}^{n}\lambda_i = 1$.

Returning to the proof of Jensen's inequality, fix $x_1, x_2, \ldots, x_n \in (a,b)$, $\lambda_i \in \mathbb{R}$ ($i = 1, \ldots, n$), and set $\xi = \sum_{i=1}^{n}\lambda_i x_i$. Let $\varphi(x) = ax + b$ be a support line of f at ξ with $f(\xi) = \varphi(\xi)$. Thus, by (7.3),

$$f\left(\sum_{i=1}^{n}\lambda_i x_i\right) = \sum_{i=1}^{n}\lambda_i \varphi(x_i) \quad \text{if and only if } \sum_{i=1}^{n}\lambda_i = 1. \tag{7.4}$$

But $\varphi(x) \leq f(x)$ for all $x \in (a,b)$. Hence

$$\sum_{i=1}^{n}\lambda_i \varphi(x_i) \leq \sum_{i=1}^{n}\lambda_i f(x_i) \quad \text{if and only if } \lambda_i \geq 0. \tag{7.5}$$

Using relations (7.4) and (7.5), we obtain Jensen's inequality. Equality holds either if $x_1 = \cdots = x_n$ or if f is linear. $\square$

Examples. (i) Taking $f(x) = -\ln x$ in Jensen's inequality, we obtain the following generalized arithmetic–geometric means inequality: for any $x_1, x_2, \ldots, x_n > 0$ and all $\lambda_i \geq 0$ ($1 \leq i \leq n$) with $\sum_{i=1}^{n}\lambda_i = 1$,

$$\lambda_1 x_1 + \cdots + \lambda_n x_n \geq x_1^{\lambda_1} \cdots x_n^{\lambda_n}.$$

(ii) Taking $f(x) = x^{q/p}$ in Jensen's inequality, we obtain the power means inequality: if $p < q$, then $M_p \leq M_q$, where

$$M_r := \left(\sum_{i=1}^{n}\lambda_i a_i^r\right)^{1/r} \quad \text{if } r \neq 0, \quad M_0 = \prod_{i=1}^{n} a_i^{\lambda_i}, \tag{7.6}$$

with $a_i > 0$ and $\sum_{i=1}^{n}\lambda_i = 1$, $\lambda_i \geq 0$. In Figure 7.1 we illustrate this property for $n = 2$ and M_r defined by $M_{-\infty} = \min(a,b) =: m$, $M_{-1} = 1/(p/a + q/b)$, $M_0 = a^p b^q$, $M_1 = pa + qb$, $M_2 = \sqrt{pa^2 + qb^2}$, $M_\infty = \max(a,b) =: M$, where $a, b \in (0, \infty)$, $p \in (0,1)$, and $q = 1 - p$.

We analyze in what follows a different framework, corresponding to $\lambda_1 = \cdots = \lambda_n = 1$, $n \geq 2$. In this case, for positive numbers $a_1, \ldots, a_n$ and $p \neq 0$, define

$$\mathscr{M}_p := \left(\sum_{i=1}^{n} a_i^p\right)^{1/p}.$$

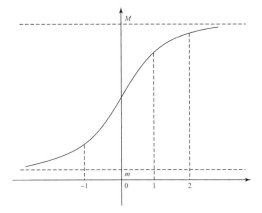

Fig. 7.1 The power mean M_r is a monotone increasing function.

It was shown (see [93]) by the German mathematician Alfred Pringsheim (1850–1941), who attributed his proof to Lüroth, and by Jensen [51, p. 192], that

$$\mathscr{M}_p > \mathscr{M}_q \quad \text{for any } 0 < p < q.$$

This inequality can be obtained directly as follows:

$$\frac{\left(\sum_{i=1}^n a_i^q\right)^{1/q}}{\left(\sum_{i=1}^n a_i^p\right)^{1/p}} = \left[\sum_{j=1}^n \frac{a_j^q}{\left(\sum_{i=1}^n a_i^p\right)^{q/p}}\right]^{1/q} = \left[\sum_{j=1}^n \left(\frac{a_j^p}{\sum_{i=1}^n a_i^p}\right)^{q/p}\right]^{1/q}$$

$$< \left(\sum_{j=1}^n \frac{a_j^p}{\sum_{i=1}^n a_i^p}\right)^{1/q} = 1.$$

Let $m = \min\{a_1,\ldots,a_n\}$ and $M = \max\{a_1,\ldots,a_n\}$. It follows that for any $p > 0$,

$$M < \mathscr{M}_p < n^{1/p} M.$$

Therefore

$$\lim_{p \to +\infty} \mathscr{M}_p = M.$$

On the other hand, since $\lim_{p \to 0+} \sum_{j=1}^n a_j^p = n$, we obtain

$$\lim_{p \to 0+} \mathscr{M}_p = +\infty.$$

Hence $\mathscr{M}_p$ decreases from $+\infty$ to M as p increases from $0+$ to $+\infty$. Moreover, since

$$\mathscr{M}_{-p} = \mathscr{M}_p,$$

it follows that $\mathcal{M}_p$ decreases from m to 0 as p increases from $-\infty$ to -0. By contrast, the mean value functions M_r defined in (7.6) increase continuously from m to M, or remain constant if $a_1 = \cdots = a_n$, as r increases from $-\infty$ to $+\infty$.

7.4 Optimization Problems

> It is not the mountain we conquer, but ourselves.
>
> Sir Edmund Hillary (1919–2008)

We have proved in Chapter 4 that any continuous function $f : [a,b] \to [a,b]$ has at least one fixed point. A natural question in applications is to provide an algorithm for finding this point. A simple and effective method given by Picard seeks to achieve what may be the next-best alternative to finding a fixed point exactly, namely, to find it approximately. More precisely, if $x_1 \in [a,b]$ is chosen arbitrarily, define $x_{n+1} = f(x_n)$, and the resulting sequence $(x_n)_{n \geq 1}$ is called the *sequence of successive approximations* of f (or a *Picard sequence* for the function f).

In geometric terms, the basic idea of the Picard method can be described as follows. First, take a point $A(x_1, f(x_1))$ on the curve $y = f(x)$. Next, consider the point $B_1(f(x_1), f(x_1))$ on the diagonal line $y = x$ and then project the point B_1 vertically onto the curve $y = f(x)$ to obtain a point $A_2(x_2, f(x_2))$. Again, project A_2 horizontally to B_2 on $y = x$ and then project B_2 vertically onto $y = f(x)$ to obtain $A_3(x_3, f(x_3))$. This process can be repeated a number of times. Often, it will weave a cobweb in which the fixed point of f, that is, the point of intersection of the curve $y = f(x)$ and the diagonal line $y = x$, gets trapped (see Figure 7.2). In fact, we will see that such trapping occurs if the slopes of tangents to the curve $y = f(x)$ are smaller (in absolute value) than the slope of the diagonal line $y = x$. When the slope condition is not met, then the points $A_1, A_2, \ldots$ may move away from a fixed point (see Figure 7.3).

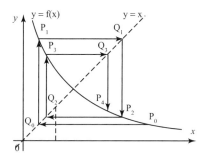

Fig. 7.2 Picard sequence converging to a fixed point.

7.4 Optimization Problems

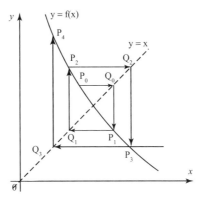

Fig. 7.3 Picard sequence diverging away from a fixed point.

A sufficient condition for the convergence of a Picard sequence, which is a formal analogue of the geometric condition of slopes mentioned above, is stated in the following result, which is also referred to as the *Picard convergence theorem*.

7.4.1. *Let $f : [a,b] \to [a,b]$ be a continuous function that is differentiable on (a,b), with $|f'(x)| < 1$ for all $x \in (a,b)$.*

(i) *Prove that f has a unique fixed point.*
(ii) *Show that any Picard sequence for f is convergent and converges to the unique fixed point of f.*

Solution. (i) By Brouwer's fixed-point theorem [Exercise 4.4.1 (i)], f has at least one fixed point. Assuming that f has two fixed points x_* and x^*, then by the Lagrange mean value theorem, there exists $\xi \in (a,b)$ such that

$$|x_* - x^*| = |f(x_*) - f(x^*)| = |f'(\xi)| \cdot |x_* - x^*| < |x_* - x^*|,$$

a contradiction. Thus, f has a unique fixed point.

We point out that the condition $|f'(x)| < 1$ for all $x \in (a,b)$ is essential for the uniqueness of a fixed point. For example, if $f : [a,b] \to [a,b]$ is defined by $f(x) = x$, then $f'(x) = 1$ for all $x \in [a,b]$ and every point of $[a,b]$ is a fixed point of f.

(ii) Let x^* denote the unique fixed point of f. Consider any $x_1 \in [a,b]$ and let $(x_n)_{n \geq 1} \subset [a,b]$ be the Picard sequence for f with its initial point x_1. Fix an integer $n \geq 1$. Thus, by the Lagrange mean value theorem, there exists ξ_n between x_n and x^* such that

$$x_{n+1} - x^* = f(x_n) - f(x^*) = f'(\xi_n)(x_n - x^*).$$

This implies that $|x_{n+1} - x^*| < |x_n - x^*|$. Next, we prove that $x_n \to x^*$ as $n \to \infty$. Since $(x_n)_{n \geq 1}$ is bounded, it suffices to show that every convergent subsequence of $(x_n)_{n \geq 1}$

converges to x^*. Let $x \in \mathbb{R}$ and $(x_{n_k})_{k \geq 1}$ be a subsequence of $(x_n)_{n \geq 1}$ converging to x. Then

$$|x_{n_{k+1}} - x^*| \leq |x_{n_k+1} - x^*| \leq |x_{n_k} - x^*|.$$

But $|x_{n_{k+1}} - x^*| \to |x - x^*|$ and

$$|x_{n_k+1} - x^*| = |f(x_{n_k}) - f(x^*)| \to |f(x) - f(x^*)| \quad \text{as } k \to \infty.$$

It follows that $|f(x) - f(x^*)| = |x - x^*|$. Now, if $x \neq x^*$, then by the Lagrange mean value theorem, there exists $\xi \in (a,b)$ such that

$$|x - x^*| = |f(x) - f(x^*)| = |f'(\xi)| \cdot |x - x^*| < |x - x^*|,$$

which is a contradiction. This proves that $x \neq x^*$.

We point out that if the condition $|f'(x)| < 1$ for all $x \in (a,b)$ is not satisfied, then f can still have a unique fixed point x^*, but the Picard sequence $(x_n)_{n \geq 1}$ with initial point $x_1 \neq x^*$ may not converge to x^*. For example, if $f : [-1,1] \to \mathbb{R}$ is defined by $f(x) = -x$, then f maps $[-1,1]$ into itself, f is differentiable, and $|f'(x)| = 1$ for all $x \in [-1,1]$. Then $x^* = 0$ is the unique fixed point of f, but if $x_1 \neq 0$ then the corresponding Picard sequence is $x_1, -x_1, x_1, -x_1, \ldots$, which oscillates between x_1 and $-x_1$ and never reaches the fixed point. In geometric terms, the cobweb that we hope to weave just traces out a square over and over again.

When the hypotheses of the Picard convergence theorem are satisfied, a Picard sequence for $f : [a,b] \to [a,b]$ with arbitrary $x_1 \in [a,b]$ as its initial point converges to a fixed point of f. It is natural to expect that if x_1 is closer to the fixed point, then the convergence rate will be better. A fixed point of f lies not only in the range of f but also in the ranges of the iterates $f \circ f$, $f \circ f \circ f$, and so on. Thus, if $\mathscr{R}_n$ is the range of the n-fold composite $f \circ \cdots \circ f$ (n times), then a fixed point is in each $\mathscr{R}_n$. If only a single point belongs to $\cap_{n=1}^{\infty} \mathscr{R}_n$, then we have found our fixed point. In fact, the Picard method amounts to starting with any $x_1 \in [a,b]$ and considering the image of x_1 under the n-fold composite $f \circ \cdots \circ f$. For example, if $f : [0,1] \to [0,1]$ is defined by $f(x) = (x+1)/4$, then

$$f \circ \cdots \circ f \ (n \text{ times}) \ (x) = \frac{3x + 4^n - 1}{3 \cdot 4^n}$$

and

$$\mathscr{R}_n = \left[\frac{1}{3}\left(1 - \frac{1}{4^n}\right), \frac{1}{3}\left(1 + \frac{2}{4^n}\right) \right].$$

Thus, $\cap_{n=1}^{\infty} \mathscr{R}_n = \{1/3\}$; hence $1/3$ is the unique fixed point of f. In general, it is not convenient to determine the ranges $\mathscr{R}_n$ for all n. So, it is simpler to use the Picard method, but this tool will be more effective if the above observations are used to some extent in choosing the initial point. □

7.4 Optimization Problems

The following elementary problem is related to the minimization of the renormalized Ginzburg–Landau energy functional (see [6]). Problems of this type arise in the study of two phenomena in quantum physics: superconductivity (discovered in 1911 by the Dutch physicist Heike Kamerlingh Ones (1853–1926), Nobel Prize in Physics 1913 "for his investigations on the properties of matter at low temperatures which led, inter alia, to the production of liquid helium") and superfluids. Superconducting material is used, for example, in magnetic resonance imaging for medical examinations and particle accelerators in physics. Knowledge about superfluid liquids can give us deeper insight into the ways in which matter behaves in its lowest and most ordered state. We point out that the Nobel Prize in Physics was awarded in 2003 to Alexei Abrikosov, Vitaly Ginzburg, and Anthony Leggett "for pioneering contributions to the theory of superconductors and superfluids." The next problem gives an idea about the location of *vortices (singularities)*, which have the tendency to be distributed in regular configurations called *Abrikosov lattices*. We refer to [68] for more details and comments.

7.4.2. *Let D be the set of complex numbers having modulus less than 1. For any fixed positive integer n and any distinct complex numbers $z_1, \ldots, z_n$ in D, define*

$$W(z_1,\ldots,z_n) = \sum_{1 \leq j < k \leq n} \log|z_j - z_k|^2 + \sum_{j,k=1}^{n} \log|1 - z_j\overline{z_k}|,$$

where $\overline{z}$ is the conjugate of the complex number z.

(i) *If $n = 2$, prove that the configuration $(z_1,\ldots,z_n) \in D^n$ that realizes the maximum of W is unique (up to a rotation) and it consists of two points that are symmetric with respect to the origin.*

(ii) *If $n = 3$, prove that the configuration that maximizes W is also unique (up to a rotation) and it consists of an equilateral triangle centered at the origin.*

Vicenţiu Rădulescu

Solution. (i) Take $n = 2$ and let z_1, z_2 be two distinct points in D. Then

$$W(z_1,z_2) = \log(|z_1|^2 + |z_2|^2 - 2|z_1|\cdot|z_2|\cdot\cos\varphi) \\ + \log(1+|z_1|^2|z_2|^2 - 2|z_1|\cdot|z_2|\cdot\cos\varphi) + \log(1-|z_1|^2) + \log(1-|z_1|^2),$$

where φ denotes the angle between the vectors $\overrightarrow{Oz_1}$ and $\overrightarrow{Oz_2}$. So, a necessary condition for the maximum of $W(z_1,z_2)$ is $\cos\varphi = -1$, that is, the points z_1, O, and z_2 are collinear, with O between z_1 and z_2. Hence one may suppose that the points z_1 and z_2 lie on the real axis and $-1 < z_2 < 0 < z_1 < 1$. Define

$$f(z_1,z_2) = 2\log(z_1 - z_2) + 2\log(1 - z_1 z_2) + \log(1 - z_1^2) + \log(1 - z_2^2).$$

Since the function $\log(1-x^2)$ is concave on $(0,+\infty)$, it follows that

$$\log(1-z_1^2) + \log(1-z_2^2) \leq 2\log\left(1 - \left(\frac{z_1-z_2}{2}\right)^2\right).$$

On the other hand, it is obvious that $1 - z_1 z_2 \leq 1 + (z_1-z_2)^2/4$. Hence

$$f(z_1,z_2) \leq f\left(\frac{z_1-z_2}{2}, \frac{z_2-z_1}{2}\right),$$

which means that the maximum of f is achieved when $z_1 = -z_2$. A straightforward computation shows that $\max W = f(5^{-1/4}, -5^{-1/4}) = 6\log 2 - 2^{-1} \cdot 5\log 5$.

(ii) For $n=3$, in order to maximize the functional $W(z_1,z_2,z_3)$, it is enough to find the maximum of

$$F(z_1,z_2,z_3) = \prod_{1\leq j<k\leq 3} |z_j - z_k|^2 \left[|z_j - z_k|^2 + (1-r_j^2)(1-r_k^2)\right] \cdot \prod_{j=1}^{3}(1-r_j^2),$$

where $r_j = |z_j|$.

Using the elementary identity

$$3\sum_{j=1}^{3} |z_j|^2 = \left|\sum_{j=1}^{3} z_j\right|^2 + \sum_{1\leq j<k\leq 3} |z_j - z_k|^2,$$

we obtain

$$3\sum_{j=1}^{3} |z_j|^2 \geq \sum_{1\leq j<k\leq 3} |z_j - z_k|^2.$$

Put $S = \sum_{j=1}^{3} r_j^2$. We try to maximize F keeping S constant. Using the above inequality, we have

$$\prod_{1\leq j<k\leq 3} |z_j - z_k|^2 \leq \left(\frac{\sum_{1\leq j<k\leq 3} |z_j - z_k|^2}{3}\right)^3 \leq \left(\sum_{j=1}^{3} |z_j|^2\right)^3 = S^3, \quad (7.7)$$

$$\prod_{j=1}^{3}(1-r_j^2) \leq \left(\frac{3-S}{3}\right)^3, \quad (7.8)$$

and

$$\prod_{1\leq j<k\leq 3}\left[|z_j - z_k|^2 + (1-r_j^2)(1-r_k^2)\right]$$

$$\leq \left[\frac{\sum_{1\leq j<k\leq 3}\left(|z_j-z_k|^2 + (1-r_j^2)(1-r_k^2)\right)}{3}\right]^3$$

$$\leq \left[\frac{\Sigma 1 - \Sigma r_j^2 - \Sigma r_k^2 + \Sigma r_j^2 r_k^2 + \Sigma |z_j - z_k|^2}{3}\right]^3 \qquad (7.9)$$

$$\leq \left(\frac{3 - 2S + \frac{S^2}{3} + 3S}{3}\right)^3 = \left(\frac{S^2 + 3S + 9}{3^2}\right)^3.$$

We have applied above the elementary inequality

$$\sum_{1 \leq j < k \leq 3} r_j^2 r_k^2 \leq \frac{1}{3}\left(\sum_{j=1}^{3} r_j^2\right)^2.$$

From (7.7), (7.8), and (7.9) we obtain

$$F \leq S^3 \cdot \left(\frac{3-S}{3}\right)^3 \cdot \left(\frac{S^2 + 3S + 9}{3^2}\right)^3 = \frac{1}{3^9}(-S^4 + 27S)^3.$$

It follows that the maximum of F (so the maximum of W) is achieved if $S = 3 \cdot 4^{-1/3}$ and $\max F = 3^6 \cdot 4^{-4}$, with equality when we have equality in (7.7), (7.8), and (7.9), that is, if and only if $z_2 = \varepsilon z_1$, $z_3 = \varepsilon^2 z_1$, where $\varepsilon = \cos(2\pi/3) + i\sin(2\pi/3)$. This implies that $\max W = 6\log 3 - 8\log 2$. □

Open problems. (i) Find the configuration that maximizes $W(z_1, \ldots, z_n)$, provided that $n \geq 4$. Is this configuration given by a regular n-gon, as for $n = 2$ or $n = 3$?

(ii) Study whether the maximal configuration "goes to the boundary" as $n \to \infty$, in the following sense: for given n, let $z = (z_1, \ldots, z_n)$ be an arbitrary configuration that realizes the maximum of $W(z_1, \ldots, z_n)$ and set $a_n = \min\{|z_j|;\ 1 \leq j \leq n\}$. Is it true that $a_n \to 1$ as $n \to \infty$?

7.5 Qualitative Results

> Nothing in life is to be feared. It is only to be understood.
>
> Marie Curie (1867–1934)

7.5.1. Let $2 < q < p + 1$ and set

$$M = \begin{cases} \frac{p}{2}(q-1) & \text{if } 1 < p \leq 2, \\ \frac{p}{2}(p-1)\left(\frac{p-1}{p+1-q}\right)^{p-2} & \text{if } p \geq 2. \end{cases}$$

Prove that for any $x \geq 0$,

$$(1+x)_+^p \leq \left(1 - \frac{q}{p+1}\right)(1+x)_+^{p+1} + (q-1)x + Mx^2 + \frac{q}{p+1}.$$

Wolfgang Reichel

Solution. Let $l(x) = (1+x)_+^p$ and

$$r(x) = \left(1 - \frac{q}{p+1}\right)(1+x)_+^{p+1} + (q-1)x + Mx^2 + \frac{q}{p+1}.$$

Note that $l(0) = r(0)$ and $l'(0) = r'(0)$. Let us first consider the case $p \geq 2$. Then our inequality holds if $l''(x) \leq r''(x)$ for all $x \in \mathbb{R}$, that is,

$$p(p-1)(1+x)_+^{p-2} \leq p(p+1-q)(1+x)_+^{p-1} + 2M. \tag{7.10}$$

Let $x_0 = (q-2)/(p+1-q)$ be the value for which $p-1 = (p+1-q)(1+x_0)$. For $x \geq x_0$ the inequality (7.10) holds automatically, while for $x \leq x_0$ it holds if

$$p(p-1)(1+x_0)^{p-2} \leq 2M,$$

which is true with equality due to the choice of M.

Now consider the case $1 < p \leq 2$. For $x \geq 0$, relation (7.10) holds if

$$p(p-1) \leq p(p+1-q) + 2M,$$

that is, if $p(q-2) \leq 2M$. For $x \leq 0$ we argue with first derivatives instead of second derivatives, that is, we show that $l'(x) \geq r'(x)$ for $x \geq 0$. This amounts to

$$p(1+x)_+^{p-1} \geq (p+1-q)(1+x)_+^p + q - 1 + 2Mx \quad \text{for } x \leq 0. \tag{7.11}$$

Combining the two inequalities

$$(1-q)(1+x)_+^p + q - 1 \leq (1-q)px \leq -2Mx \quad \text{for } x \leq 0 \text{ if } p(q-1) \leq 2M$$

and

$$(1+x)_+^p \leq (1+x)_+^{p-1} \quad \text{for } x \leq 0,$$

we obtain (7.11), provided $p(q-1) \leq 2M$. This is guaranteed by the choice of M. $\square$

7.5.2. Show that

$$1 - \frac{x}{3} < \frac{\sin x}{x} < 1.1 - \frac{x}{4} \quad \text{for all } 0 < x \leq \pi.$$

7.5 Qualitative Results

Solution. For the first inequality, consider the function $f(x) = \sin x - x + x^2/3$. Then $f'(x) = \cos x - 1 + 2x/3$ and $f''(x) = -\sin x + 2/3$. Thus, $f''(x) > 0$ except for $a < x < b$, where $\sin a = \sin b = 2/3$ and $a < \pi/2 < b$. Thus $f'(x)$ increases from 0 to a maximum at $x = a$, then decreases to a minimum at $x = b$, and then increases from b to π. The minimum value is

$$f'(b) = \cos b - 1 + \frac{2b}{3} > \frac{\sqrt{5}}{3} - 1 + \frac{\pi}{3} > 0.$$

We deduce that $f'(x) \geq 0$ for $0 \leq x \leq \pi$ and consequently $\sin x \geq x - \frac{x^2}{3}$.

For the second inequality, let $g(x) = \sin x - 1.1x + x^2/4$. Thus, $g'(x) = \cos x - 1.1 + x/2$ and $g''(x) = -\sin x + 1/2$. Then $g''(x) \geq 0$ except for $\pi/6 < x < 5\pi/6$, where $g''(x) < 0$. So $g'(x)$ increases from a value of -1.1 to a maximum of $\cos(\pi/6) - 1.1 + \pi/12 > 0$. It must be zero at a point, say $x = \alpha$, where $0 \leq \alpha \leq \pi/6$. Similarly, $g'(\beta) = 0$ for $\pi/6 \leq \beta \leq 5\pi/6$. We observe that

$$g'(\pi/4) = \frac{\sqrt{2}}{2} - 1.1 + \frac{\pi}{8} > 0$$

and

$$g'(3\pi/4) = \frac{-\sqrt{2}}{2} - 1.1 + \frac{3\pi}{8} < 0,$$

so $\pi/4 < \beta < 3\pi/4$. Then from 0, $g(x)$ decreases to a minimum at $x = \alpha$, then increases to a maximum value at $x = \beta$, and

$$g'(\beta) = 0 = \cos\beta - 1.1 + \frac{\beta}{2}.$$

Hence $\beta = 2(1.1 - \cos\beta)$. Thus

$$g(\beta) = \sin\beta - 1.1\beta + \frac{\beta^2}{4} = \sin\beta + \cos^2\beta - (1.1)^2$$
$$= (0.3 - \sin\beta)(\sin\beta - 0.7) < 0,$$

since $\sin\beta \geq \sqrt{2}/2$. We conclude that $g(x) = \sin x - 1.1 + x^2/4 \leq 0$ for $0 \leq x \leq \pi$. □

7.5.3. Prove that for any $x \in (0,1)$,

$$\frac{2}{\pi}\left(\sin\frac{\pi x}{2}\right)\arcsin x < x^2 < (\sin x)\arcsin x.$$

Solution. The first inequality is a special case of the inequality

$$f(x)f^{-1}(y) \leq xy, \qquad (7.12)$$

where $x > 0$, $y \geq f(x)$, and the positive function f is such that $f(x)/x$ increases. The inequality (7.12) follows from

$$\frac{f(x)}{x} \leq \frac{f(z)}{z} \quad \text{for } z = f^{-1}(y) \geq x.$$

The second inequality follows from the inequality

$$f(x)f^{-1}(y) \geq xy, \tag{7.13}$$

where $x, y > 0$, $y \leq f(x)$, and the function f is such that $f(x)/x$ increases. The inequality (7.13) follows from

$$\frac{f(x)}{x} \geq \frac{f(z)}{z} \quad \text{for } z = f^{-1}(y) \leq x.$$

To obtain the required inequalities, it is now enough to choose $f(x) = \arcsin x$, $x \in (0, 1)$.

This method can be applied to deduce many elementary inequalities, such as

$$(e^x - 1)\ln(1+x) > x^2 \quad \text{for } x > 0,$$
$$(\tan x)\arctan x > x^2 \quad \text{for } x \in (0, \pi/2),$$
$$[(1+x)^p - 1]\left[(1+x)^{1/p-1} - 1\right] > x^2 \quad \text{for } x > 0 \text{ and } p > 0, p \neq 1,$$
$$[(1-x)^p - 1]\left[1 - (1+x)^{1/p}\right] > x^2 \quad \text{for } x \in (0, 1) \text{ and } p < -1. \quad \square$$

7.6 Independent Study Problems

> A great truth is a truth whose opposite is also a great truth.
>
> Thomas Mann (1875–1955)

7.6.1. Prove that $m^2\sqrt{3}/4$ is the maximum area of a triangle that can be formed with the lines a, b, c, subject to the condition that $a^3 + b^3 + c^3 = 3m^3$.

C.N. Mills, Amer. Math. Monthly, Problem 3207

7.6.2. Prove that if $1 \leq x_1 \leq x_2$ and n is a positive integer, then

$$x_2^{1/n} - x_1^{1/n} \leq (x_2 - 1)^{1/n} - (x_1 - 1)^{1/n}.$$

A. Dresden, Amer. Math. Monthly, Problem 3868

7.6 Independent Study Problems

7.6.3. Show that for any real number x with $|x| < 1$,
$$\frac{|x|}{1+|x|} \leq |\ln(1+x)| \leq \frac{|x|(1+|x|)}{|1+x|}.$$

H.S. Wall, Amer. Math. Monthly, Problem 3965

7.6.4. Determine whether the following assertion is true: if x and y are real numbers such that $y \geq 0$ and $y(y+1) \leq (x+1)^2$, then $y(y-1) \leq x^2$.

7.6.5. Prove that for any real x,
$$\left|\frac{\pi}{4} - \arctan x\right| \leq \frac{\pi}{4} \frac{|x-1|}{(x^2+1)^{1/2}}.$$

7.6.6. Determine whether the inequality $a \tanh x > \sin ax$ holds for every $x > 0$ and $a > 1$.

7.6.7. Let x and y be real numbers such that $x^2 + y^2 \leq \pi$. Prove that $\cos x + \cos y \leq 1 + \cos xy$.

7.6.8. Prove that
$$x\frac{\pi - x}{\pi + x} < \sin x < \left(3 - \frac{x}{\pi}\right) x \frac{\pi - x}{\pi + x} \quad \text{for all } 0 < x < \pi.$$

7.6.9. Fix an integer $n \geq 2$ and positive numbers $a_1, \ldots, a_n$.
(i) Prove that the function $f(x) = \log\left(\sum_{i=1}^n a_i^x\right)^{1/x}$ is convex for $x \in (0, \infty)$.
(ii) Deduce that the function $g(x) = \left(\sum_{i=1}^n a_i^x\right)^{1/x}$ is convex on the interval $(0, \infty)$.

The following problem considers relations supplementing the classical inequalities:
$$\sin x < x < \tan x, \quad \ln(1+x) < x < e^x - 1, \quad \arctan x < x < \arcsin x.$$

7.6.10. Prove the following inequalities:
(i) $(1+x)\ln^2(1+x) < x^2$ for $x > -1$, $x \neq 0$;
(ii) $x/(1 + 2x/\pi) < \arctan x$ for $x > 0$;
(iii) $x^2 < \ln(1 + \tan^2 x) < (\sin x)\tan x$ for $x \in (0, \pi/2)$;
(iv) $3x - x^3 < 2\sin(\pi x/2)$ for $x \in (0,1)$;
(v) $x^3 < (\sin^2 x)\tan x$ for $x \in (0, \pi/2)$;
(vi) Is the inequality $x^{3+\varepsilon} < (\sin x)^{2+\varepsilon} \tan x$ true on $(0, \pi/2)$ for some positive number ε?

7.6.11. Let $D \subset \mathbb{R}$ and $f : D \to \mathbb{R}$ be such that $f(D) \subset D$. Prove the following generalizations and extensions of the Picard convergence theorem.

(i) **Banach Fixed Point Theorem (Contraction Principle).** If D is closed and f is a contraction (that is, there exists $\alpha \in (0,1)$ such that $|f(x) - f(y)| \leq \alpha |x - y|$ for all $x, y \in D$), then f has a unique fixed point, and any Picard sequence converges to this fixed point. Give an example to show that if f is a contraction but D is not closed, then f need not have a fixed point.

(ii) If D is closed and bounded, and f is contractive (that is, $|f(x) - f(y)| < |x - y|$ for all $x, y \in D$, $x \neq y$), then f has a unique fixed point, and any Picard sequence converges to this fixed point. Give an example to show that if f is a contractive function but D is not closed and bounded, then f need not have a fixed point.

(iii) If D is a closed and bounded interval, and f is nonexpansive (that is, $|f(x) - f(y)| \leq |x - y|$ for all $x, y \in D$), then f has a fixed point in D but it may not be unique. Give an example to show that if f is nonexpansive but D is not a closed and bounded interval, then f need not have a fixed point.

Part IV
Antiderivatives, Riemann Integrability, and Applications

Part IV is devoted to the *Riemann integral*, which is the second fundamental topic treated in calculus, the first being the *derivative*. The notion of integral arises from the need to make rigorous the intuitive concept of area of a region of the plane, though other questions lead to the same notion, typically those that require dealing with the sum of effects of a process. We are concerned with many related qualitative properties including the fundamental theorem of calculus, which states that the processes of derivation and integration are inverse to each other.

Chapter 8
Antiderivatives

> *Nature and Nature's law lay hid in night: God said, "Let Newton be!," and all was light.*
> —Alexander Pope (1688–1744), Epitaph on Newton

Abstract. The integral calculus is much older than the differential calculus, because the computation of lengths, areas, and volumes occupied the greatest mathematicians since antiquity: Archimedes, Kepler, Cavalieri, Barrow. The decisive breakthrough came when Newton and Leibniz discovered that integration is the inverse operation of differentiation, thus reducing much effort to a couple of differentiation rules.

8.1 Main Definitions and Properties

> And whereas Mr. Leibniz præfixes the letter $\int$ to the Ordinate of a curve to denote the Sum of the ordinates or area of the Curve, I did some years before represent the same thing by inscribing the Ordinate in a square... My symbols therefore... are the oldest in the kind.
>
> Sir Isaac Newton (1642–1727), Letter to Keill, 1714

Let f be a real-valued function defined on (a,b). A differentiable function F is an *antiderivative* (or a *primitive*) of f if

$$F'(x) = f(x) \quad \text{for any } x \in (a,b).$$

We shall use only the term *antiderivative*, and symbolically, we write

$$F(x) = \int f(x)dx.$$

The mean value theorem implies that any two antiderivatives of a function f differ by a constant. Thus, if F and G are antiderivatives of f then there exists $C \in \mathbb{R}$ such that $F - G = C$.

We observe that *not every function has an antiderivative* on a given interval (a,b). Indeed, if $f : (a,b) \to \mathbb{R}$ has an antiderivative F on (a,b), then by Darboux's theorem, $f = F'$ satisfies the intermediate value property. This shows that the presence of a jump discontinuity at a single point in (a,b) is enough to prevent the existence of an antiderivative F on (a,b). In other words, if f is defined on (a,b) and $f(c+)$, $f(c-)$ exist but are not both equal to $f(c)$, for some $c \in (a,b)$, then f has no antiderivative on (a,b).

A large class of functions that have antiderivatives on (a,b) is the class of all continuous functions $f : (a,b) \to \mathbb{R}$. Next, we investigate the converse of this statement: to what extent does the existence of an antiderivative F of f determine the continuity of f? We first observe that the derivative $F'(x) = f(x)$ of the function

$$F(x) = \begin{cases} x^2 \sin \frac{1}{x} & \text{if } x \neq 0, \\ 0 & \text{if } x = 0, \end{cases}$$

exists for all $x \in \mathbb{R}$, but f is not continuous at $x = 0$. This example shows that it is possible for a function f to have an antiderivative and to be discontinuous at some points.

Basic properties of antiderivatives.

(i) If f and g have antiderivatives on (a,b) and $k \in \mathbb{R}$, so do $f + g$ and kf, and

$$\int [f(x) + g(x)] dx = \int f(x) dx + \int g(x) dx$$

and

$$\int k f(x) dx = k \int f(x) dx.$$

(ii) **Integration by Parts.** If f and g are differentiable on (a,b) and $f'g$ has an antiderivative on (a,b), then so does fg', and

$$\int f(x) g'(x) dx = f(x) g(x) - \int f'(x) g(x) dx.$$

(iii) **Substitution.** If g is differentiable on (a,b) with $g(a,b) \subset (c,d)$ and $\int f(x) dx = F(x)$ on (c,d), then

$$\int f(g(x)) g'(x) dx = F(g(x)) \quad \text{for all } x \in (a,b).$$

We end this preliminary section by listing a number of useful integration formulas.

8.2 Elementary Examples

> Facts are the air of scientists. Without them you can never fly.
>
> Linus Pauling (1901–1994)

8.2.1. For $a > 0$ compute the integral

$$\int \frac{1}{x\sqrt{x^{2a} + x^a + 1}} \, dx, \quad x > 0.$$

Solution. Factor x^{2a} under the square root. The integral transforms into

$$\int \frac{1}{x^{a+1}\sqrt{1 + \frac{1}{x^a} + \frac{1}{x^{2a}}}} \, dx = \int \frac{1}{\sqrt{\left(\frac{1}{x^a} + \frac{1}{2}\right)^2 + \frac{3}{4}}} \cdot \frac{1}{x^{a+1}} \, dx.$$

With the substitution $t = 1/x^a + 1/2$ the integral becomes

$$-\frac{1}{a}\int \frac{1}{\sqrt{t^2 + \frac{3}{4}}} \, dx = -\frac{1}{a}\ln\left(t + \sqrt{t^2 + \frac{3}{4}}\right) + C$$

$$= -\frac{1}{a}\ln\left(\frac{1}{x^a} + \frac{1}{2} + \sqrt{1 + \frac{1}{x^a} + \frac{1}{x^{2a}}}\right) + C. \quad \square$$

8.2.2. Compute the integral

$$\int (1 + 2x^2)e^{x^2} \, dx.$$

Solution. Split the integral as

$$\int e^{x^2} \, dx + \int 2x^2 e^{x^2} \, dx.$$

Denote the first integral by I_1. Integrating by parts, we obtain

$$\int 2x^2 e^{x^2} \, dx = xe^{x^2} - \int e^{x^2} \, dx = xe^{x^2} - I_1.$$

We deduce that

$$\int (1 + 2x^2)e^{x^2} \, dx = xe^{x^2} + C. \quad \square$$

8.2.3. Compute the integral

$$\int \frac{x^2 + 1}{x^4 - x^2 + 1} \, dx.$$

Solution. We rewrite the integral as

$$\int \frac{x^2\left(1+\frac{1}{x^2}\right)}{x^2\left(x^2-1+\frac{1}{x^2}\right)}\,dx = \int \frac{1+\frac{1}{x^2}}{x^2-1+\frac{1}{x^2}}\,dx.$$

With the substitution $x - 1/x = t$ we have $(1+1/x^2)dx = dt$, and the integral becomes

$$\int \frac{1}{t^2+1}\,dt = \arctan t + C.$$

We conclude that

$$\int \frac{x^2+1}{x^4-x^2+1}\,dx = \arctan\left(x-\frac{1}{x}\right) + C. \quad \square$$

8.2.4. Compute

$$\int \sqrt{\frac{e^x-1}{e^x+1}}\,dx, \quad x > 0.$$

Solution. Substitute $t = \sqrt{(e^x-1)/(e^x+1)}$, $0 < t < 1$. Then $x = \ln(1+t^2) - \ln(1-t^2)$ and $dx = (2t/1+t^2 + 2t/1-t^2)dt$. The integral becomes

$$\int t\left(\frac{2t}{t^2+1}+\frac{2t}{t^2-1}\right)dt = \int\left(4-\frac{2}{t^2+1}+\frac{2t}{t^2-1}\right)dt$$

$$= 4t - 2\arctan t + \int\left(\frac{1}{t+1}+\frac{1}{1-t}\right)dt$$

$$= 4t - 2\arctan t + \ln(t+1) - \ln(t-1) + C.$$

This shows that our integral is equal to

$$4\sqrt{\frac{e^x-1}{e^x+1}} - 2\arctan\sqrt{\frac{e^x-1}{e^x+1}} + \ln\left(\sqrt{\frac{e^x-1}{e^x+1}}+1\right) - \ln\left(\sqrt{\frac{e^x-1}{e^x+1}}-1\right) + C. \quad \square$$

8.2.5. Let a be a real number. Compute the integral

$$\int \frac{1}{x^4+ax^2+1}\,dx, \quad x \in (0,\infty).$$

Solution. We only sketch the proof by pointing out the main idea, which can be applied to compute more general integrals such as

$$\int \frac{1}{ax^4+bx^2+c}\,dx.$$

The indefinite integral may be also written as

$$I = \int \frac{\frac{1}{x^2}}{x^2 + \frac{1}{x^2} + a} \, dx.$$

We associate the related indefinite integral

$$J = \int \frac{1}{x^2 + \frac{1}{x^2} + a} \, dx.$$

Then

$$J + I = \int \frac{1 + \frac{1}{x^2}}{\left(x - \frac{1}{x}\right)^2 + a + 2} \, dx = \int \frac{\left(x - \frac{1}{x}\right)'}{\left(x - \frac{1}{x}\right)^2 + a + 2} \, dx.$$

We also have

$$J - I = \int \frac{\left(x + \frac{1}{x}\right)'}{\left(x + \frac{1}{x}\right)^2 + a - 2} \, dx. \quad \square$$

Both integrals can be computed easily, and the details are left to the reader.

8.3 Existence or Nonexistence of Antiderivatives

> I would rather be a superb meteor,
> every atom of me in magnificent glow,
> than a sleepy and permanent planet.
>
> Jack London (1876–1916)

8.3.1. Prove that the function $f : \mathbb{R} \to \mathbb{R}$ defined by $f(x) = [x]$ does not have antiderivatives, where $[a]$ denotes the greatest integer less than or equal to the real number a.

Solution. Since the range of f is $f(\mathbb{R}) = \mathbb{Z}$, which is not an interval, it follows that f does not have the intermediate value property. Thus, f cannot be the derivative of a function, or equivalently, f does not admit antiderivatives. $\quad \square$

8.3.2. Prove that the function $f : \mathbb{R} \to \mathbb{R}$ defined by

$$f(x) = \begin{cases} \inf_{t \leq x}(t^2 - t + 1) & \text{if } x \leq \frac{1}{2}, \\ \sup_{t \geq x}(-t^2 + t + 1) & \text{if } x > \frac{1}{2}, \end{cases}$$

does not have antiderivatives.

Solution. A straightforward computation shows that

$$f(x) = \begin{cases} x^2 - x + 1 & \text{if } x \leq \frac{1}{2}, \\ -\frac{11}{4} & \text{if } x > \frac{1}{2}. \end{cases}$$

Since $x = 1/2$ is a discontinuity of the second kind, it follows that f does not have the intermediate value property; hence f does not admit antiderivatives. □

8.3.3. *Let $f : \mathbb{R} \to \mathbb{R}$ be a function that admits antiderivatives and assume that $g : \mathbb{R} \to \mathbb{R}$ is continuously differentiable. Prove that the function fg admits antiderivatives on $\mathbb{R}$.*

Solution. Let F denote an antiderivative of f. Since g is continuous, it has antiderivatives. Let G be an antiderivative of g. Then $(Fg)' = fg + Fg'$. Since F and g' are continuous, it follows that Fg' is continuous; hence it admits an antiderivative H. Thus, $fg = (Fg - H)'$, which shows that fg has antiderivatives. □

8.3.4. *Let $a > 0 > b$ be real numbers such that $a^2 + b \leq 0$. Assume that the function $f : \mathbb{R} \to \mathbb{R}$ satisfies*

$$(f \circ f)(x) = af(x) + bx \quad \text{for all } x \in \mathbb{R}.$$

Prove that f does not admit antiderivatives.

Solution. Arguing by contradiction, let F denote an antiderivative of f. We first prove that f is one-to-one. Indeed, if $f(x) = f(y)$ then $(f \circ f)(x) = (f \circ f)(y)$; hence $af(x) + bx = af(y) + by$, which implies $x = y$. Since f is one-to-one and has the intermediate value property, it follows that f is strictly monotone.

Let us first assume that f is decreasing and fix $x < y$. Then $f(x) > f(y)$, which implies $(f \circ f)(x) < (f \circ f)(y)$, that is, $af(x) + bx < af(y) + by$. We deduce that $0 < a(f(x) - f(y)) < b(y - x) < 0$, a contradiction.

Next, we assume that f is increasing. It follows that $f \circ f \circ f$ in increasing. On the other hand, for all $x \in \mathbb{R}$,

$$(f \circ f \circ f)(x) = a(f \circ f)(x) + bf(x) = a(af(x) + bx) + bf(x) = (a^2 + b)f(x) + abx.$$

The equality $f \circ f \circ f = (a^2 + b)f + ab\mathbf{1}_{\mathbb{R}}$ is impossible, since the left-hand side is an increasing function, while the right-hand side decreases. □

8.3.5. *Let $f : \mathbb{R} \to \mathbb{R}$ be a function satisfying $f(x) \geq 1/x$ for all $x > 0$. Prove that f does not admit antiderivatives.*

Solution. We argue by contradiction and assume that F is an antiderivative of f. By hypothesis we have $f(1/x) \geq x$, for any $x > 0$. Therefore

$$-\frac{1}{x^2} f\left(\frac{1}{x}\right) \leq -\frac{1}{x},$$

which can be rewritten as

$$\left(F\left(\frac{1}{x}\right) + \ln x\right)' \leq 0 \quad \text{for all } x > 0.$$

This shows that the function $\varphi(x) := F(1/x) + \ln x$ is nonincreasing on $(0, \infty)$. Hence

$$\varphi(1) = F(1) \geq \lim_{x \to \infty} \varphi(x) = +\infty,$$

which is a contradiction. This concludes the proof. □

8.4 Qualitative Results

> Facts are many, but the truth is one.
>
> Rabindranath Tagore (1861–1941)

8.4.1. Let $(a_n)_{n\geq 1}$ be a sequence of real numbers converging to 0 and let $f: \mathbb{R} \to \mathbb{R}$ be a function that admits antiderivatives satisfying $f(x+a_n) = f(x)$, for all $x \in \mathbb{R}$ and any integer $n \geq 1$. Prove that f is constant.

Solution. Let F be an antiderivative of f. By hypothesis, there exists a constant C such that $F(x+a_n) - F(x) \equiv C$.

We observe that for any $a \in \mathbb{R}$ there exists a sequence of **integers** (b_n) such that $\lim_{n\to\infty} a_n b_n = a$. Indeed, if b_n denotes the integer part of a/a_n, then $a = a_n b_n + r_n$, with $0 \leq r_n < a_n$. Using this and taking $r_n \to 0$ in our hypothesis, we find that $a_n b_n \to a$.

Fix a real number x_0. We have

$$F(x_0 + a_n b_n) - F(x_0) = F(x_0 + b_n a_n) - F(x_0 + (b_n-1)a_n) + \cdots + F(x_0 + a_n)$$
$$- F(x_0)$$
$$= b_n[F(x_0+a_n) - F(x_0)]$$
$$= a_n b_n \frac{F(x_0+a_n) - F(x_0)}{a_n}.$$

Since F is differentiable, we obtain, as $n \to \infty$,

$$F(x_0 + a) - F(x_0) = aF'(x_0).$$

Since a is arbitrarily chosen, we obtain for all $x \in \mathbb{R}$,

$$F(x) = F(x_0) + (x-x_0)F'(x_0) = [F(x_0) - x_0 F'(x_0)] + xF'(x_0),$$

that is, $F(x) = B + Ax$, with A and B real constants. Hence $f(x) = A = F'(x_0)$. □

8.4.2. Let I be an interval and consider a continuous function $g: I \subset \mathbb{R}$.

(a) Assume that $f: I \to \mathbb{R}^*$ admits antiderivatives. Prove that the function fg has antiderivatives.
(b) Assume that $f: I \to \mathbb{R}$ admits antiderivatives and is bounded above (or below). Prove that the function fg has antiderivatives.
(c) Deduce that the functions

$$f_1(x) = \begin{cases} e^x \cos \frac{1}{x}, & \text{if } x \neq 0, \\ 0, & \text{if } x = 0, \end{cases} \quad f_2(x) = \begin{cases} 0, & \text{if } x \leq 0, \\ \sin \frac{1}{x} \cos x, & \text{if } x > 0, \end{cases}$$

$$f_3(x) = \begin{cases} e^{-x^2} \sin \frac{1}{x}, & \text{if } x \neq 0, \\ 0, & \text{if } x = 0, \end{cases}$$

admit antiderivatives on $\mathbb{R}$.

Solution. (a) In particular, f has the intermediate value property, so $f(I)$ is an interval in $\mathbb{R}^*$. It follows that f has constant sign on I, which implies that any antiderivative F of f is strictly monotone, so one-to-one. Thus, the mapping $F : I \to F(I)$ is bijective. Moreover, since $F'(x) = f(x) \neq 0$, we deduce that the function $F^{-1} : F(I) \to I$ is differentiable. This implies that the mapping $h = g \circ F^{-1} : F(I) \to \mathbb{R}$ is continuous, so h admits antiderivatives on $F(I)$. Let H be an antiderivative of h. Hence $H \circ F : I \to \mathbb{R}$ is differentiable, and a straightforward computation shows that $H \circ F$ is an antiderivative of fg.

(b) Let us assume that f is bounded from below. Thus, there exists $m \in \mathbb{R}$ such that $f - m > 0$ on I. By (a), the function $(f - m)g =: h$ admits antiderivatives on I. It follows that the function $fg = h + mg$ admits antiderivatives on I, too.

(c) It suffices to apply the above results. □

8.4.3. Let $f : \mathbb{R} \to \mathbb{R}$ be a continuous nonconstant periodic function of period T and let F denote an antiderivative of f. Prove that there exists a T-periodic function $g : \mathbb{R} \to \mathbb{R}$ such that

$$F(x) = \left(\frac{1}{T}\int_0^T f(t)dt\right)x + g(x).$$

D. Andrica and M. Piticari

Solution. Using the relation $f(t+T) = f(t)$ for any $t \in \mathbb{R}$, it follows that

$$F(x+T) - F(x) = \int_0^T f(t)dt \quad \text{for all } x \in \mathbb{R}.$$

Considering the function $h(x) = \left(\frac{1}{T}\int_0^T f(t)dt\right)x$, we have

$$h(x+T) - h(x) = \int_0^T f(t)dt.$$

Hence $F(x+T) - F(x) = h(x+T) - h(x)$, or equivalently, $F(x+T) - h(x+T) = F(x) - h(x)$. Thus, the function defined by $g(x) = F(x) - h(x)$, $x \in \mathbb{R}$, is periodic of period T. □

8.4.4. A P-function is a differentiable function $f : \mathbb{R} \to \mathbb{R}$ with a continuous derivative f' on $\mathbb{R}$ such that $f(x + f'(x)) = f(x)$ for all x in $\mathbb{R}$.

(i) Prove that the derivative of a P-function has at least one zero.
(ii) Provide an example of a nonconstant P-function.
(iii) Prove that a P-function whose derivative has at least two distinct zeros is constant.

D. Andrica and M. Piticari

8.4 Qualitative Results

Solution. (i) If f is a P-function, and $f'(x) \neq 0$ for some $x \in \mathbb{R}$, the mean value theorem shows that f' vanishes at some point ξ between x and $x + f'(x)$: $0 = f(x + f'(x)) - f(x) = f'(x)f'(\xi)$.

(ii) Try a nonconstant polynomial function f. Identification of coefficients forces $f(x) = -x^2 + px + q$, where p and q are two arbitrarily fixed real numbers. This is not at all accidental. As shown in the comment that follows the solution, every nonconstant P-function whose derivative vanishes at a single point is of this form.

(iii) Let f be a P-function. By (a), the set $Z = \{x : x \in \mathbb{R} \text{ and } f'(x) = 0\}$ has at least one element. We now show that if it has more than one element, then it must be all of $\mathbb{R}$. The conclusion will follow. The proof is broken into three steps.

STEP 1. If f' vanishes at some point a, then $f'(x) \geq 0$ for $x \leq a$, and $f'(x) \leq 0$ for $x \geq a$. The argument is essentially the same in both cases, so we deal only with the first one. We argue by *reductio ad absurdum*. Suppose $f'(x_0) < 0$ for some $x_0 < a$ and let $\alpha = \inf\{x : x > x_0 \text{ and } f'(x) = 0\}$; clearly, this infimum exists. By continuity of f', $f'(\alpha) = 0$ and $f'(x) < 0$ for $x_0 < x < \alpha$; in particular, f is strictly monotonic (decreasing) on (x_0, α). Consider further the continuous real-valued function $g : x \mapsto x + f'(x)$, $x \in \mathbb{R}$, and note that $g(x) < x$ for $x_0 < x < \alpha$, and $g(\alpha) = \alpha$. Since $g(\alpha) = \alpha > x_0$ and g is continuous, $g(x) > x_0$ for x in (x_0, α), sufficiently close to α. Consequently, for any such x, $x_0 < g(x) < x < \alpha$, and $f(g(x)) = f(x)$, which contradicts the strict monotonicity of f on (x_0, α).

STEP 2. If f' vanishes at two points a and b, $a < b$, then f is constant on $[a,b]$. By Step 1, $f'(x) \geq 0$ for $x \leq b$ and $f'(x) \leq 0$ for $x \geq a$, so f' vanishes identically on $[a,b]$. Consequently, f is constant on $[a,b]$.

We are now in a position to conclude the proof.

STEP 3. If the set $Z = \{x : x \in \mathbb{R} \text{ and } f'(x) = 0\}$ has more than one element, then Z is all of $\mathbb{R}$. By Step 2, Z is a nondegenerate interval, and f is constant on Z: $f(x) = c$ for all x in Z. We show that $\alpha = \inf Z = -\infty$ and $\beta = \sup Z = +\infty$. Suppose, if possible, that $\alpha > -\infty$. Then α is a member of Z, by continuity of f'. Recall the function g from Step 1. By Step 1, $f'(x) > 0$ for $x < \alpha$, so $g(x) > x$, $f(x)$ is strictly monotonic (increasing), and $f(x) < c$ for $x < \alpha$. Since $f(x)$ is strictly monotonic for $x < \alpha$, the conditions $f(g(x)) = f(x)$ and $g(x) > x$ force $x < \alpha < g(x)$. Since $g(\alpha) = \alpha < \beta$, and g is continuous, it follows that $g(x) < \beta$ for $x < \alpha$, sufficiently close to α. Finally, take any such x and recall that Z is an interval to conclude that $g(x) \in Z$, so $f(x) = f(g(x)) = c$, in contradiction to $f(x) < c$ established above. Consequently, $\alpha = -\infty$. A similar argument shows that $\beta = +\infty$.

The *antiderivative test* for series, which we now state and prove, follows from the mean value theorem and the fact that a series of positive terms converges if and only if its sequence of partial sums is bounded above. □

8.4.5. Let f be positive and nonincreasing on $[k, \infty)$, where k is a positive integer, and let F be any antiderivative of f. Prove that

(i) the series $\sum_{n=k}^{\infty} f(n)$ converges if and only if F is bounded above on $[k, \infty)$;
(ii) if $\lim_{x \to \infty} F(x) = 0$, then $\sum_{n=k}^{\infty} f(n)$ converges. Furthermore, if S is the sum of $\sum_{n=k}^{\infty} f(n)$, then

$$-F(m+1) \leq \sum_{n=m+1}^{\infty} f(n) \leq -F(m) \quad \text{for } m \geq k \tag{8.1}$$

$$f(k) - F(k+1) \leq S \leq f(k) - F(k), \tag{8.2}$$

and

$$0 \leq S - \left(\sum_{n=k}^{m} f(n) - F(m+1)\right) \leq m \quad \text{for } m \geq k. \tag{8.3}$$

Solution. (i) For $n \geq k$ we apply the mean value theorem to F on $[n, n+1]$. Therefore

$$F(n+1) - F(n) = F'(x_n) = f(x_n),$$

for some $x_n \in (n, n+1)$. Since f is nonincreasing, we have

$$f(n+1) \leq f(x_n) = F(n+1) - F(n) \leq f(n) \quad \text{for } n \geq k. \tag{8.4}$$

But

$$\sum_{n=k}^{m} (F(n+1) - F(n)) = F(m+1) - F(k) \quad \text{for } m \geq k,$$

so relation (8.4) implies

$$\sum_{n=k}^{m} f(n+1) \leq F(m+1) - F(k) \leq \sum_{n=k}^{m} f(n) \quad \text{for } n \geq k. \tag{8.5}$$

Now, if $\sum_{n=k}^{\infty} f(n)$ converges, the right-hand term in (8.5) is bounded above, so that (8.5) implies that F is bounded on the set of integers that are greater than or equal to k. But $F'(x) = f(x) > 0$, and consequently F is increasing on $[k, \infty)$; hence F is bounded on $[k, \infty)$.

Conversely, if F is bounded on $[k, \infty)$, relation (8.5) implies that the sequence of partial sums for the series $\sum_{n=k}^{\infty} f(n+1)$ is bounded above. Thus, $\sum_{n=k}^{\infty} f(n+1)$ and $\sum_{n=k}^{\infty} f(n)$ are convergent.

(ii) The requirement that $\lim_{x \to \infty} F(x) = 0$ can always be met if f has an antiderivative that is bounded above. For example, if $f(x) = 1/(1+x^2)$ we would get $F(x) = \arctan x - \pi/2$ instead of $\arctan x$.

8.4 Qualitative Results

We introduce the following notation:

$$S_m = \sum_{n=k}^{m} f(n), \quad E_m = \sum_{n=m+1}^{\infty} f(n);$$

hence E_m is the error if we approximate S by S_m.

If $\lim_{x \to \infty} F(x) = 0$, then F is bounded on $[k, \infty)$, so that the series $\sum_{n=k}^{\infty} f(n)$ is convergent, by part (i). Now, if we let $m \to \infty$ in (8.5), we obtain

$$\sum_{n=k}^{\infty} f(n+1) \leq -F(k) \leq \sum_{n=k}^{\infty} f(n).$$

Since (8.5) is valid when any $k' \geq k$ is substituted for k, we have

$$\sum_{n=m}^{\infty} f(n+1) \leq -F(m) \leq \sum_{n=m}^{\infty} f(n) \quad \text{for all } m \geq k. \tag{8.6}$$

The left side of (8.6) implies that $E_m \leq -F(m)$, and the right side of (8.6) implies $-F(m+1) \leq E_m$. Thus, inequality (8.1) of (ii) holds for $m \geq k$. Inequality (8.2) follows from (8.1); just let $m = k$ in (8.1) and add $f(k)$ throughout, noticing that $S = E_k + f(k)$. Moreover, relation (8.1) implies

$$0 \leq E_m + F(m+1) = (S - S_m) + F(m+1) \leq F(m+1) - F(m)$$

for $m \geq k$; hence (8.4) yields

$$0 \leq S - (S_m - F(m+1)) \leq F(m) \quad \text{for } m \geq k,$$

which is inequality (8.3).

EXAMPLE 1. The series $\sum_{n=1}^{\infty} n^{-p}$ converges if $p > 1$ and diverges if $p < 1$. Indeed, since $F(x) = x^{1-p}/(1-p)$ for $x \geq 1$, then $F(x) \to 0$ as $x \to +\infty$ if $p > 1$, and $F(x) \to +\infty$ as $x \to +\infty$ if $p < 1$. Thus, for any $p > 1$,

(8.1) implies $\quad \dfrac{(m+1)^{1-p}}{p-1} \leq E_m \leq \dfrac{m^{1-p}}{p-1} \quad$ for $m \geq 1$,

(8.2) implies $\quad 1 + \dfrac{2^{1-p}}{p-1} \leq S \leq 1 + \dfrac{1}{p-1}$,

(8.3) implies $\quad 0 \leq S - (S_m - F(m+1)) \leq m^{-p} \quad$ for $m \geq 1$.

EXAMPLE 2. Inequality (8.1) implies that S_4 approximates $\sum_{n=1}^{\infty} n e^{-n^2}$ accurately to four places. Indeed, since $F(x) = -e^{-x^2/2}$, then $-F(m) < 0.5/10^4$ when $m \geq 4 > 2\sqrt{\ln 10}$. □

8.5 Independent Study Problems

> Order is Heaven's first law.
>
> Alexander Pope (1688–1744), *An Essay on Man IV*

8.5.1. Prove that
$$\int \sec^3 x\, dx = \frac{\sec x \tan x + \ln|\sec x + \tan x|}{2} + C, \quad x \in \left(0, \frac{\pi}{2}\right).$$

8.5.2. Find all real numbers a such that the function $f : \mathbb{R} \to \mathbb{R}$ defined by
$$f(x) = \begin{cases} \sin^2 \frac{1}{x} \cos^3 \frac{1}{x} & \text{if } x \neq 0, \\ a & \text{if } x = 0, \end{cases}$$
admits antiderivatives.

8.5.3. Prove that for any $a \in [0,1]$, the function
$$f(x) = \begin{cases} \dfrac{\cos \frac{1}{x}}{x^a} & \text{if } x \neq 0, \\ 0 & \text{if } x = 0, \end{cases}$$
admits antiderivatives.

8.5.4. Let $f : \mathbb{R} \to \mathbb{R}$ be a function that admits an antiderivative $F : \mathbb{R} \to (0, \infty)$. Prove that for any $\varepsilon > 0$ there exists $x_\varepsilon \in \mathbb{R}$ such that $|f(x_\varepsilon)| < \varepsilon$.

8.5.5. Let $F : \mathbb{R} \to \mathbb{R}$ be an antiderivative of the function $f : \mathbb{R} \to \mathbb{R}$ and assume that $f(x) \leq |x|/(1+|x|)$ for any $x \in \mathbb{R}$. Prove that there exists a single point $x_0 \in \mathbb{R}$ such that $F(x_0) = x_0$.

8.5.6. Let $f : \mathbb{R} \to \mathbb{R}$ be a differentiable function such that $\lim_{x \to -\infty} f(x)/x = \lim_{x \to +\infty} f(x)/x = 0$.

(i) Prove that the function
$$f(x) = \begin{cases} f'\left(\frac{1}{x}\right) & \text{if } x \neq 0, \\ 0 & \text{if } x = 0, \end{cases}$$
admits antiderivatives.

(iii) Deduce that the function
$$g(x) = \begin{cases} \dfrac{\cos \frac{1}{x}}{\sqrt[3]{x}} & \text{if } x \neq 0, \\ 0 & \text{if } x = 0, \end{cases}$$
admits antiderivatives.

Chapter 9
Riemann Integrability

> *If only I had the theorems! Then I should find the proofs easily enough.*
> —Bernhard Riemann (1826–1866)

Abstract. In this chapter we illustrate the definition of the definite integral of a real-valued function defined on a compact interval. We take a closer look at what kind of functions can be integrated and we develop a qualitative analysis of integrable functions, in a more precise way than is typical for calculus courses. The integral to be defined and studied here is now widely known as the *Riemann integral*.

9.1 Main Definitions and Properties

> Mathematics is very much like poetry ... what makes a good poem, a great poem; is that there is a large amount of thought expressed in very few words. In this sense formulae like $e^{\pi i} + 1 = 0$ or $\int_{-\infty}^{\infty} e^{-x^2} dx = \sqrt{\pi}$ are poems.
> —Lipman Bers (1914–1993)

Cauchy (1823) described, as rigorously as was then possible, the integral of a continuous function as the limit of a sum. Riemann (1854), merely as an a side in his celebrated Habilitation thesis on trigonometric series, defined the integral for more general functions. We describe briefly in what follows Riemann's integration theory and its extensions due to du Bois-Reymond and Darboux. A more general theory, not considered here, is due to Lebesgue (1902).

A *partition* of an interval $[a,b]$ into subintervals is a finite set of points

$$\Delta = \{x_0, x_1, x_2, \ldots, x_n\},$$

where $a = x_0 < x_1 < \cdots < x_n = b$. A partition Δ' of $[a,b]$ is called a *refinement* of Δ if it contains all the points of Δ, that is, $\Delta' \supset \Delta$.

Assume that $f : [a,b] \to \mathbb{R}$ is an arbitrary function. If $\Delta = \{x_0, x_1, x_2, \ldots, x_n\}$ is a partition of $[a,b]$, then a *selection* associated to Δ is a finite family $\xi = (\xi_1, \ldots, \xi_n)$

such that $\xi_{i-1} \le \xi_i \le \xi_i$ for $i = 1, \ldots, n$. We associate to f, Δ, and ξ the *Riemann sum* $S(f; \Delta, \xi)$ defined by

$$S(f;\Delta,\xi) = \sum_{i=1}^{n} f(\xi_i)(x_i - x_{i-1}).$$

We say that f is *Riemann integrable* on $[a,b]$ is there exists a real number I with the following property: for any $\varepsilon > 0$, there exists a partition Δ of $[a,b]$ such that $|S(f;\Delta,\xi) - I| < \varepsilon$, for every selection ξ associated to Δ. The number I is called the *integral* of f on $[a,b]$ and is denoted by $\int_a^b f(x)dx$. In particular, if we change the values of an integrable function at a finite number of points, then the function remains integrable and the value of the integral does not change.

This definition implies that if f is Riemann integrable, then f is bounded. However, not every bounded function is integrable.

Example. Recall that Dirichlet's function is defined by

$$f(x) = \begin{cases} 1 & \text{if } x \in \mathbb{Q}, \\ 0 & \text{if } x \in \mathbb{R} \setminus \mathbb{Q}. \end{cases}$$

Then f is **not** integrable on any interval $[a,b]$. Indeed, for any partition $\Delta = \{x_0, x_1, x_2, \ldots, x_n\}$ of $[a,b]$, each interval $[x_{i-1}, x_i]$ contains both rational and irrational numbers. Thus, there are two selections ξ and ξ' associated to Δ such that $S(f;\Delta,\xi) = b - a$ and $S(f;\Delta,\xi') = 0$. Hence, f is not integrable on $[a,b]$. We also point out that this function is *totally discontinuous*, that is, discontinuous at every point.

Another approach to Riemann integrability is described in what follows. We define the *lower* and *upper Darboux sums* associated to $f : [a,b] \to \mathbb{R}$ and to a division $\Delta = \{x_0, x_1, x_2, \ldots, x_n\}$ of $[a,b]$ as

$$S_-(f;\Delta) = \sum_{i=1}^{n} m_i(x_i - x_{i-1}), \quad S^+(f;\Delta) = \sum_{i=1}^{n} M_i(x_i - x_{i-1}),$$

where

$$m_i = \inf_{x_{i-1} \le x \le x_i} f(x), \quad M_i = \sup_{x_{i-1} \le x \le x_i} f(x).$$

Then $S_-(f;\Delta) \le S^+(f;\Delta)$, and moreover, $S_-(f;\Delta) \le S(f;\Delta,\xi) \le S^+(f;\Delta)$, for any selection ξ associated to the division Δ. Further, if Δ' is a refinement of Δ, then

$$S_-(f;\Delta) \le S_-(f;\Delta') \le S^+(f;\Delta') \le S^+(f;\Delta),$$

and if Δ_1 and Δ_2 are two arbitrary divisions, then

$$S_-(f;\Delta_1) \le S^+(f;\Delta_2).$$

This shows that the set of lower Darboux sums of f is majorized by every upper Darboux sum and that the set of upper Darboux sums of a certain function is

9.1 Main Definitions and Properties

minorized by any lower Darboux sum. Therefore, it makes sense to consider the supremum of the lower Darboux sums and the infimum of the upper Darboux sums. Thus, we define the *lower Darboux integral*

$$\underline{\int_a^b} f(x)dx := \sup_\Delta S_-(f;\Delta)$$

and the *upper Darboux integral*

$$\overline{\int_a^b} f(x)dx := \inf_\Delta S^+(f;\Delta).$$

The following criterion of integrability is due to Darboux.

Theorem. *A function $f : [a,b] \to \mathbb{R}$ is integrable if and only if for any $\varepsilon > 0$ there exists $\delta > 0$ such that*

$$S^+(f;\Delta) - S_-(f;\Delta) < \varepsilon,$$

for every partition $\Delta = \{x_0, x_1, x_2, \ldots, x_n\}$ with $\max_i(x_i - x_{i-1}) < \delta$.

This implies that a function $f : [a,b] \to \mathbb{R}$ is integrable if and only if the lower and upper Darboux integrals are equal. In such a case we have

$$\int_a^b f(x)dx := \underline{\int_a^b} f(x)dx = \overline{\int_a^b} f(x)dx.$$

Examples. (i) The function $f(x) = x$ is integrable on $[a,b]$. Indeed, consider the equidistant partition

$$\Delta_n = \{x_i = a + i(b-a)/n;\ i = 0, 1, 2, \ldots, n\}.$$

Then

$$S_-(f;\Delta_n) = \sum_{i=1}^n x_{i-1} \frac{b-a}{n} = \frac{b^2 - a^2}{2} - \frac{(b-a)^2}{2n}$$

and

$$S^+(f;\Delta_n) = \sum_{i=1}^n x_{i-1} \frac{b-a}{n} = \frac{b^2 - a^2}{2} + \frac{(b-a)^2}{2n}.$$

Thus, $S^+(f;\Delta_n) - S_-(f;\Delta_n) = (b-a)^2/n \to 0$ as $n \to \infty$.

(ii) Consider the Riemann function $f : [0,1] \to \mathbb{R}$ defined by

$$f(x) = \begin{cases} 0 & \text{if } x = 0 \text{ or if } x \in [0,1] \cap (\mathbb{R} \setminus \mathbb{Q}), \\ \frac{1}{n} & \text{if } x = \frac{m}{n} \in [0,1], m, n \in \mathbb{N}^*, (m,n) = 1. \end{cases}$$

We argue in what follows that f is integrable on $[0,1]$ and $\int_0^1 f(x)dx = 0$. Fix $\varepsilon > 0$. Then there is only a finite number (say, p) of numbers $x \in [0,1]$ such that $f(x) > \varepsilon$. We now choose a partition Δ of $[0,1]$ with $\max_i(x_i - x_{i-1}) < \varepsilon/p$

such that all real numbers x with $f(x) > \varepsilon$ lie in the interior of the subintervals. Then $S_-(f;\Delta) = 0$ and

$$S^+(f;\Delta) \leq \varepsilon + p \cdot \max_i(x_i - x_{i-1}) < 2\varepsilon,$$

which shows that indeed, f is integrable on $[0,1]$ and $\int_0^1 f(x)dx = 0$.

Classes of integrable functions. (i) Any *continuous* function $f : [a,b] \to \mathbb{R}$ is integrable on $[a,b]$.

(ii) Any *monotone* function $f : [a,b] \to \mathbb{R}$ is integrable on $[a,b]$.

We say that a set of real numbers A has *null measure* if for every $\varepsilon > 0$ there exists a finite collection of disjoint intervals $\{(a_j,b_j);\ 1 \leq j \leq m\}$ such that $A \subset \cup_{j=1}^m (a_j,b_j)$ and $\sum_{j=1}^m (b_j - a_j) < \varepsilon$. An important necessary and sufficient condition for the Riemann integrability of a function is stated in the following result, which is due to the French mathematician Henri Lebesgue (1875–1941).

Lebesgue's Theorem. A function $f : [a,b] \to \mathbb{R}$ is Riemann integrable if and only if f is bounded and the set of discontinuity points of f has null measure.

Properties of Riemann integrable functions.

(i) Additivity property of the integral: let $a < b < c$ and assume that $f : [a,c] \to \mathbb{R}$ is integrable on $[a,b]$ and on $[b,c]$. Then f is integrable on $[a,c]$ and

$$\int_a^c f(x)dx = \int_a^b f(x)dx + \int_b^c f(x)dx.$$

(ii) Let f be integrable on $[a,b]$ with $m \leq f(x) \leq M$ for $x \in [a,b]$. If φ is continuous on $[m,M]$, then $\varphi \circ f$ is integrable on $[a,b]$. This property implies that if f is integrable on $[a,b]$, then f^+, f^-, $|f|$, and f^2 are also integrable on $[a,b]$. Moreover,

$$\left| \int_a^b f(x)dx \right| \leq \int_a^b |f(x)|dx.$$

Newton–Leibniz Formula. Let f be integrable on $[a,b]$. If F is an antiderivative of f, then

$$\int_a^b f(x)dx = F(b) - F(a).$$

Fundamental Theorem of Calculus. Let $f : I \to \mathbb{R}$, where I is an interval, and suppose that f is integrable over any compact interval contained in I. Let $a \in I$ and define $F(x) = \int_a^x f(t)dt$, for any $x \in I$. Then F is continuous on I. Moreover, if f is continuous at $x_0 \in I$, then F is differentiable at x_0 and $F'(x_0) = f(x_0)$.

Integration by Parts. Let f and g be integrable on $[a,b]$. If F and G are antiderivatives of f and g, respectively, then

$$\int_a^b F(x)g(x)dx = F(b)G(b) - F(a)G(a) - \int_a^b f(x)G(x)dx.$$

9.2 Elementary Examples

Change of Variables in the Riemann Integral. Let φ be of class C^1 on the interval $[\alpha, \beta]$, with $a = \varphi(\alpha)$ and $b = \varphi(\beta)$. If f is continuous on $\varphi([\alpha, \beta])$ and $g = f \circ \varphi$, then

$$\int_a^b f(x)dx = \int_\alpha^\beta g(t)\varphi'(t)dt.$$

First Mean Value Theorem for Integrals. Let $f : [a,b] \to \mathbb{R}$ be a continuous function. Then there exists $\xi \in [a,b]$ such that

$$\int_a^b f(x)dx = (b-a)f(\xi).$$

The following result is sometimes called the *Weierstrass form of Bonnet's theorem*. In Bonnet's form, the monotone function g is assumed to be positive and decreasing on $[a,b]$.

Second Mean Value Theorem for Integrals. Let $f, g : [a,b] \to \mathbb{R}$ be such that f is continuous and g is monotone. Then there exists $\xi \in [a,b]$ such that

$$\int_a^b f(x)g(x)dx = g(a) \int_a^\xi f(x)dx + g(b) \int_\xi^b f(x)dx.$$

9.2 Elementary Examples

> Although to penetrate into the intimate mysteries of nature and thence to learn the true causes of phenomena is not allowed to us, nevertheless it can happen that a certain fictive hypothesis may suffice for explaining many phenomena.
>
> Leonhard Euler (1707–1783)

9.2.1. Let $0 < a < b$. Explain the curiosity

$$\int_a^b \frac{1}{x\ln x}dx = 1 \Big|_a^b + \int_a^b \frac{1}{x\ln x}dx.$$

Solution. Integrating by parts, we have

$$\int_a^b \frac{1}{x\ln x}dx = \int_a^b (\ln x)' \frac{1}{\ln x}dx = 1 \Big|_a^b + \int_a^b \frac{1}{x\ln x}dx. \quad \square$$

9.2.2. Let $f : \mathbb{R} \to \mathbb{R}$ be a continuous function such that for all $x \in \mathbb{R}$,
$$\int_0^1 f(xt)dt = 0.$$
Show that $f \equiv 0$.

Solution. We have, for any $x \neq 0$,
$$\int_0^1 f(x)dx = \int_0^x f(u)\frac{du}{x} = \frac{1}{x}\int_0^x f(u)du.$$

Hence
$$\int_0^x f(u)du = 0 \quad \text{for all } x \in \mathbb{R}.$$

This shows that
$$\frac{d}{dx}\int_0^x f(u)du = 0,$$
which shows that $f(x) = 0$ for all $x \in \mathbb{R}$. □

9.2.3. Let p, q be positive numbers. Prove that
$$\int_0^1 (1-x^p)^{1/q}dx = \int_0^1 (1-x^q)^{1/p}dx.$$

Solution. More generally, assume that f is a decreasing continuous function on $[a,b]$. Then its inverse function g exists in $[f(b), f(a)]$ and is also decreasing and continuous. Hence
$$\int_{f(b)}^{f(a)} g(y)dy = \int_a^b g(f(t))f'(t)dt = \int_b^a tf'(t)dt$$
$$= af(a) - bf(b) + \int_a^b f(t)dt.$$

If additionally we have $f(a) = b$ and $f(b) = a$, then
$$\int_a^b g(t)dt = \int_a^b f(t)dt.$$

The functions $f(x) = (1-x^q)^{1/p}$ and $g(x) = (1-x^p)^{1/q}$ represent in $[0,1]$ a special case of this situation. □

9.2.4. Let $f : [0,1] \to [0,\infty)$ be a continuous function such that
$$f^2(t) \leq 1 + 2\int_0^t f(s)ds, \quad \forall t \in [0,1].$$

Prove that $f(t) \leq 1+t$, for all $t \in [0,1]$.

9.2 Elementary Examples

Solution. Let
$$g(t) = 1 + 2\int_0^t f(s)\,ds.$$
Then $g'(t) = 2f(t) \le 2\sqrt{g(t)}$, so
$$\sqrt{g(t)} - 1 = \int_0^t \frac{g'(s)}{2\sqrt{g(s)}}\,ds \le \int_0^t ds = t.$$
Hence $f(t) \le \sqrt{g(t)} \le 1 + t$. □

9.2.5. Let $f : [1,\infty) \to \mathbb{R}$ be such that $f(1) = 1$ and
$$f'(x) = \frac{1}{x^2 + f^2(x)}.$$
Prove that $\lim_{x\to\infty} f(x)$ exists and this limit is less than $1 + \frac{\pi}{4}$.

Solution. Since $f' > 0$, our function f is increasing, so $f(t) > f(1) = 1, \forall t > 1$. Hence
$$f'(t) = \frac{1}{t^2 + f^2(t)} < \frac{1}{t^2 + 1}, \quad \forall t > 1.$$
It follows that
$$f(x) = 1 + \int_1^x f'(t)\,dt < 1 + \int_1^x \frac{1}{t^2+1}\,dt < 1 + \int_1^\infty \frac{1}{t^2+1}\,dt = 1 + \frac{\pi}{4}.$$
Thus, $\lim_{x\to\infty} f(x)$ exists and is at most $1 + \frac{\pi}{4}$. This inequality is strict because
$$\lim_{x\to\infty} f(x) = 1 + \int_1^\infty f'(t)\,dt < 1 + \int_1^\infty \frac{1}{t^2+1}\,dt = 1 + \frac{\pi}{4}. \quad \square$$

9.2.6. Prove that the function
$$F(k) = \int_0^{\pi/2} \frac{dx}{\sqrt{1-k\cos^2 x}}, \quad k \in [0,1),$$
is increasing.

Solution. Fix $0 \le k_1 < k_2 < 1$. Then, for all $x \in (0, \pi/2)$,
$$-k_1 \cos^2 x > -k_2 \cos^2 x,$$
$$\sqrt{1 - k_1 \cos^2 x} > \sqrt{1 - k_2 \cos^2 x},$$
$$\frac{1}{\sqrt{1-k_1\cos^2 x}} < \frac{1}{\sqrt{1-k_2\cos^2 x}}.$$

Hence
$$\int_0^{\pi/2} \frac{1}{\sqrt{1-k_1\cos^2 x}}dx < \int_0^{\pi/2} \frac{1}{\sqrt{1-k_2\cos^2 x}}dx. \quad \square$$

9.2.7. Let $f:[0,a]\to\mathbb{R}$ be a continuous function. Prove that
$$\int_0^x \left(\int_0^y f(t)dt\right)dy = \int_0^x (x-y)f(y)dy.$$

Solution. The integral on the left-hand side can be rewritten as
$$\int_0^x \varphi(y)\psi(y)dy,$$
where $\varphi(y)\equiv 1$, $\psi(y)=\int_0^y f(t)dt$. An integration by parts concludes the proof. $\square$

9.2.8. Let $f:[0,1]\to\mathbb{R}$ be a differentiable function such that $f'(0)<2<f'(1)$. Prove that there exists $x_0\in(0,1)$ such that $f'(x_0)=2$.

Solution. Consider the function $g(x)=f(x)-2x$. We have $g'(0)<0<g'(1)$, so $g(x)<g(0)$ for x sufficiently close to 0 and $g(x)<g(1)$ for x sufficiently close to 1. So, the minimum value of g in $[0,1]$ is achieved for some $x_0\in(0,1)$. Thus, by Fermat's theorem, $g'(x_0)=0$. It follows that $f'(x_0)=2$. $\square$

9.2.9. Prove that the following limit exists and is finite:
$$\lim_{t\searrow 0}\left(\int_0^1 \frac{dx}{(x^4+t^4)^{1/4}}+\ln t\right).$$

Solution. Let
$$I(t)=\int_0^1 \frac{dx}{(x^4+t^4)^{1/4}}+\ln t.$$

It is enough to show that this function is increasing and bounded from below. Indeed, for any $x,t\geq 0$ we have $(x+t)^4\geq x^4+t^4$. Hence
$$I(t)\geq \int_0^1 \frac{dx}{x+t}+\ln t = \int_t^{1+t}\frac{du}{u}+\ln t = \ln(1+t)\geq 0, \quad \forall t\geq 0.$$

Next, we prove that $I'(t)\geq 0$. We first observe that
$$I(t)=\int_0^t \frac{dx}{t[(x/t)^4+1]^{1/4}} + \int_t^1 \frac{dx}{t[(x/t)^4+1]^{1/4}} + \ln t.$$

By the substitution $y=x/t$ we obtain
$$I(t)=\int_0^1 \frac{dy}{(y^4+1)^{1/4}} + \int_1^{1/t}\frac{dy}{(y^4+1)^{1/4}}+\ln t.$$

Therefore
$$I'(t) = -\frac{1}{t^2(1/t^4+1)^{1/4}} + \frac{1}{t} \geq 0. \quad \square$$

9.2.10. Let $f:[0,a] \to [0,\infty)$ be a continuous function such that $f(0) = 0$ and f has a right derivative and $f'_d(0) = 0$. Moreover, we assume that
$$f(t) \leq \int_0^t \frac{f(s)}{s} ds, \quad \forall t \in [0,a].$$

Prove that $f \equiv 0$.

Solution. Let
$$F(t) = \int_0^t \frac{f(s)}{s} ds.$$

Then F is well defined because the function $f(s)/s$ can be extended by continuity in $s = 0$, since $\lim_{s \to 0} f(s)/s = f'(0) = 0$. Furthermore, we have
$$F'(t) = \frac{f(t)}{t} \leq \frac{F(t)}{t},$$

so
$$\left(\frac{F(t)}{t}\right)' \leq 0, \quad \forall t \in [0,a].$$

Thus, the mapping $F(t)/t$ is decreasing. Since $F(0) = 0$, we find that $F(t) \leq 0$. But $F(t) \geq 0$. Hence $F \equiv 0$, that is, $f \equiv 0$. $\quad \square$

9.2.11. Let $f:[0,1] \to \mathbb{R}$ be an integrable function such that
$$\int_0^1 f(x)dx = \int_0^1 xf(x)dx = 1.$$

Prove that $\int_0^1 f^2(x)dx \geq 4$.

Romanian Mathematical Olympiad, 2004

Solution. A direct application of the Cauchy–Schwarz inequality implies
$$\int_0^1 f^2(x)dx \cdot \int_0^1 x^2 dx \geq \left(\int_0^1 xf(x)dx\right)^2 = 1,$$

so $\int_0^1 f^2(x)dx \geq 3$.

In order to find a stronger bound we start from $\int_0^1 (f(x) + ax + b)^2 dx \geq 0$, for any real numbers a and b. Therefore
$$\int_0^1 f^2(x)dx \geq -\frac{a^2}{3} - b^2 - 2a - 2b - ab.$$

We are interested in finding a and b such that $E(a,b) = -a^2/3 - b^2 - 2a - 2b - ab$ is maximum. For this purpose we first fix the real parameter b and consider the mapping $\varphi(a) = -a^2/3 - (b+2)a - b^2 - 2b$, which achieves its maximum for $a_M = -3(b+2)/2$. Then $\varphi(a_M) = 4 - (b-2)^2/4$, which is maximum for $b = 2$. The above argument implies that the maximum of $E(a,b)$ is 4 and it is attained for $a = -6$ and $b = 2$, so the minimum is achieved for $f(x) = 6x - 2$. This concludes the proof. □

9.2.12. Fix real numbers a, b, c, d such that $a \neq b$ and let $g : [a,b] \to \mathbb{R}$ be a continuous function satisfying $\int_a^b g(x)dx = 0$. Find a function f such that $\int_a^b f^2(x)dx$ is a minimum subject to the conditions $\int_a^b f(x)dx = c$ and $\int_a^b f(x)g(x)dx = d$.

J.R. Hatcher, AMM E 1104, 1954

Solution. For any real numbers λ and μ we have

$$\int_a^b [f(x) - \lambda g(x) - \mu]^2 dx \geq 0.$$

Set $A = \int_a^b g^2(x)dx$. If $A > 0$ then the above inequality can be rewritten as

$$\int_a^b f^2(x)dx \geq \frac{d^2}{A} + \frac{c^2}{b-a} - \frac{(A\lambda - d)^2}{A} - \frac{[(b-a)\mu - c]^2}{b-a}.$$

This implies that the required minimum is $d^2A^{-1} + c^2(b-a)^{-1}$, and it is achieved if and only if $f(x) = dA^{-1}g(x) + c(b-a)^{-1}$.

If $A = 0$ (such a case is possible only if $d = 0$) then $g = 0$. The Cauchy–Schwarz inequality implies that the minimum of $\int_a^b f^2(x)dx$ is $c^2(b-a)^{-1}$, and it is attained if and only if $f(x) = c(b-a)^{-1}$. □

9.2.13. Find all continuous functions $f : \mathbb{R} \to \mathbb{R}$ such that for all real x and any positive integer n we have

$$n^2 \int_x^{x+1/n} f(t)dt = nf(x) + \frac{1}{2}.$$

Solution. By hypothesis we deduce that f is differentiable and, for all $n \in \mathbb{N}^*$ and any $x \in \mathbb{R}$,

$$n\left[f\left(x+\frac{1}{n}\right) - f(x)\right] = f'(x). \tag{9.1}$$

Therefore

$$n\left[f'\left(x+\frac{1}{n}\right) - f'(x)\right] = f''(x). \tag{9.2}$$

Replacing n by $2n$ in (9.1) we obtain

$$2n\left[f\left(x+\frac{1}{2n}\right) - f(x)\right] = f'(x). \tag{9.3}$$

9.2 Elementary Examples

Relation (9.1) also implies

$$2n\left[f\left(x+\frac{1}{n}\right)-f(x)\right]=2f'(x). \tag{9.4}$$

Using again (9.1) and replacing x by $x+1/n$ and n by $2n$, we have

$$2n\left[f\left(x+\frac{1}{n}\right)-f\left(x+\frac{1}{2n}\right)\right]=f'\left(x+\frac{1}{2n}\right). \tag{9.5}$$

By (9.3) and (9.4) we have

$$2n\left[f\left(x+\frac{1}{n}\right)-f\left(x+\frac{1}{2n}\right)\right]=f'(x). \tag{9.6}$$

So, by (9.5) and (9.6),

$$f'(x)=f'\left(x+\frac{1}{2n}\right), \quad \forall n\in\mathbb{N}^*, \, \forall x\in\mathbb{R}.$$

Using now (9.1), we deduce that $f''(x)=0$ for all $x\in\mathbb{R}$, so $f(x)=ax+b$, for some real numbers a and b. Using the hypothesis, a straightforward computation yields $a=1$ and b is an arbitrary real number. $\square$

9.2.14. *Let $f:[0,1]\to\mathbb{R}$ be a continuous function with the property that $xf(y)+yf(x)\le 1$, for all $x,y\in[0,1]$.*
Show that

$$\int_0^1 f(x)dx\le\pi/4.$$

Find a function satisfying the condition for which there is equality.

International Mathematics Competition for University Students, 1998

Solution. Set $I=\int_0^1 f(x)dx$ and observe that

$$I=\int_0^{\frac{\pi}{2}}f(\cos t)\sin t\,dt=\int_0^{\frac{\pi}{2}}f(\sin t)\cos t\,dt.$$

So, by hypothesis,

$$2I\le\int_0^{\frac{\pi}{2}}1dt=\frac{\pi}{2}.$$

For the second part of the problem, consider the function $f(x)=\sqrt{1-x^2}$. $\square$

9.2.15. *Let $f:[0,1]\to(0,\infty)$ be a nonincreasing function. Prove that*

$$\frac{\int_0^1 xf^2(x)dx}{\int_0^1 xf(x)dx}\le\frac{\int_0^1 f^2(x)dx}{\int_0^1 f(x)dx}.$$

Putnam Competition, 1957

Solution. Since f is positive and nonincreasing, it follows that

$$\int_0^1 \left[\int_0^1 f(x)f(y)(x-y)(f(x)-f(y))dx \right] dy \leq 0.$$

Expanding, we obtain

$$\left[\int_0^1 xf^2(x)dx \right] \int_0^1 f(y)dy - \left[\int_0^1 xf(x)dx \right] \int_0^1 f^2(y)dy$$
$$- \left[\int_0^1 f^2(x)dx \right] \int_0^1 yf(y)dy + \left[\int_0^1 yf^2(y)dy \right] \int_0^1 f(x)dx$$
$$= 2\left[\int_0^1 xf^2(x)dx \right] \int_0^1 f(x)dx - 2\left[\int_0^1 xf(x)dx \right] \int_0^1 f^2(x)dx \leq 0.$$

The proof is now complete. □

9.2.16. *For what values of $a > 1$ is*

$$\int_a^{a^2} \frac{1}{x} \ln \frac{x-1}{32}$$

minimum?

Harvard-MIT Mathematics Tournament, 2003

Solution. Define $f(a) = \int_a^{a^2} \frac{1}{x} \ln \frac{x-1}{32}$ and let $F(x)$ be an antiderivative of the function $(1, \infty) \ni x \longmapsto \frac{1}{x} \ln \frac{x-1}{32}$. So, by the Newton–Leibniz formula, $f(a) = F(a^2) - F(a)$. Hence

$$f'(a) = 2aF'(a^2) - F'(a) = \frac{2}{a} \ln \frac{a^2-1}{32} - \frac{1}{a} \ln \frac{a-1}{32}$$
$$= \frac{\ln(a-1)(a+1)^2 - \ln 32}{a}.$$

It follows that $f'(a) < 0$ if $a \in (1, 3)$ and $f'(a) > 0$, provided $a \in (3, \infty)$. This means that $f(a)$ achieves its minimum for $a = 3$. □

The following is the same problem as 9.2.15 but with a different solution.

9.2.17. *Let f be a positive nonincreasing function defined in $[0, 1]$. Prove that*

$$\frac{\int_0^1 xf^2(x)dx}{\int_0^1 xf(x)dx} \leq \frac{\int_0^1 f^2(x)dx}{\int_0^1 f(x)dx}.$$

Putnam Competition, 1957

Solution. Define $\varphi : [0, 1] \to \mathbb{R}$ by

$$\varphi(x) = \int_0^x f^2(t)dt \int_0^x tf(t)dt - \int_0^x tf^2(t)dt \int_0^x f(t)dt.$$

Then $\varphi(0) = 0$, and for all $x \in [0,1]$,

$$\varphi'(x) = f(x) \int_0^x (t-x)[f(x) - f(t)]f(t)dt \geq 0.$$

Therefore $\varphi(1) \geq 0$, which concludes the proof. $\square$

9.2.18. *Let A be an arbitrary nonnegative number and fix $p > 1$. Prove that there exists no continuous function $f : [A, \infty) \to \mathbb{R}$ ($f \not\equiv 0$) such that $|f(x)| \geq \int_A^x |f(t)|^p dt$, for all $x \geq A$.*

H. Diamond, Amer. Math. Monthly, Problem 6422

Solution. Arguing by contradiction, set $f(x) := \int_A^x |f(t)|^p dt$. Then $F(x_0) > 0$ for some $x_0 \geq A$ and $F'(x) = |f(x)|^p$ in (A, ∞). Thus, by hypothesis, $F'(x) \geq F^p(x)$, and thus $F'/F^p \geq 1$ in $[x_0, \infty)$. Integrating this inequality on $[x_0, x]$, we obtain, for all $x > x_0$,

$$\frac{F^{1-p}(x_0) - F^{1-p}(x)}{p-1} \geq x - x_0.$$

It follows that $F(x)$ is negative for x large enough, a contradiction. $\square$

9.2.19. *For $-\pi/2 < x < \pi/2$, prove the inequality*

$$\sin x \ln\left(\frac{1+\sin x}{1-\sin x}\right) \geq 2x^2.$$

M. Golomb, Math. Magazine, Problem Q887

Solution. We first observe that both sides of the inequality are even functions. Thus, it is sufficient to assume that $x \in [0, \pi/2)$. We have

$$\int_0^x \cos t \, dt = \sin x \quad \text{and} \quad \int_0^x \frac{1}{\cos t} dt = \frac{1}{2} \ln\left(\frac{1+\sin x}{1-\sin x}\right).$$

Next, we apply the Cauchy–Schwarz inequality

$$\left(\int_0^x f(t)g(t)dt\right)^2 \leq \int_0^x f^2(t)dt \cdot \int_0^x g^2(t)dt$$

with $f(t) = \sqrt{\cos t}$ and $g(t) = 1/\sqrt{\cos t}$. The conclusion follows. $\square$

9.3 Classes of Riemann Integrable Functions

Nature is not embarrassed by
difficulties of analysis.
<hr>Augustin Fresnel (1788–1827)

As an application of the mean value theorem for integrals, we know that if $f : [a,b] \to \mathbb{R}$ is a continuous function, then for all $x \in [a,b]$ there exists $c_x \in (a,x)$

such that $\int_a^x f(t)dt = f(c_x)(x-a)$. The next result gives a better understanding of the behavior of c_x as x approaches a.

9.3.1. Assume that f is a continuous function in $[a,b]$ that is differentiable in a and $f'(a+) \neq 0$. If $c_x \in (a,x)$ is such that $\int_a^x f(t)dt = f(c_x)(x-a)$, prove that

$$\lim_{x \searrow a} \frac{c_x - a}{x - a} = \frac{1}{2}.$$

B. Jacobson

Solution. Consider

$$\lim_{x \searrow a} \frac{\int_a^x f(t)dt - xf(a) + af(a)}{(x-a)^2}. \qquad (9.7)$$

Applying the mean value theorem for integrals, we obtain

$$\lim_{x \searrow a} \frac{\int_a^x f(t)dt - xf(a) + af(a)}{(x-a)^2} = \lim_{x \searrow a} \frac{f(c_x)(x-a) - f(a)(x-a)}{(x-a)^2}$$
$$= \lim_{x \searrow a} \frac{f(c_x) - f(a)}{x - a}$$
$$= \lim_{x \searrow a} \frac{f(c_x) - f(a)}{x - a} \cdot \frac{c_x - a}{x - a} = f'(a) \cdot \lim_{x \searrow a} \frac{c_x - a}{x - a}.$$

Applying L'Hôpital's rule to (9.7), we obtain

$$\lim_{x \searrow a} \frac{\int_a^x f(t)dt - xf(a) + af(a)}{(x-a)^2} = \lim_{x \searrow a} \frac{f(x) - f(a)}{2(x-a)} = \frac{f'(a)}{2}.$$

It follows that $\lim_{x \searrow a}(c_x - a)/(x-a) = 1/2$. $\square$

Exercise. Assume that f is continuous in $[a,b]$, and is k times differentiable at a with $f^{(i)}(a+) = 0$ ($i = 1, 2, \ldots, k-1$) and $f^{(k)}(a+) \neq 0$. If $c_x \in (a,x)$ is such that $\int_a^x f(t)dt = f(c_x)(x-a)$, then

$$\lim_{x \searrow a} \frac{c_x - a}{x - a} = \frac{1}{\sqrt[k]{k+1}}.$$

The following result is sometimes called the *Riemann–Lebesgue lemma*. It holds in the general framework of Riemann integrable functions $f : [a,b] \to \mathbb{R}$, but we will prove this property for continuously differentiable functions.

9.3.2. Let $f : [a,b] \to \mathbb{R}$ be a continuously differentiable function. Prove that

$$\lim_{n \to \infty} \int_a^b f(x) \sin nx \, dx = 0.$$

Solution. Integrating by parts, we obtain

$$\left| \int_a^b f(x) \sin nx \, dx \right| = \left| \frac{f(a) \cos na - f(b) \cos nb}{n} + \frac{1}{n} \int_a^b f'(x) \cos nx \, dx \right|$$

$$\leq \frac{2 \max_{x \in [a,b]} |f(x)|}{n} + \frac{1}{n} \int_a^b \max_{x \in [a,b]} |f'(x)| \, dx \to 0$$

as $n \to \infty$, since both f and f' are bounded on $[a,b]$, as continuous functions. □

9.3.3. Let $f: [-1,1] \to \mathbb{R}$ be a continuous function. Prove the following properties:

(i) If $\int_0^1 f(\sin(x+t)) dx = 0$ for all $t \in \mathbb{R}$, then $f(x) = 0$ for all $x \in [-1,1]$.
(ii) If $\int_0^1 f(\sin(nx)) dx = 0$ for all $n \in \mathbb{Z}$, then $f(x) = 0$ for all $x \in [-1,1]$.

Dorin Andrica and Mihai Piticari

Solution. (i) Taking $y = x+t$ we deduce that $\int_t^{1+t} f(\sin y) dy = 0$ for all $t \in \mathbb{R}$. Differentiating this equality, we obtain $f(\sin(1+t)) = f(\sin t)$ for all $t \in \mathbb{R}$. Taking $t = n \in \mathbb{N}$, we find that $f(\sin n) = f(\sin 0) = f(0) =: C$. Thus, $f(\sin n) = C$ for all $n \in \mathbb{N}$. We have proved in Exercise 1.3.26 that the set $\{\sin n; n \in \mathbb{N}\}$ is dense in $[-1,1]$. Using now the continuity of f, we conclude that $f(x) = C$ for all $x \in \mathbb{R}$. Returning to our hypothesis, we deduce that $C = 0$.

(ii) Taking $y = nx$, we obtain $\int_0^n f(\sin y) dy = 0$ for all $n \in \mathbb{Z}$. Setting $F(x) = \int_x^{x+1} f(\sin y) dy$, we deduce that $F(n) = 0$ for all $n \in \mathbb{Z}$. Since the function $\sin$ is 2π–periodic, we obtain $F(n + 2k\pi) = 0$ for all $n, k \in \mathbb{Z}$. But the set $\{n + 2k\pi; n, k \in \mathbb{Z}\}$ is dense in $\mathbb{R}$. So, by the continuity of F, $F(x) = 0$ for all $x \in \mathbb{R}$. Therefore $F'(x) = f(\sin(x+1)) - f(\sin x) = 0$. Taking again $x = n$, we obtain $f(\sin(n+1)) = f(\sin n) = f(0) =: C$ for all $n \in \mathbb{N}$. With the same arguments as in (i) we conclude that $f(x) = 0$ for any $x \in \mathbb{R}$. □

9.4 Basic Rules for Computing Integrals

> The profound study of nature is the most fertile source of mathematical discoveries.
>
> Joseph Fourier (1768–1830)

9.4.1. Evaluate the definite integral

$$\int_{-\pi}^{\pi} \frac{\sin nx}{(1+2^x) \sin x} dx,$$

where n is a natural number.

Solution. We have

$$I_n = \int_{-\pi}^{\pi} \frac{\sin nx}{(1+2^x)\sin x} dx$$

$$= \int_0^{\pi} \frac{\sin nx}{(1+2^x)\sin x} dx + \int_{-\pi}^0 \frac{\sin nx}{(1+2^x)\sin x} dx.$$

In the second integral we make the change of variable $y = -x$ and obtain

$$I_n = \int_0^{\pi} \frac{\sin nx}{(1+2^x)\sin x} dx + \int_0^{\pi} \frac{\sin ny}{(1+2^{-y})\sin y} dy$$

$$= \int_0^{\pi} \frac{(1+2^x)\sin nx}{(1+2^x)\sin x} dx$$

$$= \int_0^{\pi} \frac{\sin nx}{\sin x} dx.$$

For $n \geq 2$ we have

$$I_n - I_{n-2} = \int_0^{\pi} \frac{\sin nx - \sin(n-2)x}{\sin x} dx$$

$$= 2 \int_0^{\pi} \cos(n-1)x \, dx = 0.$$

Since $I_0 = 0$ and $I_1 = \pi$, we conclude that

$$I_n = \begin{cases} 0 & \text{if } n \text{ is even,} \\ \pi & \text{if } n \text{ is odd.} \end{cases} \quad \square$$

9.4.2. Let n be a positive integer. Compute

$$\int_0^{\pi} \frac{2+2\cos x - \cos(n-1)x - 2\cos nx - \cos(n+1)x}{1-\cos 2x} dx.$$

Solution. Let a_n be the value of the integral. Then $a_0 = 0$ and $a_1 = \pi$. For any $n > 0$ we have

$$a_{n+1} - 2a_n + a_{n-1} = \int_0^{\pi} \frac{\cos(-2+n)x - 2\cos nx + \cos(2+n)x}{-1+2\cos x} dx$$

$$= \int_0^{\pi} \frac{2\cos nx(-1+\cos^2 x)}{-1+\cos 2x} dx = 0.$$

Thus, $a_{n+1} - a_n = a_n - a_{n-1}$, so (a_n) is an arithmetic progression. Hence $a_n = n\pi$, for all $n \geq 0$. $\square$

9.5 Riemann Iintegrals and Limits

> If one must choose between rigor and meaning, I shall unhesitatingly choose the latter.
>
> René Thom (1923–2002)

9.5.1. Let $f : \mathbb{R} \to (0, \infty)$ be an increasing differentiable function such that $\lim_{x \to \infty} f(x) = \infty$ and f' is bounded.
Let $F(x) = \int_0^x f(t) dt$. Define the sequence $(a_n)_{n \geq 1}$ inductively by

$$a_1 = 1, \quad a_{n+1} = a_n + \frac{1}{f(a_n)},$$

and the sequence $(b_n)_{n \geq 1}$ by $b_n = F^{-1}(n)$. Prove that $\lim_{n \to \infty} (a_n - b_n) = 0$.

Solution. By hypothesis it follows that that F is increasing and $\lim_{n \to \infty} b_n = \infty$.

Using the Lagrange mean value theorem, we deduce that for all positive integers k, there exists $\xi \in (a_k, a_{k+1})$ such that

$$F(a_{k+1}) - F(a_k) = f(\xi)(a_{k+1} - a_k) = \frac{f(\xi)}{f(a_k)}.$$

By monotonicity, $f(a_k) \leq f(\xi) \leq f(a_{k+1})$; thus

$$1 \leq F(a_{k+1}) - F(a_k) \leq \frac{f(a_{k+1})}{f(a_k)} \leq 1 + \frac{f(a_{k+1}) - f(a_k)}{f(a_k)}. \tag{9.8}$$

Summing (9.8) for $k = 1, \ldots, n-1$ and substituting $F(b_n) = n$, we have

$$F(b_n) < n + F(a_1) \leq F(a_n) \leq F(b_n) + F(a_1) + \sum_{k=1}^{n-1} \frac{f(a_{k+1}) - f(a_k)}{f(a_k)}. \tag{9.9}$$

From the first two inequalities we deduce that $a_n > b_n$ and $\lim_{n \to \infty} a_n = \infty$.

Fix $\varepsilon > 0$. Choose an integer K_ε such that $f(a_{K_\varepsilon}) > 2/\varepsilon$. If n is sufficiently large, then

$$F(a_1) + \sum_{k=1}^{n-1} \frac{f(a_{k+1}) - f(a_k)}{f(a_k)} = F(a_1) + \sum_{k=1}^{K_\varepsilon - 1} \frac{f(a_{k+1}) - f(a_k)}{f(a_k)}$$
$$+ \sum_{k=K_\varepsilon}^{n-1} \frac{f(a_{k+1}) - f(a_k)}{f(a_k)} \tag{9.10}$$
$$< O_\varepsilon(1) + \frac{1}{f(a_{K_\varepsilon})} \sum_{k=K_\varepsilon}^{n-1} (f(a_{k+1}) - f(a_k))$$
$$< O_\varepsilon(1) + \frac{\varepsilon}{2}(f(a_n) - f(a_{K_\varepsilon})) < \varepsilon f(a_n).$$

Inequalities (9.9) and (9.10) together say that for any positive ε, if n is sufficiently large,
$$F(a_n) - F(b_n) < \varepsilon f(a_n).$$

Applying again the Lagrange mean value theorem, we obtain a real number $\xi \in (b_n, a_n)$ such that
$$F(a_n) - F(b_n) = f(\xi)(a_n - b_n) > f(b_n)(a_n - b_n).$$

Therefore
$$f(b_n)(a_n - b_n) < \varepsilon f(a_n). \tag{9.11}$$

Let B be an upper bound for f'. Since $f(a_n) < f(b_n) + B(a_n - b_n)$, relation (9.11) yields
$$f(b_n)(a_n - b_n) < \varepsilon(f(b_n) + B(a_n - b_n));$$
hence
$$(f(b_n) - \varepsilon B)(a_n - b_n) < \varepsilon f(b_n).$$

Since $\lim_{n \to \infty} f(b_n) = \infty$, the first factor is positive, and we have
$$0 < a_n - b_n < \varepsilon \frac{f(b_n)}{f(b_n) - \varepsilon B} < 2\varepsilon,$$

provided n is sufficiently large. This concludes the proof. $\square$

We have established in Chapter 2 that the harmonic series $\sum_{n=1}^{\infty} 1/n$ diverges, but we have asserted that the sequence $(a_n)_{n \geq 1}$ defined by
$$a_n = \sum_{k=1}^{n} \frac{1}{k} - \ln n$$
converges to a positive number γ, called Euler's constant. We are now in position to prove rigorously this important result.

9.5.2. *Prove that the sequence $(a_n)_{n \geq 1}$ defined by*
$$a_n = \sum_{k=1}^{n} \frac{1}{k} - \ln n$$
converges to a positive number $\gamma \in (0, 1)$.

Solution. For any $n \geq 2$, set
$$x_n = \ln n - \ln(n-1) - \frac{1}{n}.$$

It follows that $x_2 + \cdots + x_n = 1 - a_n$, so it is sufficient to show that the series $\sum_{n=2}^{\infty} x_n$ converges to a number $\ell \in (0, 1)$. We observe that
$$\int_0^1 \frac{x}{n(n-x)} dx = \int_0^1 \left(\frac{1}{n-x} - \frac{1}{n} \right) dx = x_n.$$

Therefore
$$0 < x_n = \int_0^1 \frac{x}{n(n-x)}dx < \frac{1}{n(n-1)} = \frac{1}{n-1} - \frac{1}{n}.$$
We deduce that the series $\sum_{n=2}^\infty x_n$ is convergent and
$$0 < \sum_{n=2}^\infty x_n < 1,$$
which concludes the proof. □

9.5.3. Let $f : [0,1] \to [0,\infty)$ be a continuous function. Define, for all $n \geq 1$,
$$I_n(f) = n\int_0^1 [f(t)]^n dt.$$

Find the limit of the sequence $(I_n(f))_{n\geq 1}$ in the following cases:

(i) if $\max_{x\in[0,1]} f(x) < 1$;
(ii) if $\max_{x\in[0,1]} f(x) > 1$.

Solution. If $\max_{x\in[0,1]} f(x) < 1$, then there exists $\varepsilon > 0$ such that $f(t) \leq 1-\varepsilon < 1$, so
$$|I_n| \leq n(1-\varepsilon)^n \to 0 \quad \text{as } n\to\infty.$$

If $\max_{x\in[0,1]} f(x) > 1$, then there exists $t_0 \in [0,1]$ such that $f(t_0) > 1$. Assume $t_0 \in (0,1)$. Then, by the continuity of f, there exists $\eta > 0$ such that $f(t) \geq 1+\varepsilon > 1$, for all $t \in [t_0-\eta, t_0+\eta]$. Therefore
$$I_n \geq 2n\eta(1+\varepsilon)^n \to \infty \quad \text{as } n\to\infty.$$

Similar arguments can be applied if $t_0 \in \{0,1\}$. □

9.5.4. Let $f, g : \mathbb{R} \to \mathbb{R}$ be continuous functions such that $f(x+1) = f(x)$ and $g(x+1) = g(x)$, for all $x \in \mathbb{R}$. Prove that
$$\lim_{n\to\infty} \int_0^1 f(x)g(nx)dx = \left(\int_0^1 f(x)dx\right)\left(\int_0^1 g(x)dx\right).$$

Solution. Since $\int_0^1 f(x)dx = \sum_{k=0}^{n-1} \int_{k/n}^{(k+1)/n} f(x)dx$, we have
$$\int_0^1 f(x)g(nx)dx = \frac{1}{n}\sum_{k=0}^{n-1} \int_0^1 f((x+k)/n)g(x)dx$$
$$= \int_0^1 g(x) \sum_{k=0}^{n-1} f((x+k)/n) \cdot \frac{1}{n} \longrightarrow \left(\int_0^1 f(x)dx\right)\cdot\left(\int_0^1 g(x)dx\right). \quad □$$

9.5.5. Let $f : [0,1] \to \mathbb{R}$ be a continuous function. Compute

(a) $\int_0^1 x^n f(x)dx.$

(b) $\int_0^1 x^n f(x) dx.$

Solution. (a) Fix $\varepsilon > 0$. Let

$$L = \max_{x \in [0,1]} (|f(x)| + 1) \quad \text{and} \quad 0 < \delta < \min\left\{\frac{\varepsilon}{2L}, 1\right\}.$$

We have

$$\left|\int_{1-\delta}^1 x^n f(x) dx\right| \leq \int_{1-\delta}^1 x^n |f(x)| dx \leq L\delta \leq \frac{\varepsilon}{2}$$

and

$$\left|\int_0^{1-\delta} x^n f(x) dx\right| \leq \int_0^{1-\delta} (1-\delta)^n |f(x)| dx \leq L(1-\delta)^{n+1}.$$

It follows that

$$\lim_{n \to \infty} \int_0^1 x^n f(x) dx = 0.$$

(b) We claim that

$$\lim_{n \to \infty} n \int_0^1 x^n (f(x) - f(1)) dx = 0.$$

For $\varepsilon > 0$ fixed, let $\delta > 0$ be such that $|f(x) - f(1)| < \varepsilon/2, \forall x \in [1-\delta, 1]$. Hence

$$\left|n \int_{1-\delta}^1 x^n (f(x) - f(1)) dx\right| \leq n \int_{1-\delta}^1 x^n |f(x) - f(1)| dx \leq n \int_{1-\delta}^1 x^n \frac{\varepsilon}{2} dx \leq \frac{\varepsilon}{2}.$$

Taking $L = \sup_{x \in [0,1]} |f(x) - f(1)|$, we have

$$\left|n \int_0^{1-\delta} x^n (f(x) - f(1)) dx\right| \leq n \int_0^{1-\delta} x^n L dx = n \frac{(1-\delta)^{n+1}}{n+1},$$

which concludes the proof of our claim.

Finally, it is enough to observe that

$$n \int_0^1 x^n f(x) dx = n \int_0^1 x^n (f(x) - f(1)) dx + n \int_0^1 f(1) x^n dx \to f(1). \quad \square$$

9.5.7. Let $f : [0,1] \to \mathbb{R}$ be a differentiable function such that

$$\sup_{x \in (0,1)} |f'(x)| = M < \infty.$$

Let n be a positive integer. Prove that

$$\left|\sum_{j=0}^{n-1} \frac{f(j/n)}{n} - \int_0^1 f(x) dx\right| \leq \frac{M}{2n}.$$

Solution. We have

$$\left|\sum_{j=0}^{n-1}\frac{f(j/n)}{n}-\int_0^1 f(x)dx\right|=\left|\sum_{j=0}^{n-1}\left(\frac{f(j/n)}{n}-\int_{j/n}^{(j+1)/n}f(x)dx\right)\right|$$

$$\leq \sum_{j=0}^{n-1}\int_{j/n}^{(j+1)/n}|f(j/n)-f(x)|dx.$$

On the other hand, by the Lagrange mean value theorem, we obtain that for any $x \in (j/n, (j+1)/n)$ there exists $c_x \in (j/n, x)$ such that

$$f'(c_x) = \frac{f(x) - f(j/n)}{x - j/n}.$$

By the hypothesis that f' is bounded we deduce that

$$|f(x) - f(j/n)| \leq M(x - j/n), \quad \forall x \in (j/n - (j+1)/n).$$

Hence

$$\left|\sum_{j=0}^{n-1}\frac{f(j/n)}{n}-\int_0^1 f(x)dx\right| \leq \sum_{j=0}^{n-1}\int_{j/n}^{(j+1)/n} M(x - j/n)dx$$

$$= M\sum_{j=0}^{n-1}\left(\frac{(j+1)^2}{2n^2} - \frac{j^2}{2n^2} - \frac{j}{n^2}\right) = M\sum_{j=0}^{n-1}\frac{1}{2n^2} = \frac{M}{2n}. \quad \square$$

9.5.8. Let $f : [0,\infty) \to [0,\infty)$ be a continuous function with the property that $\lim_{x\to\infty} f(x)\int_0^x f(t)dt = \ell \in (0,\infty)$. Prove that $\lim_{x\to\infty} f(x)\sqrt{x}$ exists and is finite, and evaluate it in terms of ℓ.

C. Popescu and D. Schwartz

Solution. Let $F : [0,\infty) \to [0,\infty)$ be defined by $F(x) = \int_0^x f(t)dt$. Since f takes on nonnegative values, it follows that F is increasing, so

$$\lim_{x\to\infty} F(x) = \sup\{F(x) : x \geq 0\} > 0.$$

We now prove that $\lim_{x\to\infty} F(x) = \infty$. Indeed, if this limit is finite and equals $L > 0$, then $\lim_{x\to\infty} f(x) = \ell/L$. Hence $f(x) \geq \ell/(2L)$ for $x \geq x_0 > 0$. So, for all $x \geq x_0$, $F(x) = F(x_0) + \int_{x_0}^x f(t)dt \geq \ell(x-x_0)/(2L)$, which contradicts the assumption that $\lim_{x\to\infty} F(x)$ is finite.

Applying l'Hôpital's rule, we obtain

$$\lim_{x\to\infty}\frac{F^2(x)}{x} = 2\lim_{x\to\infty} f(x)F(x) = 2\ell.$$

It follows that $\lim_{x\to\infty} \frac{F(x)}{\sqrt{x}} = \sqrt{2\ell}$. Finally, we write

$$f(x)\sqrt{x} = \frac{f(x)F(x)}{\frac{F(x)}{\sqrt{x}}},$$

for x large enough, to conclude that $\lim_{x\to\infty} f(x)\sqrt{x}$ exists, is finite, and equals $\sqrt{\ell/2}$. □

9.5.9. *Assume that f is a real-valued continuously differentiable function on some interval $[a,\infty)$ such that $f'(x) + \alpha f(x)$ tends to zero as x tends to infinity, for some $\alpha > 0$. Prove that $f(x)$ tends to zero as x tends to infinity.*

Solution. Set $g := f' + \alpha f$. Arguing by contradiction, there exist $C_1 > 0$ and a sequence $(x_n)_{n\geq 1}$ such that $\lim_{n\to\infty} x_n = \infty$ and $|f(x_n)| \geq C_1$, for all $n \geq 1$. Without loss of generality, we can assume that $f(x_n) \geq C_1$, for all $n \geq 1$. Hence $f'(x_n) = g(x_n) - \alpha f(x_n) \leq g(x_n) - \alpha C_1$. So, since $g(x) \to 0$ as $x \to \infty$ and $\alpha > 0$, there exists $N_0 \in \mathbb{N}$ such that $f'(x_n) \leq -\alpha C_1/2 \equiv -C_2$, for all $n \geq N_0$. Therefore

$$\frac{f'(x_n)}{f(x_n)} \leq -\frac{C_2}{C_1} \equiv -C_3,$$

for all $n \geq N_0$, where $C_3 > 0$. By integration on $[a, x_n]$ in the above inequality we obtain

$$\ln f(x_n) \leq -C_3 x_n + C_4,$$

or equivalently,

$$f(x_n) \leq e^{-C_3 x_n + C_4},$$

for all $n \geq N_0$. This implies $f(x_n) \to 0$ as $n \to \infty$, which contradicts our assumption $f(x_n) \geq C_1 > 0$, for all $n \geq 1$. In conclusion, $\lim_{x\to\infty} f(x) = 0$. It is obvious that the statement does not remain true if $\alpha \leq 0$. Give an example! □

9.5.10. *(i) Prove that for any positive integer n,*

$$(n-1)! \leq n^n e^{-n} e \leq n!.$$

(ii) Deduce that

$$\frac{(n!)^{1/n}}{n} = \left[\frac{n!}{n^n}\right]^{1/n}$$

approaches e^{-1} as n tends to infinity.

Solution. (i) A direct proof uses an induction argument based on the inequalities

$$\left(1 + \frac{1}{n}\right)^n < e < \left(1 + \frac{1}{n}\right)^{n+1}.$$

An alternative proof using the integral calculus uses the evaluation and comparison of $\int_1^n \ln x\, dx$ with the upper and lower Darboux sums associated with the partition $(1, 2, \ldots, n)$ of the interval $[1, n]$.

(ii) Taking the nth root of the right inequality in (i), we obtain
$$ne^{-1}e^{1/n} \leq (n!)^{1/n}.$$

Dividing by n yields
$$\frac{1}{e}e^{1/n} \leq \frac{(n!)^{1/n}}{n}.$$

On the other hand, we multiply the first inequality in (i) by n and take the nth root. Therefore
$$(n!)^{1/n} \leq ne^{-1}e^{1/n}n^{1/n}.$$

Dividing by n yields
$$\frac{(n!)^{1/n}}{n} \leq \frac{1}{e}e^{1/n}n^{1/n}.$$

But both $n^{1/n}$ and $e^{1/n}$ approach 1 as n becomes large. Thus our quotient is squeezed between two numbers approaching e^{-1}, and must therefore approach e^{-1}. □

9.5.11. *Let $g : [0, 1] \to \mathbb{R}$ be a continuous function such that $\lim_{x \to 0+} g(x)/x$ exists and is finite. Prove that*
$$\lim_{n \to \infty} \int_0^1 g(x^n) dx = \int_0^1 \frac{g(x)}{x} dx.$$

D. Andrica and M. Piticari

Solution. Define the function $h : [0, 1] \to \mathbb{R}$ by
$$h(t) = \begin{cases} \frac{g(t)}{t} & \text{if } t \in (0, 1], \\ \lim_{x \to 0, x > 0} \frac{g(x)}{x} & \text{if } t = 0. \end{cases}$$

Then h is continuous, and we can set
$$H(x) = \int_0^x h(t) dt.$$

We have
$$n \int_0^1 g(x^n) dx = n \int_0^1 x^n h(x^n) dx = x H(x^n)\Big|_0^1 - \int_0^1 H(x^n) dx$$
$$= H(1) - \int_0^1 H(x^n) dx = \int_0^1 \frac{g(x)}{x} dx - \int_0^1 H(x^n) dx.$$

If $0 < a < 1$, then
$$\left|\int_0^1 H(x^n) dx\right| \leq \int_0^1 |H(x^n)| dx = \int_0^a |H(x^n)| dx + \int_a^1 |H(x^n)| dx \qquad (9.12)$$
$$\leq a|H(\alpha_n^n)| + (1-a)M,$$

where $\alpha_n \in [0, a]$ and $M = \max_{t \in [0,1]} |H(t)|$.

Consider $\varepsilon > 0$ such that $a > 1 - \varepsilon/(2M)$. Since $\lim_{n\to\infty} |H(\alpha_n^n)| = 0$, it follows that $a|H(\alpha_n^n)| < \frac{\varepsilon}{2}$ for all positive integers $n \geq N(\varepsilon)$. Relation (9.12) yields

$$\left|\int_0^1 H(x^n)dx\right| \leq \frac{\varepsilon}{2} + (1-a)M < \frac{\varepsilon}{2} + \left(1 - 1 + \frac{\varepsilon}{2M}\right)M = \varepsilon.$$

Hence $\lim_{n\to\infty} \int_0^1 H(x^n)dx = 0$ and the conclusion follows. □

9.5.12. Let $g : [0,1] \to \mathbb{R}$ be a continuous function such that $\lim_{x\to 0+} g(x)/x$ exists and is finite. Prove that for any function $f : [0,1] \to \mathbb{R}$ of class C^1,

$$\lim_{n\to\infty} n \int_0^1 f(x)g(x^n)dx = f(1)\int_0^1 \frac{g(x)}{x}dx.$$

D. Andrica and M. Piticari

Solution. For any $x \in [0,1]$, define $G(x) = \int_0^x g(t)/t\, dt$, and observe that

$$n\int_0^1 f(x)g(x^n)dx = n\int_0^1 x^n f(x)\frac{g(x^n)}{x^n}dx$$

$$= G(x^n)xf(x)\Big|_0^1 - \int_0^1 [xf'(x) + f(x)]G(x^n)dx \quad (9.13)$$

$$= G(1)f(1) - \int_0^1 [xf'(x) + f(x)]G(x^n)dx$$

$$= f(1)\int_0^1 \frac{g(x)}{x}dx - \int_0^1 [xf'(x) + f(x)]G(x^n)dx.$$

We will prove that

$$\lim_{n\to\infty} \int_0^1 [xf'(x) + f(x)]G(x^n)dx = 0.$$

Indeed, by considering $M = \max_{x\in[0,1]} |xf'(x) + f(x)|$, we have

$$\left|\int_0^1 [xf'(x) + f(x)]G(x^n)dx\right| \leq \int_0^1 |xf'(x) + f(x)||G(x^n)|dx$$

$$\leq M\int_0^1 |G(x^n)|dx.$$

Using that $\lim_{n\to\infty} \int_0^1 |G(x^n)|dx = 0$ (see Exercise 9.4.11), our conclusion follows from (9.13).

Some direct consequences of the above property are the following:

(i) If $f : [0,1] \to \mathbb{R}$ is a continuous function, then

$$\lim_{n\to\infty} n\int_0^1 \frac{x^n f(x)}{1+x^{2n}}dx = \frac{\pi}{4}f(1).$$

(ii) If $a > 0$, then

$$\lim_{n \to \infty} n \int_0^1 \frac{x^n}{a + x^n} dx = \ln \frac{a+1}{a}.$$

Romanian National Olympiad, 2001

(iii) If $f : [0, 1] \to \mathbb{R}$ is a continuous function, then

$$\lim_{n \to \infty} n \int_0^1 f(x) \ln(1 + x^n) dx = \frac{\pi^2}{12} f(1).$$

The properties

$$\lim_{n \to \infty} n \int_1^a \frac{dx}{x^n + 1} = \ln 2$$

and

$$\lim_{n \to \infty} n \int_0^1 \frac{x^{n-2}}{x^{2n} + x^n + 1} dx = \frac{\pi}{3\sqrt{3}}$$

are direct consequences of the more general property stated in the following exercise. □

9.5.13. Let $f : [1, +\infty) \to \mathbb{R}$ be a continuous function such that the limit $\lim_{x \to \infty} x f(x)$ exists and is finite. Prove that $\lim_{t \to \infty} \int_1^t \frac{f(x)}{x} dx$ exists and

$$\lim_{n \to \infty} n \int_1^a f(x^n) dx = \lim_{t \to \infty} \int_1^t \frac{f(x)}{x} dx,$$

for any $a > 1$.

D. Andrica and M. Piticari

Solution. Define $\lim_{x \to \infty} x f(x) = \ell$, where $\ell \in \mathbb{R}$. Thus, we can find a real number $x_0 > 1$ such that for any $x \geq x_0$,

$$\frac{\ell - 1}{x^2} \leq \frac{f(x)}{x} \leq \frac{\ell + 1}{x^2}.$$

Choose $m > 0$ satisfying $\ell - 1 + m \geq 0$. Then, for any $x \geq x_0$,

$$0 \leq \frac{\ell - 1 + m}{x^2} \leq \frac{f(x)}{x} + \frac{m}{x^2} \leq \frac{\ell + 1 + m}{x^2}. \tag{9.14}$$

Define the function $J : [1, \infty) \to \mathbb{R}$ by

$$J(t) = \int_1^t \left(\frac{f(x)}{x} + \frac{m}{x^2} \right) dx.$$

Then J is differentiable and

$$J'(t) = \frac{f(t)}{t} + \frac{m}{t^2} \geq 0.$$

Therefore J is an increasing function on the interval $[x_0, +\infty)$. Moreover, using the last inequality in (9.14), we get by integration

$$J(t) = \int_1^{x_0} \left(\frac{f(x)}{x} + \frac{m}{x^2}\right) dx + \int_{x_0}^t \left(\frac{f(x)}{x} + \frac{m}{x^2}\right) dx$$

$$\leq \int_1^{x_0} \left(\frac{f(x)}{x} + \frac{m}{x^2}\right) dx + (\ell + 1 + m) \int_{x_0}^t \frac{dx}{x^2}$$

$$\leq \int_1^{x_0} \left(\frac{f(x)}{x} + \frac{m}{x^2}\right) dx + \frac{\ell + 1 + m}{x_0},$$

for any $t \geq x_0$. It follows that $\lim_{t \to \infty} J(t)$ is finite. On the other hand,

$$J(t) = \int_1^t \frac{f(x)}{x} dx + m\left(1 - \frac{1}{t}\right);$$

hence

$$\lim_{t \to \infty} \int_1^t \frac{f(x)}{x} dx = \lim_{t \to \infty} J(t) - m,$$

which is finite.

For a fixed number $a > 1$, define

$$J(t) = t \int_1^a f(x^t) dx \quad \text{and} \quad U(t) = \int_1^{a^t} \frac{f(x)}{x} dx.$$

Because the function $g : [1, +\infty) \to \mathbb{R}$ defined by $g(x) = xf(x)$ is continuous and $\lim_{x \to \infty} g(x)$ is finite, it follows that g is bounded. Thus, there exists $M > 0$ such that

$$|g(x)| \leq M, \quad x \in [1, \infty). \tag{9.15}$$

Changing the variable $x = u^t$, we obtain $dx = tu^{t-1} du$, hence

$$U(t) = t \int_1^a \frac{f(u^t)}{u} du. \tag{9.16}$$

From (9.15) and (9.16) we obtain

$$|J(t) - U(t)| = t \left| \int_1^a f(x^t) dx - \int_1^a \frac{f(x^t)}{x} dx \right|$$

$$= t \left| \int_1^a \left(f(x^t) - \frac{f(x^t)}{x}\right) dx \right| \leq t \int_1^a |f(x^t)| \frac{x-1}{x} dx \tag{9.17}$$

$$= t \int_1^a x^t |f(x^t)| \frac{x-1}{x^{t+1}} dx \leq tM \int_1^a \frac{x-1}{x^{t+1}} dx$$

$$= Mt \left[\frac{1}{1-t}(a^{-t+1} - 1) - \frac{1}{t}(1 - a^{-t})\right], \quad t > 0.$$

Because

$$\lim_{t \to \infty} \left[\frac{1}{1-t}(a^{-t+1} - 1) - \frac{1}{t}(1 - a^{-t})\right] = 0,$$

from (9.17) it follows that

$$\lim_{t \to \infty} J(t) = \lim_{t \to \infty} U(t).$$

Therefore

$$\lim_{t \to \infty} \int_1^t \frac{f(x)}{x} dx = \lim_{t \to \infty} \int_1^a f(x^t) dx,$$

and the desired result follows. □

9.6 Qualitative Results

> I listen only to Bach, Beethoven or Mozart. Life is too short to waste on other composers.
>
> John Edensor Littlewood (1885–1977)

9.6.1. Let $\mathscr{E}$ be the set of all continuous functions $u : [0,1] \to \mathbb{R}$ such that

$$|u(x) - u(y)| \leq |x - y|, \quad 0 \leq x, y \leq 1, \quad u(0) = 0.$$

Define the function $\varphi : \mathscr{E} \to \mathbb{R}$ by

$$\varphi(u) = \int_0^1 (u^2(x) - u(x)) dx.$$

Prove that φ achieves its maximum on $\mathscr{E}$.

Solution. We have

$$|u(x)| = |u(x) - u(0)| \leq |x|$$

and

$$|u^2(x) - u(x)| = |u(x)| \cdot |u(x) - 1| \leq |x|(|x| + 1).$$

Hence

$$|\varphi(u)| \leq \int_0^1 |u^2(x) - u(x)| \leq \int_0^1 x(x+1) dx = \frac{5}{6}.$$

Equality holds if $|u(x)| = x$ and $|u(x) - 1| = x + 1$. This happens for $u \in \mathscr{E}$ defined by $u(x) = -x$. □

9.6.2. Evaluate the limit

$$\lim_{x \to 0+} \int_x^{2x} \frac{\sin^m t}{t^n} dt, \quad \text{where } m, n \in \mathbb{N}.$$

Solution. We use the fact that $\sin t / t$ is decreasing in the interval $(0, \pi)$ and $\lim_{x \to 0+} \sin t / t = 1$. For all $x \in (0, \pi/2)$ and $t \in [x, 2x]$ we have

$\sin 2x/(2x) < \sin t/t < 1$, thus

$$\left(\frac{\sin 2x}{2x}\right)^m \int_x^{2x} \frac{t^m}{t^n} dt < \int_x^{2x} \frac{\sin^m t}{t^n} dt < \int_x^{2x} \frac{t^m}{t^n} dt.$$

On the other hand,

$$\int_x^{2x} \frac{t^m}{t^n} dt = x^{m-n+1} \int_1^2 u^{m-n} du.$$

The factor $\sin 2x/(2x)^m$ tends to 1 as $x \to 0$. If $m - n + 1 < 0$, the limit of x^{m-n+1} is infinity, while if $m - n + 1 > 0$, then it equals 0. If $m - n + 1 = 0$, then $x^{m-n+1} \int_1^2 u^{m-n} du = \ln 2$. Hence,

$$\lim_{x \to 0+} \int_x^{2x} \frac{\sin^m t}{t^n} dt = \begin{cases} 0, & \text{if } m \geq n, \\ \ln 2, & \text{if } n - m = 1, \\ +\infty, & \text{if } n - m > 1. \end{cases} \quad \square$$

9.6.3. Let $f : \mathbb{R} \to \mathbb{R}$ be a function of class C^1 such that

$$\lim_{x \to \infty} f(x) = a \quad \text{and} \quad \lim_{x \to \infty} f'(x) = b.$$

Prove that $b = 0$.

Solution. By addition and multiplication by positive constants we can assume that $a = 0 \leq b$. Fix $\varepsilon > 0$. Choose $R \geq 1$ such that $|f(x)| \leq \varepsilon$ and $f'(x) > b/2 \geq 0$, $\forall x \geq R$. On the other hand,

$$f(x) = f(R) + \int_R^x f'(t) dt.$$

It follows that

$$2\varepsilon \geq f(x) - f(R) \geq \int_R^x \frac{b}{2} dx = \frac{(x-R)b}{2}.$$

For $x = 5R$ we obtain $b \leq \varepsilon/R \leq \varepsilon$. Since $\varepsilon > 0$ is chosen arbitrarily, it follows that $b = 0$. $\square$

9.6.4. Let $f : \mathbb{R} \to \mathbb{R}$ be a nontrivial function of class C^∞ such that $\lim_{|x| \to \infty} f(x) = 0$ and

$$f(x)f(y) = f\left(\sqrt{x^2 + y^2}\right), \quad \forall x, y \in \mathbb{R}.$$

(a) Prove that f is even and $f(0) = 1$.
(b) Show that $f'(x) = f''(0)x f(x)$ and find all functions with these properties.

Solution. (a) Let $x = y = 0$. So $f^2(0) = f(0)$, and we have either $f(0) = 0$ or $f(0) = 1$. If $f(0) = 0$ then $0 = f(\sqrt{x^2})$ for all x, so $f(x) = 0$ for any $x > 0$. If there exists $y < 0$ such that $f(y) \neq 0$, then $f(x)f(y) = 0$ for all x. In conclusion, if $f(0) = 0$ then $f(x) = 0$, for all x. Since f is nontrivial, this is a contradiction. Therefore $f(0) = 1$. Taking $y = 0$, we obtain $f(x) = f(\sqrt{x^2}) = f(-x)$, so f is even.

9.6 Qualitative Results

(b) Let $r = \sqrt{x^2 + y^2}$. Differentiating with respect to y, we obtain

$$f(x)f'(y) = f'(r) \cdot \frac{y}{r}.$$

Again by differentiation we obtain

$$f(x)f''(y) = f''(r) \cdot \frac{y^2}{r^2} + f'(r) \cdot \frac{x^2}{r^3}.$$

Taking $y = 0$, we obtain $f'(x) = f''(0)xf(x)$, so

$$\frac{f'(x)}{f(x)} = f''(0)x.$$

By integration we deduce that $f(x) = e^{f''(0)x^2/2}$. Since $\lim_{|x|\to\infty} f(x) = 0$, we obtain $f''(0)/2 = -\gamma < 0$. Hence $f(x) = e^{-\gamma x^2}$, for some positive constant γ. □

9.6.5. Let $0 \leq a \leq 1$. Find all continuous functions $f : [0,1] \to [0,\infty)$ such that

$$\int_0^1 f(x)dx = 1, \quad \int_0^1 xf(x)dx = a, \quad \int_0^1 x^2 f(x)dx = a^2.$$

Solution. Let f be such a function. Applying the Cauchy–Schwarz inequality, we have

$$a = \int_0^1 xf(x)dx \leq \left(\int_0^1 x^2 f(x)dx\right)^{1/2} \left(\int_0^1 f(x)dx\right)^{1/2} \leq a.$$

So, all the above inequalities are, in fact, equalities. Taking into account the conditions that ensure equality in the Cauchy–Schwarz inequality, we deduce that there exists a nonnegative constant C such that

$$x\sqrt{f(x)} = C\sqrt{f(x)}, \quad \forall x \in [0,1].$$

Hence $\sqrt{f(x)} \equiv 0$, which contradicts $\int_0^1 f(x)dx = 1$. In conclusion, there is no function satisfying our hypotheses. □

Alternative proof. Multiplying the three equalities in our statement by $\alpha^2, -2\alpha$, resp. 1, we obtain

$$\int_0^1 f(x)(\alpha - x)^2 dx = 0, \quad \forall \alpha \in \mathbb{R}.$$

This implies $f \equiv 0$, which contradicts $\int_0^1 f(x)dx = 1$. Thus, there does not exist such a function.

9.6.6. Let $f : [0,\infty) \to [0,\infty)$ be a continuous function satisfying $f(0) = 1$ and such that $\lim_{t\to\infty} \int_x^t f(s)ds \leq f(x)$, for all $x \geq 0$. Find the maximum value of $f(x)$, for $x > 0$.

Solution. The maximum value of $f(x)$ is 1 if $0 \leq x \leq 1$ and e^{1-x} if $x > 1$. Let $F(x) = \lim_{t \to \infty} \int_x^t f(s)ds$. By hypothesis we deduce that $F(0) \leq 1$ and $F(x) \leq -F'(x)$, for all x. Hence

$$(\ln F(x))' = \frac{F'(x)}{F(x)} \leq -1, \quad \forall x \in \mathbb{R}.$$

By integration from 0 to x we obtain $\ln(F(x)/F(0)) \leq -x$. Since $F(0) \leq 1$, it follows that $F(x) \leq e^{-x}$, for all $x > 0$. Using now the monotony of f, we obtain $f(x) \leq 1$, for any $x \geq 0$. We also have

$$f(x) \leq \int_{x-1}^x f(t)dt \leq F(x-1) \leq e^{1-x}, \quad \forall x > 1.$$

To prove that this bound is achieved, fix $x \geq 0$. For any $x \in [0,1]$, define

$$f^*(t) = \begin{cases} 1 & \text{if } 0 \leq t \leq 1, \\ 0 & \text{if } t > 1. \end{cases}$$

The function f^* satisfies our assumptions and $f^*(x) = 1$. For any $x > 1$, define

$$f^*(t) = \begin{cases} e^{-t} & \text{if } 0 \leq t < x-1, \\ e^{1-x} & \text{if } x-1 \leq t \leq x, \\ 0 & \text{if } t > x. \end{cases}$$

Then f^* satisfies our hypotheses and $f^*(x) = e^{1-x}$. $\square$

9.6.7. Assume that $g : (0,1) \to \mathbb{R}$ is a continuous function and let $f : \mathbb{R} \to \mathbb{R}$ be a continuous function such that the limits (finite or infinite) $f(-\infty) = \lim_{t \to -\infty} f(t)$ and $f(+\infty) = \lim_{t \to +\infty} f(t)$ exist, and $f(+\infty) < f(t) < f(-\infty)$, for all $t \in \mathbb{R}$.

Prove that there exists a function u satisfying $u'' + f(u) - g(x) = 0$ in $(0,1)$ and $u'(0) = u'(1) = 0$ if and only if

$$f(+\infty) < \int_0^1 g(x)dx < f(-\infty).$$

Ph. Korman, SIAM Review, Problem 00-001

Solution. (Michael Renardy) The necessity of the condition follows by integrating the differential equation on $(0,1)$. We obtain that if such a function u exists, then

$$f(+\infty) < \int_0^1 f(u(x))dx = \int_0^1 g(x)dx < f(-\infty).$$

To prove the sufficiency, consider the initial value problem

$$u'' + f(u) - g(x) = 0, \quad u(0) = a, \quad u'(0) = 0.$$

9.6 Qualitative Results

We obtain that if $a \to \infty$,

$$u'(1) = \int_0^1 u''(x)dx = \int_0^1 [g(x) - f(u(x))]dx \longrightarrow \int_0^1 g(x)dx - f(+\infty).$$

Similarly, as $a \to -\infty$,

$$u'(1) \longrightarrow \int_0^1 g(x)dx - f(-\infty). \quad \square$$

Using the hypothesis in combination with the intermediate value property, we obtain a value of $a \in \mathbb{R}$ for which $u'(1) = 0$.

9.6.8. Let a and b be positive real numbers. Let f and g be functions from $\mathbb{R}$ into $\mathbb{R}$, twice differentiable, with initial conditions $f(0) = a$, $f'(0) = 0$, $g(0) = 0$, $g'(0) = b$, and satisfying the differential equations

$$f'' = -f(1 - f^2 - g^2), \quad g'' = -g(1 - f^2 - g^2).$$

(a) Show that there is a nontrivial polynomial function $E(X,Y)$ such that for all $a, b > 0$, $E(f^2(t) + g^2(t), f'^2(t) + g'^2(t))$ is independent of t.
(b) Show that if f and g are both periodic in t, with period T, and if at $t = 0$, $f^2(t) + g^2(t)$ is not at a local minimum, then $a \leq 1$, $b^2 \leq a^2(1 - a^2)$, and $T > 2\pi$.
(c) Give an example of f and g satisfying the premises of part (b).
(d) Prove that there exist choices of a and b such that the resulting (f, g) is periodic, and $\min(f^2 + g^2) < (1/2) \max(f^2 + g^2)$.

Vicenţiu Rădulescu, Amer. Math. Monthly, Problem 11073

Solution. (a) This statement is a kind of energy conservation law. Let $r = \sqrt{f^2 + g^2}$ and $s = \sqrt{f'^2 + g'^2}$. Then

$$(f^2 + g^2)(1 - (1/2)(f^2 + g^2)) + (f'^2 + g'^2) = r^2(1 - r^2/2) + s^2.$$

Define $E(X,Y) = X - X^2/2 + Y$. It follows that

$$\frac{d}{dt} E(f^2(t) + g^2(t), f'^2(t) + g'^2(t)) = 2\left[(ff' + gg')(1 - f^2 - g^2) + (f'f'' + g'g'')\right](t).$$

Using now the differential equation satified by f and g, we obtain

$$\frac{d}{dt} E(f^2(t) + g^2(t), f'^2(t) + g'^2(t)) = 2(ff' + gg')(1 - f^2 - g^2)(t)$$

$$- 2\left[f'f(1 - f^2 - g^2) + g'g(1 - f^2 - g^2)\right](t)$$

$$= 0,$$

for any $t \in \mathbb{R}$. Hence

$$E(f^2(t)+g^2(t), f'^2(t)+g'^2(t)) \equiv E(f^2(0)+g^2(0), f'^2(0)+g'^2(0)) = a^2 - a^4/2 + b^2.$$

$\square$

Alternative proof of (a). We multiply the differential equation $f'' = -f(1 - f^2 - g^2)$ by f' and then we integrate on $[0,t]$. Using the assumptions $f(0) = a$ and $f'(0) = 0$, we obtain

$$f^2(t) - \frac{f^4(t)}{2} + f'^2(t) - \int_0^t f(s)f'(s)g^2(s)ds = a^2 - \frac{a^4}{2}.$$

Similarly, using the differential equation satisfied by g, we obtain

$$g^2(t) - \frac{g^4(t)}{2} + g'^2(t) - \int_0^t f^2(s)g(s)g'(s)ds = b^2.$$

By addition, we obtain, for any $t \in \mathbb{R}$,

$$f^2(t) + g^2(t) - \frac{(f^2(t)+g^2(t))^2}{2} + f'^2(t) + g'^2(t) = a^2 - \frac{a^4}{2} + b^2.$$

(b) We first prove that $a \leq 1$. Indeed, arguing by contradiction, let us assume the contrary. Set $u := f^2 + g^2$. The assumption $a > 1$ enables us to choose $M > 1$ such that

$$\min\{u(x); x \in \mathbb{R}\} < M^2 < a^2.$$

Let $I \subset \mathbb{R}$ be a bounded interval such that $u > M^2$ in I, and $u = M^2$ on the boundary of I. But

$$u'' = 2u(u-1) + 2(f'^2 + g'^2) \geq 2u(u-1) > 0 \quad \text{in } I.$$

So, u is convex in I and $u = M^2$ on the boundary of I. Hence $u \leq M^2$, which contradicts the choice of I.

Applying Taylor's formula we have

$$f(x) = a - \frac{a(1-a^2)}{2}x^2 + O(x^3), \quad \text{as } x \to 0$$

and

$$g(x) = bx + O(x^3), \quad \text{as } x \to 0.$$

So

$$u(x) = a^2 + [b^2 - a^2(1-a^2)]x^2 + O(x^3), \quad \text{as } x \to 0.$$

Since $x = 0$ is a local maximum point of u, it follows that $b^2 \leq a^2(1-a^2)$.

An alternative proof of this statement is based on the fact that $u''(0) \leq 0$ (since $x = 0$ is a local maximum point of u) combined with $u''(0) = 2[b^2 + a^2(a^2-1)]$.

The above arguments also show that $a < 1$. Indeed, if $a = 1$, then $u''(0) = 2b^2 > 0$, a contradiction to the fact that the origin is a local maximum point of u.

9.6 Qualitative Results

Let us now prove that $T > 2\pi$. We first notice that f (or g) cannot have the same sign on an unbounded interval. Indeed, in this case, f'' (or g'') would have the same sign. But due to the periodicity, this is possible only for constant functions, which is impossible in our case.

Let x_1, x_2 be two consecutive zeros of f. We can assume that $f > 0$ in (x_1, x_2), so that $f'(x_1) > 0$ and $f'(x_2) < 0$. Denote by x_3 the smallest real number greater than x_2 such that $f(x_3) = 0$. Hence $f < 0$ in (x_2, x_3). If we prove that $x_2 - x_1 > \pi$, it will also follow that $x_3 - x_2 > 2\pi$ and there does not exist $x \in (x_1, x_3)$ such that $f(x) = 0$ and $f'(x) > 0$. This implies that the principal period of f must be greater than 2π. For our purpose, we multiply by $\varphi(x) := \sin \frac{\pi(x-x_1)}{x_2-x_1}$ in $f'' + f(1 - f^2 - g^2) = 0$ and then we integrate on $[x_1, x_2]$. Hence

$$\left(\frac{\pi}{x_2 - x_1}\right)^2 \int_{x_1}^{x_2} f(x)\varphi(x)dx = \int_{x_1}^{x_2} f(x)\left(1 - f^2(x) - g^2(x)\right)\varphi(x)dx$$
$$< \int_{x_1}^{x_2} f(x)\varphi(x)dx.$$

It follows that $x_2 - x_1 > \pi$.

Alternative proof of (b). Define $u : \mathbb{R} \to [0, \infty)$ by $u(x) = f^2(x) + g^2(x)$, $x \in \mathbb{R}$. Clearly, u is a T-periodic function of class $C^2(\mathbb{R})$, and

$$u''(x) = 2(u(x)(u(x) - 1) + (f'(x))^2 + (g'(x))^2 \tag{9.18}$$

for all real x. In particular, $u''(0) = 2\left[a^2(a^2 - 1) + b^2\right]$. Since u has a local maximum at the origin, it follows that $u''(0) \leq 0$, which yields immediately $b^2 \leq a^2(1 - a^2)$ by the preceding. This establishes the first inequality in (ii).

The proof of the second is a bit trickier, so let us first outline the strategy. The main idea consists in showing that the distance between two "consecutive" zeros of f must exceed π. To be more precise, we shall prove the following claim:

Claim 1: *If $x_1 < x_2$ are real numbers such that $f(x_1) = f(x_2) = 0$, but $f(x) \neq 0$ for $x_1 < x < x_2$, then $x_2 - x_1 > \pi$.*

Therefore, if we are able to produce two disjoint open subintervals, say I and J, of an interval of length T, with the property that f does not vanish on $I \cup J$, but $f = 0$ on the boundary of $I \cup$ on the boundary of J, then the length of each of these subintervals exceeds π, by Claim 1, and we are done: $T \geq \text{length}\,I + \text{length}\,J > \pi + \pi = 2\pi$.

The easiest way to produce such intervals consists in showing that f must take on values of either sign on each interval of length T:

Claim 2: *The function f takes on values of either sign on each interval of length T.*

Assuming this, let us see how it applies to produce the desired subintervals. First, let $I = (\alpha, \beta)$ be the connected component of the open set $\{x : f(x) > 0\}$ that contains the origin (recall that $f(0) = a > 0$). Then $f(\alpha) = f(\beta) = 0$ by continuity of f and maximality of I. Next, by Claim 2, $f(x_0) < 0$ for some $x_0 \in [0, T]$. But since $f(T) = f(0) = a > 0$, x_0 must be an interior point of $[0, T]$. Now let $J = (\gamma, \delta)$ be the connected component of the open set $\{x : 0 < x < T \text{ and } f(x) < 0\}$ that contains x_0. Again, $f(\gamma) = f(\delta) = 0$ by continuity of f and maximality of J. Finally, observe that $\beta \leq \gamma$ and $\delta \leq \alpha + T$ (here we use the T-periodicity of f) to conclude that I and J are indeed disjoint open subintervals of $[\alpha, \alpha + T]$ satisfying the required conditions.

So all that remains to prove is Claims 1 and 2 above. Both proofs rely on the next claim

Claim 3: $u(x) \leq 1$ for all $x \in \mathbb{R}$.

Proof of Claim 3. Recall the inequality $b^2 \leq a^2(1 - a^2)$, proved in the first part. Since $b > 0$, we deduce that $a^2 < 1$, so $u(0) = a^2 < 1$. Now we argue by reductio ad absurdum. Suppose, by contradiction, that $u(x_0) > 1$ for some real x_0. Then $U = \{x : u(x) > 1\}$ is an open nonempty set. Let K denote the connected component of U that contains x_0. Since u is periodic and $u(0) = a^2 < 1$, it follows that $\alpha = \inf K > -\infty$ and $\beta = \sup K < \infty$. Observe now that $u(\alpha) = u(\beta) = 1$, by continuity of u and maximality of K, while $u(x) > 1$ for $\alpha < x < \beta$. This leads to the following two contradictory facts: on the one hand, by virtue of (9.18),

$$u''(x) \geq 2u(x)(u(x) - 1) > 0 \tag{9.19}$$

for $\alpha < x < \beta$; on the other hand, since u is continuous, it must attain a *maximum* value on the compact set $[\alpha, \beta]$ at some *interior* point x_1, so $u''(x_1) \leq 0$, thus contradicting (9.19) at x_1. Consequently, $u(x) \leq 1$ for any $x \in \mathbb{R}$.

Proof of Claim 1. To make a choice, let $f(x) > 0$ for $x_1 < x < x_2$; in the case that $f(x) < 0$ for $x_1 < x < x_2$, we merely replace f by $-f$ everywhere. Now set $\varphi(x) = \sin(\pi(x - x_1)/(x_2 - x_1))$ for $x_1 \leq x \leq x_2$, and note that $\varphi(x_1) = \varphi(x_2) = 0$ and $\varphi(x) > 0$ for $x_1 < x < x_2$. Then

$$\int_{x_1}^{x_2} f(x)\varphi(x)dx > \int_{x_1}^{x_2} f(x)(1 - u(x))\varphi(x)dx \quad \text{(by Claim 3)}$$

$$= -\int_{x_1}^{x_2} f''(x)\varphi(x)dx \quad \text{for } f'' + f(1-u) = 0$$

$$= \frac{\pi}{x_2 - x_1}\int_{x_1}^{x_2} f'(x)\cos\frac{\pi(x - x_1)}{x_2 - x_1}dx \quad \text{for } \varphi(x_1) = \varphi(x_2) = 0$$

$$= \left(\frac{\pi}{x_2 - x_1}\right)^2 \int_{x_1}^{x_2} f(x)\varphi(x)dx \text{ for } f(x_1) = f(x_2) = 0,$$

which shows that $x_2 - x_1$ is indeed strictly greater than π.

Proof of Claim 2. Since f is T-periodic, there is no loss in considering the interval $[0, T]$. Clearly, f takes on positive values around the origin, for $f(0) = a > 0$ by hypothesis. To prove that it also takes on negative values, we argue by reductio ad absurdum. Suppose, if possible, that $f(x) \geq 0$ for all $x \in [0, T]$. By Claim 3, it then follows that $f''(x) = -f(x)(1 - u(x)) \leq 0$ for all $x \in [0, T]$, that is, f' is decreasing on $[0, T]$. But f' is itself T-periodic, so it must be constant on $[0, T]$, which implies in turn that f must be constant on $[0, T]$, by T-periodicity. Thus, f'' vanishes on

9.6 Qualitative Results

$[0,T]$, and since $f'' + f(1 - f^2 - g^2) = 0$, we deduce that g is itself constant on $[0,T]$; that is, g' vanishes on $[0,T]$, thus contradicting the hypothesis $g'(0) = b > 0$. We conclude that f must take on values of either sign on $[0,T]$.

(c) Choose $a, b > 0$ such that $b^2 = a^2(1 - a^2)$. Define $f(x) = a\cos\frac{bx}{a}$ and $g(x) = a\sin\frac{bx}{a}$. It follows that $T = 2\pi a/b$. In particular, we observe that $T > 2\pi$.

(d) We shall prove the following more general result.

Fix arbitrarily $\delta > 0$. Then there exist choices of a and b such that the resulting (f,g) is periodic, and $\min(f^2 + g^2) < \delta \max(f^2 + g^2)$.

Define
$$\Omega := \{(a,b) \in (0,1] \times [0,1];\ b^2 \leq a^2(1 - a^2)\}.$$

Let $(a,b) \in \text{Int}\,\Omega$ and set $v(x) := f(x) + ig(x)$. Since $v(0) = a \neq 0$, it follows that for small x,
$$v(x) = e^{i\varphi(x)} r(x), \quad r(x) = \sqrt{f^2(x) + g^2(x)},$$
where $\varphi(0) = 0$ and $r > 0$. Then r satisfies
$$\begin{cases} -r'' = r(1 - r^2) - \frac{a^2 b^2}{r^3}, \\ r(0) = a,\ r'(0) = 0, \end{cases} \tag{9.20}$$

while φ is given by
$$\varphi' = \frac{ab}{r^2}, \quad \varphi(0) = 0.$$

Hence, if the problem (9.20) has a global positive solution, it follows that v is global. Moreover, if r is periodic of period T_0, then
$$v(nT_0 + x) = e^{in\varphi(T_0)} e^{i\varphi(x)} r(x), \quad \forall 0 \leq x < T_0,\ \forall n \in \mathbb{N},$$

so that (9.20) gives a periodic solution if and only if $\varphi(T_0) \in \pi\mathbb{Q}$.

We prove in what follows the global existence. More precisely, if $(a,b) \in \text{Int}\,\Omega$, then (9.20) has a global positive periodic solution. We first observe that the assumption made on (a,b) implies $r''(0) < 0$. So, multiplying in (9.20) by r', we obtain, for small $x > 0$,
$$r'^2 = -r^2 + \frac{r^4}{2} - \frac{a^2 b^2}{r^2} + a^2 - \frac{a^4}{2} + b^2 \tag{9.21}$$

and
$$r' = -\sqrt{-r^2 + \frac{r^4}{2} - \frac{a^2 b^2}{r^2} + a^2 - \frac{a^4}{2} + b^2}. \tag{9.22}$$

Now, relation (9.21) implies that r and r' are bounded if a solution exists and moreover, that
$$\inf\{r(x);\ r \text{ exists}\} > 0.$$

It follows that r is a global solution.

Set
$$t_0 := \sup\{x > 0; \ r'(y) < 0 \ \text{ for all } 0 < y < x\}.$$

Note that (9.22) is valid if $0 < x < t_0$.

Let $0 < c < a$ be the **unique** root of

$$\psi(x) := \frac{x^4}{2} - x^2 - \frac{a^2 b^2}{x^2} + a^2 - \frac{a^4}{2} + b^2 = 0.$$

Since $\psi(x) < 0$ if $x \in (0, c)$ or if $x > a$, x close to a, it follows from relation (9.21) that

$$c \le r(x) \le a \quad \text{for all } x \in \mathbb{R}. \tag{9.23}$$

Claim 4. We have $\lim_{x \nearrow t_0} r(x) = c$.

Proof of Claim 4. If $t_0 < \infty$, it follows that $r'(t_0) = 0$. Now, relation (9.21) in conjunction with the definitions of t_0 and c shows that $r(t_0) = c$. If $t_0 = +\infty$, then we have $\lim_{x \to \infty} r(x) \ge c$. If $\lim_{x \to \infty} r(x) > c$, then there exists a constant $M > 0$ such that $r'(x) \le -M$ for each $x > 0$. The last inequality contradicts (9.23) for large x.

For any $0 < x < t_0$, relation (9.22) yields

$$x = \int_{r(x)}^{a} \frac{dt}{\sqrt{-t^2 + \frac{t^4}{2} - \frac{a^2 b^2}{t^2} + a^2 - \frac{a^4}{2} + b^2}}.$$

Therefore

$$t_0 = \int_{c}^{a} \frac{dt}{\sqrt{-t^2 + \frac{t^4}{2} - \frac{a^2 b^2}{t^2} + a^2 - \frac{a^4}{2} + b^2}} < \infty.$$

It follows by a reflection argument that $r(2t_0) = r(0) = a$, $r'(2t_0) = r'(0) = 0$, so that r is $(2t_0)$-periodic. This concludes the proof of Claim 4.

Define $\Psi(x) := 2x^2 \psi(x) = x^6 - 2x^4 - 2a^2 b^2 + (2a^2 - a^4 + 2b^2)x^2$. Taking into account Claim 4 and observing that $\Psi(0) < 0$, it follows that it is enough to show that $\Psi(\varepsilon) > 0$, for some $\varepsilon > 0$ and for a convenient choice of $(a, b) \in \text{Int}\,\Omega$. But

$$\Psi(\varepsilon) > \varepsilon^2[\varepsilon^4 - 2\varepsilon^2 + a^2(2 - a^2)] - 2a^2 b^2.$$

For $a = 2^{-1/2}$ we obtain

$$\Psi(\varepsilon) > \varepsilon^2[\varepsilon^4 - 2\varepsilon^2 + 3 \cdot 4^{-1}] - b^2 > \frac{\varepsilon^2}{2} - b^2 > 0,$$

provided that $\varepsilon > 0$ is sufficiently small and $b = \varepsilon/2$. It is obvious that this choice of a and b guarantees $(a, b) \in \text{Int}\,\Omega$.

9.6.9. Let a be a positive real number. Assume that f is a twice differentiable function from $\mathbb{R}$ into $\mathbb{R}$ such that $f(0) = a$ and satisfying the differential equation $f'' = -f(1 - f^2)$. Moreover, we suppose that f is periodic with principal period $T = T(a)$ and that f^2 achieves its maximum at the origin.

9.6 Qualitative Results

(a) Prove that $a \leq 1$ and deduce that f is well defined on the real axis.
(b) Prove that $T > 2\pi$.
(c) Define $t_0(a) := \sup\{t > 0;\ f > 0 \text{ in } (0,t)\}$. Show that f is decreasing on $(0, t_0(a))$ and express T in terms of $t_0(a)$.
(d) Prove that $a^2 + (2\pi/T)^2 > 1$.
(e) Show that the mapping $t_0 : (0,1) \to \mathbb{R}$ is increasing and compute $\lim_{a \searrow 0} t_0(a)$ and $\lim_{a \nearrow 1} t_0(a)$.
(f) Deduce that for any $T > 2\pi$, there exists a unique periodic function $f : \mathbb{R} \to \mathbb{R}$ with principal period T such that $f'' = -f(1 - f^2)$ and f^2 achieves its maximum at the origin.

Vicenţiu Rădulescu, Amer. Math. Monthly, Problem 11104

Solution. (a) Arguing by contradiction, let us assume that $a > 1$. Using the differential equation satisfied by f, it follows that $f''(0) > 0$, which contradicts our hypothesis that f^2 achieves its maximum at the origin. If $a = 1$, we get only the trivial solution $f \equiv 1$. That is why we shall assume in what follows that $a \in (0,1)$. Multiplying by f' in the differential equation satisfied by f and integrating, we obtain

$$f'^2 = -f^2 + \frac{1}{2}f^4 + a^2 - \frac{1}{2}a^4. \tag{9.24}$$

It follows that if a function f with the required properties exists, we have $|f(x)| \leq a$ and $|f'(x)| \leq \left(a^2 - a^4/2\right)^{1/2}$, for all $x \in \mathbb{R}$. Hence f is globally defined.

(b) We first observe that f cannot be positive (resp, negative) on an infinite interval, if f is periodic. Indeed, in that case, f would be a periodic concave (resp., convex) function, that is, a constant function. But this is impossible, due to our choice of a.

Let x_1, x_2 be two consecutive zeros of f. We may suppose that $f(x) > 0$ if $x_1 < x < x_2$, so that $f'(x_1) > 0$ and $f'(x_2) < 0$. If x_3 denotes the smallest $x > x_2$ such that $f(x_3) = 0$, it follows that $f(x) < 0$, for any $x \in (x_2, x_3)$. If we prove that $x_2 - x_1 > \pi$, it will also follow that $x_3 - x_1 > 2\pi$ and that there is no $x \in (x_1, x_3)$ such that $f(x) = 0$ and $f'(x) > 0$. This means that the principal period of f **must** be greater than 2π. This will be done in the following auxiliary result. □

Lemma 1. *Let $\Psi : \mathbb{R} \to [0,1]$ be such that the set $\{x;\ \Psi(x) = 0 \text{ or } \Psi(x) = 1\}$ contains only isolated points. Let f be a real function such that $f(x_1) = f(x_2) = 0$, and $f > 0$ in (x_1, x_2). Assume that $-f'' = f\Psi$ in $[x_1, x_2]$. Then $x_2 - x_1 > \pi$.*

Proof of Lemma. We may assume that $x_1 = 0$. Multiplying by $\varphi(x) := \sin \pi x/x_2$ in the differential equation $-f'' = f\Psi$ and integrating by parts, we obtain

$$\int_0^{x_2} f\varphi\, dx > \int_0^{x_2} f\Psi\varphi\, dx = \frac{\pi^2}{x_2^2} \int_0^{x_2} f\varphi\, dx,$$

that is, $x_2 > \pi$.

(c) Since $f'(0) = 0$ and $f''(0) < 0$, it follows that f decreases for small $x > 0$. Moreover, $f'(x) < 0$ for $0 < x < t_0(a)$. Indeed, suppose the contrary. Then, taking into account relation (9.24), we obtain the existence of some $\tau > 0$ with $\tau < t_0(a)$ and such that $f(\tau) = a$. If we consider the smallest $\tau > 0$ such that the above equality holds, then $f(x) < a$ for any $0 < x < \tau$. Since $f(0) = f(\tau) = a$, it follows that there exists some $0 < t_1 < \tau$ such that $f'(t_1) = 0$, which is the desired contradiction. Hence

$$f' = -\sqrt{a^2 - \frac{a^4}{2} - f^2 + \frac{f^4}{2}} < 0 \quad \text{in } (0, t_0(a)).$$

It follows that for any $0 < x < t_0(a)$,

$$\int_{f(x)}^{a} \frac{dt}{\sqrt{\frac{1}{2}t^4 - t^2 + a^2 - \frac{1}{2}a^4}} = x, \tag{9.25}$$

which yields

$$t_0(a) = \int_0^a \frac{dt}{\sqrt{\frac{1}{2}t^4 - t^2 + a^2 - \frac{1}{2}a^4}} = \int_0^1 \frac{d\xi}{\sqrt{(1-\xi^2)[1 - \frac{a^2}{2}(1+\xi^2)]}}. \tag{9.26}$$

Taking into account the differential equation satisfied by f, we first deduce that

$$f(t_0(a) + x) = -f(t_0(a) - x). \tag{9.27}$$

Indeed, both functions $g(x) = f(t_0(a) + x)$ and $h(x) = -f(t_0(a) - x)$ are solutions of the Sturm–Liouville problem

$$\begin{cases} z'' = -z(1 - z^2), & \text{in } (0, t_0(a)), \\ z(0) = 0, \ z'(0) = f'(t_0(a)). \end{cases}$$

Using now the uniqueness of the solution to the above boundary value problem, we deduce relation (9.27). Next, similar arguments imply $f(2t_0(a) - x) = -f(x)$ and $f(4t_0(a) + x) = f(x)$. It follows that f is periodic and its principal period is $T(a) = 4t_0(a)$.

We observe that **(b)** easily follows from the above results. Indeed, relation (9.26) yields

$$t_0(a) > \int_0^1 (1-\xi^2)^{-1/2} d\xi = \frac{\pi}{2}.$$

So, by $T(a) = 4t_0(a)$, we obtain **(b)**.

We may give the following alternative proof in order to justify that f decreases on the interval $(0, t_0(a))$. Using the differential equation $f'' = -f(1 - f^2)$ in conjunction with $f > 0$ on $(0, t_0(a))$ and $f^2 \leq a^2 < 1$, it follows that $f'' < 0$ on $(0, t_0(a))$. Hence f' is decreasing on $(0, t_0(a))$, that is, $f'(x) < f'(0) = 0$ for any $x \in (0, t_0(a))$.

9.6 Qualitative Results

(d) Since $T(a) = 4t_0(a)$, it is enough to show that $\sqrt{1-a^2}\, t_0(a) < \frac{\pi}{2}$. Relation (9.26) yields

$$\sqrt{1-a^2}\, t_0(a) = \int_0^1 \frac{\sqrt{1-a^2}}{\sqrt{(1-\xi^2)[1-\frac{a^2}{2}(1+\xi^2)]}}\, d\xi < \int_0^1 \frac{1}{\sqrt{1-\xi^2}}\, d\xi = \frac{\pi}{2}.$$

(e) Relation (9.26) implies that the mapping $a \longmapsto t_0(a)$ is increasing and $\lim_{a \searrow 0} t_0(a) = \frac{\pi}{2}$, $\lim_{a \nearrow 1} t_0(a) = +\infty$.

(f) Since $t_0'(a) > 0$, it follows that the mapping $T(a) \longmapsto a := a(T)$ is analytic. Taking into account relation (9.25), we conclude the proof. Moreover, we have $\lim_{T \searrow 2\pi} a(T) = 0$ and $\lim_{T \nearrow +\infty} a(T) = 1$. □

9.6.10. Let F and G be twice differentiable functions from $\mathbb{R}^2$ into $\mathbb{R}$, and satisfying the partial differential equations

$$\Delta F = -F(1 - F^2 - G^2), \quad \Delta G = -G(1 - F^2 - G^2).$$

Assume that there exist $\omega_1, \omega_2 \in \mathbb{R}^2$ that are with respect to the standard basis and such that F and G are periodic, with periods ω_1 and ω_2.

(a) Show that $F^2 + G^2 \leq 1$.
(b) Prove that if $|\omega_1|$ and $|\omega_2|$ are sufficiently small, then F and G are constant.

Solution. (a) Set

$$P = \{\lambda \omega_1 + \mu \omega_2;\, 0 \leq \lambda, \mu \leq 1\} \subset \mathbb{R}^2; \quad P^+ = \{x \in P;\, F^2(x) + G^2(x) \geq 1\} \subset \mathbb{R}^2.$$

Let $\zeta \in C^\infty(\mathbb{R}^2)$ be a nonnegative function with compact support such that $\zeta = 0$ in a neighborhood of the origin. For any $n \geq 1$, set $\zeta_n(x) = n^{-2}\zeta(x/n)$. Multiplying the equality satisfied by F (resp., by G) by $F(F^2 + G^2 - 1)^+ \zeta_n$ (resp., by $G(F^2 + G^2 - 1)^+ \zeta_n$) and integrating by parts, we obtain, by addition and passing to the limit as $n \to \infty$,

$$\int_{P^+} \left[(|\nabla F|^2 + |\nabla G|^2)(F^2 + G^2 - 1) + |\nabla(F^2 + G^2)|^2 \right]$$
$$\leq -\int_{P^+} (F^2 + G^2)(F^2 + G^2 - 1)^2.$$

This inequality implies **(a)**. □

Alternative proof of (a). Define $u : \mathbb{R}^2 \to \mathbb{R}$ by $u(x) = F^2(x) + G^2(x) - 1$. Clearly, u is an (ω_1, ω_2)-periodic function of class $C^2(\mathbb{R}^2)$. Thus, its values are determined by the values it takes on on the fundamental cell

$$P = \{\lambda \omega_1 + \nu \omega_2 : 0 \leq \lambda, \nu \leq 1\}.$$

In particular,
$$\sup\{u(x) : x \in \mathbb{R}^2\} = \sup\{u(x) : x \in P\}.$$
But the fundamental cell is clearly compact, so there exists $x_0 \in P$ such that
$$u(x_0) = \sup\{u(x) : x \in P\},$$
by continuity of u; whence
$$u(x_0) = \sup\{u(x) : x \in \mathbb{R}^2\},$$
by the preceding remark. Suppose now by way of contradiction that u takes a (strictly) positive value at some point. Then $u(x_0) > 0$, so $u(x) > 0$ for all $x \in B(x_0, r)$, some $r > 0$, by continuity. Here and here in after, $B(x_0, r)$ denotes the open r-ball centered at x_0, that is, the set of all points x in the plane whose distance from x_0 is (strictly) less than r. It should be clear that
$$u(x_0) = \sup\{u(x) : x \in B(x_0, r)\}; \tag{9.28}$$
that is, u assumes an interior maximum value on $B(x_0, r)$. Now a straightforward calculation yields
$$\Delta u = 2((\nabla F)^2 + (\nabla G)^2) + 2u(u+1) \geq 2u(u+1).$$
(Use was made here of the condition $\Delta F + F(1 - F^2 - G^2) = \Delta G + G(1 - F^2 - G^2) = 0$.) This shows that
$$\Delta u > 0 \text{ on } B(x_0, r). \tag{9.29}$$
Thus, u is a subharmonic function on $B(x_0, r)$, assuming an interior maximum value, by (9.28). The strong maximum principle now applies to show that u must be constant on $B(x_0, r)$. But then clearly $\Delta u = 0$ on $B(x_0, r)$, thus contradicting (9.29). We conclude that $u(x) \leq 0$, i.e., $(F(x))^2 + (G(x))^2 \leq 1$, for all $x \in \mathbb{R}^2$.

(b) Consider the Hilbert space H of doubly periodic smooth functions $u : P \to \mathbb{R}$ of periods ω_1 and ω_2 endowed with the norm
$$\|u\|_H = \left(\int_P (u^2(x) + |\nabla u(x)|^2) dx\right)^{1/2}.$$

Let $(\varphi_n)_{n \geq 0}$ be an orthonormal basis of eigenfunctions of the Laplace operator $(-\Delta)$ in H and denote by $(\lambda_n)_{n \geq 0}$ the corresponding eigenvalues. We can suppose that $\varphi_0 = 1$, so that $\lambda_n > 0$ for any $n \geq 1$. If P is "small," so if $|\omega_1|$ and $|\omega_2|$ are sufficiently small, then $\lambda_n > 2$, for any $n \geq 1$. If $u = (F, G) : \mathbb{R}^2 \to \mathbb{R}^2$ is a doubly periodic mapping that solves the above system, then we obtain the Fourier developments
$$u = \sum_{n \geq 0} c_n \varphi_n$$

9.6 Qualitative Results

and
$$u|u|^2 = \sum_{n\geq 0} d_n \varphi_n.$$

Integrating the relations $\Delta F + F(1 - F^2 - G^2) = 0$ and $\Delta G + G(1 - F^2 - G^2) = 0$ over P, we obtain $c_0 = d_0$. Next, multiplying the same identities by φ_n (for $n \geq 1$), we obtain, after integration, $|d_n| = (\lambda_n - 1)|c_n| > |c_n|$, provided that $c_n \neq 0$. On the other hand, since $F^2 + G^2 \leq 1$, it follows that
$$\int_P (F^2(x) + G^2(x))dx \geq \int_P (F^2(x) + G^2(x))^3 dx,$$

which can be rewritten, by Parseval's identity,
$$\sum_{n\geq 0} |c_n|^2 \geq \sum_{n\geq 0} |d_n|^2.$$

This last inequality implies $c_n = d_n = 0$ for any $n \geq 1$, so u is a constant mapping.

Alternative proof of (b). We shall prove that if the diameter δ of the open fundamental cell
$$\Omega = \{\lambda_1 \omega_1 + \lambda_2 \omega_2 : 0 < \lambda_1, \lambda_2 < 1\}$$

does not exceed $1/2$ (e.g., if $|\omega_k| \leq 1/4$, $k \in 1,2$), then F and G are both constant. To begin with, notice that Ω is a relatively compact domain with a piecewise C^1 boundary. Next, define $u,v: \Omega \to \mathbb{C}$ by $u(x) = F(x) + iG(x)$ and $v(x) = u(x)|u(x)|^2$, respectively. Clearly, u and v are both of class $C^2(\Omega)$. But by **(a)**, $|u(x)| \leq 1$ for all $x \in \Omega$, so $\|u\|_{L^2(\Omega)} \geq \|v\|_{L^2(\Omega)}$.

Let us show that if $\|u\|_{L^2(\Omega)} = \|v\|_{L^2(\Omega)}$, then F and G are both constant on Ω, so throughout the plane. Indeed, if $\|u\|_{L^2(\Omega)} = \|v\|_{L^2(\Omega)}$, then
$$\int_\Omega |u|^2(1 - |u|^2) = 0,$$

so, with reference again to **(a)** above, $|u|(1 - |u|) = 0$ on all of Ω, by continuity. Since Ω is connected and $|u|$ is continuous, it follows that either $|u| = 0$ on all of Ω or $|u| = 1$ on all of Ω. But
$$\Delta |u|^2 = 2((\nabla F)^2 + (\nabla G)^2) + 2|u|^2(|u|^2 - 1),$$

so $\nabla F = \nabla G = 0$ on all of Ω, and F and G are indeed both constant throughout Ω.

Assume now once and for all that F and G are not constant. Then
$$\|u\|_{L^2(\Omega)} > \|v\|_{L^2(\Omega)} \tag{5}$$

by the preceding. Next, recall that there exists a sequence $\{\varphi_n\}_{n\in\mathbb{Z}_+}$ of functions in $C^\infty(\Omega)$ and an unbounded increasing sequence $\{\lambda_n\}_{n\in\mathbb{Z}_+}$ of positive real numbers with the following properties:

(a) $-\Delta \varphi_n = \lambda_n \varphi_n$ on Ω, for each index n;

(b) the φ_n are orthonormal in the Hilbert space $L^2(\Omega)$, endowed with the standard scalar product

$$\langle \varphi, \psi \rangle_{L^2(\Omega)} = \int_\Omega \varphi \bar{\psi},$$

and their $\mathbb{C}$-linear span is dense in $L^2(\Omega)$;

(c) the elements $\varphi_n/\sqrt{\lambda_n}$ are orthonormal in the Sobolev space $H_0^1(\Omega)$, endowed with the alternative scalar product

$$\langle \varphi, \psi \rangle_{H_0^1(\Omega)} = \int_\Omega (\nabla \varphi)(\nabla \bar{\psi}),$$

and their $\mathbb{C}$-linear span is dense in $H_0^1(\Omega)$.

Recall the notation δ for the diameter of Ω. The Poincaré inequality for φ_m,

$$\|\varphi_m\|_{L^2(\Omega)} \leq \sqrt{2}\delta \|\varphi_m\|_{H_0^1(\Omega)},$$

in conjunction with the orthonormality of the φ_n in $L^2(\Omega)$ and of the $\varphi_n/\sqrt{\lambda_n}$ in $H_0^1(\Omega)$, yields $\delta\sqrt{2\lambda_m} \geq 1$ for all m. Thus, if $\delta \leq 1/2$, then $\lambda_n \geq 2$ for all n. Assume henceforth that $\delta \leq 1/2$, so $\lambda_n \geq 2$ for all n, and consider the Fourier expansions of u and v in $L^2(\Omega)$ in terms of the φ_n:

$$u = \sum_{n \in \mathbb{Z}_+} a_n \varphi_n, \text{ with } a_n = \int_\Omega u \bar{\varphi}_n,$$

and

$$v = \sum_{n \in \mathbb{Z}_+} b_n \varphi_n, \text{ with } b_n = \int_\Omega v \bar{\varphi}_n.$$

Since $u \in C^2(\Omega) \subset L^2(\Omega)$, it follows that $\Delta u \in C(\Omega) \subset L^2(\Omega)$, so the Fourier expansion of Δu in $L^2(\Omega)$, in terms of the φ_n, is given by

$$\Delta u = -\sum_{n \in \mathbb{Z}_+} a_n \lambda_n \varphi_n;$$

of course, use was made here of condition (a) above. But $\Delta u + u = v$ in Ω, so $b_n = (1 - \lambda_n)a_n$ for all n, by the preceding. Recalling that $\lambda_n \geq 2$ for all n, it follows that $|b_n| \geq |a_n|$ for all n, whence

$$\sum_{n \in \mathbb{Z}_+} |b_n|^2 \geq \sum_{n \in \mathbb{Z}_+} |a_n|^2.$$

Finally, rewrite (5) in terms of Parseval's identity to get

$$\sum_{n \in \mathbb{Z}_+} |a_n|^2 > \sum_{n \in \mathbb{Z}_+} |b_n|^2,$$

and thus reach a blatant contradiction. Consequently, if the diameter of the open fundamental cell does not exceed $1/2$, then both F and G must indeed be constant throughout the plane.

Remark. This problem is related to the study of stationary periodic solutions of the coupled Allen–Cahn/Cahn–Hilliard system that describes various phase transition phenomena. The framework corresponds to doubly periodic solutions in several dimensions, and the results formulated in this problem are related to those proved in the one-dimensional case.

9.7 Independent Study Problems

> As the sun eclipses the stars by its brilliancy, so the man of knowledge will eclipse the fame of others in assemblies of the people if he proposes problems, and still more if he solves them.
>
> Brahmagupta (598–668)

9.7.1. Prove that if n is a positive integer, then

$$\int_0^{\pi/2} \frac{\sin(2n+1)x}{\sin x}\,dx = \frac{\pi}{2}$$

and

$$\int_0^{\pi/2} \frac{\sin 2nx}{\sin x}\,dx = 2\left(1 - \frac{1}{3} + \frac{1}{5} - \frac{1}{7} + \cdots + \frac{(-1)^{n-1}}{2n-1}\right).$$

L. Richardson, Amer. Math. Monthly, Problem 2758

9.7.2. Let f be a real-valued continuous function on $[a,b]$ that satisfies $0 \le f(x) \le M$ for all $x \in [a,b]$. Prove that

$$0 < \left[\int_a^b f(x)\,dx\right]^2 - \left[\int_a^b f(x)\cos x\,dx\right]^2 - \left[\int_a^b f(x)\sin x\,dx\right]^2 \le \frac{M^2(b-a)^4}{12}.$$

O. Dunkel, Amer. Math. Monthly, Problem 3104

9.7.3. Let $f : \mathbb{R} \to \mathbb{R}$ be a function that is locally integrable, that is, f is integrable on every interval $[a,b] \subset \mathbb{R}$. Assume that

$$f(x+y) = f(x) + f(y), \quad \forall x, y \in \mathbb{R}.$$

(a) Prove that

$$\int_0^x f(t)\,dt + \int_0^y f(t)\,dt = \int_0^{x+y} f(t)\,dt.$$

(b) Prove that there exists $c \in \mathbb{R}$ such that $f(x) = cx$ for all $x \in \mathbb{R}$.

9.7.4. Let $f : [0,1] \to \mathbb{R}$ be a monotone function and $a \in \mathbb{R}$. Prove that
$$\int_0^1 |f(x) - a| dx \geq \int_0^1 \left| f(x) - f\left(\frac{1}{2}\right) \right| dx.$$

9.7.5. Let $f : (0, \infty) \to \mathbb{R}$ be a differentiable function such that $f'(x)$ tends monotonically to 0 as $x \to \infty$. Prove that the following limit exists:
$$\lim_{n \to \infty} \left(\frac{1}{2} f(1) + f(2) + f(3) + \cdots + f(n-1) + \frac{1}{2} f(n) - \int_1^n f(x) dx \right).$$

Let s be the value of this limit. Prove that if $f'(x)$ is increasing, then
$$\frac{1}{8} f'(n) < \frac{1}{2} f(1) + f(2) + f(3) + \cdots + f(n-1) + \frac{1}{2} f(n) - \int_1^n f(x) dx - s < 0.$$

Particular cases: $f(x) = 1/x$ and $f(x) = -\ln x$.

9.7.6. Let $f : [a,b] \to \mathbb{R}$ be an integrable function. Define $f_{kn} := f(a + k\delta_n)$, where $\delta_n = (b-a)/n$.

Prove that
$$\lim_{n \to \infty} (1 + f_{1n} \delta_n)(1 + f_{2n} \delta_n) \cdots (1 + f_{nn} \delta_n) = e^{\int_a^b f(x) dx}.$$

9.7.7. Let $f : [a,b] \to \mathbb{R}$, $\varphi : [a, b+d] \to \mathbb{R}$, $d > 0$, be integrable functions. Show that
$$\lim_{\delta \searrow 0} \int_a^b f(x) \varphi(x + \delta) dx = \int_a^b f(x) \varphi(x) dx.$$

9.7.8. Prove that
$$\int_0^a e^{-t^2/2} dt = \left(\frac{\pi}{2} - aI \right)^{1/2},$$
where
$$I = \int_{a^2}^{2a^2} \frac{e^{-s/2}}{s \sqrt{s^2 - a^2}} ds.$$

9.7.9. Let $f : \mathbb{R} \to \mathbb{R}$ be an integrable function and $a < b$. Prove that
$$\lim_{t \to 0} \int_a^b |f(x+t) - f(x)| dx = 0.$$

Deduce that $\lim_{t \to 0} \int_a^b f(x+t) dx = \int_a^b f(x) dx$.

9.7 Independent Study Problems

9.7.10. Let $I_n = \int_0^{2\pi} \cos x \cos 2x \cdots \cos nx\, dx$. Find all integers $1 \le n \le 10$ such that $I_n \ne 0$.

9.7.11. Compute
$$\int_2^4 \frac{\sqrt{\ln(9-x)}}{\sqrt{\ln(9-x)} + \sqrt{\ln(x+3)}}\, dx.$$

9.7.12. Let a and b be positive numbers. Compute
$$\int_0^a \left(\int_0^b e^{\max\{b^2 x^2, a^2 y^2\}}\, dy \right) dx.$$

9.7.13. Compute
$$\max_{y \in [0,1]} \int_0^y \sqrt{x^4 + (y - y^2)^2}\, dx.$$

9.7.14. Prove that
$$\int_{-100}^{-10} \left(\frac{x^2 - x}{x^3 - x + 1} \right)^2 dx + \int_{\frac{1}{101}}^{\frac{1}{11}} \left(\frac{x^2 - x}{x^3 - x + 1} \right)^2 dx + \int_{\frac{101}{100}}^{\frac{11}{10}} \left(\frac{x^2 - x}{x^3 - x + 1} \right)^2 dx$$
is a rational number.

9.7.15. Let $1 < p \le q$ be integers. For all $x = \lim_{n \to \infty} \sum_{k=1}^n x_k/p^k$, $x_k \in \{0, 1, \ldots, p-1\}$, define the function $f(x) = \lim_{n \to \infty} \sum_{k=1}^n x_k/q^k$. Compute $\int_0^1 f(x)\, dx$.
Answer. $(p-1)/2(q-1)$.

9.7.16. Let a and b be relatively prime positive numbers. Prove that
$$\int_0^1 \left(\{ax\} - \frac{1}{2} \right) \left(\{bx\} - \frac{1}{2} \right) dx = \frac{1}{12ab},$$
where $\{u\}$ denotes the fractional part of the real number u.

9.7.17. Prove that for any positive integer n there exists a unique polynomial $B_n(x)$ satisfying $\int_x^{x+1} B_n(t)\, dt = x^n$. Find $B_n(x)$. The numbers $B_n = B_n(0)$ are called Bernoulli numbers. Show that $B_{2k+1} > 0$ and $B_{2k}B_{2k+1} < 0$, for all integers $k > 0$. Find an explicit formula for $s_n(m) = 1^n + 2^n + \cdots + m^n$.
Hint. Prove that $B_n(0) = B_n(1)$. This yields $B_n(x) = \sum_{k=0}^n C_n^k B_k x^{n-k}$ and $s_n(m) = (B_{n+1}(m) - B_{n+1})/(n+1)$.

9.7.18. Let $a_1, \ldots, a_n$ be real numbers. Prove that
$$\left(\sum_{k=1}^n \frac{a_k}{k} \right)^2 \le \sum_{k=1}^n \sum_{j=1}^n \frac{a_k a_j}{k+j-1}.$$
Hint. We have $\sum_{k=1}^n a_k/k = \int_0^1 (\sum_{k=1}^n a_k x^{k-1})\, dx$.

9.7.19. Fix n real numbers $a_1, \ldots, a_n$ such that $a_k > -1$ for all $k = 1, \ldots, n$. Assume that
$$\frac{a_1}{1+a_1 x} + \frac{a_2}{1+a_2 x} + \cdots + \frac{a_n}{1+a_n x} \leq 1, \quad \forall x \in [0,1].$$
Prove that $(1+a_1) \cdots (1+a_n) \leq e$.
Hint. Integrate the inequality in the hypothesis.

9.7.20. Let $p > 1$ and $a \geq 0$. Prove that
$$\frac{2}{\pi} \int_a^{ap} \left(\frac{\sin t}{t}\right)^2 dt \leq 1 - \frac{1}{a}.$$

9.7.21. Compute
$$\int_{-1}^{1} \frac{dx}{(e^x + 1)(x^2 + 1)}.$$
Hint. Apply the identity $(e^x + 1)^{-1} + (e^{-x} + 1)^{-1} = 1$. The value of the integral is $\pi/4$.

9.7.22. Let $f : [0, \pi] \to \mathbb{R}$ be a continuous function such that
$$\int_0^\pi f(t) \sin t \, dt = \int_0^\pi f(t) \cos t \, dt = 0.$$
Prove that f has at least two zeros in $(0, \pi)$.
Hint. By hypothesis, f has at least one zero c. If this were be the unique zero of f, then the function $f(x) \sin(x - c)$ would have constant sign.

9.7.23. Let $f : [a,b] \to \mathbb{R}$ be a continuous function, $p_1, \ldots, p_n > 0$. Prove that for all $x_1, \ldots, x_n \in [a,b]$ there exists $x \in [a,b]$ such that
$$p_1 \int_x^{x_1} f(t) dt + \cdots + p_n \int_x^{x_n} f(t) dt = 0.$$

9.7.24. Let $f, g : [0,1] \to \mathbb{R}$, $g > 0$, be continuous functions. Prove that
$$\lim_{n \to \infty} \frac{\int_0^1 x^n f(x) dx}{\int_0^1 x^n g(x) dx} = \frac{f(1)}{g(1)}.$$

9.7.25. (**The du Bois-Reymond lemma**) Let $f, g : [0,1] \to \mathbb{R}$ be continuous functions such that for all $\varphi \in C^1[0,1]$ satisfying $\varphi(0) = \varphi(1) = 0$ we have
$$\int_0^1 (f(x) \varphi'(x) + g(x) \varphi(x)) dx = 0.$$
Prove that $f \in C^1[0,1]$ and $f' = g$.

9.7 Independent Study Problems

9.7.26. Let $f,g : [0.\infty) \to \mathbb{R}$ be continuous functions that equal 1 identically on $[0,1]$, are differentiable in $(1,\infty)$, and satify

$$xf'(x) = -f(x-1), \ xg'(x) = g(x-1), \quad \forall x > 1.$$

Prove that $\int_0^x f(x-t)g(t)dt = x$, for all $x \geq 0$.

9.7.27. Compute $\lim_{n \to \infty} \int_0^{\sqrt{n}} (1 - x^2/n)^n dx$.

9.7.28. Let A_n and G_n denote the arithmetic and the geometric means of the numbers $C_n^0, C_n^1, \ldots, C_n^n$. Show that $\lim_{n \to \infty} A_n^{1/n} = 2$ and $\lim_{n \to \infty} G_n^{1/n} = \sqrt{e}$.

9.7.29. Let $0 < a < b$. Compute

$$\lim_{t \searrow 0} \left(\int_0^1 (bx + a(1-x))^t dx \right)^{1/t}.$$

Hint. Use the substitution $u = bx + a(1-x)$. The limit is $(b^b/a^a)^{1/(b-a)}/e$.

9.7.30. (**The d'Alembert equation**) Find all continuous functions $f : \mathbb{R} \to \mathbb{R}$ such that

$$2f(x)f(y) = f(x+y) + f(x-y), \quad \forall x, y \in \mathbb{R}.$$

Hint. Integrate f on $[-t,t]$ and deduce a differential equation satisfied by this integral. We obtain the following situations: $f \equiv 0$, $f \equiv 1$, $f(x) = \cos kx$, $f(x) = \cosh kx$, where k is a real number.

9.7.31. Let $f : [0, \pi] \to \mathbb{R}$ be a continuous function such that

$$\int_0^\pi f(x) \sin x \, dx = \int_0^\pi f(x) \cos x \, dx = 0.$$

Prove that f has at least two distinct zeros in $[0, \pi]$.

The results stated in the following problem are special cases of the Fredholm alternative, a celebrated property in the theory of differential and integral equations.

9.7.32. Let $a(x)$ and $f(x)$ be periodic continuous functions with period 2π.

(a) Prove that the equation $u'' = f$ has a 2π-periodic solution if and only if $\int_0^{2\pi} f(x) \, dx = 0$.
(b) Prove that the equation $u'' + u = f$ has a 2π-periodic solution if and only if both $\int_0^{2\pi} f(x) \sin x \, dx = 0$ and $\int_0^{2\pi} f(x) \cos x \, dx = 0$.
(c) Prove that the equation $u'' + au = f$ has a 2π-periodic solution if and only if $\int_0^{2\pi} f(x)z(x) \, dx = 0$ for all 2π-periodic solutions of the differential equation $z'' + a(x)z = 0$.

9.7.33. Let α be a real number, $\alpha \neq \pm 1$. Prove that the Dini integral

$$\int_0^\pi (1 - 2\alpha \cos x + \alpha^2)\, dx$$

equals 0 if $|\alpha| < 1$ and $2\pi \ln |\alpha|$ if $|\alpha| > 1$.

9.7.34. Show that $\int_0^\pi f(\sin x) \cos x\, dx = 0$ for any continuous function f on $[0,1]$.

9.7.35. Find all continuous functions f on $[0,1]$ such that

$$\int_0^x f(t)\, dt = \int_x^1 f(t)\, dt \quad \text{for every } x \in (0,1).$$

9.7.36. Let f be of class C^1 on $[a,b]$, with $f(a) = f(b) = 0$.

(i) Prove that

$$\int_a^b x f(x) f'(x)\, dx = -\frac{1}{2} \int_a^b [f(x)]^2\, dx.$$

(ii) Additionally, assume that $\int_a^b [f(x)]^2\, dx = 1$. Show that

$$\int_a^b [f'(x)]^2\, dx \cdot \int_a^b [xf(x)]^2\, dx > \frac{1}{4}.$$

Chapter 10
Applications of the Integral Calculus

> *As are the crests on the heads of peacocks, as are the gems on the hoods of cobras, so is mathematics, at the top of all sciences.*
> —The Yajurveda, circa 600 B.C.

Abstract. The determination of areas and volumes has exercised the curiosity of mathematicians since Greek antiquity. Two of the greatest achievements of Archimedes (287–212 B.C.) were the computations of the areas of the parabola and of the circle. The early seventeenth century then saw the computation of areas under the curve $y = x^a$ with either integer or arbitrary values of a (Cavalieri, Fermat). In this chapter, we are concerned with various applications of the Riemann integral.

10.1 Overview

> I always prefer to believe the best of everybody—it saves so much trouble.
> —Rudyard Kipling (1865–1936)

In what follows we consider certain integrals of the form $\int_a^\infty f(x)dx$ or $\int_a^b f(x)dx$, where f is not bounded on $[a,b]$. These integrals are no longer defined as Riemann integrals and involve an extra limiting process.

Let $f:[a,\infty) \to \mathbb{R}$ be integrable on $[a,b]$, for every $b > a$. We say that the *improper integral* $\int_a^\infty f(x)dx$ converges if

$$I := \lim_{b \to \infty} \int_a^b f(x)dx$$

exists, and we write $\int_a^\infty f(x)dx = I$ in this case.

Let $f:(a,b] \to \mathbb{R}$ be integrable on $[c,b]$, for every $c \in (a,b)$. If

$$I := \lim_{c \to a+} \int_c^b f(x)dx$$

exists, we say that $\int_a^b f(x)dx$ converges, and equals I.

Examples. (i) The integral $\int_0^1 x^p dx$ exists for $p \geq 0$, since the mapping $f(x) = x^p$ is continuous on $[0,1]$, but fails to exist for $p < 0$, since f is unbounded. However, for $p \neq -1$,

$$\lim_{\delta \to 0+} \int_\delta^1 x^p dx = \lim_{\delta \to 0+} \frac{1 - \delta^{p+1}}{p+1} = \begin{cases} \frac{1}{p+1} & \text{if } p > -1, \\ +\infty & \text{if } p < -1. \end{cases}$$

Thus, the improper integral $\int_0^1 x^p dx$ converges for $p > -1$ and diverges for $p < -1$. We propose to the reader to check that this integral also diverges for $p = -1$.

(ii) Similarly, the improper integral $\int_1^\infty x^p dx$ converges for $p < -1$ and diverges for $p \geq -1$.

An important application of improper integrals concerns the convergence of series.

Integral Test for Series. Suppose that $f : [1, \infty) \to [0, \infty)$ is nonincreasing. Then the improper integral $\int_1^\infty f(x) dx$ and the series $\sum_{n=1}^\infty f(n)$ are both convergent or both divergent.

10.2 Integral Inequalities

> You may have to fight a battle more than once to win it.
>
> Margaret Thatcher

The following result is due to H. Brezis and M. Willem [13].

10.2.1. Let $a \geq 0$, $b > 0$, $L > 0$ and $f \in C^1[0,L]$ be such that $f(0) = 0$ and, for $0 \leq x \leq L$,

$$f'(x)^2 \leq a^2 + 2b^2 \int_0^x |f(s) f'(s)| ds.$$

Prove that for any $0 \leq x \leq L$,

$$|f(x)| \leq \frac{a}{b}(e^{bx} - 1)$$

and

$$|f'(x)| \leq a e^{bx}.$$

Solution. Define, on $[0, L]$, $F(x) = \int_0^x |f'(s)| ds$, so that $F' = |f'|$ and $|f| \leq F$. By assumption, we have

$$F'(x)^2 \leq a^2 + 2b^2 \int_0^x F(s) F'(s) ds = a^2 + b^2 F(x)^2. \tag{10.1}$$

10.2 Integral Inequalities

Hence we obtain
$$F'(x) \leq a + bF(x),$$
or
$$\left(e^{-bx}F(x)\right)' \leq ae^{-bx}.$$

We find after integration
$$e^{-bx}F(x) \leq \frac{a}{b}\left(1 - e^{-bx}\right),$$
or
$$|f(x)| \leq F(x) \leq \frac{a}{b}\left(e^{bx} - 1\right).$$

Inserting this into (10.1) yields $|f'(x)| \leq ae^{bx}$. □

10.2.2. Let f be a continuous function on $[0,1]$ such that
$$\int_x^1 f(t)\,dt \geq \frac{1-x^2}{2}, \quad \text{for every } x \in [0,1].$$

Show that $\int_0^1 f^2(t)\,dt \geq \frac{1}{3}$.

Solution. From the inequality
$$0 \leq \int_0^1 (f(x) - x)^2\,dx = \int_0^1 f^2(x)\,dx - 2\int_0^1 xf(x)\,dx + \int_0^1 x^2\,dx$$

we deduce that
$$\int_0^1 f^2(x)\,dx \geq 2\int_0^1 xf(x)\,dx - \int_0^1 x^2\,dx = 2\int_0^1 xf(x) - \frac{1}{3}.$$

From the hypotheses we have
$$\int_0^1 \left(\int_x^1 f(t)\,dt\right) dx \geq \int_0^1 \frac{1-x^2}{2}\,dx,$$

that is,
$$\int_0^1 tf(t)\,dt \geq \frac{1}{3}.$$

This completes the proof. □

10.2.3. Let $f : [0,1] \to \mathbb{R}$ be a continuous function such that for any $x, y \in [0,1]$,
$$xf(y) + yf(x) \leq 1.$$

(i) Show that
$$\int_0^1 f(x)\,dx \leq \frac{\pi}{4}.$$

(ii) Find a function satisfying the condition for which there is equality.

Solution. (i) We observe that

$$\int_0^1 f(x)dx = \int_0^{\frac{\pi}{2}} f(\sin\theta)\cos\theta\, d\theta = \int_0^{\frac{\pi}{2}} f(\cos\theta)\sin\theta\, d\theta.$$

So, by hypothesis,

$$2\int_0^1 f(x)dx \le \int_0^{\frac{\pi}{2}} 1\, d\theta = \frac{\pi}{2}.$$

(ii) Let $f(x) = \sqrt{1-x^2}$. If $x = \sin\theta$ and $y = \sin\Phi$, then

$$xf(y) + yf(x) = \sin\theta\cos\Phi + \sin\Phi\cos\theta = \sin(\theta+\Phi) \le 1.$$

On the other hand, $\int_0^1 \sqrt{1-x^2}\, dx = \pi/4$. □

10.2.4. Let $f : [a,b] \to \mathbb{R}$ be a function of class C^1 such that $\int_a^b f(t)dt = 0$.

(i) For any integer $n \ge 1$ define the function $f_n(t) = \cos(2\pi nt)$. Prove that f_n satisfies the above hypotheses and compute

$$\sup_{n\ge 1} \frac{\int_0^1 |f_n(t)|^2 dt}{\int_0^1 |f_n'(t)|^2 dt}.$$

(ii) Show that

$$\int_a^b f^2(t)dt \le (b-a)^2 \int_a^b (f'(t))^2 dt.$$

(iii) Let $g : [a,b] \to \mathbb{R}$ be a function of class C^1 such that $g(a) = g(b) = 0$ (in this case, $\int_a^b g(t)dt$ is not necessarily zero). Prove that

$$\int_a^b g^2(t)dt \le \frac{(b-a)^2}{8} \int_a^b (g'(t))^2 dt.$$

Solution. (i) The required value is $4\pi^{-1}$.

(ii) Since f is continuous and $\int_a^b f(t)dt = 0$, there exists $c \in (a,b)$ such that $f(c) = 0$. Hence

$$f(t) = f(c) + \int_c^t f'(u)du = \int_c^t f'(u)du, \quad \forall t \in [a,b].$$

By the Cauchy–Schwarz inequality we obtain for all $t \in [a,b]$,

$$|f(t)| \le \left|\int_c^t f'(u)du\right| \le \sqrt{|t-c|}\left(\int_c^t (f'(u))^2 du\right)^{1/2}.$$

Therefore

$$\int_a^b |f(t)|^2 dt \le \left(\int_a^b (f'(u))^2 du\right)\int_a^b |t-c|dt.$$

10.2 Integral Inequalities

To evaluate the second term, we observe that $|t-c| \leq b-a$, so $\int_a^b |t-c|dt \leq (b-a)^2$, and the proof is concluded. An alternative proof is based on the remark that

$$\int_a^b |t-c|dt = c^2 - c(a+b) + \frac{a^2+b^2}{2}.$$

This is a quadratic function in c, whose minimum is achieved at $c = (a+b)/2$. The maximum is attained either at a or at b, and this value equals $(b-a)^2/2$. We conclude that

$$\int_a^b |f(t)|^2 dt \leq \frac{(b-a)^2}{2} \left(\int_a^b (f'(u))^2 du \right).$$

(iii) We intend to apply the above result for $f = g'$. In this case, we would find $\int_a^b f(t)dt = g(b) - g(a) = 0$. However, this in not a good idea, since it does not yield our conclusion. We remark that the points a and b play the same role as c. This implies that it is better to restart the proof, but considering the different cases $t \leq (b+a)/2$ and $t \geq (b+a)/2$.

If $t \leq (b+a)/2$ we have

$$f(t) = \int_a^t f'(u)du,$$

so

$$|f(t)|^2 \leq (t-a) \int_a^t (f'(u))^2 du \leq (t-a) \int_a^{(b+a)/2} (f'(u))^2 du.$$

By integration we obtain

$$\int_a^{(b+a)/2} |f(t)|^2 dt \leq \frac{(b-a)^2}{8} \int_a^{(b+a)/2} (f'(t))^2 dt.$$

If $t \geq (b+a)/2$, we obtain

$$\int_{(b+a)/2}^b |f(t)|^2 dt \leq \frac{(b-a)^2}{8} \int_{(b+a)/2}^b (f'(t))^2 dt.$$

Adding the above two inequalities, we obtain

$$\int_a^b |f(t)|^2 dt \leq \frac{(b-a)^2}{8} \int_a^b (f'(t))^2 dt,$$

which concludes the proof. □

We give in what follows an elementary proof of the Young inequality in a particular case.

10.2.5. Let $f : [0,+\infty) \to \mathbb{R}$ be a strictly increasing function with a continuous derivative such that $f(0) = 0$. Let f^{-1} denote the inverse of f. Prove that for all positive numbers a and b, with $b < f(a)$,

$$ab < \int_0^a f(x)dx + \int_0^b f^{-1}(y)dy.$$

Solution. We first observe that if f is a strictly increasing function with a continuous derivative, then integration by parts yields

$$\int_a^b f(x)\,dx = bf(b) - af(a) - \int_a^b x f'(x)\,dx.$$

Let $y = f(x)$, $x = f^{-1}(y)$. We obtain

$$\int_a^b f(x)\,dx = bf(b) - af(a) - \int_{f(a)}^{f(b)} f^{-1}(y)\,dy. \tag{10.2}$$

Now if $u = f(a)$, $v = f(b)$, relation (10.2) becomes

$$\int_{f^{-1}(u)}^{f^{-1}(v)} f(x)\,dx = v f^{-1}(v) - u f^{-1}(u) - \int_u^v f^{-1}(y)\,dy.$$

Thus, we can always express $\int f^{-1}(y)\,dy$ in terms of $\int f(x)\,dx$.

Returning to our problem, assume that $0 < r < a$. Then

$$(a-r)f(r) < \int_r^a f(x)\,dx.$$

Hence

$$af(r) - \int_0^a f(x)\,dx < rf(r) - \int_0^r f(x)\,dx.$$

Applying relation (10.2) to $\int_0^r f(x)\,dx$, we obtain

$$af(r) - \int_0^a f(x)\,dx < \int_0^{f(r)} f^{-1}(y)\,dy.$$

Since $0 < b < f(a)$, we can take $r = f^{-1}(b)$, and our conclusion follows. □

10.2.6. Let $f : \mathbb{R} \to [0, \infty)$ be a continuously differentiable function. Prove that

$$\left| \int_0^1 f^3(x)\,dx - f^2(0) \int_0^1 f(x)\,dx \right| \leq \max_{0 \leq x \leq 1} |f'(x)| \left(\int_0^1 f(x)\,dx \right)^2.$$

Solution. Let $M = \max_{0 \leq x \leq 1} |f'(x)|$. We have

$$-M f(x) \leq f(x) f'(x) \leq M f(x), \quad \text{for all } x \in [0,1].$$

Therefore

$$-M \int_0^x f(t)\,dt \leq \frac{1}{2} f^2(x) - \frac{1}{2} f^2(0) \leq M \int_0^x f(t)\,dt, \quad \text{for all } x \in [0,1].$$

We deduce that

$$-M f(x) \int_0^x f(t)\,dt \leq \frac{1}{2} f^3(x) - \frac{1}{2} f^2(0) f(x) \leq M f(x) \int_0^x f(t)\,dt.$$

10.2 Integral Inequalities

Integrating the last inequality on $[0,1]$, it follows that

$$-M\left(\int_0^1 f(x)\,dx\right)^2 \leq \int_0^1 f^3(x)\,dx - f^2(0)\int_0^1 f(x)\,dx \leq M\left(\int_0^1 f(x)\,dx\right)^2,$$

which concludes the proof. $\square$

10.2.7. (**Poincaré's inequality, the smooth case**). *Let $f:[0,1]\to\mathbb{R}$ be a function of class C^1 with $f(0)=0$. Show that*

$$\sup_{0\leq x\leq 1} |f(x)| \leq \left(\int_0^1 (f'(x))^2\,dx\right)^{1/2}.$$

Solution. Let $x \in [0,1]$. Using $f(0)=0$ and applying the Cauchy–Schwarz inequality, we obtain

$$|f(x)| = \left|\int_0^x f'(t)\,dt\right|$$

$$\leq \left(\int_0^x |f'(t)|^2\,dt\right)^{1/2}\left(\int_0^x 1^2\,dt\right)^{1/2}$$

$$\leq \left(\int_0^x |f'(t)|^2\,dt\right)^{1/2},$$

which concludes the proof. $\square$

10.2.8. (**Poincaré's inequality, the L^p variant**). *Let $p\geq 1$ be a real number, a, b real numbers, and let $f:(a,b)\to\mathbb{R}$ be an arbitrary function such that $f(0)=0$ and $\int_a^b |f(x)|^p\,dx < \infty$. Prove that there exists a positive constant C depending only on $b-a$ such that*

$$\int_a^b |f(x)|^p\,dx \leq C\int_a^b |f'(x)|^p\,dx,$$

for all $f \in W_0^{1,p}(a,b)$.

Solution. By the Cauchy–Schwarz inequality we obtain

$$|f(x)| = |f(x) - f(a)| = \left|\int_a^x f'(t)\,dt\right| \leq \int_a^b |f'(t)|\,dt,$$

for any $x \in (a,b)$. So, by the Hölder inequality,

$$\int_a^b |f(x)|^p\,dx \leq \left(\int_a^b |f'(t)|\,dt\right)^p$$

$$\leq (b-a)^{p/p'}\cdot \int_a^b |f'(x)|^p\,dx$$

$$= (b-a)^{p-1} \cdot \int_a^b |f'(x)|^p dx.$$

The conclusion follows with $C = (b-a)^{p-1}$. □

10.2.9. Let $f : [0,1] \to \mathbb{R}$ be a function of class C^1 such that $f(1) = 0$. Prove that

$$\int_0^1 f^2(x)dx \leq 2\sqrt{\int_0^1 x^2 f^2(x)dx} \cdot \sqrt{\int_0^1 f'^2(x)dx}.$$

Solution. Integrating by parts and using $f(1) = 0$, we obtain

$$\int_0^1 f^2(x)dx = \int_0^1 x' f^2(x)dx = -\int_0^1 2xf(x)f'(x)dx.$$

Applying the Cauchy–Schwarz inequality, we have

$$\left|\int_0^1 xf(x)f'(x)dx\right| \leq \sqrt{\int_0^1 x^2 f^2(x)dx} \sqrt{\int_0^1 f'^2(x)dx}. \quad \Box$$

10.2.10. Let $f : [0,a] \to [0,\infty)$ be a continuous function and consider $K > 0$ such that

$$f(t) \leq K \int_0^t f(s)ds, \quad \text{for all } t \in [0,a].$$

Prove that $f \equiv 0$.

Solution. The most direct argument is based on iteration. By hypothesis and a recurrence argument we obtain

$$f(t) \leq \frac{K^n}{(n-1)!} \int_0^t (t-s)^{n-1} f(s)ds, \quad \forall t \in [0,a], \ \forall n \geq 1.$$

Let $M = \max_{t \in [0,a]} f(t)$. Hence

$$f(t) \leq \frac{MK^n}{(n-1)!} \int_0^t (t-s)^{n-1} ds = \frac{Mt^n K^n}{n!} \leq \frac{M(aK)^n}{n!} \longrightarrow 0, \quad \forall t \in [0,a],$$

which shows that $f(t) = 0$, for all $t \in [0,a]$.

The following alternative proof can be used in similar situations. Let

$$F(t) = K \int_0^t f(s)ds.$$

By hypothesis, we have
$$F'(t) = Kf(t) \leq KF(t).$$

Hence
$$\left(e^{-Kt} F(t)\right)' \leq 0.$$

10.2 Integral Inequalities

Thus, the mapping $t \longmapsto e^{-Kt}F(t)$ is decreasing in $[0,a]$. Since $F(0)=0$, it follows that $F(t) \leq 0$. On the other hand, $f(t) \geq 0$, so $F(t) \geq 0$. Therefore $F(t) \equiv 0$, that is, $f \equiv 0$. □

10.2.11. *Fix a real number $p > 1$ and let $f : \mathbb{R} \to \mathbb{R}$ be a continuous function. Prove that for any $x \in \mathbb{R}$ and $\lambda > 0$ there exists a unique real y such that for all $z \in \mathbb{R}$,*

$$\int_{x-\lambda}^{x+\lambda} |y-f(t)|^p dt \leq \int_{x-\lambda}^{x+\lambda} |z-f(t)|^p dt.$$

École Normale Supérieure, Paris, 2003

Solution. For any fixed $(x,\lambda) \in \mathbb{R} \times (0,+\infty)$ we define the continuous function $g(z) = \int_{x-\lambda}^{x+\lambda} |z-f(t)|^p dt$. Let us observe that for any $z \in \mathbb{R}$ with $|z| \geq \max\{|f(t)|;\ x-\lambda \leq t \leq x+\lambda\} =: M$ we have $g(z) \geq 2\lambda(|z|-M)^p$, which implies $g(z) \to +\infty$ as $|z| \to +\infty$. Since g is continuous, it follows that g achieves its minimum.

On the other hand, by the strict convexity of the mapping $[0,+\infty) \ni u \longmapsto u^p$ we deduce that g is strictly convex, too. In particular, this implies that the minimum point of g is unique.

The important inequality we prove in what follows does not belong to the theory of linear differential equations per se, but is very useful to establish the global existence of solutions of homogeneous differential linear systems. □

10.2.12. (**Gronwall's inequality, the differential form**) *(a) Consider the differentiable function $f : [0,T] \to [0,\infty)$ satisfying*

$$f'(x) \leq a(x)f(x) + b(x), \quad \forall x \in [0,T],$$

where $a,b : [0,T] \to [0,\infty)$ are continuous functions. Prove that

$$f(x) \leq e^{\int_0^x f(t)dt}\left[f(0) + \int_0^x f(t)dt\right], \quad \forall x \in [0,T].$$

(b) In particular, if $f'(x) \leq a(x)f(x)$ pe $[0,T]$ and $f(0) = 0$, prove that $f \equiv 0$ in $[0,T]$.

Solution. (a) By hypothesis it follows that

$$\left(f(x)e^{-\int_0^x a(t)dt}\right)' = e^{-\int_0^x a(t)dt}(f'(x) - a(x)f(x)) \leq e^{-\int_0^x a(t)dt}b(x),$$

for any $x \in [0,T]$. Hence

$$f(x)e^{-\int_0^x a(t)dt} \leq f(0) + \int_0^x e^{-\int_0^s a(t)dt} b(s)ds \leq f(0) + \int_0^x b(s)ds,$$

which concludes the proof. □

(b) Follows from (a).

10.2.13. (**Gronwall's inequality, the integral form**) *(a) Consider the continuous function* $f : [0,T] \to [0,\infty)$ *satisfying*

$$f(x) \leq C_1 \int_0^x f(t)dt + C_2, \quad \forall x \in [0,T],$$

where $C_1, C_2 \geq 0$ *are constants.*
  Prove that

$$f(x) \leq C_2\left(1 + C_1 x e^{C_1 x}\right), \quad \forall 0 \leq x \leq T.$$

(b) In particular, if

$$f(x) \leq C_1 \int_0^x f(t)dt, \quad \forall x \in [0,T],$$

prove that $f \equiv 0$ *în* $[0,T]$.

Solution. (a) Let $g(x) = \int_0^x f(t)dt$. Then $g' \leq C_1 g + C_2$ în $[0,T]$. Applying the Gronwall inequality in differential form, we obtain

$$g(x) \leq e^{C_1 x}(g(0) + C_2 x) = C_2 x e^{C_1 x}.$$

Using now the hypothesis, we obtain

$$f(x) \leq C_1 g(x) + C_2 \leq C_2(1 + C_1 x e^{C_1 x}). \quad \square$$

Independent Study. (i) What can we say about a continuous function $f : \mathbb{R} \to [0,\infty)$ if

$$f(x) \leq \int_0^x f(t)dt \quad \text{for all } x \in [a,b]?$$

(ii) Assume that for all $t \in [a,b]$,

$$f(x) \leq C_1(x-a) + C_2 \int_a^x f(t)dt + C_3,$$

where f is a nonnegative continuous function on $[a,b]$, and $C_1 \geq 0$, $C_2 > 0$, and $C_3 \geq 0$ are constants. Prove that for all $x \in [a,b]$,

$$f(x) \leq \left(\frac{C_1}{C_2} + C_3\right) e^{C_2(x-a)} - \frac{C_1}{C_2}.$$

The following inequality was proved by G.H. Hardy in 1920. He was looking for a nice proof of Hilbert's double series inequality. Some six years later, Landau was able to supply the best constant in this inequality. The interest illustrated for this result is reflected by the fact that twenty papers giving new proofs or modifying slightly the inequality appeared before 1934. One of these modified inequalities is the Carleman inequality, which in some sense is a limiting version of Hardy's

10.2 Integral Inequalities

inequality. We also point out that special cases of the following result become Poincaré's inequality, Friedrich's inequality, or the weighted Sobolev's inequalities. All these inequalities have many applications in embedding theories and in differential equations theory, both ordinary and partial. We state below the Hardy inequality in its simplest and original form.

10.2.14. (Hardy's inequality). *Let $p > 1$ and let $f : [0, \infty) \to \mathbb{R}$ be a differentiable increasing function such that $f(0) = 0$ and $\int_0^\infty (f'(x))^p dx$ is finite. Show that*

$$\int_0^\infty x^{-p} |f(x)|^p dx \leq \left(\frac{p}{p-1}\right)^p \int_0^\infty (f'(x))^p dx,$$

and equality holds if and only if f vanishes identically. Moreover, the best constant is given by $p^p(p-1)^{-p}$.

Solution. If f vanishes identically, then equality holds. So, we will assume that $f \not\equiv 0$. For any integer $n \geq 1$, set $g_n = \min\{f', n\}$, $G_n(x) = \int_0^x g_n(t)dt$. Let A be sufficiently large that $f' \not\equiv 0$ on $(0, X)$, for all $X \geq A$. It follows that the same property is satisfied by g_n and G_n. After integration by parts, we obtain

$$\int_0^X \left(\frac{G_n(x)}{x}\right)^p dx = -\frac{1}{p-1} \int_0^X G_n^p(x)(x^{1-p})' dx$$

$$= \left[-\frac{x^{1-p} G_n^p(x)}{p-1}\right]_0^X + \frac{p}{p-1} \int_0^X \left(\frac{G_n(x)}{x}\right)^{p-1} g_n(x) dx$$

$$\leq \frac{p}{p-1} \int_0^X \left(\frac{G_n(x)}{x}\right)^{p-1} g_n(x) dx.$$

We have used above the fact that $G_n(x) = o(x)$ for $x \to 0$. So, by Hölder's inequality,

$$\int_0^X \left(\frac{G_n(x)}{x}\right)^p dx \leq \frac{p}{p-1} \left(\int_0^X \left(\frac{G_n(x)}{x}\right)^p dx\right)^{1/p'} \left(\int_0^X g_n^p(x) dx\right)^{1/p}. \quad (10.3)$$

Since the left-hand side is positive and finite, it follows from (10.3) that

$$\int_0^X \left(\frac{G_n(x)}{x}\right)^p dx \leq \left(\frac{p}{p-1}\right)^p \int_0^X g_n^p(x) dx.$$

Passing to the limit as $n \to \infty$, it follows that

$$\int_0^X \left(\frac{f(x)}{x}\right)^p dx \leq \left(\frac{p}{p-1}\right)^p \int_0^X (f'(x))^p dx.$$

Next, taking $X \to \infty$, we obtain

$$\int_0^\infty \left(\frac{f(x)}{x}\right)^p dx \leq \left(\frac{p}{p-1}\right)^p \int_0^\infty (f'(x))^p dx. \quad (10.4)$$

This is exactly the Hardy inequality in the weak form, that is, for $<$ replaced by $\leq$. It remains to show that the inequality is strict, under the hypothesis that $f \neq 0$. Indeed, taking $n \to \infty$ and $X \to \infty$ in (10.3), we obtain

$$\int_0^\infty \left(\frac{f(x)}{x}\right)^p dx \leq \frac{p}{p-1} \left(\int_0^\infty \left(\frac{f(x)}{x}\right)^p dx\right)^{1/p'} \left(\int_0^\infty (f'(x))^p dx\right)^{1/p}. \quad (10.5)$$

Using now (10.4) in the right-hand side of (10.5), we deduce that

$$\int_0^\infty \left(\frac{f(x)}{x}\right)^p dx < \left(\frac{p}{p-1}\right)^p \int_0^\infty (f'(x))^p dx,$$

with equality if and only if $x^{-p} f^p(x)$ and $(f'(x))^p$ are proportional. This situation is impossible, since it would imply that f' is a certain power of x, so $\int_0^\infty (f'(x))^p dx$ diverges, contradiction.

The fact that $p^p(p-1)^{-p}$ is the best constant in Hardy's inequality follows by choosing the family of functions

$$f_\varepsilon(x) = \begin{cases} 0, & \text{if } 0 \leq x \leq 1, \\ \frac{p}{p-1-p\varepsilon} \left(x^{(p-1-p\varepsilon)/p} - 1\right), & \text{if } x > 1, \end{cases}$$

for $\varepsilon > 0$ sufficiently small.

The following inequality due to Hardy is closely related to the uncertainty principle.

Let $f : [a,b] \to \mathbb{R}$ be a continuously differentiable function such that $f(a) = f(b) = 0$. Prove that

$$\int_a^b \frac{f^2(x)}{4d(x)^2} dx \leq \int_a^b f'(x)^2 dx, \quad (10.6)$$

where $d(x) = \min\{|x-a|, |x-b|\}$.

It is sufficient to prove that

$$\int_a^c \frac{f^2(x)}{4(x-a)^2} dx \leq \int_a^c f'(x)^2 dx,$$

where $2c = a+b$, and a similar inequality for the other half-interval. Without loss of generality, we consider only the case $a = 0$. We then have

$$\int_0^c f'(x)^2 dx = \int_0^c \left\{x^{1/2}(x^{-1/2}f(x))' + \frac{f(x)}{2x}\right\}^2 dx$$

$$\geq \int_0^c \left\{x^{-1/2}f(x)(x^{-1/2}f(x))' + \frac{f^2(x)}{4x^2}\right\} dx$$

$$= \left[\frac{1}{2}(x^{-1/2}f(x))^2\right]_0^c + \int_0^c \frac{f^2(x)}{4x^2}dx$$
$$= \frac{f^2(c)}{2c} + \int_0^c \frac{f^2(x)}{4x^2}dx$$
$$\geq \int_0^c \frac{f^2(x)}{4x^2}dx,$$

as required. □

Comments. The N-dimensional version of the Hardy's inequality can be stated as follows. For this purpose, if Ω is a region in $\mathbb{R}$ and $u \in \mathbb{R}^N$ such that $|u| = 1$, we define
$$d_u(x) = \min\{|u|;\ x + tu \notin \Omega\}.$$
This definition implies that $0 \leq d_u(x) \leq +\infty$ and $d_u(x) = 0$ if and only if $x \notin \Omega$. Defining the vectors
$$e_1 = (1,0,\ldots,0),\ e_2 = (0,1,\ldots,0),\ \ldots, e_2 = (0,0,\ldots,1),$$
we put $d_i(x) = d_{e_i}(x)$, for any $1 \leq i \leq n$.

Let Ω be an arbitrary open set in $\mathbb{R}^N$ and assume that f is a continuously differentiable real-valued function with compact support in Ω. Then
$$\sum_{i=1}^N \int_\Omega \frac{f^2(x)}{4d_i(x)^2} dx \leq \int_\Omega |\nabla f(x)|^2 dx, \tag{10.7}$$
where $\nabla f(x) = (f_{x_1}(x),\ldots, f_{x_N}(x))$.

Indeed, applying Hardy's inequality (10.6), we deduce that for any $1 \leq i \leq n$,
$$\int_\Omega \frac{f^2(x)}{4d_i(x)^2} dx \leq \int_\Omega |f_{x_i}(x)|^2 dx.$$

By summation from $i = 1$ to N we deduce (10.7).

Remark. Opic and Kufner [87] were looking at higher-dimensional versional of Hardy's inequality. If Ω is a bounded set in $\mathbb{R}^N$, they established the inequality
$$\left(\int_\Omega |f(x)|^q v(x) dx\right)^{1/q} \leq C \left(\sum_{i=1}^N \int_\Omega \left|\frac{\partial f}{\partial x_i}(x)\right|^p w(x) dx\right)^{1/p}.$$

Here the constant C is proved to exist if and only if (sample result for $N = 1$ and $\Omega = (a,b)$)
$$\sup_{x \in (a,b)} \left(\int_x^b v(t) dt\right)^{1/q} \left(\int_a^x w(t)^{1-p'} dt\right)^{1/p'} < \infty,$$
where p' denotes the conjugate exponent of p.

10.2.15. Let f be a C^1 function on $[a,b]$ and let g be a C^1 convex function on $[a,b]$ such that $f(a) = g(a)$, $f(b) = g(b)$, and $f(x) \leq g(x)$ for all $x \in [a,b]$. Prove that
$$\int_a^b \sqrt{1+f'^2(x)}dx \geq \int_a^b \sqrt{1+g'^2(x)}dx.$$

International Mathematics Competition for University Students, 2004

Solution. Consider the function $h(t) := \sqrt{1+t^2}$, $t \in \mathbb{R}$. Since h is convex, then for all $t, s \in \mathbb{R}$,
$$h(t) \geq h(s) + (t-s)h'(s).$$
Thus, for all $x \in [a,b]$,
$$h(f'(x)) \geq h(g'(x)) + [f'(x) - g'(x)] h'(g'(x)).$$
Therefore
$$\int_a^b \sqrt{1+f'^2(x)}dx \geq \int_a^b \sqrt{1+g'^2(x)}dx + \int_a^b [f'(x) - g'(x)] h'(g'(x))dx.$$
So, it is enough to prove that
$$\int_a^b [f'(x) - g'(x)] h'(g'(x))dx \geq 0.$$
This follows by the second mean value theorem for Riemann integrals. Indeed, since g and h are convex and of class C^1, the function $h' \circ g'$ is nondecreasing and continuous. It follows that there exists $c \in [a,b]$ such that
$$\int_a^b [f'(x) - g'(x)] h'(g'(x))dx = h'(g'(a)) \int_a^c (f'(x) - g'(x))dx$$
$$+ h'(g'(b)) \int_c^b (f'(x) - g'(x))dx.$$
But $f(a) = g(a)$ and $f(b) = g(b)$. Hence
$$\int_a^b [f'(x) - g'(x)] h'(g'(x))dx = [f(c) - g(c)] [h'(g'(a)) - h'(g'(b))],$$
and the conclusion follows after using the hypothesis $f \leq g$ and the fact that $h' \circ g'$ is nondecreasing. □

The above proof also shows that equality occurs if and only if $f = g$.

10.2.16. Prove that there is no positive continuously differentiable function f on $[0, \infty)$ such that $f'(x) \geq f^2(x)$, for all $x \geq 0$.

Walter Rudin, Amer. Math. Monthly, Problem E 3331

10.2 Integral Inequalities

Solution. If f is such a function, we obtain the contradiction

$$\frac{1}{f(0)} = \int_0^{1/f(0)} dx \le \int_0^{1/f(0)} \frac{f'(x)}{f^2(x)} dx = \frac{1}{f(0)} - \frac{1}{f(1/f(0))} < \frac{1}{f(0)}.$$

The following result is related to the previous one, and it yields a nonexistence result in a framework involving the second derivative. □

10.2.17. Let $a \in \mathbb{R}$, $\alpha > 0$, and $C > 0$ be constants. Prove that there is no twice-differentiable function $f : [a, \infty) \to (0, \infty)$ that satisfies $f''(x) > f^{1+\alpha}(x)$ and $f'(x) > C$.

Richard Stong

Solution. Suppose that such a function exists. Since $f'(x) > C > 0$, $f(x)$ increases without bound. Integrating the first inequality in the form

$$f'(x) f''(x) > f^{1+\alpha}(x) f'(x)$$

gives

$$(f'(x))^2 \ge \frac{2}{2+\alpha} f^{2+\alpha}(x) + C_1,$$

for some constant C_1. Since f increases without bound, there exists some $b \ge a$ such that for $x \ge b$ we have

$$(f'(x))^2 \ge \frac{1}{2+\alpha} f^{2+\alpha}(x),$$

which can be rewritten as

$$\frac{f'(x)}{f^{1+\alpha/2}(x)} \ge \frac{1}{\sqrt{2+\alpha}} > 0.$$

Integrating again gives the inequality

$$-\frac{2}{\alpha f^{\alpha/2}(x)} \ge \frac{x}{\sqrt{2+\alpha}} + C_2.$$

This is a contradiction, because the left-hand side is always negative, but the right-hand side will be positive for x large enough. □

10.2.18. Prove that

$$\int_0^\pi e^{\sin x} dx > \pi e^{2/\pi}.$$

Solution. We have $e^{\sin x} = \sum_{n=0}^\infty \sin^n x / n!$. On the other hand, an integration by parts yields $(n+2) I_{n+2} = (n+1) I_n$, where $I_n := \int_0^\pi \sin^n x \, dx$. Since $I_0 = \pi$ and $I_1 = 2$, an elementary argument by induction yields $I_n \ge \pi (2/\pi)^n$, for all $n \ge 0$, with strict inequality if $n \ge 2$. We conclude that $\int_0^\pi e^{\sin x} dx > \pi \sum_{n=0}^\infty (2/\pi)^n = \pi e^{2/\pi}$. □

Exercise. Using Stirling's formula, find a sharper lower bound for $\int_0^\pi e^{\sin x} dx$.

10.2.19. Lyapunov's Inequality. Let p be a real-valued continuous function on $[a,b]$ ($p \not\equiv 0$) and let f be a nontrivial function of class C^2 such that $f''(x) + p(x)f(x) = 0$ for all $x \in [a,b]$ and $f(a) = f(b) = 0$. Prove that

$$\int_a^b \left|\frac{f''(x)}{f(x)}\right| dx > \frac{4}{b-a}.$$

Solution. Set $M := \sup_{a \leq x \leq b} |f(x)| < \infty$. We have

$$\int_a^b \left|\frac{f''(x)}{f(x)}\right| dx > \frac{\int_a^b |f''(x)| dx}{M} > \frac{\left|\int_c^d f''(x) dx\right|}{M} = \frac{|f'(d) - f'(c)|}{M}, \qquad (10.8)$$

for arbitrary $a \leq c < d \leq b$. Take $x_0 \in (a,b)$ such that $f(x_0) = M$. Then, by Rolle's theorem, there exist $\xi_1 \in (a,x_0)$ and $\xi_2 \in (x_0,b)$ such that $M = (x_0 - a)f'(\xi_1) = -(b - x_0)f'(\xi_2)$. Combining this with (10.8), we obtain

$$\int_a^b \left|\frac{f''(x)}{f(x)}\right| dx > \frac{1}{x_0 - a} + \frac{1}{b - x_0} > \frac{4}{b-a},$$

the last inequality following by the minimization of the mapping $(a,b) \ni x \longmapsto (x-a)^{-1} + (b-x)^{-1}$. □

Exercise. Assume that $q: [a,b] \to (0,\infty)$ is a continuous function. Prove that under the same assumptions as in the statement of the above exercise,

$$\int_a^b q(x) p_+(x) dx > \alpha(b-a),$$

where $\alpha = \inf_{x \in [a,b]} (x-a)^{-1}(b-x)^{-1} q(x)$ and $p_+(x) = \max\{p(x), 0\}$. This inequality generalizes Lyapunov's inequality, provided that $q(x) = 1$, for all $x \in [a,b]$.

10.2.20. Let a be a positive real number and let $f: [0,\infty) \to [0,a]$ be a function that has the intermediate value property on $[0,\infty)$ and is continuous on $(0,\infty)$. Assume that $f(0) = 0$ and

$$xf(x) \geq \int_0^x f(t) dt, \quad \text{for all } x \in (0,\infty).$$

Prove that f has antiderivatives on $[0,\infty)$.

D. Andrica and M. Piticari, Romanian Mathematical Olympiad, 2007

Solution. Since f is continuous on $(0,\infty)$ and bounded, it follows that f is integrable on $[0,x]$, for every $x \geq 0$. The function $F: [0,\infty) \to \mathbb{R}$, defined by

$$F(x) = \int_0^x f(t) dt,$$

is therefore differentiable on $(0,\infty)$, and $F'(x) = f(x)$ for all $x > 0$. Define a map $g : (0,\infty) \to \mathbb{R}$ by letting $g(x) = F(x)/x$ for $x > 0$. Since

$$g'(x) = \frac{xf(x) - F(x)}{x^2} \geq 0,$$

for all $x > 0$, g is nondecreasing, so $\lim_{x \to 0} g(x)$ exists. Let ℓ denote this limit.

Since f has the intermediate value property, there exists a sequence (a_n) of positive real numbers such that $a_n \to 0$ and $f(a_n) \to f(0) = 0$. By hypothesis, $f(a_n) \geq g(a_n) \geq 0$, whence $\lim_{n \to \infty} g(a_n) = 0$, so $\ell = 0$. Since $F(0) = 0$ and $g(x) = F(x)/x$, we get

$$F'(0) = \lim_{x \to 0} \frac{F(x) - F(0)}{x} = \lim_{x \to 0} g(x) = \ell = 0,$$

so $F'(0) = f(0)$ and F is an antiderivative of f on $[0, \infty)$. □

10.2.21. Let $\mathscr{C}$ be the class of all differentiable functions $f : [0,1] \to \mathbb{R}$ with a continuous derivative f' on $[0,1]$, and $f(0) = 0$ and $f(1) = 1$. Determine the minimum value that the integral $\int_0^1 (1+x^2)^{1/2}(f'(x))^2 dx$ can assume as f runs through all of $\mathscr{C}$, and find all functions in $\mathscr{C}$ that achieve this minimum value.

D. Andrica and M. Piticari, Romanian Mathematical Olympiad, 2008

Solution. Apply the Cauchy–Schwarz inequality to get

$$1 = f(1) - f(0) = \int_0^1 f'(x)dx$$
$$= \int_0^1 (1+x^2)^{-1/4}[(1+x^2)^{1/4} f'(x)] dx$$
$$\leq \left(\int_0^1 (1+x^2)^{-1/2} dx \right)^{1/2} \left(\int_0^1 (1+x^2)^{1/2} (f'(x))^2 dx \right)^{1/2}$$
$$= \left(\ln(1+\sqrt{2}) \right)^{1/2} \left(\int_0^1 (1+x^2)^{1/2} (f'(x))^2 dx \right)^{1/2}.$$

Consequently,

$$\int_0^1 (1+x^2)^{1/2} (f'(x))^2 dx \geq \frac{1}{\ln(1+\sqrt{2})},$$

for all $f \in \mathscr{C}$. Equality holds if and only if $f'(x) = k(1+x^2)^{-1/2}$; that is,

$$f(x) = k \ln\left(x + \sqrt{1+x^2}\right) + c.$$

The conditions $f(0) = 0$ and $f(1) = 1$ yield

$$f(x) = \frac{1}{\ln(1+\sqrt{2})} \ln\left(x + \sqrt{1+x^2}\right), \quad 0 \leq x \leq 1,$$

a function that clearly belongs to $\mathscr{C}$. □

10.3 Improper Integrals

> We know what we are, but know not what we may be.
>
> William Shakespeare (1564–1616)

For any $t > 0$ we define *Euler's gamma function* by

$$\Gamma(t) = \int_0^\infty x^{t-1} e^{-x} dx.$$

10.3.1. Prove that $\int_0^\infty x^{t-1} e^{-x} dx$ converges for any $t > 0$.

Solution. To show that this integral is convergent, there are two difficulties: the integrated function is unbounded in a neighborhood of the origin, and the integration interval is infinite. Therefore we split the integral into

$$\int_0^\infty x^{t-1} e^{-x} dx = \int_0^1 x^{t-1} e^{-x} dx + \int_1^\infty x^{t-1} e^{-x} dx.$$

Using the estimate

$$x^{t-1} e^{-x} \leq x^{t-1}$$

and

$$\int_0^1 x^{-\alpha} = \begin{cases} (1-\alpha)^{-1} & \text{if } \alpha < 1, \\ +\infty & \text{if } \alpha \geq 1, \end{cases}$$

we deduce that $\int_0^1 x^{t-1} e^{-x} dx$ converges for $t > 0$. Next, we use the estimate

$$x^{t-1} e^{-x} = x^{t-1} e^{-x/2} e^{-x/2} \leq M e^{-x/2},$$

provided x is sufficiently large. Thus, $\int_1^\infty x^{t-1} e^{-x} dx$ converges for any $t > 0$. A direct computation also shows that

$$\Gamma(t+1) = t \Gamma(t) \quad \text{for any } t > 0. \quad \square$$

10.3.2. Prove that

$$\int_1^\infty \frac{\sin x}{x} dx$$

converges, but

$$\int_1^\infty \left| \frac{\sin x}{x} \right| dx$$

diverges.

Solution. Set $f(x) = \sin x / x$, for any $x \geq 1$. Integrating by parts, we have

$$\int_1^R \frac{\sin x}{x} dx = \cos 1 - \frac{\cos R}{R} - \int_1^R \frac{\cos x}{x^2} dx.$$

10.3 Improper Integrals

Since $|\cos x/x^2| \leq 1/x^2$ and $\int_1^\infty 1/x^2 \, dx$ converges, we deduce that $\int_1^\infty \sin x/x \, dx$ converges.

To show that $\int_1^\infty |\sin x/x| \, dx$ diverges, we observe that

$$\int_\pi^{(n+1)\pi} \frac{|\sin x|}{x} dx = \sum_{k=1}^n \int_{k\pi}^{(k+1)\pi} \frac{|\sin x|}{x} dx$$

$$\geq \sum_{k=1}^n \frac{1}{(k+1)\pi} \int_{k\pi}^{(k+1)\pi} |\sin x| dx = \sum_{k=1}^n \frac{2}{(k+1)\pi},$$

which diverges to $+\infty$ as $n \to \infty$. $\square$

10.3.3. Using the fact that

$$\int_{-\infty}^\infty e^{-x^2} dx = \sqrt{\pi},$$

compute $f'(t)$, where

$$f(t) = \int_{-\infty}^\infty e^{-tx^2} dx, \quad \forall t > 0.$$

Solution. By the change of variables $y = x\sqrt{t}$ we obtain, for all $t > 0$,

$$f(t) = \int_{-\infty}^\infty e^{-tx^2} = \int_{-\infty}^\infty e^{-y^2} \frac{dy}{\sqrt{t}} = \frac{1}{\sqrt{t}} \int_{-\infty}^\infty e^{-y^2} dy = \sqrt{\frac{\pi}{t}}.$$

It follows that

$$f'(t) = -\frac{\sqrt{\pi}}{2} t^{-3/2}. \quad \square$$

10.3.4. Let $f : [0, \infty) \to [0, \infty)$ be a decreasing function such that the integral $\int_0^\infty f(x) dx$ is convergent. Prove that $\lim_{x \to \infty} x f(x) dx = 0$.

Solution. We argue by contradiction. Assume that there exist $\varepsilon > 0$ and $x_n \to \infty$ such that $x_n f(x_n) \geq \varepsilon$. Since f is decreasing, it follows that there exists $A > 0$ such that $f(x) \geq \varepsilon/x$, $\forall x \geq A$. This implies $\lim_{t \to \infty} \int_0^t f(x) dx = +\infty$, a contradiction. $\square$

10.3.5. Let $f : \mathbb{R} \to [0, \infty)$ be a continuous function such that the integral $\int_0^\infty f(x) dx$ is convergent. Prove that

$$\lim_{n \to \infty} \frac{1}{n} \int_0^n x f(x) dx = 0.$$

Solution. Fix $\varepsilon \in (0, 1)$. From $\int_0^\infty f(x) dx := \lim_{t \to \infty} \int_0^t f(x) dx < \infty$, it follows that there exists $N \in \mathbb{N}$ such that

$$\int_n^\infty f(x) dx < \varepsilon, \quad \forall n > N.$$

Thus, for all $n \in \mathbb{N}$ such that $n > \left[\frac{N}{\varepsilon}\right]$, we have

$$\int_0^n \frac{x}{n} f(x) dx = \int_0^{n\varepsilon} \frac{x}{n} f(x) dx + \int_{n\varepsilon}^n \frac{x}{n} f(x) dx$$

$$< \varepsilon \int_0^{n\varepsilon} f(x) dx + \int_{n\varepsilon}^n f(x) dx$$

$$< \varepsilon \int_0^{n\varepsilon} f(x) dx + \varepsilon < \varepsilon \left(\int_0^\infty f(x) dx + 1\right).$$

Since the above inequality holds for any $\varepsilon > 0$ and for all n sufficiently large, we obtain

$$\lim_{n \to \infty} \frac{1}{n} \int_0^n x f(x) dx = 0. \quad \square$$

10.3.6. Find all positive numbers a, b such that the integral

$$\int_b^\infty \left(\sqrt{\sqrt{x+a} - \sqrt{x}} - \sqrt{\sqrt{x} - \sqrt{x-b}}\right) dx$$

is convergent.

Solution. We use the Landau notation and denote by $O(x^2)$ a quantity obtained by multiplication of a bounded function and x^2. Thus, we can write

$$(1+x^2)^{1/2} = 1 + \frac{x}{2} + O(x^2), \quad \text{for } |x| < 1.$$

Hence

$$\sqrt{x+a} - \sqrt{x} = x^{1/2}\left(\sqrt{1 + \frac{a}{x}} - 1\right) = x^{1/2}\left(1 + \frac{a}{2x} + O(x^{-2})\right).$$

It follows that

$$\sqrt{\sqrt{x+a} - \sqrt{x}} = x^{1/4} x^{1/4}\left(1 + \frac{a}{4x} + O(x^{-2})\right),$$

and similarly,

$$\sqrt{\sqrt{x} - \sqrt{x-b}} = x^{1/4} x^{1/4}\left(1 + \frac{b}{4x} + O(x^{-2})\right).$$

So, our integral becomes

$$\int_b^\infty x^{1/4} \left(\frac{a-b}{4x} + O(x^{-2})\right) dx.$$

The term $x^{1/4} O(x^{-2})$ is bounded by the product of a constant and $x^{-7/4}$, whose integral converges.

10.3 Improper Integrals

So, we have only to decide whether

$$\lim_{A\to\infty} \int_b^A x^{-3/4} \frac{a-b}{4} dx$$

exists and is finite.

The function $x^{-3/4}$ yields a divergent integral, so the above integral converges if and only if $a = b$. □

10.3.7. Let $\omega : [0,a] \to [0,\infty)$ be a continuous increasing function such that $\omega(0) = 0$, $\omega(t) > 0$, for all $t > 0$ and

$$\lim_{\delta \searrow 0} \int_\delta^a \frac{1}{\omega(s)} ds = +\infty.$$

Let $f : [0,a] \to [0,\infty)$ be a continuous function such that

$$f(t) \le \int_0^t \omega(f(s)) ds, \quad \forall t \in (0,a].$$

Prove that $f \equiv 0$.

Solution. Let $g(t) = \max_{0 < s \le t} f(s)$ and assume that $g(t) > 0$ for all $0 < t \le a$. Then $f(t) \le g(t)$, and for all t, there exists $t_1 \le t$ such that $f(t_1) = g(t)$. Therefore

$$g(t) = f(t_1) \le \int_0^{t_1} \omega(f(s)) ds \le \int_0^{t_1} \omega(g(s)) ds \le \int_0^t \omega(g(s)) ds.$$

It follows that g satisfies the same inequality as f. Let

$$G(t) = \int_0^t \omega(g(s)) ds.$$

Hence $G(0) = 0$, $g(t) \le G(t)$, $G'(t) = \omega(g(t))$. It follows that $G'(t) \le \omega(G(t))$. So, on the one hand,

$$\int_\delta^a \frac{G'(t)}{\omega(G(t))} dt < a,$$

and on the other hand,

$$\int_\delta^a \frac{G'(t)}{\omega(G(t))} dt = \int_\Delta^A \frac{du}{\omega(u)},$$

where $G(a) = A$ and $G(\delta) = \Delta$. But by our hypotheses, the last integral tends to $+\infty$ as $\delta \searrow 0$. This contradiction shows that g cannot be positive, that is, $f \equiv 0$. □

10.3.8. Let $f : [0,\infty) \to [0,\infty)$ be a continuous function with the property that

$$\lim_{x\to\infty} f(x) \int_0^x f(t) dt = \ell > 0, \quad \ell \text{ finite.}$$

Prove that $\lim_{x\to\infty} f(x)\sqrt{x}$ exists and is finite, and evaluate it in terms of ℓ.

Solution. Define $F(x) = \int_0^x f(t)dt$, for any $x \geq 0$. Since f takes on nonnegative values, F is increasing, and by our hypothesis,

$$\lim_{x \to \infty} F(x) = \sup\{F(x) : x \geq 0\} > 0.$$

We first prove that
$$\lim_{x \to \infty} F(x) = +\infty.$$

Assuming that this limit is some finite $L > 0$, it follows that
$$\lim_{x \to \infty} f(x) = \ell/L.$$

Hence $f(x) \geq \ell/(2L)$ for $x \geq x_0$, for some $x_0 \geq 0$, so

$$F(x) = F(x_0) + \int_{x_0}^x f(t)dt \geq \ell(x - x_0)/(2L), \quad x \geq x_0,$$

which would contradict the assumption that $\lim_{x \to \infty} F(x)$ is finite. This establishes that $\lim_{x \to \infty} F(x) = +\infty$. Therefore, l'Hôpital's rule yields

$$\lim_{x \to \infty} \frac{F^2(x)}{x} = 2 \lim_{x \to \infty} f(x) F(x) = 2\ell.$$

It follows that
$$\lim_{x \to \infty} \frac{F(x)}{\sqrt{x}} = \sqrt{2\ell}.$$

Finally, writing
$$f(x)\sqrt{x} = \frac{f(x)F(x)}{\frac{F(x)}{\sqrt{x}}},$$

for x large enough, we conclude that $\lim_{x \to \infty} f(x)\sqrt{x}$ exists, is finite, and equal to $\sqrt{\ell/2}$. □

10.3.9. (**Barbălat's lemma, [5]**). *Let $f : [0, \infty) \to \mathbb{R}$ be uniformly continuous and Riemann integrable. Prove that $f(x) \to 0$ as $x \to \infty$.*

Solution. Arguing by contradiction, there exist $\varepsilon > 0$ and a sequence $(x_n)_{n \geq 0} \subset [0, \infty)$ such that $x_n \to \infty$ as $n \to \infty$ and $|f(x_n)| \geq \varepsilon$ for all integers $n \geq 0$. By the uniform continuity of f, there exists $\delta > 0$ such that for any $n \in \mathbb{N}$ and for all $x \in [0, \infty)$,

$$|x_n - x| \leq \delta \implies |f(x_n) - f(x)| \leq \frac{\varepsilon}{2}.$$

Therefore, for all $x \in [x_n, x_n + \delta]$ and for any $n \in \mathbb{N}$, $f(x) \geq |f(x_n)| - |f(x_n) - f(x)| \geq \varepsilon/2$. This yields, for all $n \in \mathbb{N}$,

$$\left| \int_0^{x_n+\delta} f(x)dx - \int_0^{x_n} f(x)dx \right| = \left| \int_{x_n}^{x_n+\delta} f(x)dx \right| = \int_{x_n}^{x_n+\delta} |f(x)|dx \geq \frac{\varepsilon\delta}{2} > 0.$$

10.3 Improper Integrals

By hypothesis, the improper Riemann integral $\int_0^\infty f(x)dx$ exists. Thus, the left-hand side of the inequality converges to 0 as $n\to\infty$. This contradicts the above relation, and the proof is concluded. □

10.3.10. Let $f : [0,\infty) \to \mathbb{R}$ be a decreasing function such that $\lim_{x\to\infty} f(x) = 0$. Prove that $\int_0^\infty f(x)dx$ converges.

Solution. We first observe that our assumptions imply that $f > 0$ on $(0,\infty)$ and f is integrable on $[0,T]$ for any $T > 0$. Fix T and choose $n \in \mathbb{N}$ such that $n\pi < T \leq (n+1)\pi$. Then

$$\int_0^T f(x)\sin x\,dx = \sum_{k=0}^{n-1} \int_{k\pi}^{(k+1)\pi} f(x)\sin x\,dx + \int_{n\pi}^T f(x)\sin x\,dx$$

$$= \sum_{k=0}^{n-1} (-1)^k \int_{k\pi}^{(k+1)\pi} |f(x)\sin x|dx + \int_{n\pi}^T f(x)\sin x\,dx.$$

We consider the limit as $T\to\infty$, directly for last term and by Leibniz's alternating series test for the first term. First,

$$\left|\int_{n\pi}^T f(x)\sin x\,dx\right| \leq \int_{n\pi}^{(n+1)\pi} |f(x)||\sin x|dx \leq f(n\pi)\int_{n\pi}^{(n+1)\pi} |\sin x|dx = 2f(n\pi),$$

and the $f(x)\to 0$ hypothesis shows that the last term above has limit 0 as $n\to\infty$, that is, as $T\to\infty$.

Next, the limit of the first term exists if and only if $\sum_{k=0}^\infty (-1)^k a_k$ converges, where $a_k = \int_{k\pi}^{(k+1)\pi} f(x)|\sin x|dx$. But

$$a_k = \int_{k\pi}^{(k+1)\pi} f(x)|\sin x|dx \geq \int_{(k+1)\pi}^{(k+2)\pi} f(x)|\sin x|dx = a_{k+1},$$

because f is decreasing, and for the same reason,

$$a_k \leq f(k\pi)\int_{k\pi}^{(k+1)\pi} |\sin x|dx = 2f(k\pi) \to 0 \quad \text{as } k\to\infty.$$

The alternating series thus converges, and the proof is complete. □

This exercise may be used to deduce the convergence of many improper integrals. For instance, $\int_0^\infty (\sin x)/x\,dx$ converges (we observe that 0 is not a "bad" point, since $\sin x/x \to 1$ as $x\to 0$. An alternative proof of this result follows from the following exercise:

Exercise. Prove that if $0 < a < R < \infty$, then

$$\left|\int_a^R \frac{\sin x}{x}dx\right| < \frac{2}{a}.$$

Next, we are concerned with the *Gaussian integral* (or *probability integral*) $\int_{-\infty}^{+\infty} e^{-x^2} dx$.

10.3.11. Prove that
$$\int_0^\infty e^{-x^2} dx = \frac{\sqrt{\pi}}{2}.$$

Solution. We define
$$f(x) = \int_0^1 \frac{e^{-x(1+t^2)}}{1+t^2} dt.$$

Since f is differentiable for all $x \in \mathbb{R}$, we have
$$f(0) = \arctan 1 = \frac{\pi}{4}, \quad f(\infty) := \lim_{x \to +\infty} f(x) = 0, \tag{10.9}$$

the latter because when $x > 0$,
$$0 < f(x) = e^{-x} \int_0^1 \frac{e^{-xt^2}}{1+t^2} dt < \frac{e^{-x}\pi}{4}.$$

Applying the Newton–Leibniz rule, we obtain $x > 0$,
$$f'(x) = -\int_0^1 e^{-x(1+t^2)} dt = -e^{-x} \int_0^1 e^{-xt^2} dt$$
$$= -\frac{e^{-x}}{\sqrt{x}} \int_0^{\sqrt{x}} e^{-u^2} du = -\frac{e^{-x}}{\sqrt{x}} g(\sqrt{x}), \tag{10.10}$$

where
$$g(z) = \int_0^z e^{-u^2} du.$$

Integrating (10.10) from 0 to ∞, then making the substitution $z = \sqrt{x}$, we deduce that
$$f(\infty) - f(0) = -\int_0^\infty \frac{e^{-x}}{\sqrt{x}} g(\sqrt{x}) dx = -2 \int_0^\infty e^{-z^2} g(z) dz$$
$$= -2 \int_0^\infty g'(z) g(z) dz = [g(0)]^2 - [g(\infty)]^2. \tag{10.11}$$

Because of (10.9), we read relation (10.11) as $-(\pi/4) = -(\int_0^\infty e^{-x^2} dx)^2$. We conclude that $\int_0^\infty e^{-x^2} dx = \sqrt{\pi}/2$. □

Exercise 10.3.11. shows that $\int_{-\infty}^{+\infty} e^{-x^2} dx = 2\int_0^{+\infty} e^{-x^2} dx = \sqrt{\pi}$. After a change of variable we obtain
$$\int_0^{+\infty} e^{-x^2} x^{-1/2} dx = \Gamma\left(\frac{1}{2}\right).$$

10.3 Improper Integrals

This shows why the factorial of a half-integer is a rational multiple of $\sqrt{\pi}$. More generally, for any $a, b > 0$,

$$\int_0^{+\infty} e^{-ax^b} dx = a^{-1/b} \Gamma\left(a + \frac{1}{b}\right).$$

In the following exercise we give a simple method to evaluate the Fresnel integrals (named after the French mathematician Augustin-Jean Fresnel (1788–1827), who applied them in diffraction phenomena) $\int_0^\infty \sin x^2 dx$ and $\int_0^\infty \cos x^2 dx$.

10.3.11. Prove that

$$\int_0^\infty \sin x^2 dx = \int_0^\infty \cos x^2 dx = \frac{\sqrt{2\pi}}{4}.$$

Solution. Define

$$F = \int_0^\infty \cos x^2 \, dx, \quad G = \int_0^\infty \sin x^2 \, dx. \tag{10.12}$$

We introduce the auxiliary functions

$$\alpha(x) = \int_0^1 \frac{\cos xt^2}{1+t^2} dt, \quad \beta(x) = \int_0^1 \frac{\sin xt^2}{1+t^2} dt.$$

We observe that α and β are continuously differentiable functions satisfying

$$\alpha(0) = \frac{\pi}{4}, \quad \beta(0) = 0. \tag{10.13}$$

We will also show that

$$\alpha(\infty) := \lim_{x \to +\infty} \alpha(x) = 0, \quad \beta(\infty) := \lim_{x \to +\infty} \beta(x) = 0. \tag{10.14}$$

Applying the Newton–Leibniz formula to both integrals in (10.12), we deduce that

$$\alpha'(x) - \beta(x) = -q(x), \quad \beta'(x) + \alpha(x) = p(x), \tag{10.15}$$

where for $x > 0$,

$$\begin{cases} p(x) = \int_0^1 \cos xt^2 dt = \frac{1}{\sqrt{x}} \int_0^{\sqrt{x}} \cos y^2 dy = \frac{1}{\sqrt{x}} u(\sqrt{x}), \\ q(x) = \int_0^1 \sin xt^2 dt = \frac{1}{\sqrt{x}} \int_0^{\sqrt{x}} \sin y^2 dy = \frac{1}{\sqrt{x}} v(\sqrt{x}), \end{cases} \tag{10.16}$$

and

$$u(z) := \int_0^z \cos y^2 dy, \quad v(z) = \int_0^z \sin y^2 dy. \tag{10.17}$$

We treat the system of differential equations (10.15) as follows: (i) we multiply the first equation by $-\cos x$, the second one by $\sin x$, then add the products; (ii) we multiply the first equation by $\sin x$, the second one by $\cos x$, then add. The two results can be written, respectively,

(i) $\dfrac{d}{dx}\{\beta(x)\sin x - \alpha(x)\cos x\} = p(x)\sin x + q(x)\cos x,$

(ii) $\dfrac{d}{dx}\{\beta(x)\cos x + \alpha(x)\sin x\} = p(x)\cos x - q(x)\sin x.$ (10.18)

Integrating the above relations from 0 to ∞, we obtain, by means of relations (10.13) and (10.14),

$$\int_0^\infty [p(x)\sin x + q(x)\cos x]dx = \frac{\pi}{4}, \quad \int_0^\infty [p(x)\cos x - q(x)\sin x]dx = 0. \quad (10.19)$$

Now, using (10.16) and (10.17), the substitution $z = \sqrt{x}$, and relation (10.17), we rewrite (10.19) as

$$\frac{\pi}{4} = 2\int_0^\infty [u(z)\sin z^2 + v(z)\cos z^2]dz$$
$$= 2\int_0^\infty [u(z)v'(z) + v(z)u'(z)]dz \quad (10.20)$$
$$= 2\int_0^\infty \frac{d}{dz}[u(z)v(z)]dz = 2FG$$

and

$$0 = 2\int_0^\infty [u(z)\cos z^2 - v(z)\sin z^2]dz = 2\int_0^\infty [u(z)u'(z) - v(z)v'(z)]dz$$
$$= \int_0^\infty \frac{d}{dz}\{[u(z)]^2 - [v(z)]^2\}dz = F^2 - G^2. \quad (10.21)$$

To show that G is positive (and with it F, by (10.20)), we write

$$G = \frac{1}{2}\int_0^\infty \frac{\sin x}{\sqrt{x}}dx = \frac{1}{2}\sum_{k=0}^\infty \int_{\pi k}^{\pi(k+1)} \frac{\sin x}{\sqrt{x}}dx.$$

This is an alternating series of diminishing terms whose kth term approaches 0 as $k \to \infty$ and whose $k = 0$ term is positive. It follows that $G > 0$. From (10.20) and (10.21), therefore, $F = G = (1/2)\sqrt{\pi}/2$.

To prove the second assertion in (10.14), we substitute $y = xt^2$ in (10.12) and we deduce that

$$\beta(x) = \frac{1}{2\sqrt{x}}\int_0^x \frac{\sin y}{\sqrt{y}[1+(y/x)]}dy. \quad (10.22)$$

10.3 Improper Integrals

We partition $[0,x]$ into the subintervals $[0,\pi], [\pi, 2\pi], \ldots, [n\pi, x]$ for the appropriate positive integer n, and write (10.22) as a sum of integrals over the subintervals. We obtain a sum of diminishing alternating-sign terms, the first one being positive. Thus, for $x > \pi$,

$$0 < \beta(x) < \frac{1}{2\sqrt{x}} \int_0^\pi \frac{\sin y}{\sqrt{y}[1+(y/x)]} dy < \frac{1}{2\sqrt{x}} \int_0^\pi \frac{\sin y}{\sqrt{y}} dy,$$

hence $\beta(\infty) = 0$.

It is almost as obvious that a like partition of the interval $[0,x]$ for evaluation of

$$\alpha(x) = \frac{1}{2\sqrt{x}} \int_0^x \frac{\cos y}{\sqrt{y}[1+(y/x)]} dy.$$

Thus, for any $x > \pi$,

$$0 < \alpha(x) < \frac{1}{2\sqrt{x}} \int_0^\pi \frac{\cos y}{\sqrt{y}[1+(y/x)]} dy$$

$$< \frac{1}{2\sqrt{x}} \int_0^{\pi/2} \frac{\cos y}{\sqrt{y}[1+(y/x)]} dy < \frac{1}{2\sqrt{x}} \int_0^{\pi/2} \frac{\cos y}{\sqrt{y}} dy,$$

which implies that $\alpha(\infty) = 0$. This completes the proof of (10.14). □

In Exercise 10.2.2 we have argued that $\int_0^\infty (\sin x)/x\, dx$ converges. In what follows we show that the exact value of this integral is $\pi/2$. As in the previous exercises, this computation is difficult, since the antiderivative of $f(x) = (\sin x)/x$ cannot be expressed as a finite combination of elementary functions.

10.3.12. Prove that

$$\int_0^\infty \frac{\sin x}{x} dx = \frac{\pi}{2}.$$

Solution. Integrating on $[0, \pi]$ the identity

$$\frac{1}{2} + \cos x + \cos 2x + \cdots + \cos nx = \frac{\sin\left(n+\frac{1}{2}\right)x}{2\sin\frac{x}{2}},$$

we obtain

$$\int_0^\infty \frac{\sin\left(n+\frac{1}{2}\right)x}{2\sin\frac{x}{2}} = \frac{\pi}{2}. \tag{10.23}$$

Define the function

$$f(x) = \begin{cases} \frac{2\sin\frac{x}{2} - x}{2x\sin\frac{x}{2}} & \text{if } x \in (0, \pi], \\ 0 & \text{if } x = 0. \end{cases}$$

Then f is continuously differentiable. Thus, by the Riemann–Lebesgue lemma (Exercise 9.2.2.),

$$\lim_{n \to \infty} \int_0^\pi \left(\frac{1}{x} - \frac{1}{2\sin\frac{x}{2}} \right) \sin\left(n + \frac{1}{2}\right) dx = 0.$$

Taking into account relation (10.23), we deduce that

$$\lim_{n \to \infty} \int_0^\pi \frac{\sin\left(n + \frac{1}{2}\right)}{x} dx = \frac{\pi}{2}.$$

Using the substitution $t = (n + \frac{1}{2})x$, we obtain

$$\lim_{n \to \infty} \int_0^{(n+1/2)\pi} \frac{\sin t}{t} dt = \frac{\pi}{2}. \tag{10.24}$$

Since

$$\int_0^\infty \frac{\sin t}{t} dt$$

is convergent, relation (10.24) yields $\int_0^\infty (\sin x)/x \, dx = \pi/2$.

A more direct approach uses the Laplace transform F of $(\sin x)/x$:

$$F(x) = \int_0^\infty \frac{\sin x}{x} e^{-sx} dx.$$

As s becomes large, the *kernel* e^{-sx} converges to 0 as $s \to \infty$, for any $x > 0$. Since $|\sin x/x| < 1$ for $x > 0$, we have $|F(s)| < 1/s$, which gives $\lim_{s \to \infty} F(s) = 0$.

Assuming that differentiation with respect to s may be performed either before or after the integration, we have

$$F'(s) = -\int_0^\infty (\sin x) e^{-sx} dx = -\frac{1}{s^2 + 1}.$$

Hence $F(s) = -\arctan s + C$, where C is a constant. But $\lim_{s \to \infty} F(s) = 0$, so C must be $\pi/2$. Thus, since $C = F(0)$,

$$\int_0^\infty \frac{\sin x}{x} dx = \frac{\pi}{2}.$$

A generalized version of the Riemann–Lebesgue lemma is the following: Assume that $f : [a,b] \to \mathbb{R}$ is a continuous function (where $0 \le a < b$) and let $g : [0, \infty) \to \mathbb{R}$ be continuous and periodic of period T. Then

$$\lim_{n \to \infty} \int_a^b f(x) g(nx) dx = \frac{1}{T} \int_0^T g(x) dx \cdot \int_a^b f(x) dx.$$

10.3 Improper Integrals

Some direct consequences of the generalized Riemann–Lebesgue lemma are

$$\lim_{n\to\infty} \int_{\pi}^{2\pi} \frac{|\sin nx|}{x}\,dx = \frac{2\ln 2}{\pi},$$

$$\lim_{n\to\infty} n \int_{2\pi}^{4\pi} \frac{\sin nx}{x^2}\,dx = \frac{3}{16\pi^2},$$

$$\lim_{n\to\infty} \int_0^{\pi} \frac{\sin x}{1+\cos^2 nx}\,dx = \sqrt{2}.$$

The following generalization of the Riemann–Lebesgue lemma to aperiodic functions is due to D. Andrica and M. Piticari. □

10.3.13. Let $0 \leq a < b$ be real numbers and let $f : [a,b] \to \mathbb{R}$ be a function of class C^1. Assume that $g : [0,\infty) \to \mathbb{R}$ is a continuous function such that $\lim_{x\to\infty} \frac{1}{x}\int_0^x g(t)\,dt = L$ exists and is finite. Prove that

$$\lim_{n\to\infty} \int_a^b f(x)g(nx)\,dx = L \int_a^b f(x)\,dx.$$

Solution. Set $G(x) = \int_0^x g(t)\,dt$ and define $\omega(x) = G(x) - Lx$, for any $x \geq 0$. Then ω is differentiable and $\lim_{x\to\infty} \frac{\omega(x)}{x} = 0$. We have

$$\int_a^b f(x)g(nx)\,dx = \frac{1}{n}[G(nb)f(b) - f(a)G(na)] - \frac{1}{n}\int_a^b f'(x)G(nx)\,dx$$

$$= \frac{1}{n}[G(nb)f(b) - f(a)G(na)] - \frac{1}{n}\int_a^b f'(x)[Lnx + \omega(nx)]\,dx$$

$$= \frac{1}{n}[G(nb)f(b) - f(a)G(na)] - L\int_a^b f'(x)x\,dx - \frac{1}{n}\int_a^b \omega(nx)f'(x)\,dx$$

$$= \frac{1}{n}[G(nb)f(b) - f(a)G(na)] - L[bf(b) - af(a)]$$

$$+ L \int_a^b f(x)\,dx - \frac{1}{n}\int_a^b \omega(nx)f'(x)\,dx.$$

We show that

$$\lim_{n\to\infty} \frac{1}{n}\int_a^b \omega(nx)f'(x)\,dx = 0. \tag{10.25}$$

Indeed, we first observe that

$$\left|\frac{1}{n}\int_a^b \omega(nx)f'(x)\,dx\right| \leq \frac{1}{n}\int_a^b |\omega(nx)||f'(x)|\,dx = \frac{|\omega(c_n)|}{n}\int_a^b |f'(x)|\,dx,$$

where $na < c_n < nb$. So, by Exercise 4.6.15, $\lim_{n\to\infty} \frac{|\omega(c_n)|}{n} = 0$, which implies relation (10.25).

Next, we observe that

$$\lim_{n\to\infty} \frac{1}{n}[G(nb)f(b) - G(na)f(a)] = \lim_{n\to\infty}\left(b\frac{G(nb)}{nb}f(b) - a\frac{G(na)}{na}f(a)\right)$$
$$= L[bf(b) - af(a)]. \qquad (10.26)$$

Using relations (10.25) and (10.26), we conclude the proof. □

10.3.14. *Evaluate*

$$\lim_{y\to+\infty} \int_0^\infty \frac{\sin x^2}{x^3} \sin(xy)\,dx.$$

N.I. Alexandrova

Solution. Define the auxiliary function

$$f(x) = \begin{cases} \frac{\sin x^2}{x^3} - \frac{1}{x} & \text{if } x \in (0,1), \\ \frac{\sin x^2}{x^3} & \text{if } x \in [1,\infty). \end{cases}$$

Since $-x^3/6 < (\sin x^2)/x^3 - 1/x < 0$ for $x \in (0,1)$, we deduce that $\int_0^\infty f(x)dx$ exists. Hence, we may write

$$\int_0^\infty \frac{\sin x^2}{x^3} \sin(xy)\,dx = \int_0^\infty f(x)\sin(xy)\,dx + \int_0^1 \sin(xy)\frac{dx}{x}.$$

The first term on the right tends to zero as $y\to\infty$ by the Riemann–Lebesgue lemma, while the second term equals

$$\int_0^y \sin t\, \frac{dt}{t},$$

which tends to $\pi/2$ as $y\to\infty$. We conclude that the value of the limit is $\pi/2$. □

10.4 Integrals and Series

> The mathematics of the 20th century mostly followed the road shown by Poincaré, the main difficulty being—as A. Weil once told me—the fact that too many good mathematicians have appeared, whereas all valuable mathematicians personally knew each other at Poincaré's time.
>
> Vladimir I. Arnold

10.4.1. *Establish the nature of the series*

$$\sum_{n=2}^\infty \frac{1}{n\ln n}.$$

Solution. Trying to apply the comparison tests for series, we have

$$\frac{1}{n^2} < \frac{1}{n \ln n} < \frac{1}{n} \quad \text{for all } n \geq 3.$$

However, the series $\sum_{n=3}^{\infty} 1/n^2$ converges and the series $\sum_{n=3}^{\infty} 1/n$ diverges; hence neither the first nor the second comparison test for series can be applied.

Using the integral test for series, we observe that the function $f(n) = 1/(x \ln x)$ is decreasing on $[2, \infty)$ and

$$\int_2^R \frac{dx}{x \ln x} = \int_{\ln 2}^{\ln R} \frac{dy}{y} = \ln(\ln R) - \ln(\ln 2) \to +\infty \quad \text{as } R \to +\infty.$$

Thus, the series $\sum_{n=2}^{\infty} 1/(n \ln n)$ diverges. □

A more general result is established in the following exercise.

10.4.2. Prove that the series

$$\sum_{n=2}^{\infty} \frac{1}{n(\ln n)^a}$$

converges if and only if $a > 1$.

Solution. Let $f(x) = 1/[x(\ln x)^a]$. Then

$$\int_1^x f(t)\,dt = \int_{\ln 2}^{\ln(x+1)} \frac{1}{s^a}\,ds = \begin{cases} \frac{[\ln(x+1)]^{1-a} - (\ln 2)^{1-a}}{1-a} & \text{if } a \neq 1, \\ \ln \ln(x+1) - \ln \ln 2 & \text{if } a = 1. \end{cases}$$

Letting $x \to \infty$, we see that

$$\int_1^{\infty} f(t)\,dt = \frac{1}{(a-1)(\ln 2)^{a-1}} \quad \text{if } a > 1$$

and $\int_1^{\infty} f(t)\,dt$ is divergent if $p \leq 1$. Thus, by the integral test for series,

$$\sum_{n=2}^{\infty} \frac{1}{n(\ln n)^a}$$

is convergent if $a > 1$ and it diverges if $p \leq 1$. Moreover, if $a > 1$, then

$$\frac{1}{(a-1)(\ln 2)^{a-1}} \leq \sum_{n=2}^{\infty} \frac{1}{n(\ln n)^a} \leq \frac{1}{(\ln 2)^{a-1}} \left[\frac{1}{2\ln 2} + \frac{1}{a-1} \right]. \quad □$$

10.4.3. *Compute*

$$\int_0^{\infty} \left(x - \frac{x^3}{2} + \frac{x^5}{2 \cdot 4} - \frac{x^7}{2 \cdot 4 \cdot 6} + \cdots \right) \left(1 + \frac{x^2}{2^2} + \frac{x^4}{2^2 \cdot 4^2} + \cdots \right) dx.$$

Solution. We use the classical development

$$e^t = 1 + \frac{t}{1} + \frac{t^2}{1\cdot 2} + \frac{t^3}{1\cdot 2\cdot 3} + \cdots, \quad \forall t \in \mathbb{R}.$$

This shows that the sum in the first set of parantheses under the integral sign equals $xe^{-x^2/2}$. On the other hand, after integration by parts, we obtain

$$\int_0^\infty x^{2n+1} e^{-x^2/2} dx = 2n \int_0^\infty x^{2n-1} e^{-x^2/2} dx.$$

An induction argument leads to

$$\int_0^\infty x^{2n+1} e^{-x^2/2} dx = (2n)!!.$$

It follows that the value of the integral is

$$\sum_{n=0}^\infty \frac{1}{2^n n!} = \sqrt{e}.$$

A particular case of the following general abstract result is

$$\lim_{x \searrow 0} x^{-1} \int_0^x \sin\frac{1}{t} dt = \lim_{x \searrow 0} x^{-1} \int_0^x \cos\frac{1}{t} dt = 0. \quad \square$$

10.4.4. Let $a_1, \ldots, a_n$ be real numbers. Prove the inequality

$$\sum_{i,j=1}^n \frac{a_i a_j}{i+j} \geq 0.$$

Solution. Consider the function $f : \mathbb{R} \to \mathbb{R}$ defined by $f(x) = \sum_{k=1}^n a_k e^{kx}$. The conclusion follows after observing that

$$\sum_{i,j=1}^n \frac{a_i a_j}{i+j} = \int_{-\infty}^0 f^2(x) dx.$$

This identity also shows that equality holds in our inequality if and only if $a_1 = \cdots = a_n = 0$.

We point out that the above inequality is in conjunction with the celebrated Hilbert inequality

$$\sum_{i,j=1}^\infty \frac{a_i a_j}{i+j} \leq \pi \sum_{k=1}^\infty a_k^2,$$

for any real numbers a_k, $k \geq 1$. $\quad \square$

10.4.5. Prove that there does not exist a subset of natural numbers A such that the function $f(x) = \sum_{n \in A} \frac{x^n}{n!}$ has the property that $x^2 e^{-x} f(x)$ tends to 1 as x tends to infinity.

Solution. We first observe that A cannot be finite. Indeed, in this case, we have $\lim_{x\to\infty} x^2 e^{-x} f(x) = 0$, which contradicts our hypothesis. Next, we suppose that A is infinite. Our argument is based on the fact that for any $n \in \mathbb{N}$,

$$\int_0^\infty \frac{x^n e^{-x}}{n!} dx = 1.$$

Let B an arbitrary **finite** subset of A. By hypothesis, there exists a positive constant C such that for all $x \geq 1$,

$$\sum_{n \in A} \frac{x^n e^{-x}}{n!} \leq \frac{C}{x^2}.$$

Then

$$\operatorname{Card} B = \int_0^\infty \sum_{n \in B} \frac{x^n e^{-x}}{n!} dx \leq \int_0^\infty \sum_{n \in A} \frac{x^n e^{-x}}{n!} dx$$

$$\leq \int_0^1 \sum_{n \in A} \frac{x^n e^{-x}}{n!} dx + \int_1^\infty \frac{C}{x^2} dx = K,$$

where K is a universal constant. We have obtained that if A is infinite, then the cardinality of **any** finite subset of A is bounded by the same constant, contradiction.

In what follows we use the second mean value theorem for integrals and the Dirichlet kernel

$$D_n(x) = \frac{1}{2} + \sum_{k=1}^n \cos kx = \frac{\sin\frac{(2n+1)x}{2}}{\sin\frac{x}{2}} \tag{10.27}$$

to deduce the value of $\zeta(2) = \sum_{n=1}^\infty 1/n^2$. □

10.4.6. *(i) Prove that*

$$\sum_{k=1}^{2n-1} \int_0^\pi x \cos kx \, dx = -2 \sum_{k=1}^n \frac{1}{(2k-1)^2}.$$

(ii) Deduce that

$$\sum_{k=1}^\infty \frac{1}{(2k-1)^2} = \frac{\pi^2}{8} \quad \text{and} \quad \sum_{k=1}^\infty \frac{1}{k^2} = \frac{\pi^2}{6}.$$

Solution. (i) For any integer $1 \leq k \leq n$ we have

$$\int_0^\pi x \cos kx \, dx = \frac{1}{k} \int_0^\pi x(\sin kx)' dx = -\frac{1}{k} \int_0^\pi \sin kx \, dx = \frac{(-1)^k - 1}{k^2}$$

$$= \begin{cases} 0 & \text{if } k \text{ is even,} \\ -\frac{2}{k^2} & \text{if } k \text{ is odd.} \end{cases}$$

This concludes the proof of (i).

(ii) By (i), we have

$$\int_0^\pi x\left(\frac{1}{2}+\sum_{k=1}^{2n-1}\cos kx\right)dx = \frac{\pi^2}{4} - 2\sum_{k=1}^{n}\frac{1}{(2k-1)^2}.$$

On the other hand, by (10.27),

$$\int_0^\pi x\left(\frac{1}{2}+\sum_{k=1}^{2n-1}\cos kx\right)dx = \int_0^\pi \frac{\frac{x}{2}}{\sin\frac{x}{2}}\sin\frac{(4n-1)x}{2}dx$$

$$= 2\left[1+\left(\frac{\pi}{2}-1\right)\cos\frac{(4n-1)\xi}{2}\right]\frac{1}{4n-1} = O\left(\frac{1}{n}\right),$$

as $n\to\infty$. In the above representation we have applied the second mean value formula for integrals for

$$f(x) = \sin\frac{(4n-1)x}{2} \quad \text{and} \quad g(x) = \begin{cases} \frac{x/2}{\sin(x/2)} & \text{if } x \in (0,\pi], \\ 1 & \text{if } x = 0. \end{cases}$$

We conclude that $\sum_{k=1}^{\infty} 1/(2k-1)^2 = \frac{\pi^2}{8}$. The value of $\sum_{k=1}^{\infty} 1/k^2$ is immediate after observing that

$$\sum_{k=1}^{\infty}\frac{1}{k^2} = \sum_{k=1}^{\infty}\frac{1}{(2k-1)^2} + \frac{1}{4}\sum_{k=1}^{\infty}\frac{1}{k^2}. \quad \square$$

10.5 Applications to Geometry

> I will not define time, space, place, and motion, as being well known to all.
>
> Sir Isaac Newton (1642–1727)

Let $f:[a,b]\to\mathbb{R}$ be a differentiable function and assume that its derivative is continuous. Then the *length* of the curve $y = f(x)$ between a and b is given by

$$\int_a^b \sqrt{1+f'(x)^2}\,dx.$$

Assuming that $f:[a,b]\to[0,\infty)$ is continuous, the *volume* of the solid obtained by revolving the curve $y = f(x)$ around the x-axis is

$$\pi\int_a^b f(x)^2\,dx.$$

Assuming now that $f:[a,b]\to[0,\infty)$ is a differentiable function with continuous derivative, the area of the surface of revolution of the graph of f around the x-axis is

10.5 Applications to Geometry

$$2\pi \int_a^b f(x)\sqrt{1+f'(x)^2}\,dx.$$

An important example in mathematical physics is played by the following sequence of functions. For any positive integer n, define the Dirac sequence[1] $(f_n)_{n\geq 1}$ by

$$f_n(x) = \alpha_n(1-x^2)^n,$$

where

$$\alpha_n = \frac{(2n+1)!!}{(2n)!!}.$$

Then $\int_{-1}^1 f_n(x)\,dx = 1$. Moreover, these functions concentrate (see Figure 10.1), for increasing n, more and more of their "mass" at the origin, in the following sense: for any $\varepsilon > 0$ there exist $\delta \in (0,1)$ and an integer N such that for all $n \geq N$,

$$1 - \varepsilon < \int_{-\delta}^\delta f_n(x)\,dx < 1 \quad \text{and} \quad \int_{-1}^{-\delta} f_n(x)\,dx + \int_\delta^1 f_n(x)\,dx < \varepsilon.$$

10.5.1. Let $f, g : [a,b] \to [0,\infty)$ be continuous and nondecreasing functions such that for each $x \in [a,b]$,

$$\int_a^x \sqrt{f(t)}\,dt \leq \int_a^x \sqrt{g(t)}\,dt \quad \text{and} \quad \int_a^b \sqrt{f(t)}\,dt = \int_a^b \sqrt{g(t)}\,dt.$$

Prove that $\int_a^b \sqrt{1+f(t)}\,dt \geq \int_a^b \sqrt{1+g(t)}\,dt$.

Solution. Let $F(x) = \int_a^x \sqrt{f(t)}\,dt$ and $G(x) = \int_a^x \sqrt{g(t)}\,dt$. The functions F and G are convex, $F(a) = 0 = G(a)$, and $F(b) = G(b)$, by the hypothesis. We have to prove that

$$\int_a^b \sqrt{1+(F'(t))^2}\,dt \geq \int_a^b \sqrt{1+(G'(t))^2}\,dt. \quad \square$$

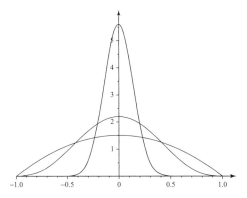

Fig. 10.1 Mass concentration of the Dirac sequence ($n = 1, 2, 3$).

---

[1] Introduced by the British scientist Paul Adrien Maurice Dirac (1902–1984), Nobel Prize in Physics 1933, "for the discovery of new productive forms of atomic theory."

This is equivalent to showing that the arc length of the graph of F is greater than or equal to the arc length of the graph of G. This is clear, since both functions are convex, their graphs have common endpoints, and the graph of F is below the graph of G.

10.5.2. Let f be a real-valued function with continuous nonnegative derivative. Assume that $f(0) = 0$, $f(1) = 1$, and let ℓ be the length of f on the interval $[0,1]$. Prove that
$$\sqrt{2} \leq \ell < 2.$$

Solution. We apply the elementary inequalities
$$\frac{\sqrt{2}}{2}(a+b) \leq \sqrt{a^2+b^2} \leq a+b,$$
for any $a, b \geq 0$.

We have
$$\ell = \int_0^1 \sqrt{1+(f'(x))^2}\, dx.$$

Therefore, since $f'(x) \geq 0$,
$$\sqrt{2} = \frac{\sqrt{2}}{2}\int_0^1 (1+f'(x))\, dx \leq \ell \leq \int_0^1 (1+f'(x))\, dx = 2.$$

Thus, $\sqrt{2} \leq \ell \leq 2$. If $\ell = 2$, then necessarily,
$$\sqrt{1+(f'(x))^2} = 1+f'(x), \quad \text{for all } x \in [0,1].$$

This implies $f'(x) = 0$, for any $x \in [0,1]$; hence f is constant. The latter contradicts the conditions $f(0) = 0$, $f(1) = 1$. □

10.5.3. Let $f : [0,1] \to [0,1]$ be a concave function of class C^1 such that $f(0) = f(1) = 0$. Prove that the length of the graph of f does not exceed 3.

Solution. By Rolle's theorem, there exists $c \in (0,1)$ such that $f'(c) = 0$. Since f is concave, it follows that f' is decreasing, so f increases in $(0,c)$ and is decreasing in $(c,1)$. The length of the graph on f in $[0,c]$ is given by
$$L_{[0,c]} = \int_0^c \sqrt{1+f'(x)^2}\, dx = \lim_{n \to \infty} \frac{c}{n}\sum_{k=0}^{n-1} \sqrt{1+f'(\xi_k)^2},$$
where $\xi_k \in (kc/n, (k+1)c/n)$. By the Lagrange mean value theorem, we can assume that
$$f'(\xi_k) = \frac{f((k+1)c/n) - f(kc/n)}{c/n}.$$

Hence

$$L_{[0,c]} = \lim_{n\to\infty} \sum_{k=0}^{n-1} \sqrt{(c/n)^2 + (f((k+1)c/n) - f(kc/n))^2}$$

$$\leq \lim_{n\to\infty} \sum_{k=0}^{n-1} [(c/n) + (f((k+1)c/n) - f(kc/n))] = c + f(c),$$

since f is increasing. A similar argument shows that $L_{[c,1]} \leq 1 - c + f(c)$. Therefore $L_{[0,1]} \leq c + f(c) + 1 - c + f(c) \leq 3$. □

10.6 Independent Study Problems

> Be not afraid of going slowly; be afraid only of standing still.
>
> Chinese Proverb

10.6.1. For any ordered pair of real numbers (x,y), define the sequence $(a_n(x,y))_{n\geq 0}$ by

$$a_0(x,y) = x, \quad a_{n+1}(x,y) = \frac{(a_n(x,y))^2 + y^2}{2}, \quad \forall n \geq 0.$$

Find the area of the region defined by

$$\{(x,y) \in \mathbb{R}^2; \text{ the sequence } (a_n(x,y))_{n\geq 0} \text{ is convergent}\}.$$

10.6.2. Let $I \subset \mathbb{R}$ be an open interval and $\rho : I \to (0, \infty)$ a function of class C^1. Fix $a, b \in I$ and define

$$\mathscr{F}_{a,b} = \{f : [0,1] \to I; \ f \in C^1[0,1], \ f(0) = a, \ f(1) = b\}.$$

For any C^1 function $f : [0,1] \to I$ set

$$E_1(f) = \int_0^1 f'(t)^2 dt$$

and

$$E_\rho(f) = \int_0^1 f'(t)^2 \rho(f(t)) dt.$$

(i) Prove that if $g : [0,1] \to \mathbb{R}$ is a continuous function, then

$$\int_0^1 g^2(t) dt \geq \left(\int_0^1 g(t) dt\right)^2$$

and establish when equality holds.

(ii) Find
$$\min\left\{\sqrt{E_1(g)};\ g\in\mathscr{F}_{a,b}\right\}$$
and show that there exists $f\in\mathscr{F}_{a,b}$ such that the minimum is achieved.

(iii) Prove that there exists an increasing function $\psi:I\to\mathbb{R}$ of class C^2 such that for all $f\in\mathscr{F}_{a,b}$,
$$E_\rho(f)=\int_0^1 [(\psi\circ f)'(t)]^2 dt.$$

(iv) Fix such a function ψ. Prove that for any $a,b\in I$, we have
$$\min\left\{\sqrt{E_\rho(f)};\ f\in\mathscr{F}_{a,b}\right\}=|\psi(a)-\psi(b)|.$$
Show that the minimum is achieved by a unique function $f_{a,b}$ and find this function.

(v) Prove that $f_{a,b}$ is of class C^2 and satisfies
$$2f_{a,b}''(x)\rho(f_{a,b}(x))+(f_{a,b}'(x))^2\rho'(f_{a,b}(x))=0,\quad \forall x\in[0,1].$$

10.6.3. Compute $\lim_{t\to\infty}\int_0^\pi x^t\sin x\, dx$.

10.6.4. Find a function f of the form $f(x)=(x+b)/(cx+d)$ such that the expression $\int_0^{\pi/2}|\tan x-f(x)|dx$ achieves its maximum.
Answer. $b=\pi(\pi-2)/2(4-\pi),\ c=\pi/(\pi-4),\ d=-\pi^2/2(\pi-4)$.

10.6.5. Let $f:[0,1]\to(0,\infty)$ be a decreasing function. Show that
$$\frac{\int_0^1 xf^2(x)dx}{\int_0^1 xf(x)dx}\leq\frac{\int_0^1 f^2(x)dx}{\int_0^1 f(x)dx}.$$

Hint. Consider the function
$$\int_0^x f^2(t)dt\cdot\int_0^x tf(t)dt-\int_0^x tf^2(t)dt\cdot\int_0^x f(t)dt.$$

10.6.6. (**Generalized Gronwall Inequality.**) Consider the continuous functions $a,b,c,f:[0,\infty)\to[0,\infty)$ such that
$$f(x)\leq\int_0^x [a(t)f(t)+b(t)]dt+c(x),\quad \forall x\geq 0.$$

Prove that
$$f(x)\leq\left[\int_0^x b(t)dt+\max_{0\leq t\leq x} c(t)\right]e^{\int_0^x a(t)dt},\quad \forall x\geq 0.$$

10.6 Independent Study Problems

10.6.7. (Nagumo's theorem) Let $f : [0,1] \to [0,\infty)$ be a continuous function such that $f(0) = 0$, $\lim_{t \searrow 0} f(t)/t = 0$ and $tf(t) \leq \int_0^t f(s) s^{-1} ds$, for all $t > 0$. Show that $f \equiv 0$.

10.6.8. Find all values of the real parameter a such that the function
$$f(x) = \int_0^x |t|^a \sin(1/t)^{|1/t|} dt$$
is differentiable at the origin.

10.6.9. Let $f : [0,1] \to \mathbb{R}$ be a continuous function. Prove that
$$\left(\int_0^1 x^2 f(x) dx \right)^2 \leq \frac{1}{3} \int_0^1 x^2 f^2(x) dx.$$

10.6.10. Let $f : [a,b] \to \mathbb{R}$ be a function of class C^1 such that $f(a) = 0$ and $0 \leq f'(x) \leq 1$. Prove that
$$\int_a^b f^3(x) dx \leq \left(\int_a^b f(x) dx \right)^2.$$

Hint. We have
$$\left(\int_a^b f(x) dx \right)^2 - \int_a^b f^3(x) dx = 2 \int_a^b \left(\int_a^y f(y) f(v)[1 - f'(v)] dv \right) dy.$$

10.6.11. Let $f : [0,1] \to \mathbb{R}$ be a function of class C^1 such that $f(0) = f(1) = 0$. Prove that
$$\left(\int_0^1 f(x) dx \right)^2 \leq \frac{1}{12} \int_0^1 (f'(x))^2 dx$$
and find for what functions the equality holds.

10.6.12. Let $f : [a,b] \to [0,\infty)$ be a differentiable function such that $f(a) = f(b) = 0$. Prove that there exists $c \in [a,b]$ such that
$$|f'(c)| \geq \frac{4}{(b-a)^2} \int_a^b f(x) dx.$$

10.6.13. Let $f : [0,1] \to \mathbb{R}$ be a continuous function such that $\int_0^1 x^k f(x) dx = 0$ for $k = 0, 1, \ldots, n-1$ and $\int_0^1 x^n f(x) dx = 1$. Show that there exists $x_0 \in [0,1]$ such that $|f(x_0)| \geq 2^n(n+1)$.

Hint. Observe that $\int_0^1 (x - 1/2)^n f(x) dx = 1$.

10.6.14. Let $a \le 0$, $b \ge 2$ and let $f : [a,b] \to \mathbb{R}$ be a function of class C^2. Prove that

$$\int_a^b (f''(x))^2 dx \ge \frac{3}{2}(f(0) - 2f(1) + f(2))^2$$

and that the equality holds if and only if

$$f(x) = A + Bx + C(x_+^3 - 2(x-1)_+^3 + (x-2)_+^3),$$

where $t_+ = \max(0,t)$.

Hint. Integrate by parts in $\int gf''$, where $g(x) = x_+ - 2(x-1)_+ + (x-2)_+$.

10.6.15. Let $K : [0,a] \to (0,\infty)$ be a continuous function and let $f : [0,a] \to [0,\infty)$ be a continuous function satisfying

$$f(t) \le \int_0^t K(s) f(s) ds, \quad \forall t \in [0,a].$$

Prove that $f \equiv 0$.
Hint. Apply the same ideas as in the proof of the solved problem No. 10.1.10.

10.6.16. Let $f : [0,a] \to [0,\infty)$ be a continuous function such that

$$f(t) \le e^{\int_0^t f(s) ds} - 1, \quad \forall t \in [0,a].$$

Prove that $f \equiv 0$.

10.6.17. Let $f : [0,a] \to [0,\infty)$ be a continuous function such that

$$f(t) \le \left(\int_0^t f(s) ds \right)^2, \quad \forall t \in [0,a].$$

Can we assert that $f \equiv 0$? The same question if the exponent 2 is replaced by $1/2$.

10.6.18. Let $f : [0,1] \to \mathbb{R}$ be a continuous function.

(a) Show that $\lim_{\lambda \to \infty} \int_0^1 f(x) \sin(\lambda x) dx = 0$.
(b) Prove the following generalization of the above result. If $g : \mathbb{R} \to \mathbb{R}$ is continuous with period $T > 0$, show that

$$\lim_{\lambda \to \infty} \int_0^1 f(x) g(\lambda x) dx = \bar{g} \int_0^1 f(x) dx,$$

where $\bar{g} := T^{-1} \int_0^T g(x) dx$ signifies the average of g over one period.

10.6.19. Let $\varphi : (0,1] \to [0,\infty)$ be a twice differentiable decreasing function such that φ' is increasing on $(0,1]$, $\lim_{x \searrow 0} \varphi(x) = \infty$, and $\lim_{x \searrow 0} x\varphi'(x) = -\infty$. Let $g : [0,\infty) \to \mathbb{R}$ be a continuous bounded function that admits a bounded antiderivative $G(x) = \int_0^x f(t) dt$ on $[0,\infty)$ and such that $\lim_{x \to \infty} g(x)$ does not exist.

Prove that
$$\lim_{x \searrow 0} x^{-1} \int_0^x f(t)dt = 0.$$

T.G. Feeman and O. Marrero

10.6.20. Let $f : [0, \infty) \to \mathbb{R}$ be a differentiable function satisfying
$$f'(x) = -3f(x) + 6f(2x), \quad \text{for all } x > 0.$$
Assume that $|f(x)| \leq e^{-\sqrt{x}}$, for all $x \geq 0$. For any $n \geq 0$ we define the nth-order momentum of f by
$$\mu_n = \int_0^\infty x^n f(x) dx.$$

(a) Express μ_n in terms of μ_0.
(b) Prove that the sequence $(\mu_n 3^n / n)_{n \geq 1}$ is convergent and moreover, its limit is 0 if and only if $\mu_0 = 0$.

10.6.21. Let $f : [a,b] \to \mathbb{R}$ be a continuous function that is twice differentiable in (a,b) and such that $f(a) = f(b) = 0$. Prove that
$$\int_a^b |f(x)| dx \leq \frac{(b-a)^3}{12} \sup_{a<x<b} |f''(x)|.$$
If, in addition, $f((a+b)/2) = 0$, show that in the above inequality we can replace $1/12$ by $1/24$.
Hint. Apply Rolle's theorem twice to the function $(t-a)(t-b)f(x) - (x-a)(x-b)f(t)$ for some fixed $a < x < b$. We obtain $|f(x)| \leq M(x-a)(b-x)/2$.

10.6.22. Let a be a fixed real number and define
$$\mathscr{F} = \{f \in C^2[0,1]; \ f(0) = 0, \ f'(0) = a, \ f(1) = 0\}.$$
Find $\min_{f \in \mathscr{F}} \int_0^1 [f''(x)]^2 dx$ and the function that realizes the minimum.

10.6.23. Using the ideas developed in the proof of Exercise 10.2.13, show that
$$\lim_{y \to +\infty} \int_0^\infty x^{-(1+r)} \sin x^r \sin(xy) dx = \frac{\pi}{2}$$
for any fixed positive number r.

Klaus Schürger

10.6.24. Let $a \in (0, \pi)$ and assume that n is a positive integer. Prove that
$$\int_0^\pi \frac{\cos nx - \cos na}{\cos x - \cos a} dx = \pi \frac{\sin na}{\sin a}.$$

10.6.25. Consider the functions $f, g : [0, 1) \to \mathbb{R}$ defined by
$$f(x) = \frac{1}{2} \ln \frac{1+x}{1-x} - x, \quad g(x) = f(x) - \frac{x^3}{3(1-x^2)}.$$

(i) Prove that $f(x) > 0$ and $g(x) < 0$ for all $x \in (0, 1)$.

(ii) Deduce that for $x \in [0, 1)$ we have
$$0 \leq \frac{1}{2} \ln \frac{1+x}{1-x} - x \leq \frac{x^3}{3(1-x^2)}.$$

(iii) Show that for any positive integer n,
$$0 \leq \left(n + \frac{1}{2}\right) \ln \frac{n+1}{n} - 1 \leq \frac{1}{12} \left(\frac{1}{n} - \frac{1}{n+1}\right).$$

(iv) Define
$$a_n = \frac{n^{n+\frac{1}{2}} e^{-n}}{n!} \quad \text{and} \quad b_n = a_n e^{1/(12n)}.$$

Prove that $a_{n+1} \geq a_n$ and $b_{n+1} \leq b_n$.

(v) Deduce Stirling's formula
$$n! = c^{-1} n^{n+\frac{1}{2}} e^{-n} e^{\theta/(12n)},$$
for some $\theta \in (0, 1)$, where c is the unique real number such that $a_n \leq c \leq b_n$ for all n.

10.6.26. A function $\delta : \mathbb{R} \to \mathbb{R}$ is a Dirac delta function if it is nonnegative and
$$\int_{-\infty}^{+\infty} \delta(x) f(x) dx = f(0),$$
for all continuous functions $f : \mathbb{R} \to [0, \infty)$. Show that there is no such function.

10.6.27. (i) Using the recurrence formulas for the integrals of powers of the sine, prove that
$$\int_0^{\pi/2} \sin^{2n} x\, dx = \frac{2n-1}{2n} \cdot \frac{2n-3}{2n-2} \cdots \frac{1}{2} \cdot \frac{\pi}{2},$$
$$\int_0^{\pi/2} \sin^{2n+1} x\, dx = \frac{2n}{2n+1} \cdot \frac{2n-2}{2n-1} \cdots \frac{2}{3}.$$

(ii) Deduce that
$$1 \leq \frac{\int_0^{\pi/2} \sin^{2n-1} x\, dx}{\int_0^{\pi/2} \sin^{2n+1} x\, dx} \leq 1 + \frac{1}{2n}.$$

(iii) Taking the ratio of the integrals of $\sin^{2n} x$ and $\sin^{2n+1} x$ between 0 and $\pi/2$, deduce Wallis's formula
$$\frac{\pi}{2} = \lim_{n \to \infty} \left(\frac{2}{1} \cdot \frac{4}{3} \cdots \frac{2n}{2n-1} \cdot \frac{1}{\sqrt{2n+1}}\right)^2.$$

Part V
Appendix

The purpose of this appendix is to recall some basic notions in elementary set theory and topology of the real line. We are concerned with the following notions: direct and inverse image of a set under a function; one-to-one and surjective mappings; finite and infinite (countable or uncountable) sets; neighborhood of a point; open and closed subsets of $\mathbb{R}$; distinguished points (interior, isolated, and accumulation points) of a set. Several examples and diagrams illustrate these abstract notions.

Appendix A
Basic Elements of Set Theory

> *General statements are simpler than their particular cases. A mathematical idea should not be petrified in a formalized, axiomatic setting, but should be considered instead as flowing as a river.*
> —James Joseph Sylvester (1814–1897)

This appendix provides a short introduction to those concepts from elementary set theory that we have used in this volume.

A.1 Direct and Inverse Image of a Set

Let A and B be two sets and let f be a mapping of A into B. If $C \subset A$, the *direct image* of C under f is defined by

$$f(C) = \{f(x);\ x \in C\}.$$

The set $f(A)$ (see Figure A.1) is called the *range* of f. Then $f(A) \subset B$. If $f(A) = B$, we say that f maps A *onto* B, that is, f is a *surjective* function.

If $D \subset B$, the *inverse image* of D under f is defined by

$$f^{-1}(D) = \{a \in A;\ f(x) \in D\}.$$

The inverse image of the set B is depicted in Figure A.2.

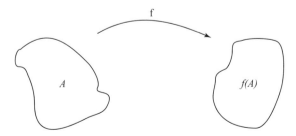

Fig. A.1 The direct image of the set A under the function f.

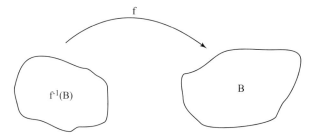

Fig. A.2 The inverse image of the set B under the function f.

In particular, if $y \in B$, then $f^{-1}(y)$ is the set of all $x \in A$ such that $f(x) = y$. If for each $y \in B$, $f^{-1}(y)$ consists of at most one element of A, then f is said to be a *one-to-one* mapping of A into B. This may be also expressed as follows: f is a one-to-one mapping of A into B provided that $f(x_1) \neq f(x_2)$ whenever $x_1 \neq x_2$, $x_1 \in A$, $x_2 \in A$.

A function that is one-to-one and surjective is said to be *bijective*.

A.2 Finite, Countable, and Uncountable Sets

An arbitrary set A is *finite* if there is a bijective map between A and $\{1,\ldots,n\}$, for some positive integer n (the empty set is also considered to be finite).

The set A is *infinite* if it is not finite.

The set A is said to be *countable* if there is a bijective function between A and $\mathbb{N}$.

The set A is *at most countable* if it is either finite or countable.

The set A is said to be *uncountable* if it is neither finite or countable.

Examples. (i) The set $\mathbb{Z}$ of integers is countable, since the function $f : \mathbb{N} \to \mathbb{Z}$ defined by

$$f(n) = \begin{cases} \dfrac{n}{2} & \text{if } n \text{ is even,} \\ -\dfrac{n-1}{2} & \text{if } n \text{ is odd,} \end{cases}$$

is bijective.

(ii) The set of all rational numbers $\mathbb{Q}$ is countable.
(iii) Both the set of real numbers $\mathbb{R}$ and the set of irrational numbers $\mathbb{R} \setminus \mathbb{Q}$ are uncountable sets.
(iv) The set of all sequences whose elements are the digits 0 or 1 is uncountable.

Appendix B
Topology of the Real Line

> *Mathematicians are born, not made.*
> —Henri Poincaré (1854–1912)

We recall in what follows the main topological definitions and properties that we have used in this volume.

B.1 Open and Closed Sets

An *open interval* in $\mathbb{R}$ (see Figure B.1) is any set of the form

$$(a,b) = \{x \in \mathbb{R};\ a < x < b\}.$$

Fig. B.1 An open interval on the real axis.

A *closed interval* in $\mathbb{R}$ (see Figure B.2) is any set of the form

$$[a,b] = \{x \in \mathbb{R};\ a \leq x \leq b\}.$$

Fig. B.2 A closed interval on the real axis.

The intersection of two open intervals is either empty or is another open interval. The union of two open intervals is either another open interval or is just two disjoint open intervals. If I is an open interval and $x \in I$, then there exists $\varepsilon > 0$ such that $(a - \varepsilon, a + \varepsilon) \subset I$.

A subset U of $\mathbb{R}$ is called *open* if whenever $x \in U$, there exists $\varepsilon > 0$ such that $(x - \varepsilon, x + \varepsilon) \subset U$.

In $\mathbb{R}$, any open set U is the countable union of disjoint open intervals.

The union of any number (finite or infinite) of open sets is open. The intersection of finitely many open sets is open. This property is no longer valid for an arbitrary intersection of open sets.

If a is real number, then a neighborhood of a is a set V that contains an open set U such that $a \in U$.

A subset F of $\mathbb{R}$ is called *closed* if its complement $\mathbb{R} \setminus F$ is open.

The intersection of any number (finite or infinite) of closed sets is closed. The union of finitely many (but in general, not of infinitely many) closed sets is closed.

A set S of real numbers is closed if and only if the limit of any convergent sequence $(a_n)_{n \geq 1} \subset S$ is also an element of S.

Let S be a subset of $\mathbb{R}$. A real number a is called a *boundary point* of S if every nontrivial neighborhood $(a - \varepsilon, a + \varepsilon)$ of a contains both points of S and points of $\mathbb{R} \setminus S$.

A boundary point of S might lie either in S or in the complement of S.

B.2 Some Distinguished Points

Let S be a set of real numbers.

A point a of S is called an *interior point* of S if there exists $\varepsilon > 0$ such that the interval $(a - \varepsilon, a + \varepsilon)$ lies in S. A point $b \in S$ is called an *isolated point* of S if there exists $\varepsilon > 0$ such that the intersection of the interval $(b - \varepsilon, b + \varepsilon)$ with S is just the singleton $\{b\}$. A set consisting only of isolated points is called *discrete*. A point $c \in R$ is called an *accumulation point* of S (see Figure B.3) if every neighborhood of c contains infinitely many distinct elements of S.

Fig. B.3 Accumulation points of a set of real numbers.

A real number c is an accumulation point of the set S if it is the limit of a nonconstant sequence in S. In such a case c is either a boundary point of S or an interior point of S, but it cannot be an isolated point of S.

An important result in classical analysis (which is a reformulation of the Bolzano–Weierstrass theorem) asserts that every bounded, infinite subset of $\mathbb{R}$ has an accumulation point.

Glossary

Abel's Summation Formula. The relation

$$\sum_{k=1}^{n} a_k b_k = A_n b_n - \sum_{k=1}^{n-1} A_k(b_{k+1} - b_k),$$

for $n \geq 2$, where $A_k = \sum_{i=1}^{k} a_i$.

Abel's Test. Let $\sum_{n=1}^{\infty} a_n$ be a convergent series of real numbers. Then for any bounded monotone sequence $(b_n)_{n \geq 1}$, the series $\sum_{n=1}^{\infty} a_n b_n$ is also convergent.

Accumulation Point. Let S be a subset of $\mathbb{R}$. A point x is called an *accumulation point* of S if every neighborhood of x contains infinitely many distinct elements of S.

Arithmetic–Geometric Means Inequality (AM–GM Inequality). If n is a positive integer and $a_1, a_2, \ldots, a_n$ are nonnegative real numbers, then

$$\frac{1}{n} \sum_{k=1}^{n} a_k \geq \sqrt[n]{a_1 a_2 \cdots a_n},$$

with equality if and only if $a_1 = a_2 = \cdots = a_n$.

Arithmetic–Harmonic Means Inequality. If n is a positive integer and $a_1, a_2, \ldots, a_n$ are positive numbers, then

$$\frac{1}{n} \sum_{k=1}^{n} a_k \geq \frac{1}{\frac{1}{n} \sum_{k=1}^{n} \frac{1}{a_k}},$$

with equality if and only if $a_1 = a_2 = \cdots = a_n$.

Banach Fixed-Point Theorem (Contraction Principle). Let $D \subset \mathbb{R}$ be a closed set. Then any contraction $f : D \to D$ has a unique fixed point.

Barbălat's Lemma. Let $f : [0, \infty) \to \mathbb{R}$ be uniformly continuous and Riemann integrable. Then $f(x) \to 0$ as $x \to \infty$.

Bernoulli's Inequality. Given any $r > 0$, we have
$$(1+x)^r \geq 1 + rx \quad \text{for any } x > -1.$$

Bertrand Series. Let α and β be real numbers. Then the Bertrand series
$$\sum_{n=2}^{\infty} \frac{1}{n^\alpha (\ln n)^\beta}$$
converges if and only if either $\alpha > 1$ or $\alpha = 1$ and $\beta > 1$.

Bolzano–Weierstrass Theorem. Every bounded sequence in $\mathbb{R}$ has a convergent subsequence with limit in $\mathbb{R}$.

Boundary Point. A real number a is called a *boundary point* of a set of real numbers S if every nontrivial neighborhood $(a - \varepsilon, a + \varepsilon)$ of a contains both points of S and points of $\mathbb{R} \setminus S$.

Brouwer Fixed-Point Theorem. Any continuous function $f : [a, b] \to [a, b]$ has at least one fixed point.

Carleman's Inequality. If $a_1, a_2, \ldots, a_n, \ldots$ are positive real numbers, then
$$\sum_{n=1}^{\infty} (a_1 a_2 \cdots a_n)^{1/n} \leq e \sum_{n=1}^{\infty} a_n,$$
where e denotes the base of the natural logarithm $2.71828\ldots$.

Cauchy's Condensation Criterion for Series. Suppose that $a_1 \geq a_2 \geq \cdots \geq 0$. Then the series $\sum_{n=1}^{\infty} a_n$ is convergent if and only if the series
$$\sum_{n=0}^{\infty} 2^n a_{2^n} = a_1 + 2a_2 + 4a_4 + 8a_8 + \cdots$$
is convergent.

Cauchy's Criterion for Infinite Products. Let $(a_n)_{n \geq 1}$ be a sequence of real numbers such that $a_n > -1$ for all n. Assume that $\lim_{n \to \infty} \sum_{k=1}^{n} a_k$ exists. Then $\lim_{n \to \infty} \prod_{k=1}^{n} (1 + a_k)$ exists, too. Moreover, this limit is zero if and only if the series $\sum_{n=1}^{\infty} a_n^2$ diverges.

Cauchy's Criterion for Sequences. A sequence of real numbers is convergent if and only if it is a Cauchy sequence.

Cauchy's Criterion for Series. A series $\sum_{n=1}^{\infty} a_n$ is convergent if and only if for each $\varepsilon > 0$, there is a positive integer N such that
$$\left| \sum_{k=n}^{m} a_k \right| < \varepsilon \quad \text{for all } m \geq n \geq N.$$

Cauchy Sequence. A sequence $(a_n)_{n\geq 1}$ of real numbers is called a *Cauchy sequence* if for every $\varepsilon > 0$ there is a natural number N_ε such that $|a_m - a_n| < \varepsilon$, for all $m, n \geq N_\varepsilon$.

Cauchy's Mean Value Theorem. Let $f, g : [a,b] \to \mathbb{R}$ be two functions that are continuous on $[a,b]$ and differentiable on (a,b). Then there exists a point $\xi \in (a,b)$ such that
$$(f(b) - f(a))g'(\xi) = (g(b) - g(a))f'(\xi).$$

Cauchy–Schwarz Inequality (discrete version). For any real numbers $a_1, a_2, \ldots, a_n$ and $b_1, b_2, \ldots, b_n$,
$$(a_1^2 + a_2^2 + \cdots + a_n^2)(b_1^2 + b_2^2 + \cdots + b_n^2) \geq (a_1 b_1 + a_2 b_2 + \cdots + a_n b_n)^2,$$
with equality if and only if a_k and b_k are proportional, $k = 1, 2, \ldots, n$.

Cauchy–Schwarz Inequality (integral version). Let $f, g : I \to \mathbb{R}$ be two nonnegative and integrable functions defined in a possible unbounded interval I. Then
$$\left(\int_a^b f^2(x)dx\right)^{1/2} \left(\int_a^b g^2(x)dx\right)^{1/2} \geq \int_a^b f(x)g(x)dx.$$
If f and g are continuous, then equality holds if and only if f and g are proportional.

Cesàro's Lemma. Let $(a_n)_{n\geq 1}$ be a sequence of positive numbers. Then the series $\sum_{n\geq 1} a_n$ diverges if and only if for any sequence of real numbers $(b_n)_{n\geq 1}$ that admits a limit ℓ in $\overline{\mathbb{R}}$, the sequence $\left(\frac{a_1 b_1 + \cdots + a_n b_n}{a_1 + \cdots + a_n}\right)_{n\geq 1}$ tends to ℓ, too.

Change of Variables in the Riemann Integral. Let φ be of class C^1 on the interval $[\alpha, \beta]$, with $a = \varphi(\alpha)$ and $b = \varphi(\beta)$. If f is continuous on $\varphi([\alpha, \beta])$ and $g = f \circ \varphi$, then
$$\int_a^b f(x)dx = \int_\alpha^\beta g(t)\varphi'(t)dt.$$

Closed Set. A subset F of $\mathbb{R}$ is called *closed* if its complement $\mathbb{R} \setminus F$ is open.

Continuous Function. A function f defined on an interval (a,b) is *continuous* at some point $c \in (a,b)$ if for each $\varepsilon > 0$, there exists $\delta > 0$ depending on both ε and c such that $|f(x) - f(c)| < \varepsilon$ whenever $|x - c| < \delta$.

Contraction. This is a mapping $f : D \subset \mathbb{R} \to \mathbb{R}$ for which there exists $\alpha \in (0,1)$ such that $|f(x) - f(y)| \leq \alpha |x - y|$ for all $x, y \in D$.

Convexity. A function f is *convex* (resp., *concave*) on $(a,b) \subset \mathbb{R}$ if the graph of f lies under (resp., over) the line connecting $(a_1, f(a_1))$ and $(b_1, f(b_1))$ for all $a < a_1 < b_1 < b$.

Coriolis Test. If $(a_n)_{n\geq 1}$ is a sequence of real numbers such that $\sum_{n=1}^{\infty} a_n$ and $\sum_{n=1}^{\infty} a_n^2$ are convergent, then $\prod_{n=1}^{\infty}(1 + a_n)$ converges.

Countable and Uncountable Sets. A set S is *countable* if it can be put into one-to-one correspondence with the set of natural numbers. Otherwise, S is *uncountable*. Examples: the sets $\mathbb{N}$, $\mathbb{Z}$, and $\mathbb{Q}$ are countable, while the sets $\mathbb{R}\setminus\mathbb{Q}$, $\mathbb{R}$, and $\mathbb{C}$ are uncountable.

Croft Lemma. Let $f : \mathbb{R} \to \mathbb{R}$ be a continuous function such that $\lim_{n\to\infty} f(n\delta) = 0$ for all $\delta > 0$. Then $\lim_{x\to\infty} f(x) = 0$.

Darboux Sums. We define the *lower* and *upper Darboux sums* associated to the function $f : [a,b] \to \mathbb{R}$ and to a partition $\Delta = \{x_0, x_1, x_2, \ldots, x_n\}$ of $[a,b]$ as

$$S_-(f;\Delta) = \sum_{i=1}^n m_i(x_i - x_{i-1}), \quad S^+(f;\Delta) = \sum_{i=1}^n M_i(x_i - x_{i-1}),$$

where

$$m_i = \inf_{x_{i-1} \leq x \leq x_i} f(x), \quad M_i = \sup_{x_{i-1} \leq x \leq x_i} f(x).$$

Darboux's Criterion. A function $f : [a,b] \to \mathbb{R}$ is Riemann integrable if and only if for any $\varepsilon > 0$ there exists $\delta > 0$ such that for every partition $\Delta = \{x_0, x_1, x_2, \ldots, x_n\}$ of $[a,b]$ with $\max_i(x_i - x_{i-1}) < \delta$, we have $S^+(f;\Delta) - S_-(f;\Delta) < \varepsilon$ (here, $S^+(f;\Delta)$ and $S_-(f;\Delta)$ denote the associated upper and lower Darboux sums).

Darboux's Theorem. Let $f : I \to \mathbb{R}$ be a differentiable function, where I is an open interval. Then f' has the intermediate value property.

Denjoy's Theorem. Let $f : I \to \mathbb{R}$ be a function that admits one-sided derivatives at any point of $I \setminus A$, where I is an interval and A is at most countable. Then f admits a derivative at any point of I, excepting a set that is at most countable.

Denjoy–Bourbaki Theorem. Let E be a normed vector space and consider the continuous function $f : [a,b] \to E$. Let $\varphi : [a,b] \to \mathbb{R}$ be a continuous nondecreasing function. Assume that both f and φ admit a right derivative at every point of $[a,b) \setminus A$, where the set A is at most countable, and moreover, for all $x \in [a,b) \setminus A$, we have $\|f'(x+)\| \leq \varphi'(x+)$. Then $\|f(b) - f(a)\| \leq \varphi(b) - \varphi(a)$.

Differentiation Inverse Functions Theorem. Suppose f is a bijective differentiable function on the interval $[a,b]$ such that $f'(x) \neq 0$ for all $x \in [a,b]$. Then f^{-1} exists and is differentiable on the range of f, and moreover, $(f^{-1})'[f(x)] = 1/f'(x)$ for all $x \in [a,b]$.

Dirac Sequence. This is the sequence of functions $(f_n)_{n \geq 1}$ that is defined by $f_n(x) = \alpha_n(1-x^2)^n$ for all $n \geq 1$, where $\alpha_n = \prod_{k=1}^n (2k+1)/(2k)$. These functions concentrate their "mass" at the origin, in the following sense: for any $\varepsilon > 0$ there exists $\delta \in (0,1)$ and an integer N such that for all $n \geq N$,

$$1 - \varepsilon < \int_{-\delta}^{\delta} f_n(x)dx < 1 \quad \text{and} \quad \int_{-1}^{-\delta} f_n(x)dx + \int_{\delta}^{1} f_n(x)dx < \varepsilon.$$

Dirichlet's Test. Let $\sum_{n=1}^{\infty} a_n$ be a series of real numbers whose partial sums $s_n = \sum_{k=1}^{n} a_k$ form a bounded sequence. If $(b_n)_{n\geq 1}$ is a decreasing sequence of nonnegative numbers converging to 0, then the series $\sum_{n=1}^{\infty} a_n b_n$ converges.

Discontinuity Points. Let f be a function with domain I. Let $a \in I$ and assume that f is discontinuous at a. Then there are two ways in which this discontinuity can occur:

(i) If $\lim_{x \to a-} f(x)$ and $\lim_{x \to a+} f(x)$ exist, but either do not equal each other or do not equal $f(a)$, then we say that f has a *discontinuity of the first kind* at the point a.
(ii) If either $\lim_{x \to a-} f(x)$ does not exist or $\lim_{x \to a+} f(x)$ does not exist, then we say that f has a *discontinuity of the second kind* at the point a.

Euler's Formula. If ζ denotes Riemann's zeta function, then

$$\zeta(x) = \prod_{n=1}^{\infty} \frac{1}{1-(p_n^x)^{-1}} \quad \text{for all } x > 1,$$

where $(p_n)_{n\geq 1}$ is the sequence of prime numbers ($p_1 = 2, p_2 = 3, p_3 = 5, \ldots$).

Euler's Gamma Function. This is the function defined by

$$\Gamma(t) = \int_0^{\infty} x^{t-1} e^{-x} dx \quad \text{for all } t > 0.$$

Fermat's Theorem. Let $f : I \to \mathbb{R}$ be a function and let x_0 be an interior point of I that is a relative maximum point or a relative minimum point for f. If f is differentiable at x_0, then $f'(x_0) = 0$.

Fibonacci Sequence. This sequence is defined by $F_0 = 1, F_1 = 1$, and $F_{n+1} = F_n + F_{n-1}$ for every positive integer n.

First Comparison Test for Series. Let $\sum_{n=1}^{\infty} a_n$ and $\sum_{n=1}^{\infty} b_n$ be two series of nonnegative numbers and suppose that $a_n \leq b_n$, for all $n \in \mathbb{N}$. Then the following properties are true:

(i) If $\sum_{n=1}^{\infty} b_n$ is convergent, then $\sum_{n=1}^{\infty} a_n$ is convergent, too.
(ii) If $\sum_{n=1}^{\infty} a_n$ is divergent, then $\sum_{n=1}^{\infty} b_n$ is divergent, too.

First Mean Value Theorem for Integrals. Let $f : [a,b] \to \mathbb{R}$ be a continuous function. Then there exists $\xi \in [a,b]$ such that

$$\int_a^b f(x) dx = (b-a) f(\xi).$$

Froda's Theorem. The set of discontinuity points of the first kind of any function $f : \mathbb{R} \to \mathbb{R}$ is at most countable.

Fundamental Theorems of Calculus. Let $f : I \to \mathbb{R}$, where I is an interval, and suppose that f is integrable over any compact interval contained in I. Let $a \in I$ and define $F(x) = \int_a^x f(t)dt$, for any $x \in I$. Then F is continuous on I. Moreover, if f is continuous at $x_0 \in I$, then F is differentiable at x_0 and $F'(x_0) = f(x_0)$.

Gauss's Test for Series. Let $(a_n)_{n \geq 1}$ be a sequence of positive numbers such that for some constants $r \in \mathbb{R}$ and $p > 1$, we have

$$\frac{a_{n+1}}{a_n} = 1 - \frac{r}{n} + O\left(\frac{1}{n^p}\right) \quad \text{as } n \to \infty.$$

Then the series $\sum_{n=1}^{\infty} a_n$ converges if $r > 1$ and diverges if $r \leq 1$.

Generalized Arithmetic–Geometric Means Inequality. For any $x_1, x_2, \ldots, x_n > 0$ and all $\lambda_i \geq 0$ ($1 \leq i \leq n$) with $\sum_{i=1}^{n} \lambda_i = 1$,

$$\lambda_1 x_1 + \cdots + \lambda_n x_n \geq x_1^{\lambda_1} \cdots x_n^{\lambda_n}.$$

Green–Tao Theorem. The set of prime numbers contains arbitrarily long arithmetic progressions.

Gronwall's Inequality (differential form). Let f be a nonnegative differentiable function on $[0, T]$ that satisfies the differential inequality

$$f'(x) \leq a(x)f(x) + b(x),$$

where a and b are nonnegative continuous functions on $[0, T]$. Then

$$f(x) \leq e^{\int_0^x a(t)dt} \left[f(0) + \int_0^x b(t)dt \right] \quad \text{for all } 0 \leq x \leq T.$$

Gronwall's Inequality (integral form). Let f be a nonnegative continuous function on $[0, T]$ that satisfies the integral inequality

$$f(x) \leq C_1 \int_0^x f(t)dt + C_2$$

for constants $C_1, C_2 \geq 0$. Then

$$f(x) \leq C_2 \left(1 + C_1 x e^{C_1 x}\right) \quad \text{for all } 0 \leq x \leq T.$$

Hardy's Inequality (discrete version). Assume that $p > 1$ and let $(a_n)_{n \geq 1}$ be a sequence of nonnegative numbers. Then

$$\sum_{n=1}^{\infty} \left(\frac{1}{n} \sum_{k=1}^{n} a_k \right)^p \leq \left(\frac{p}{p-1} \right)^p \sum_{n=1}^{\infty} a_n^p,$$

with equality if and only if $a_n = 0$ for every $n \geq 1$. Moreover, the constant $p^p(p-1)^{-p}$ is the best possible.

Hardy's Inequality (integral version). Assume that $p > 1$ and let $f : [0, \infty) \to [0, \infty)$ be a continuous function such that $\int_0^\infty f^p(x)\,dx := \lim_{x \to \infty} \int_0^x f^p(t)\,dt$ exists and is finite. Then

$$\int_0^\infty \left[\frac{1}{x}\int_0^x f(t)\,dt\right]^p dx \leq \left(\frac{p}{p-1}\right)^p \int_0^\infty f^p(x)\,dx,$$

with equality if and only if $f \equiv 0$. Moreover, the constant $p^p(p-1)^{-p}$ is the best possible.

Heine's Criterion. Let $f : I \to \mathbb{R}$ be a function defined on an interval I and let x_0 be an accumulation point of I. Then $f(x) \to \ell$ as $x \to x_0$ if and only if $f(x_n) \to \ell$ as $n \to \infty$, for any sequence $(x_n)_{n \geq 1} \subset I$ converging to x_0.

Heine–Borel Theorem. A set of real numbers is compact if and only if it is closed and bounded.

Hilbert's Double Series Theorem. Assume $p > 1$, $p' = p/(p-1)$ and consider $A := \sum_{n=1}^\infty a_n^p$, $B := \sum_{k=1}^\infty b_k^{p'}$, where $(a_n)_{n \geq 1}$ and $(b_n)_{n \geq 1}$ are sequences of nonnegative numbers. Then

$$\sum_{n=1}^\infty \sum_{k=1}^\infty \frac{a_n b_k}{n+k} < \frac{\pi}{\sin(\pi/p)} A^{1/p} B^{1/p'},$$

with equality if and only if either $A = 0$ or $B = 0$.

Hölder's Inequality (discrete version). Let $a_1, a_2, \ldots, a_n$ and $b_1, b_2, \ldots, b_n$ be positive numbers. If p and q are positive numbers such that $p^{-1} + q^{-1} = 1$, then

$$(a_1^p + a_2^p + \cdots + a_n^p)^{1/p} (b_1^q + b_2^q + \cdots + b_n^q)^{1/q} \geq a_1 b_1 + a_2 b_2 + \cdots + a_n b_n,$$

with equality if and only if a_k and b_k are proportional, $k = 1, 2, \ldots, n$.

Hölder's Inequality (integral version). Let f and g be nonnegative and integrable functions on $[a,b] \subset \mathbb{R}$. If p and q are positive numbers such that $p^{-1} + q^{-1} = 1$, then

$$\int_a^b f(x)g(x)\,dx \leq \left(\int_a^b f^p(x)\,dx\right)^{1/p} \left(\int_a^b g^q(x)\,dx\right)^{1/q}.$$

L'Hôpital's Rule. Let $f, g : (a,b) \to \mathbb{R}$ and $x_0 \in [a,b]$ be such that

(i) f and g are differentiable in $(a,b) \setminus \{x_0\}$;
(ii) $g'(x) \neq 0$ in $(a,b) \setminus \{x_0\}$;
(iii) f and g both tend either to 0 or to $\pm\infty$ as $x \to x_0$;
(iv) $f'(x)/g'(x) \to \ell \in \overline{\mathbb{R}}$, as $x \to x_0$.

Then

$$\lim_{x \to x_0} \frac{f(x)}{g(x)} = \ell.$$

Horizontal Chord Theorem. Let $f : [0,1] \to \mathbb{R}$ be a continuous function that has a horizontal chord of length λ. Then f has horizontal chords of lengths λ/n, for every integer $n \geq 2$, but horizontal chords of any other length cannot exist.

Increasing Function Theorem. If f is differentiable on an open interval I, then f is increasing on I if and only if $f'(x) \geq 0$ for all $x \in I$. If $f'(x) > 0$ for all $x \in I$, then f is strictly increasing in I.

Infimum. The *infimum* (or *greatest lower bound*) of a set $A \subset \mathbb{R}$ is an element $\alpha \in \mathbb{R} \cup \{-\infty\}$ that is a lower bound of A and such that no $\alpha_0 > \alpha$ is a lower bound of A. Notation: $\alpha = \inf A$.

Integral Test for Series. Suppose that $f : [1, \infty) \to [0, \infty)$ is nonincreasing. Then the improper integral $\int_1^\infty f(x)dx$ and the series $\sum_{n=1}^\infty f(n)$ are both convergent or both divergent.

Integration by Parts. Let f and g be integrable on $[a,b]$. If F and G are antiderivatives of f and g, respectively, then

$$\int_a^b F(x)g(g)dx = F(b)G(b) - F(a)G(a) - \int_a^b f(x)G(x)dx.$$

Interior Point. Let S be a subset of $\mathbb{R}$. A point x is called an *interior point* of S if there exists $\varepsilon > 0$ such that the interval $(x - \varepsilon, x + \varepsilon)$ is contained in S.

Intermediate Value Property. Let $I \subset \mathbb{R}$ be an arbitrary interval. A function $f : I \to \mathbb{R}$ is said to have the *intermediate value property* if for any $a, b \in I$ the function f takes on all the values between $f(a)$ and $f(b)$.

Isolated Point. Let S be a subset of $\mathbb{R}$. A point x is called an *isolated point* of S if there exists $\varepsilon > 0$ such that the intersection of the interval $(x - \varepsilon, x + \varepsilon)$ with S is just the singleton $\{x\}$.

Jensen's Inequality. Let $f : (a,b) \to \mathbb{R}$ be a convex function and assume that $\lambda_1, \lambda_2, \ldots, \lambda_n$ are nonnegative numbers with sum equal to 1. Then

$$\lambda_1 f(x_1) + \lambda_2 f(x_2) + \cdots + \lambda_n f(x_n) \geq f(\lambda_1 x_1 + \lambda_2 x_2 + \cdots + \lambda_n x_n)$$

for any $x_1, x_2, \ldots, x_n$ in the interval (a,b). If the function f is concave, then inequality is reversed.

Knaster Fixed Point Theorem. Any nondecreasing function $f : [a,b] \to [a,b]$ has at least a fixed point.

Kolmogorov's Inequality. Let $f : \mathbb{R} \to \mathbb{R}$ be a function of class C^3. Assume that both f and f''' are bounded and set

$$M_0 = \sup_{x \in \mathbb{R}} |f(x)|, \quad M_3 = \sup_{x \in \mathbb{R}} |f'''(x)|.$$

Then f' is bounded and

$$\sup_{x\in\mathbb{R}}|f'(x)| \leq \frac{1}{2}\left(9M_0^2 M_3\right)^{1/3}.$$

Kronecker Theorem. Let α be an irrational real number. Then the set

$$A = \{m+n\alpha;\ m,n \in \mathbb{Z}\}$$

is dense in $\mathbb{R}$.

Kummer's Test for Series. Let $(a_n)_{n\geq 1}$ and $(b_n)_{n\geq 1}$ be two sequences of positive numbers. Suppose that the series $\sum_{n=1}^{\infty} 1/b_n$ diverges and let $x_n = b_n - (a_{n+1}/a_n)b_{n+1}$. Then the series $\sum_{n=1}^{\infty} a_n$ converges if there is some $h > 0$ such that $x_n \geq h$ for all n (equivalently, if $\liminf_{n\to\infty} x_n > 0$) and diverges if $x_n \leq 0$ for all n (which is the case if, e.g., $\limsup_{n\to\infty} x_n > 0$).

Lagrange's Mean Value Theorem. Let $f:[a,b]\to\mathbb{R}$ be a function that is continuous on $[a,b]$ and differentiable on (a,b). Then there exists a point $\xi \in (a,b)$ such that

$$\frac{f(b)-f(a)}{b-a} = f'(\xi).$$

Geometrically, this theorem states that there exists a suitable point $(\xi, f(\xi))$ on the graph of $f:[a,b]\to\mathbb{R}$ such that the tangent to the curve $y = f(x)$ is parallel to the straight line through the points $(a, f(a))$ and $(b, f(b))$.

Landau's Inequality. Let $f:\mathbb{R}\to\mathbb{R}$ be a function of class C^2. Assume that both f and f'' are bounded and set

$$M_0 = \sup_{x\in\mathbb{R}}|f(x)|,\quad M_2 = \sup_{x\in\mathbb{R}}|f''(x)|.$$

Then f' is bounded and

$$\sup_{x\in\mathbb{R}}|f'(x)| \leq 2\sqrt{M_0 M_2}.$$

Landau–Kolmogorov Generalized Inequality. Let $f:\mathbb{R}\to\mathbb{R}$ be a nonconstant function of class C^n such that both f and $f^{(n)}$ are bounded. Then all the derivatives $f^{(k)}$ are bounded, $1 \leq k \leq n-1$.

For any integer $0 \leq k \leq n$, set $M_k = \sup_{x\in\mathbb{R}}|f^{(k)}(x)|$. Then, for all $0 \leq k \leq n$,

$$M_k \leq 2^{k(n-k)/2} M_0^{1-k/n} M_n^{k/n}.$$

Lebesgue's Theorem. A function $f:[a,b]\to\mathbb{R}$ is Riemann integrable if and only if f is bounded and the set of discontinuity points of f has null measure.

Leibniz's Test for Series. Let $(a_n)_{n\geq 1}$ be a decreasing sequence of positive numbers. Then the alternating series $\sum_{n=1}^{\infty}(-1)^n a_n$ is convergent.

Liminf of a sequence. Let $(a_n)_{n\geq 1}$ be a sequence of real numbers. The *limit infimum* of this sequence (denoted by $\liminf_{n\to\infty} a_n$) is the least limit of all subsequences of the given sequence. More rigorously, for each n let

$$A_n = \inf\{a_n, a_{n+1}, a_{n+2}, \ldots\}.$$

Then $(A_n)_{n\geq 1}$ is a monotone increasing sequence, so it has a limit. We define

$$\liminf_{n\to\infty} a_n := \lim_{n\to\infty} A_n \in \mathbb{R} \cup \{\pm\infty\}.$$

Limsup of a sequence. Let $(a_n)_{n\geq 1}$ be a sequence of real numbers. The *limit supremum* of this sequence (denoted by $\limsup_{n\to\infty} a_n$) is the greatest limit of all subsequences of the given sequence. More rigorously, for each n let

$$B_n = \sup\{a_n, a_{n+1}, a_{n+2}, \ldots\}.$$

Then $(B_n)_{n\geq 1}$ is a monotone decreasing sequence, so it has a limit. We define

$$\limsup_{n\to\infty} a_n := \lim_{n\to\infty} B_n \in \mathbb{R} \cup \{\pm\infty\}.$$

Limit Comparison Test for Series. Let $(a_n)_{n\geq 1}$ and $(b_n)_{n\geq 1}$ be two sequences of positive numbers such that $\ell := \lim_{n\to\infty}(a_n/b_n)$ exists.

(i) If $\ell > 0$, then $\sum_{n=1}^{\infty} a_n$ converges if and only if $\sum_{n=1}^{\infty} b_n$ converges.
(ii) If $\ell = 0$ and $\sum_{n=1}^{\infty} b_n$ converges, then $\sum_{n=1}^{\infty} a_n$ converges.

Limit of a Function. Let $f : I \to \mathbb{R}$ and assume that $x_0 \in \overline{\mathbb{R}}$ is an accumulation point of I. We say that f has *limit* $\ell \in \overline{\mathbb{R}}$ as $x \to x_0$ if for every neighborhood V of ℓ there exists a neighborhood U of x_0 such that for every $x_0 \in U \cap I$, $x \neq x_0$, we have $f(x) \in V$.

Lipschitz Function. Let $I \subset \mathbb{R}$ be an interval. A function $f : I \to \mathbb{R}$ is Lipschitz (or satisfies a Lipschitz condition) if there is a constant $L > 0$ such that for any $x, y \in I$,

$$|f(x) - f(y)| \leq L|x - y|.$$

The constant L is then called a Lipschitz constant for f.

Lower Bound. Let A be a set of real numbers. If there exists $m \in \mathbb{R}$ such that $x \geq m$ for every $x \in A$, we say that A is bounded below, and call m a *lower bound* of A.

Lyapunov's Inequality. Let p be a real-valued continuous function on $[a,b]$ ($p \not\equiv 0$) and let f be a nontrivial function of class C^2 such that $f''(x) + p(x)f(x) = 0$ for all $x \in [a,b]$ and $f(a) = f(b) = 0$. Then

$$\int_a^b \left|\frac{f''(x)}{f(x)}\right| dx > \frac{4}{b-a}.$$

Minkowski's Inequality (for numbers). Let $a_1, a_2, \ldots, a_n$ and $b_1, b_2, \ldots, b_n$ be positive numbers. If $p \geq 1$, then

$$\left(\sum_{k=1}^{n}(a_k+b_k)^p\right)^{1/p} \leq \left(\sum_{k=1}^{n}a_k^p\right)^{1/p} + \left(\sum_{k=1}^{n}b_k^p\right)^{1/p}.$$

Minkowski's Inequality (for functions). Let f and g be nonnegative and integrable functions on $[a,b] \subset \mathbb{R}$. If $p \geq 1$, then

$$\left(\int_a^b (f+g)^p(x)dx\right)^{1/p} \leq \left(\int_a^b f^p(x)dx\right)^{1/p} + \left(\int_a^b g^p(x)dx\right)^{1/p}.$$

Monotone Convergence Theorem. Let $(a_n)_{n\geq 1}$ be a bounded sequence that is monotone. Then $(a_n)_{n\geq 1}$ is a convergent sequence. If increasing, then $\lim_{n\to\infty} a_n = \sup_n a_n$, and if decreasing, then $\lim_{n\to\infty} a_n = \inf_n a_n$.

Monotone Function. Let f be a real function on (a,b). Then f is said to be *nondecreasing* (resp., *increasing*) on (a,b) if $a < x < y < b$ implies $f(x) \leq f(y)$ (resp., $f(x) < f(y)$). If $-f$ is nondecreasing (resp., increasing), then f is said to be *nonincreasing* (resp., *decreasing*) on (a,b). The class of *monotone* functions on (a,b) consists of all functions that are either nondecreasing or nonincreasing on (a,b).

Neighborhood. If a is real number, then a *neighborhood* of a is a set V that contains an open set U such that $a \in U$.

Nested Intervals Theorem. Suppose that $I_n = [a_n, b_n]$ are closed intervals such that $I_{n+1} \subset I_n$, for all $n \geq 1$. If $\lim_{n\to\infty}(b_n - a_n) = 0$, then there is a unique real number that belongs to every I_n.

Newton's Binomial. For all $a, b \in \mathbb{R}$ and for all $n \in \mathbb{N}$ we have

$$(a+b)^n = \sum_{k=0}^{n} \binom{n}{k} a^{n-k} b^k.$$

It seems that this formula was found by the French mathematician Blaise Pascal (1623–1662) in 1654. One of Newton's brilliant ideas in his *anni mirabiles*,[1] inspired by the work of Wallis was to try to interpolate by the polynomials $(1+x)^n$ ($n \geq 1$), in order to obtain a series for $(1+x)^a$, where a is a real number. Thus, Newton found the following generalized binomial theorem: for any $a \in \mathbb{R}$ and all $x \in \mathbb{R}$ with $|x| < 1$,

$$(1+x)^a = \sum_{k=0}^{n} \frac{a(a-1)\cdots(a-k+1)}{k!} x^k + s_n(x),$$

---

[1] *All this was in the two plague years 1665 and 1666, for in those days I was in the prime of my age for invention, and minded mathematics and philosophy more than at any other time since.* (Newton, quoted from Kline [58] 1972, p. 357.)

where $s_n(x)\to 0$ as $n\to\infty$. This is the formula that was engraved on Newton's gravestone at Westminster Abbey.

Newton–Leibniz Formula. Let f be integrable on $[a,b]$. If F is an antiderivative of f, then
$$\int_a^b f(x)dx = F(b) - F(a).$$

Newton's Method. Given a function f on $[a,b]$ and a point $x_0 \in [a,b]$, the iterative sequence $(x_n)_{n\geq 0}$ given by
$$x_{n+1} = x_n - \frac{f(x_n)}{f'(x_n)}, \quad n \geq 0,$$
determines Newton's method (or Newton's iteration) with initial value x_0.

Nonexpansive Function. This is a function $f: D \subset \mathbb{R} \to \mathbb{R}$ such that $|f(x)-f(y)| \leq |x-y|$ for all $x, y \in D$.

Open Set. A subset U of $\mathbb{R}$ is called *open* if whenever $x \in U$, there exists $\varepsilon > 0$ such that $(x-\varepsilon, x+\varepsilon) \subset U$.

Osgood Property. Let $(U_n)_{n\geq 1}$ be a sequence of open and dense subsets in $\mathbb{R}$. Then their intersection $\cap_{n=1}^{\infty} U_n$ is also dense in $\mathbb{R}$.

Pell Equation. This is the Diophantine equation $x^2 - Dy^2 = m$, where D is a nonsquare positive integer and m is an integer.

Picard Convergence Theorem. Let $f: [a,b] \to [a,b]$ be a continuous function that is differentiable on (a,b), with $|f'(x)| < 1$ for all $x \in (a,b)$. Then any Picard sequence for f is convergent and converges to the unique fixed point of f.

Pigeonhole Principle (Dirichlet's Principle). If $n+1$ pigeons are placed in n pigeonholes, then some pigeonhole contains at least two of the pigeons.

Pinching Principle. Let $(a_n)_{n\geq 1}$, $(b_n)_{n\geq 1}$, and $(c_n)_{n\geq 1}$ be sequences of real numbers satisfying
$$a_n \leq b_n \leq c_n$$
for every n. If
$$\lim_{n\to\infty} a_n = \lim_{n\to\infty} c_n = \ell$$
for some real number ℓ, then
$$\lim_{n\to\infty} b_n = \ell.$$

Poincaré's Inequality. Let $f: [0,1] \to \mathbb{R}$ be a function of class C^1 with $f(0) = 0$. Then
$$\sup_{0\leq x\leq 1} |f(x)| \leq \left(\int_0^1 (f'(x))^2 dx\right)^{1/2}.$$

Power Mean Inequality. Let $a_1, a_2, \ldots, a_n$ be any positive numbers for which $a_1 + a_2 + \cdots + a_n = 1$. For positive numbers $x_1, x_2, \ldots, x_n$ we define

$$M_{-\infty} = \min\{x_1, x_2, \ldots, x_n\},$$
$$M_{\infty} = \max\{x_1, x_2, \ldots, x_n\},$$
$$M_0 = x_1^{a_1} x_2^{a_2} \cdots x_n^{a_n},$$
$$M_t = \left(a_1 x_1^t + a_2 x_2^t + \cdots + a_n x_n^t\right)^{1/t},$$

where t is a nonzero real number. Then

$$M_{-\infty} \leq M_s \leq M_t \leq M_{\infty}$$

for $s \leq t$. The arithmetic–geometric means inequality and the arithmetic–harmonic means inequality are particular cases of the power mean inequality.

Raabe's Test for Series. Let $(a_n)_{n \geq 1}$ be a sequence of positive numbers. Then the series $\sum_{n=1}^{\infty} a_n$ converges if $a_{n+1}/a_n \leq 1 - r/n$ for all n, where $r > 1$ (equivalently, if $\liminf_{n \to \infty} n(1 - a_{n+1}/a_n) > 1$) and diverges if $a_{n+1}/a_n \geq 1 - 1/n$ for all n (which is the case if, e.g., $\limsup_{n \to \infty} n(1 - a_{n+1}/a_n) < 1$).

Racetrack Principle. Let $f, g : [a,b] \to \mathbb{R}$ be differentiable functions. If $f'(x) \leq g'(x)$ on $[a,b]$, then $f(x) - f(a) \leq g(x) - g(a)$ for all $x \in [a,b]$.

Rademacher Theorem. Let $f : I \to \mathbb{R}$ be a convex function, where I is an interval. Then f is locally Lipschitz. Furthermore, if $f : I \to \mathbb{R}$ is locally Lipschitz, then f is differentiable almost everywhere.

Ratio Test for Series. Let $\sum_{n=1}^{\infty} a_n$ be a series such that $a_n \neq 0$ for all n. Then the following properties are true:

(i) The series $\sum_{n=1}^{\infty} a_n$ converges if $\limsup_{n \to \infty} |a_{n+1}/a_n| < 1$.
(ii) The series $\sum_{n=1}^{\infty} a_n$ diverges if there exists $m \in \mathbb{N}$ such that $|a_{n+1}/a_n| \geq 1$ for all $n \geq m$.
(iii) If $\liminf_{n \to \infty} |a_{n+1}/a_n| \leq 1 \leq \limsup_{n \to \infty} |a_{n+1}/a_n|$, then the test is inconclusive.

Relative Extremum Point. Assume that $I \subset \mathbb{R}$ is an interval and let $f : I \to \mathbb{R}$ be a function. A point $x_0 \in I$ is said to be a *relative* or *local maximum point* (resp., *relative* or *local minimum point*) if there exists $\delta > 0$ such that $f(x) \leq f(x_0)$ (resp., $f(x) \geq f(x_0)$) whenever $x \in I$ and $|x - x_0| < \delta$.

Riemann–Lebesgue Lemma. Let $0 \leq a < b$ and assume that $f : [0,b] \to \mathbb{R}$ is a continuous function and $g : [0,\infty) \to \mathbb{R}$ is continuous and periodic of period T. Then

$$\lim_{n \to \infty} \int_a^b f(x) g(nx) dx = \frac{1}{T} \int_0^T g(x) dx \cdot \int_a^b f(x) dx.$$

Riemann's ζ Function. The Riemann zeta function is defined by

$$\zeta(x) = \sum_{n=1}^{\infty} \frac{1}{n^x}, \quad \text{for any } x > 1.$$

Rolle's Theorem. Let $f : [a,b] \to \mathbb{R}$ be a function that is continuous on $[a,b]$ and differentiable on (a,b). If $f(a) = f(b)$, then there exists a point $\xi \in (a,b)$ such that $f'(\xi) = 0$.

Rolle's Theorem (Polar Form). Let $f : [\theta_1, \theta_2] \to \mathbb{R}$ be a continuous real-valued function, nowhere vanishing in $[\theta_1, \theta_2]$, differentiable in (θ_1, θ_2), and such that $f(\theta_1) = f(\theta_2)$. Then there exists $\theta_0 \in (\theta_1, \theta_2)$ such that the tangent line to the graph $r = f(\theta)$ at $\theta = \theta_0$ is perpendicular to the radius vector at that point.

Root Test for Series. Given a series $\sum_{n=1}^{\infty} a_n$, define $\ell = \limsup_{n \to \infty} \sqrt[n]{|a_n|} \in [0, +\infty]$. Then the following properties are true:

(i) If $\ell < 1$ then the series $\sum_{n=1}^{\infty} a_n$ is convergent.
(ii) If $\ell > 1$ then the series $\sum_{n=1}^{\infty} a_n$ is divergent.
(iii) If $\ell = 1$ then the test is inconclusive.

Schwartzian Derivative. Let $f : I \to \mathbb{R}$ and assume that $f'''(x)$ exists and $f'(x) \neq 0$ for all $x \in I$. The Schwartzian derivative of f at x is defined by

$$\mathscr{D}f(x) := \frac{f'''(x)}{f'(x)} - \frac{3}{2}\left[\frac{f''(x)}{f'(x)}\right]^2.$$

Second Comparison Test for Series. Let $\sum_{n=1}^{\infty} a_n$ and $\sum_{n=1}^{\infty} b_n$ be two series of positive numbers such that $\sum_{n=1}^{\infty} a_n$ is convergent and $\sum_{n=1}^{\infty} b_n$ is divergent. Given a series $\sum_{n=1}^{\infty} x_n$ of positive numbers, we have:

(i) If the inequality $x_{n+1}/x_n \leq a_{n+1}/a_n$ is true for all $n \geq 1$, then $\sum_{n=1}^{\infty} x_n$ is convergent.
(ii) If the inequality $x_{n+1}/x_n \geq b_{n+1}/b_n$ is true for all $n \geq 1$, then $\sum_{n=1}^{\infty} x_n$ is divergent.

Second Mean Value Theorem for Integrals. Let $f, g : [a,b] \to \mathbb{R}$ be such that f is continuous and g is monotone. Then there exists $\xi \in [a,b]$ such that

$$\int_a^b f(x)g(x)dx = g(a)\int_a^\xi f(x)dx + g(b)\int_\xi^b f(x)dx.$$

Sierpiński's Theorem. Let I be an interval of real numbers. Then any function $f : I \to \mathbb{R}$ can be written as $f = f_1 + f_2$, where f_1 and f_2 have the intermediate value property.

Squeezing and Comparison Test. Let f, g, h be three functions defined on the interval I and let x_0 be an accumulation point of I. Assume that

$$g(x) \leq f(x) \leq h(x) \quad \text{for all } x \in I.$$

If $g(x) \to \ell$ and $h(x) \to \ell$ as $x \to x_0$, then $f(x) \to \ell$ as $x \to x_0$.

Glossary

Stirling's Formula. The limit
$$\lim_{n\to\infty} \frac{n!}{n^n e^{-n}\sqrt{2\pi n}}$$
exists and equals 1. In particular, the value of $n!$ is asymptotically equal to
$$n^n e^{-n}\sqrt{2\pi n}$$
as n becomes large. More precisely,
$$n! = \frac{\sqrt{2\pi n}\, n^n}{e^n} \cdot \exp\left(\frac{1}{12n} - \frac{1}{360n^3} + \frac{1}{1260n^5} - \frac{1}{1680n^7} + O(n^{-8})\right) \quad \text{as } n\to\infty.$$

Stolz–Cesàro Lemma. Let $(a_n)_{n\geq 1}$ and $(b_n)_{n\geq 1}$ two sequences of real numbers.

(i) Assume that $a_n\to 0$ and $b_n\to 0$ as $n\to\infty$. Suppose, moreover, that $(b_n)_{n\geq 1}$ is decreasing for all sufficiently large n and
$$\lim_{n\to\infty} \frac{a_{n+1}-a_n}{b_{n+1}-b_n} =: \ell \in \mathbb{R}$$
exists. Then $\lim_{n\to\infty} a_n/b_n$ exists, and moreover, $\lim_{n\to\infty} a_n/b_n = \ell$.

(ii) Assume that $b_n \to +\infty$ as $n\to\infty$ and that $(b_n)_{n\geq 1}$ is increasing for all sufficiently large n. Suppose that
$$\lim_{n\to\infty} \frac{a_{n+1}-a_n}{b_{n+1}-b_n} =: \ell \in \mathbb{R}$$
exists. Then $\lim_{n\to\infty} a_n/b_n$ exists, and moreover, $\lim_{n\to\infty} a_n/b_n = \ell$.

Strong Maximum Principle. Let $f : [a,b] \to \mathbb{R}$ be a twice differentiable convex function such that $f(a) = f(b) = 0$. Then the following alternative holds: either

(i) $f \equiv 0$ in $[a,b]$

or

(ii) $f < 0$ in (a,b), and moreover, $f'(a) < 0$ and $f'(b) > 0$.

Supremum. The *supremum* (or *least upper bound*) of a set $A \subset \mathbb{R}$ is an element $\beta \in \mathbb{R} \cup \{+\infty\}$ that is an upper bound of A and such that no $\beta_0 < \beta$ is an upper bound of A. Notation: $\beta = \sup A$.

Taylor's Formula. Let n be a nonnegative integer and suppose that f is an $(n+1)$-times continuously differentiable function on an open interval $I = (a-\varepsilon, a+\varepsilon)$. Then, for $x \in I$,
$$f(x) = \sum_{k=0}^{n} f^{(k)}(a) \frac{(x-a)^k}{k!} + \int_a^x f^{(n+1)}(t) \frac{(x-a)^n}{n!} dt.$$

Uniformly Continuous Function. A function f is uniformly continuous on a set D if for any $\varepsilon > 0$, there exists $\delta > 0$ such that $|f(x) - f(y)| < \varepsilon$ whenever $x, y \in D$ and $|x - y| < \delta$.

Upper Bound. Let A be a set of real numbers. If there exists $M \in \mathbb{R}$ such that $x \leq M$ for every $x \in A$, we say that A is bounded above, and call M an *upper bound* of A.

Young's Inequality (for numbers). If a, b, p, and q are positive numbers such that $p^{-1} + q^{-1} = 1$, then
$$ab \leq \frac{a^p}{p} + \frac{b^q}{q}.$$

Young's Inequality (for functions). Let $f : [0, +\infty) \to \mathbb{R}$ be a strictly increasing function with a continuous derivative such that $f(0) = 0$. Then for all $a, b \geq 0$,
$$ab \leq \int_0^a f(x) dx + \int_0^b f^{-1}(y) dy.$$

Volterra's Theorem. If two real continuous functions defined on the real axis are continuous on dense subsets of $\mathbb{R}$, then the set of their common continuity points is dense in $\mathbb{R}$, too.

Wallis' Formula. As $n \to \infty$,
$$\frac{2}{1} \cdot \frac{4}{3} \cdots \frac{2n}{2n-1} \cdot \frac{1}{\sqrt{2n+1}} \longrightarrow \sqrt{\frac{\pi}{2}}.$$

Weak Maximum Principle. Let $f : [a,b] \to \mathbb{R}$ be a continuous convex function. Then f attains its maximum on $[a,b]$ either in a or in b. In particular, if $f(a) \leq 0$ and $f(b) \leq 0$, then $f \leq 0$ in $[a,b]$.

Weierstrass's Nowhere Differentiable Function. The continuous function
$$f(x) = \sum_{n=1}^{\infty} b^n \cos(a^n x) \quad (0 < b < 1)$$
is nowhere differentiable, provided $ab > 1 + 3\pi/2$.

Weierstrass's Theorem. Every real-valued continuous function on a closed and bounded interval attains its maximum and its minimum.

References

1. Aigner, M., Ziegler, G.: Proofs from the Book. Springer-Verlag, Berlin Heidelberg (2001)
2. Arnold, V.I.: Evolution processes and ordinary differential equations. Kvant **1986**, No. 2, 13–20 (1986)
3. Bailey, D.F.: Krasnoselski's theorem on the real line. Amer. Math. Monthly **81**, 506–507 (1974)
4. Banerjee, C.R., Lahiri, B.K.: On subseries of divergent series. Amer. Math. Monthly **71**, 767–768 (1964)
5. Bărbălat, I.: Systèmes d'équations différentielles d'oscillations non linéaires. Rev. Roumaine Math. Pures Appl. **IV**, 267–270 (1959)
6. Bethuel, F., Brezis, H., Hélein, F.: Ginzburg-Landau Vortices. Progress in Nonlinear Differential Equations and Their Applications, Vol. 13, Birkhäuser, Boston (1994)
7. Biler P., Witkowski, A.: Problems in Mathematical Analysis. Marcel Dekker, New York (1990)
8. Blomer, V.: The theorem of Green-Tao. Newsletter European Math. Soc. **67**, 13–16 (2008)
9. Boas, R.P.: Inequalities for a collection. Math. Mag. **52**, 28–31 (1979)
10. Boas, R.P.: Counterexamples to l'Hôpital's rule. Amer. Math. Monthly **93**, 644–645 (1986)
11. Boju, V., Funar, L.: The Math Problems Notebook. Birkhäuser, Boston (2007)
12. Brezis, H., Oswald, L.: Remarks on sublinear elliptic equations. Nonlinear Analysis **10**, 55–64 (1986)
13. Brezis, H., Willem, M.: On some nonlinear equations with critical exponents. Preprint (2008)
14. Burnside, R., Convexity and Jensen's inequality. Amer. Math. Monthly **82**, 1005 (1975)
15. Carleman T.: Sur les fonctions quasi-analytiques. Conférences faites au cinquième congrès des mathématiciens scandinaves, Helsinki, 181–196 (1923)
16. Cauchy, A.-L.: Analyse algébrique. Reprint of the 1821 edition. Cours d'Analyse de l'École Royale Polytechnique. Éditions Jacques Gabay, Sceaux (1989)
17. Chapman, R.: Evaluating $\zeta(2)$. Manuscript, 1999 (corrected 2003), 13 pp., available electronically at http://www.secamlocal.ex.ac.uk/~rjc/etc/zeta2.pdf
18. Clouet, J.-F., Després, B., Ghidaglia, J.-M., Lafitte, O.: L'Épreuve de Mathématique en PSI (Concours d'entrée à l'École Polytechnique et à l'École normale supérieure de Cachan 1997). Springer, Berlin, Heidelberg (1998)
19. Costara, C., Popa, D.: Berkeley Preliminary Exams. Ex Ponto, Constanţa (2000)
20. Darboux, G.: Mémoire sur les fonctions discontinues. Ann. Sci. École Norm. Sup. **4**, 57–112 (1875)
21. Darboux, G: Sur la composition des forces en statique. Bull. Sci. Math. **9**, 281–299 (1875)
22. Dieudonné, J.: Abrégé d'histoire des mathématiques. Hermann, Paris (1978)
23. Dini, U.: Sulle serie a termini positivi. Ann. Univ. Toscana **9**, 41–76 (1867)
24. Erickson M., Flowers, J.: Principles of Mathematical Problem Solving. Prentice Hall (1999)

25. Euler, L.: Vollständige Anleitung zur Algebra. Zweyter Theil, Kays. Acad. der Wissenschaften, St. Petersburg (1770); Opera Mathematica. ser. I, vol. 1, B.G. Teubner, Leipzig (1991). English translation: Elements of Algebra. Springer, New York (1984)
26. Euler, L.: De summis serierum reciprocarum (On the sum of series of reciprocals). Comm. Acad. Sci. Petrop. **7**, 123–134 (1734/35)
27. Feeman, T.G., Marrero, O.: The right-hand derivative of an integral. Amer. Math. Monthly **109**, 565–568 (2002)
28. Froda, A.: Sur la distribution des propriétés de voisinage des fonctions de variables réelles. Bull. Math. Soc. Roum. Sci. **32**, 105–202 (1929)
29. Gelbaum, B.: Problems in Analysis. Problem Books in Mathematics, Springer, Berlin Heidelberg (1982)
30. Gelbaum, B., Olmsted, J.M.H.: Counterexamples in Analysis. Holden-Day, Inc., San Francisco, London, Amsterdam (1964)
31. Ghorpade, S., Limaye, B.: A Course in Calculus and Real Analysis. Springer, New York (2006)
32. Gelca, R., Andreescu, T.: Putnam and Beyond. Springer, New York (2007)
33. Giaquinta, M., Modica, G.: Mathematical Analysis: Functions of One Variable. Birkhäuser, Boston (2003)
34. Grabiner, J.V.: The Origins of Cauchy's Rigorous Calculus. MIT Press, Cambridge (1981)
35. Green, B., Tao., T.: The primes contain arbitrarily long arithmetic progressions. Ann. of Math. **167**, 481–547 (2008)
36. Hairer, E., Wanner, G.: Analysis by Its History. Undergraduate Texts in Mathematics. Springer-Verlag, New York (1996)
37. Halmos, P.: A Hilbert Space Problem Book. Van Nostrand, New York (1967)
38. Halperin, I.: A fundamental theorem of the Calculus. Amer. Math. Monthly **61**, 122–123 (1954)
39. Hamel, G: Eine Basis aller Zahlen und die unstetigen Lösungen der Funktionalgleichung $f(x+y) = f(x) + f(y)$. Math. Ann. **60**, 459–462 (1905)
40. Hardy, G.H.: Note on a theorem of Hilbert. Math. Zeitschr. **6**, 314–317 (1920)
41. Hardy, G.H.: A Course of Pure Mathematics. Cambridge University Press (1952)
42. Hardy, G.H., Littlewood, J.E.: Some integral inequalities connected with Calculus of Variations. Quart. J. Math. Oxford **3**, 241–252 (1932)
43. Hardy, G.H., Littlewood, J.E., Pólya, G.: Inequalities. Cambridge University Press (1967)
44. Havil, J.: Gamma. Princeton University Press (2003)
45. Heine, E.: Die Elemente der Funktionenlehre. J. Reine Angew. Math. (Crelle) **74**, 172–188 (1872)
46. Helmberg, G.: A construction concerning $(l^p)' \subset l^q$. Amer. Math. Monthly **111**, 518–520 (2004)
47. Herzog, G.: C^1–solutions of $x = f(x')$ are convex or concave. Amer. Math. Monthly **105**, 554–555 (1998)
48. Hille, E.: Lectures on Ordinary Differential Equations. Addison-Wesley Publ. Comp., Reading Massachussets (1969)
49. Hofbauer, J.: A simple proof of $1 + \frac{1}{2^2} + \frac{1}{3^2} + \cdots = \frac{\pi^2}{6}$. Amer. Math. Monthly **109**, 196–200 (2002)
50. Jacobson, B.: On the mean value theorem for integrals. Amer. Math. Monthly **89**, 300–301 (1982)
51. Jensen, J.L.W.V.: Sur les fonctions convexes et les inégalités entre les valeurs moyennes. Acta Math. **30**, 175–193 (1906)
52. Jost, J.: Postmodern Analysis. Springer, Berlin Heidelberg New York (1998)
53. Jungck, G.: An alternative to the integral test. Math. Magazine **56**, 232–235 (1983)
54. Kaczor, W.J., Nowak, M.T.: Problems in Mathematical Analysis I: Real Numbers, Sequences and Series. Student Mathematical Library, Vol. 4, American Mathematical Society, Providence, RI, (2000)

55. Kaczor, W.J., Nowak, M.T.: Problems in Mathematical Analysis II: Continuity and Differentiation. Student Mathematical Library, Vol. 12, American Mathematical Society, Providence, RI, (2001)
56. Kedlaya, K.: Proof of a mixed arithmetic-mean, geometric-mean inequality. Amer. Math. Monthly **101**, 355–357 (1994)
57. Kedlaya, K.: A weighted mixed-mean inequality. Amer. Math. Monthly **106**, 355–358 (1999)
58. Kline, M.: Mathematics Thought from Ancient to Modern Times. Oxford University Press (1972)
59. Knopp, K.: Theory and Application of Infinite Series, 2nd ed. Dover, Mineola, NY (1990)
60. Krantz, S.G.: A Handbook of Real Variables. With Applications to Differential Equations and Fourier Analysis. Birkhäuser (2004)
61. Krasnoselski, M.A.: Two remarks on the method of successive approximations. Math. Nauk (N. S.) **10**, 123–127 (1955)
62. Kummer, E.E.: Über die Convergenz und Divergenz der unendlichen Reihen. Journal für die Reine und Angewandte Mathematik **13**, 171–184 (1835)
63. Landau, E.: Einige Ungleichungen für Zweimal Differenzierbare Funktionen. Proc. London Math. Soc. **13**, 43–49 (1914)
64. Landau, E.: A note on a theorem concerning series of positive terms. J. London Math. Soc. **1**, 38–39 (1926)
65. Landau, E.: Elementary Number Theory. Chelsea, 2nd ed. (1966)
66. Larson, L.: Problem-Solving Through Problems. Problem Books in Mathematics, Springer, New York (1983)
67. Lax, P.: A curious functional equation. J. d'Analyse Mathématique **105**, 383–390 (2008)
68. Lefter, C., Rădulescu, V.: Minimization problems and corresponding renormalized energies. Diff. Integral Equations **9**, 903–918 (1996)
69. Lenstra, H.W.: Solving the Pell equation. Notices Amer. Math. Soc. **49**, 182–192 (2002)
70. Lewin, J., Lewin, M.: A simple test for the nth term of a series to approach zero. Amer. Math. Monthly. **95**, 942 (1988)
71. Lupu, T.: Probleme de Analiză Matematică. Calculul Integral. Gil Press, Zalău (1996)
72. Lüroth, J.: Bemerkung über gleichmässige Stetigkeit. Math. Ann. **6**, 319–320 (1873)
73. Makarov, B.M., Goluzina, M.G., Lodkin, A.A., Podkorytov, A.N., Selected Problems in Real Analysis. Translated from the Russian by H.H. McFaden. Translations of Mathematical Monographs, Vol. 107, American Mathematical Society, Providence, RI, (1992)
74. Maligranda, L.: A simple proof of the Hölder and the Minkowski inequality. Amer. Math. Monthly **102**, 256–259 (1995)
75. Mashaal, M.: Bourbaki. A Secret Society of Mathematicians. American Mathematical Society, Providence, RI, (2006)
76. Mawhin, J.: Analyse. Fondements, Techniques, Évolution. DeBoeck Université, Paris Bruxelles (1997)
77. Mitrinović, D.S.: Analytic Inequalities. Springer, Heidelberg (1970)
78. Mond, B., Pečarić, J.: A mixed means inequality. Austral. Math. Soc. **23**, 67–70 (1996)
79. Monier, V.: Analyse. Dunod, Paris (1990)
80. Neuser, D.A., Wayment, S.G.: A note on the intermediate value property. Amer. Math. Monthly **81**, 995–997 (1974)
81. Newman, D.: A Problem Seminar. Problem Books in Mathematics, Springer, New York (1982)
82. Nicolescu, M.: Analiză Matematică II. Editura Tehnică, București (1958)
83. Niculescu, C.P.: Fundamentele Analizei Matematice. Analiza pe Dreapta Reală. Editura Academiei, București (1996)
84. Niven, I., Zuckerman, H.S., Montgomery, H.L.: An Introduction to the Theory of Numbers. John Wiley & Sons, New York (1991)
85. Olivier, L.: Remarques sur les séries infinies et leur convergence. J. Reine Angew. Math. **2**, 31–44 (1827)

86. Olsen, L.: A new proof of Darboux's theorem. Amer. Math. Monthly **111**, 713–715 (2004)
87. Opic, B., Kufner, A.: Hardy-type Inequalities. Pitman Res. Notes Math. Ser. 219, Longman Scientific and Technical, Harlow, UK (1990)
88. Oxtoby, J.: Horizontal chord theorems. Amer. Math. Monthly **79**, 468–475 (1972)
89. Poincaré, H.: Sur les équations aux dérivées partielles de la physique mathématique. Amer. J. Math. **12**, 211–294 (1890)
90. Pólya, G.: How to Solve It. A New Aspect of Mathematical Method. Princeton University Press (1945)
91. Pólya, G., Szegő, G.: Problems and Theorems in Analysis I. Springer, Berlin, Heidelberg (1972)
92. Pólya, G., Szegő, G.: Problems and Theorems in Analysis II. Springer, Berlin, Heidelberg (1976)
93. Pringsheim, A.: Zur theorie der ganzen transzendenten Funktionen. Sitzungsberichte der Mathematisch-Physikalischen Klasse der Königlich Bayerischen Akademie der Wissenschaften **32**, 163–192 (1902)
94. Protter, M., Weinberger, H.: Maximum Principles in Differential Equations. Prentice-Hall, Inc., Englewood Cliffs (1967)
95. Prus-Wiśniowski, F.: A refinement of Raabe's test. Amer. Math. Monthly **115**, 249–252 (2008)
96. Rajwade, A.R., Bhandari, A.K.: Surprises and Counterexamples in Real Function Theory. Texts and Readings in Mathematics, Vol. 42, Hindustan Book Agency, New Delhi, (2007)
97. Rassias, T.: Survey on Classical Inequalities. Kluwer Acad. Publ., Dordrecht (2000)
98. Rădulescu, V.: A Liouville-type property for differential inequalities. SIAM Problems and Solutions. Problem 06-005 (2005)
99. Rădulescu, V.: Abrikosov lattices in superconductivity. SIAM Problems and Solutions. Problem 06-006 (2006)
100. Richmond, D.E.: An elementary proof of a theorem of calculus. Amer. Math. Monthly **92**, 589–590 (1985)
101. Robertson, R.: An improper application of Green's theorem. College Math. J. **38**, 142–145 (2007)
102. Rudin, W.: Principles of Mathematical Analysis. McGraw-Hill, New York (1976)
103. Shklyarsky, D.O., Chentsov, N.N., Yaglom, I.M.: Selected Problems and Theorems in Elementary Mathematics. English translation, Mir Publishers (1979)
104. Sohrab, H.: Basic Real Analysis. Birkhäuser, Boston (2003)
105. Souza, P.N., Silva, J.N.: Berkeley Problems in Mathematics. Problem Books in Mathematics, Springer, New York, Berlin, Heidelberg (1998)
106. Sperb, R.: Maximum Principles and Their Applications. Mathematics in Science and Engineering, Vol. 157, Academic Press, New York London (1981)
107. Steele, J.M.: The Cauchy-Schwarz Master Class. An Introduction to the Art of Mathematical Inequalities. MAA Problem Books Series. Mathematical Association of America, Washington, DC; Cambridge University Press, Cambridge (2004)
108. Stolz, O.: Ueber die Grenzwerthe der Quotienten. Math. Ann. **15**, 556–559 (1879)
109. Stromberg, K.: An Introduction to Classical Real Analysis. Wadsworth International Group, Belmont, CA (1981)
110. Székely, G.: Contests in Higher Mathematics. Miklós Schweitzer Competitions 1962–1991. Problem Books in Mathematics, Springer, New York (1996)
111. Tolpygo, A.K.: Problems of the Moskow Mathematical Competitions. Prosveshchenie, Moskow (1986)
112. Vasil'ev, N.B., Egorov, A.A.: The Problems of the All-Soviet-Union Mathematical Competitions. Nauka, Moskow (1988)
113. Volterra, V.: Alcune osservasioni sulle funzioni punteggiate discontinue. Giornale di Matematiche **19**, 76–86 (1881)

114. Young, G.S.: The linear functional equation. Amer. Math. Monthly **65**, 37–38 (1958)
115. Ward, M.: A mnemonic for γ. Amer. Math. Monthly **38**, 522 (1931)
116. Wildenberg, G.: Convergence-Preserving Functions. Amer. Math. Monthly **95**, 542–544 (1988)
117. Willem, M.: Analyse fonctionnelle élémentaire. Cassini, Paris (2003)
118. Willem, M.: Principes d'analyse fonctionnelle. Cassini, Paris (2007)

Index

(a, b), open interval, 419
$S(f; \Delta, \xi)$, Riemann sum associated to the function f, 326
$S^+(f; \Delta)$, upper Darboux sum associated to the function f, 326
$S_-(f; \Delta)$, lower Darboux sum associated to the function f, 326
$[a, b]$, closed interval, 419
e, base of natural logarithms, 4
$\int_a^b f(x)dx$, Riemann integral of the function f, 326
$\lim_{n \to \infty} a_n$, limit of the sequence $(a_n)_{n \geq 1}$, 4
$\lim_{x \to x_0+} f(x)$, limit to the right of the function f at x_0, 116
$\lim_{x \to x_0-} f(x)$, limit to the left of the function f at x_0, 116
$\lim_{x \to x_0} f(x)$, limit of the function f at x_0, 116
$\liminf_{n \to \infty} a_n$, limit infimum of the sequence $(a_n)_{n \geq 1}$, 5
$\limsup_{n \to \infty} a_n$, limit supremum of the sequence $(a_n)_{n \geq 1}$, 5
$\arccos x, arccosine at x \in [-1, 1]$, 195
$\arcsin x, arcsine at x \in [-1, 1]$, 195
$\arctan x, arctangent at x \in \mathbb{R}$, 196
$\cosh x, hyperbolic cosine at x \in \mathbb{R}$, 196
$\cos x, cosine at x \in \mathbb{R}$, 195
$\sinh x, hyperbolic sine at x \in \mathbb{R}$, 196
$\sin x, sine at x \in \mathbb{R}$, 195
$\tanh x, hyperbolic tangent at x \in \mathbb{R}$, 196
$\tan x, tangent at x \in \mathbb{R}$, 195
$\sum_{n=1}^{\infty} a_n$, 60
$f'(x_0)$, derivative of the function f at x_0, 184
$f'(x_0+)$, right-hand derivative of the function f at x_0, 185
$f'(x_0-)$, left-hand derivative of the function f at x_0, 185

Abel Prize, 30, 217
Abel's summation formula, 63, 421
Abel's test, 63, 421
Abel, N.H., 30
Abrikosov lattice, 303
Abrikosov, A., 303
absolute value, 210
accumulation point, 4, 41, 45, 115, 129, 420, 421
additive function, 163
additivity property of the integral, 328
affine function, 263
algorithm
    Euclidean, 39
Allen–Cahn system, 367
alternative
    Fredholm, 371
AM–GM inequality, 7, 8, 77, 107, 295, 421
anagram, 183
angular point, 185
anni mirabiles, 431
antiderivative, 313, 319, 320, 328, 336, 412, 428
antiderivative test, 321
Archimedes, 38, 373
Aristotle, 3
arithmetic mean test, 83
arithmetic progression, 34, 52, 55, 103, 340
arithmetic–geometric means inequality, 7, 8, 77, 107, 264, 265, 421
arithmetic–harmonic means inequality, 421
Arnold, V.I., 238, 402
astrodynamics, 37
asymptotic estimate, 123, 195
asymptotic property, 4
asymptotic regularity, 158
at most countable set, 171, 418, 425

443

Atiyah, M.F., 30
axiom of choice, 163

Baire lemma, 152, 153
Banach fixed point theorem, 310, 421
Banach space, 90, 239
Barbălat's lemma, 394, 422
Barone theorem, 158
Barrow, I., 238
Basel problem, 67
basis
  orthonormal, 364
Bernoulli numbers, 67, 369
Bernoulli's inequality, 422
Bernoulli, J., 67, 196
Bertrand series, 81, 422
best local linear approximation, 183
bijective function, 197, 418
binomial coefficients, 97, 196
Boas, R.P., 174
Bolzano, B., 4
Bolzano–Weierstrass theorem, 5, 17, 146, 147, 420, 422
boundary point, 420, 422
bounded function, 132, 165
bounded sequence, 4–6, 11, 14, 22, 26, 27, 37, 422, 431
bounded set, 310
Bourbaki, 222, 424
Brezis, H., 249, 374
Brouwer fixed-point theorem, 154, 218, 301, 422
Brouwer, L.E.J., 154, 218
Browder, F., 158
bump function, 216

Cahn–Hilliard system, 367
Cantor principle, 154
Carleman's inequality, 30, 45–51, 382, 422
Carleman, T., 47
Carleson, L., 30
Cauchy functional equation, 163, 217
Cauchy mean value theorem, 192, 255
Cauchy sequence, 5, 248, 423
Cauchy's condensation criterion, 61, 422
Cauchy's criterion, 5
Cauchy's criterion for infinite products, 65, 422
Cauchy's criterion for sequences, 157, 158, 422
Cauchy's criterion for series, 61, 422
Cauchy's theorem, 423
Cauchy, A.-L., 192

Cauchy–Schwarz inequality, 78, 91, 93, 284, 333, 334, 337, 353, 376, 379, 380, 423
Cavalieri, B., 373
celestial mechanics, 37
Cesàro lemma, 89, 423
Cesàro, E., 7
change of variable, 117
change of variables in Riemann integral, 329, 423
characteristic function, 118
chord
  horizontal, 172
Clay Mathematics Institute, 7
closed interval, 419
closed set, 140, 310, 420, 421, 423
cobweb, 300, 302
combinatorics, 265
compact set, 140, 146
compact support, 216
complete metric space, 152
completely monotonic function, 191
concave function, 221, 229, 246, 263, 279, 280, 423
concentration of mass, 407, 424
conjecture
  Erdős–Turán, 103
constancy of sign, 117
constant
  Euler, 65, 66, 112, 342
  Lipschitz, 274, 430
continued fraction, 38
continuous function, 139, 423
continuous process, 3
contraction, 310, 421, 423
contraction principle, 310, 421
contractive function, 161, 310
convergence-preserving function, 108
convergent sequence, 422, 431
convex function, 204, 206, 208, 209, 225, 229, 235, 246, 263–265, 268, 273, 274, 283, 407, 423, 433
  support line, 275, 297, 298
convex set, 266, 267
cooperative recurrence, 19
Coriolis test, 65, 423
countable set, 115, 151, 208, 209, 222, 252, 418, 423, 424
countable union, 420
criterion
  Cauchy, 5, 61, 157, 158, 422
  Cauchy's condensation, 61, 422
  Darboux, 424
  Heine, 117, 427
critical point theory, 187

Index 445

Croft lemma, 174, 424
cusp point, 185

d'Alembert equation, 371
d'Alembert, J., 62
Darboux sum, 282, 424
Darboux's criterion, 424
Darboux's theorem, 192–194, 226, 227, 314, 424
Darboux, G., 143, 192
decreasing sequence, 7
decreasing function, 21, 148, 201, 204, 205, 210, 229, 244, 245, 255, 285, 291, 431
decreasing sequence, 3–5, 22, 40, 41, 48, 53
Denjoy theorem, 208, 209, 424
Denjoy, A., 209, 222
Denjoy–Bourbaki theorem, 222, 424
dense subset, 152, 153
derivative
    discontinuous, 143
    one-sided, 185, 208, 209, 236, 252, 424
    Schwartzian, 258, 434
    symmetric, 224
    to the right, 222, 237
devil's staircase, 191
diagonal line, 300
difference equation, 18, 38
difference quotient, 183
differential equation
    nonlinear, 244
differential calculus, 183
differential equation, 235, 238, 244–251
    linear, 244, 250, 251
    nonlinear, 245–250
    singular, 245
differential inequality, 223, 235–238, 242, 256, 257
differentiation inverse function theorem, 197, 424
diffraction, 397
digital signal processing, 123
Dini integral, 372
Dini, U., 62
Diophantine equation, 38, 432
Dirac delta function, 414
Dirac sequence, 407, 424
Dirac, P.A.M., 407
direct image, 140
direct image of a set, 417
Dirichlet kernel, 405
Dirichlet's function, 198, 326
Dirichlet's principle, 45, 432
Dirichlet's test, 63, 84, 424
discontinuity
    of the first kind, 141, 151, 277, 425
    of the second kind, 141, 151, 318, 425
discontinuity point, 328, 429
discontinuous derivative, 143
discrete process, 3
discrete set, 420
divergent sequence, 6, 8, 11, 14, 26, 45, 46, 53
division of an interval, 325
double periodic solution, 367
du Bois-Reymond lemma, 370

eccentric anomaly, 37
eigenfunction, 364
eigenvalue, 364
elliptic integral, 19
epigraph, 265
equation
    d'Alembert, 371
    difference, 18
    differential, 238, 251
    diophantine, 38, 432
    Ginzburg–Landau, 245, 248
    gravity, 250
    Kepler, 37
    Pell, 38, 432
equidistant division, 327
equilateral triangle, 303
Erdős, P., 103
Erdős–Szekeres theorem, 53
estimate
    asymptotic, 123
Euclidean algorithm, 39
Euler constant, 65, 66, 112, 342
Euler's formula, 67, 425
Euler's function, 19
Euler's gamma function, 66, 390, 425
Euler, L., 19, 38, 67
even function, 133
expansion
    Taylor, 195
extremum point, 433

Fermat's theorem, 187, 193, 201, 226, 332, 425
Fermat, P., 187
Fibonacci recurrence, 18
Fibonacci sequence, 18, 20, 23, 425
Fibonacci, L., 18
Fields Medal, 103
finite differences, 19
finite set, 418
first comparison test, 61, 74, 425
first mean value theorem for integrals, 329, 425

fixed point, 154, 156–160, 177–180, 218, 300, 301, 310, 422, 428
formula
   Abel's summation, 63, 421
   change of variables in Riemann integral, 329, 423
   Euler, 67, 425
   integration by parts, 314, 328, 339, 428
   Newton–Leibniz, 328, 336, 396, 397, 432
   Stirling, 4, 69, 96, 387, 414, 435
   substitution, 314
   Taylor, 239, 257, 259, 276, 277, 435
   Viète, 88
   Wallis, 414, 436
Fredholm alternative, 371
French Academy of Sciences, 238
Fresnel integrals, 397
Fresnel, A., 397
Friedrich's inequality, 383
Froda's theorem, 152, 425
Froda, A., 152
function
   sinc, 123
   additive, 163
   affine, 263
   antiderivative, 313
   arc cosine, 195
   arc sine, 195
   arc tangent, 195
   best local linear approximation, 183
   bijective, 197, 418
   bounded, 132, 165
   bump, 216
   characteristic, 118
   completely monotonic, 191
   concave, 221, 229, 246, 263, 279, 280, 423
   continuous, 139, 423
   contraction, 310, 421, 423
   contractive, 161, 310
   convergence preserving, 108
   convex, 204, 206, 208, 209, 225, 229, 235, 246, 263–265, 268, 273, 274, 283, 407, 423, 433
     support line, 275, 297, 298
   cosine, 195
   decreasing, 21, 148, 201, 204, 205, 210, 229, 244, 245, 255, 285, 291, 431
   Dirac delta, 414
   Dirichlet, 198, 326
   Euler, 19
   even, 133
   exponential, 195
   gamma, 66, 390, 425
   hyperbolic
     cosine, 196
     sine, 196
     tangent, 196
   increasing, 21, 23, 24, 189, 200, 203–207, 211, 221, 223–225, 227, 229, 238, 242, 245, 257, 270, 273, 279, 280, 285, 286, 289, 428, 431
   inverse, 143, 197
   limit, 115, 430
   Lipschitz, 161, 187, 189, 207, 208, 274, 430
   locally bounded, 117
   locally Lipschitz, 274
   logarithm, 195
   monotone, 164, 431
   nondecreasing, 431
   nonexpansive, 310, 432
   nonincreasing, 431
   nowhere differentiable, 186, 214, 436
   one-to-one, 147, 152, 166, 277, 418
   periodic, 120, 125, 172, 173
   primitive, 313
   range, 140
   Riemann, 152, 327
   Riemann integrable, 326–328, 429
   Riemann zeta, 6, 67, 77, 112, 434
   sine, 195
   strictly concave, 263
   strictly convex, 263
   subadditive, 210
   subexponential, 258
   subharmonic, 364
   superlinear, 132, 258, 287
   surjective, 150, 171, 258, 417
   tangent, 195
   totally discontinuous, 163, 326
   trigonometric, 228
   uniformly continuous, 143, 146, 180, 436
   with compact support, 216
functional equation, 164, 167, 178, 179
   Cauchy, 163, 179, 217
fundamental theorem of calculus, 328, 426

gamma function, 66, 390, 425
Gauss, C., 19
Gauss test, 63, 67, 75, 87, 426
Gaussian integral, 396
Gelfand, I., 103
generalized AM–GM inequality, 298, 426
generalized arithmetic–geometric means inequality, 298, 426
generalized means inequality, 264
geometric means inequality, 8
geometric progression, 40, 52
geometric series, 39

Index 447

Ginzburg, V., 303
Ginzburg–Landau equation, 245
Ginzburg–Landau theory, 237
global property, 140
golden ratio, 18
graph of a function, 408
gravity equation, 250
greatest lower bound, 428
Green, B., 103
Gronwall's inequality, 410, 426
Gronwall's lemma, 249, 381, 382
growth
    superlinear, 242

Hölder's inequality, 107, 296, 427
Hamel base, 163
Hardy's inequality, 45–47, 50, 383–385, 426
Hardy's notation, xx
Hardy, G.H., 238
Hardy–Littlewood theorem, 256
harmonic series, 6, 61, 66, 93, 99, 342
Hauptlehrsatz, 140
Hawking, S., 183
Heine's criterion, 117, 427
Heine, E., 146
Heine–Borel theorem, 427
Hilbert problems, 163
Hilbert space, 364, 366
Hilbert's double series inequality, 382
Hilbert's fifth problem, 163
Hillam fixed-point theorem, 159
Hincin, A., 209
horizontal chord, 172
horizontal chord theorem, 173, 174, 428
Huygens, C., 238
hypergeometric series, 75

identity
    Abel, 421
    Parseval, 365, 366
image
    direct, 417
    inverse, 417
improper integral, 373, 395, 428
increasing sequence, 6
increasing function, 21, 23, 24, 189, 200, 203–207, 211, 221, 223–225, 227, 229, 238, 242, 245, 257, 270, 273, 279, 280, 285, 286, 289, 428, 431
increasing function theorem, 189, 428
increasing sequence, 3–5, 7, 11, 35, 40–42, 48, 53, 55, 207
indeterminate form, 117
inequality

arithmetic–geometric means, 7, 8, 77, 107, 264, 265, 295, 421
arithmetic–harmonic means, 421
Bernoulli, 422
Carleman, 30, 45–51, 382, 422
Cauchy–Schwarz, 78, 91, 93, 284, 333, 334, 337, 353, 376, 379, 380, 423
differential, 223, 235–238, 242, 256, 257
Friedrich, 383
generalized arithmetic–geometric means, 298, 426
generalized means, 264
Gronwall, 249, 381, 382, 410, 426
Hölder, 107, 296, 427
Hardy, 45–47, 50, 383–385, 426
harmonic mean, 8
Hilbert, 281
Hilbert's double series, 382
Jensen, 264, 265, 268, 270, 272, 273, 283, 285, 297, 428
Knopp, 47
Kolmogorov, 240, 428
Landau, 238, 240, 241, 429
Landau–Kolmogorov generalized, 241, 429
Lyapunov, 388, 430
mean value, 222, 269, 280
Minkowski, 296, 430, 431
Poincaré, 366, 379, 383, 432
Popoviciu, 287
power mean, 298, 432
Sobolev, 383
Young, 211, 265, 295, 297, 377, 436
infimum, 428
infinite product
    absolutely convergent, 64
infinite process, 115
infinite product, 63
    Cauchy's criterion, 65, 422
    convergent, 63
    Coriolis test, 65, 423
    divergent, 63
    unconditionally convergent, 64
infinite set, 418
information theory, 286
initial value problem, 354
integral
    Dini, 372
    elliptic, 19
    Fresnel, 397
    Gaussian, 396
    improper, 373, 395, 428
    probability, 396
    Riemann, 325, 373
integral of a function, 326

integral test for series, 374, 403, 428
integration by parts, 314, 328, 339, 428
interior point, 273, 276, 420, 428
intermediate value property, 155
intermediate value property, 143, 144, 147–151, 155, 158, 167, 168, 170, 172, 173, 179, 355, 424, 428
interpolation theory, 19
interval
  closed, 419
  open, 419
inverse function, 143, 197
inverse image, 140
inverse image of a set, 417
Ishikawa iteration process, 90
isolated point, 420, 428
iterative property, 148, 158

Jensen's inequality, 264, 265, 268, 270, 272, 273, 283, 285, 297, 428
Jensen, J., 264

Kamerlingh Ones, H., 303
Kepler's equation, 37
Kepler's second law, 37
Kepler, J., 37
kernel, 400
  Dirichlet, 405
Knaster-fixed point theorem, 154, 428
Knopp inequality, 47
Kolmogorov's inequality, 240, 428
Kolmogorov, A.N., 239
Krasnoselski's property, 160
Kronecker density theorem, 45
Kronecker theorem, 44, 429
Kronecker, L., 44
Kummer's test, 62, 429
Kummer, E., 62

L'Hôpital's rule, 196, 197, 202, 205, 214, 227, 229, 232, 427
Lévy, P., 173
Lagrange mean value theorem, 188, 189, 191–193, 218–223, 226–228, 231, 282, 289, 301, 302, 341, 342, 429
Lagrange, J.-L., 19
Laguerre's theorem, 257
Lalescu, T., 9
Landau's inequality, 238, 240, 241, 429
Landau's notations, xx, 195, 392
Landau, L., 103
Landau–Kolmogorov generalized inequality, 241, 429
Laplace transform, 400

lateral limit, 116
lattice
  Abrikosov, 303
law
  Kepler's second, 37
  of planetary motion, 37
  of universal gravitation, 37
Lax, P., 30, 217
least upper bound, 435
Lebesgue's theorem, 328, 429
Lebesgue, H., 328
Leggett, A., 303
Leibniz's test, 63, 96, 97, 395, 429
lemma
  Baire, 152, 153
  Barbălat, 394, 422
  Cesàro, 89, 423
  Croft, 174, 424
  du Bois-Reymond, 370
  Gronwall, 249, 381, 382
  Riemann–Lebesgue, 338, 400–402, 433
  rising sun, 181
  Stolz–Cesàro, 7–9, 13, 22, 29, 121, 196, 435
  straddle, 234
  three chords, 267
length of a curve, 406
limit
  lateral, 116
  to the left, 116, 139
  to the right, 116, 139
  uniqueness, 116
limit comparison test, 430
limit infimum, 15, 429
limit of a function, 115, 430
limit point, 17, 18
limit supremum, 15, 430
linear recurrence, 10, 24
Lipschitz condition, 161, 162, 178, 430
Lipschitz constant, 274, 430
Lipschitz function, 161, 187, 189, 207, 208, 274, 430
liquid helium, 303
Littlewood, J.E., 238
local maximum point, 187, 201, 202, 246, 256
local minimum point, 187, 201, 202
local property, 140
locally bounded function, 117
locally Lipschitz function, 274
logarithmic convexity, 246
logarithmic mean, 270
Lorentz mean, 269
low temperatures, 303

lower bound, 430
lower Darboux integral, 327
Lyapunov's inequality, 388, 430

Maclaurin
　polynomial, 194
　series, 194
mass
　concentration, 407, 424
mathematical induction, 18
maximum principle, 235–237, 256
　strong, 235, 435
　weak, 235, 436
mean value inequality, 222, 269, 280
method
　Newton, 258, 259, 432
　Picard, 300, 302
metric space
　complete, 152
Millennium Problems, 7
Minkowski's inequality, 296, 430, 431
monotone convergence theorem, 5, 431
monotone function, 164, 431
monotone sequence, 5, 42, 53

Nagumo theorem, 411
natural base, 71
neighborhood, 4, 42, 116, 117, 127, 420, 431
nested intervals theorem, 5, 289, 431
Newton quotient, 184
Newton's binomial, 71, 101, 131, 292, 431
Newton's iteration, 432
Newton's method, 258, 259, 432
Newton, I., 19
Newton–Leibniz formula, 328, 336, 396, 397, 432
Nobel Prize, 30, 103, 303, 407
nondecreasing sequence, 3, 43
nonexpansive function, 310, 432
nonincreasing sequence, 3, 27
nonlinear differential equation, 244, 246
nonlinear recurrence, 19
norm, 239
normed vector space, 222, 424
notations
　Hardy, xx
　Landau, xx, 195, 392
numbers
　of Bernoulli, 67, 369

one-side derivative, 185, 208, 209, 236, 252, 424
one-to-one function, 147, 152, 166, 277
open interval, 419

open set, 140, 420, 432
orthonormal basis, 364
oscillation, 142
Osgood property, 152, 432
Oswald, L., 249

Pólya, G., 50
paradox
　Zeno, 39
paradox of dichotomy, 3
Parseval's identity, 366
Parseval's identity, 365
partial sum of a series, 60
Pascal, B., 431
Pell equation, 38, 432
Pell, J., 38
periodic function, 120, 125, 172, 173
permutation, 42
phase transition phenomena, 367
Pi Day, 60
Picard convergence theorem, 301, 302, 310, 432
Picard method, 300, 302
Picard sequence, 156, 300–302, 310, 432
Picard, E., 156
pigeonhole principle, 45, 432
pinching principle, 150, 432
Poincaré's inequality, 366, 379, 383, 432
Poincaré, H., 187
point
　accumulation, 4, 41, 45, 115, 129, 420, 421
　angular, 185
　boundary, 420, 422
　cusp, 185
　fixed, 154, 156, 160, 300, 301, 310
　interior, 273, 276, 420, 428
　isolated, 420, 428
　local maximum, 187, 201, 202, 246, 256
　local minimum, 187, 201, 202
　relative extremum, 433
　strict local minimum, 277
polynomial recurrence, 40
Popescu, C., 14
Popoviciu's inequality, 287
Popoviciu, T., 287
power mean, 269
power mean inequality, 298, 432
primitive, 313
principle
　Cantor, 154
　contraction, 310, 421
　Dirichlet, 45, 432
　pigeonhole, 45, 432
　pinching, 150, 432

racetrack, 189, 289, 433
uncertainty, 384
uniform boundedness, 91
Pringsheim, A., 299
probability integral, 396
probability theory, 4
problem
　Basel, 67
　rabbit, 18
　Sturm–Liouville, 362
product
　infinite, 63
progression
　arithmetic, 34, 52, 55, 103, 119, 340
　geometric, 40, 52
property
　asymptotic, 4
　global, 140
　intermediate value, 158
　intermediate value, 149, 167
　intermediate value, 143, 144, 147, 148, 150, 151, 155, 158, 168, 170, 172, 173, 179, 355, 424, 428
　iterative, 148, 158
　Krasnoselski, 160
　local, 140
　logarithmic convexity, 246
　Osgood, 152, 432

quadratic recurrence, 10

Raabe's test, 63, 75, 433
rabbit problem, 18
racetrack principle, 189, 289, 433
Rademacher theorem, 207, 274, 433
Rademacher, H., 207
range of a function, 140, 417
ratio test, 62, 71, 75, 433
recurrence
　cooperative, 19
　Fibonacci, 18
　linear, 10, 24
　nonlinear, 19
　polynomial, 40
　quadratic, 10
refinement of a division, 325
regular n–gon, 122
relative extremum point, 433
removable singularity, 123
renormalized Ginzburg–Landau energy, 303
Riemann discontinuous function, 143
Riemann function, 67, 77, 112, 152, 327, 434
Riemann hypothesis, 6
Riemann integrable function, 326–328, 429

Riemann integral, 325, 373
Riemann series, 93
Riemann zeta function, 6
Riemann, B., 6, 67
Riemann–Lebesgue lemma, 338, 400–402, 433
right derivative, 222, 237
rising sun lemma, 181
Rolle theorem, 188, 218, 225, 227, 388, 408, 413, 434
Rolle theorem (polar form), 188, 434
Rolle, M., 188
root test, 62, 74, 434
rotation, 303
Royal Society, 183
Rudin, W., 386
rule
　L'Hôpital's, 196, 197, 202, 205, 214, 227, 229, 232, 427

sawtooth curve, 153
scalar product, 366
Scandinavian Congress of Mathematics, 47
Schwartzian derivative, 258, 434
second comparison test, 74, 434
second mean value theorem for integrals, 329, 386, 405, 406, 434
sequence
　bounded, 4–6, 11, 14, 22, 26, 27, 37, 422, 431
　Cauchy, 5, 248, 423
　Cauchy's condensation criterion, 157
　convergent, 422, 431
　decreasing, 3–5, 7, 22, 40, 41, 48, 53
　Dirac, 407, 424
　divergent, 6, 8, 11, 14, 26, 45, 46, 53
　Fibonacci, 18, 20, 23, 425
　increasing, 3–7, 11, 35, 40–42, 48, 53, 55, 207
　monotone, 5, 42, 53
　monotone convergence theorem, 5
　nondecreasing, 3, 43
　nonincreasing, 3, 27
　Picard, 156, 300–302, 310, 432
　successive approximations, 156, 300
series
　Abel's test, 63, 421
　arithmetic mean test, 83
　Bertrand, 81, 422
　binomial coefficients, 97
　Cauchy's condensation criterion, 61, 422
　Cauchy's criterion, 61, 422
　convergent, 60
　Dirichlet's test, 63, 84, 424
　divergent, 60

Index 451

first comparison test, 61
first comparison test, 74, 425
Gauss test, 63, 67, 75, 87, 426
geometric, 39
harmonic, 6, 61, 66, 93, 99, 342
hypergeometric, 75
integral test, 374
Kummer's test, 62, 429
Leibniz's test, 63, 96, 97, 395, 429
limit comparison test, 430
Maclaurin, 194
partial sum, 60
Raabe's test, 63, 75, 433
ratio test, 62, 71, 75, 433
Riemann, 93
root test, 62, 74, 434
second comparison test, 74, 434
set
  at most countable, 151, 152, 171, 277, 418, 425
  bounded, 310
  bounded above, 436
  bounded below, 430
  closed, 140, 310, 420, 421, 423
  compact, 140, 146
  complementary, 152
  convex, 266, 267
  countable, 115, 208, 209, 222, 252, 418, 423, 424
  dense, 152, 153, 214, 432
  discrete, 420
  finite, 418
  infinite, 418
  null measure, 328, 429
  open, 140, 420, 432
  uncountable, 153, 418, 423
Shafrir, I., 249
Sierpiński's theorem, 155, 434
Sierpiński, W., 144
sinc function, 123
Singer, I.M., 30
singular differential equation, 245
singularities, 303
sinus cardinalis, 123
slope, 161, 183, 185, 188, 267, 290, 300
Sobolev inequalities, 383
Sobolev space, 366
solution
  double periodic, 367
  weak, 187
space
  Banach, 90, 239
  Hilbert, 364, 366
  normed vector, 222, 424

Sobolev, 366
squeezing and comparison test, 117, 123, 434
Stirling's formula, 4, 69, 96, 387, 414, 435
Stirling, J., 19
Stolz, O., 7
Stolz–Cesàro lemma, 7–9, 13, 22, 29, 121, 196, 435
Stong, R., 387
straddle lemma, 234
strict local minimum point, 277
strictly concave function, 263
strictly convex function, 263
strong maximum principle, 235, 435
strongly accretive operator, 90
Sturm–Liouville problem, 362
subadditive function, 210
subexponential function, 258
subharmonic function, 364
subsequence
  convergent, 5
subseries, 98
successive approximation, 156, 158
successive approximations sequence, 156, 300
sum
  Darboux, 282, 424
superconductivity, 237
superconductors, 303
superfluid liquids, 303
superlinear function, 132, 258, 287
superlinear growth, 242
support line, 275, 297, 298
supremum, 435
surface of revolution, 406
surjective function, 150, 171, 258
symmetric
  derivative, 186, 209, 210
symmetric derivative, 186, 209, 210, 224
system
  Allen–Cahn, 367
  Cahn–Hilliard system, 367

tangent
  vertical, 185
Tao, T., 103
Taylor expansion, 195
Taylor's formula, 239, 257, 259, 276, 277, 435
test
  Abel, 63, 421
  antiderivative, 321
  arithmetic mean, 83
  Coriolis, 65, 423
  Dirichlet, 63, 84, 424
  first comparison, 61, 74, 425
  Gauss, 63, 67, 75, 87, 426

integral for series, 374, 403, 428
  Kummer, 62, 429
  Leibniz, 63, 96, 97, 395, 429
  limit comparison, 430
  Raabe, 63, 75, 433
  ratio, 62, 71, 75, 433
  root, 62, 74, 434
  second comparison, 74, 434
  squeezing and comparison, 117, 123, 434
theorem
  Banach fixed point, 310, 421
  Barone, 158
  Bolzano–Weierstrass, 5, 17, 146, 147, 420, 422
  Brouwer fixed-point, 154, 218, 301, 422
  Cauchy, 192, 255, 423
  Darboux, 192–194, 226, 227, 314, 424
  Denjoy, 208, 209, 424
  Denjoy–Bourbaki, 222, 424
  differentiation inverse function, 197, 424
  Erdős–Szekeres, 53
  Fermat, 187, 193, 201, 226, 332, 425
  first mean value for integrals, 329, 425
  fixed-point, 159
  Froda, 152, 425
  fundamental of calculus, 328, 426
  Green–Tao, 103, 426
  Hardy–Littlewood, 256
  Heine–Borel, 427
  Hilbert double series, 281, 427
  horizontal chord, 173, 174, 428
  increasing function, 189, 428
  Knaster fixed point, 154, 428
  Kronecker, 44, 429
  Kronecker density, 45
  Lagrange, 188, 189, 191–193, 218–223, 226–228, 231, 282, 289, 301, 302, 341, 342, 429
  Laguerre, 257
  Lebesgue, 328, 429
  monotone convergence, 5, 431
  Nagumo, 411
  nested intervals, 5, 289, 431
  Picard convergence, 301, 302, 310, 432
  Rademacher, 207, 274, 433
  Rolle, 188, 218, 225, 227, 388, 408, 413, 434
  Rolle (polar form), 188, 434
  second mean value for integrals, 329, 386, 405, 406, 434
  Sierpiński, 155, 434
  Volterra, 153, 436
  Weierstrass, 140, 147, 436
theory
  Ginzburg–Landau, 237
  interpolation, 19
  probability, 4

Thompson, J.G., 30
three chords lemma, 267
Tits, J., 30
totally discontinuous function, 163, 326
Tower of London, 183
trigonometric function, 228
trigonometric series, 325
Turán, P., 103
twin primes, 103

uncertainty principle, 384
uncountable set, 153, 418, 423
uniform boundedness principle, 91
uniformly continuous function, 143, 146, 180, 436
union
  countable, 420
uniqueness of the limit, 116
universal gravitation, 37
upper bound, 436
upper contour set, 265
upper Darboux integral, 327

van der Waerden, B.L., 103
Vandermonde determinant, 130
Varadhan, S., 30
vertical tangent, 185
Viète formula, 88
Viète, F., 64
Volterra theorem, 153, 436
Volterra, V., 152
volume of a solid, 406
vortices, 303

Wallis's formula, 414, 436
Wallis, J., 64
weak maximum principle, 235, 436
weak solution, 187
Weierstrass form of Bonnet's theorem, 329
Weierstrass's nowhere differentiable function, 186, 436
Weierstrass's theorem, 140, 147, 436
Weierstrass, K., 5
Westminster Abbey, 432
Willem, M., 374

Young's inequality, 211
Young's inequality (for functions), 377, 436
Young's inequality (for numbers), 265, 295, 297, 436

Zeno, 3
Zeno's paradox, 39
zero
  of a function, 199, 231, 255

Printed in the United States of America